Two Manuals Combined into One Book



# Military Police

# And

# Military Working Dogs

Department of Defense

Pentagon
Publishing

**FM 3-19.4** (FM 19-4)

Field Manual
No. 3-19.4

Headquarters
Department of the Army
Washington, DC, 4 March 2002

# Military Police Leaders' Handbook

## Contents



**Distribution Restriction:** Approved for public release; distribution is unlimited.

*This publication supersedes FM 19-4, 7 May 1993.

Page

Page

# Preface

This field manual (FM) addresses military police (MP) maneuver and mobility support (MMS), area security (AS), internment and resettlement (I/R), law and order (L&O), and police intelligence operations (PIO) across the full spectrum of Army operations. Although this manual includes a discussion of corps and division MP elements, it primarily focuses on the principles of platoon operations and the tactics, techniques, and procedures (TTP) the platoon uses to accomplish its mission.

This FM provides the capabilities and organization of the MP, demonstrates the flexibility and diversity of MP in adapting to any mission throughout the full spectrum of Army operations, and characterizes the MP as a combat-force multiplier. Additionally, this manual identifies the fact that the Army will not conduct operations alone and defines the role of the MP in support of joint, multinational, and interagency operations.

The MP TTP are organized by the MP functions of MMS, AS, I/R, LO, and PIO with supporting tasks, both individual and collective, to help illustrate the functions.

**NOTE: United States (US) policy regarding the use and employment of antipersonnel land mines (APLs) outlined in this FM is subject to the convention on certain conventional weapons and executive orders (EOs). Current US policy limits the use of non-self-destructing APLs to (1) defending the US and its allies from armed aggression across the Korean demilitarized zone and (2) training personnel engaged in demining and countermine operations. The use of the M18A1 claymore in the command-detonation mode is not restricted under international law or EO.**

*Appendix A* complies with current Army directives which state that the metric system will be incorporated into all new publications. *Appendix B* deals with media relations.

The proponent of this publication is Headquarters (HQ) United States Army Training and Doctrine Command (TRADOC). Send comments and recommendations on Department of the Army (DA) Form 2028 directly to Commandant, US Army Military Police School, ATTN: ATSJ-MP-TD, 401 MANSCEN Loop, Suite 2060, Fort Leonard Wood, Missouri 65473-8926.

Unless this publication states otherwise, masculine nouns and pronouns do not refer exclusively to men.

# Chapter 1

# Military Police Overview

This chapter provides information about how MP are organized, equipped, and trained to provide combat support (CS) across the full spectrum of Army operations.

## INTRODUCTION

1-1. As a flexible economy-of-force organization, MP provide a wide range of diverse support because of their agility and versatility to adapt to any mission or environment. As a combat multiplier, they support the maneuver commander through the five MP functions. MP contribute to the commander's overall combat power by integrating efforts with those of other combat, CS, and combat service support (CSS) elements.

1-2. In addition to single-service operations, MP also support joint, multinational, and interagency activities. MP support air base defense in concert with Air Force Security Forces, operate joint and multinational checkpoints, conduct combined police patrols, and exchange police information and criminal intelligence with the host nation (HN), military, and civilian police agencies.

1-3. MP have the capability to expedite the movement of combat resources, provide critical asset security and protection, conduct I/R, contribute to force protection efforts through L&O operations, and gather and disseminate police information and intelligence.

# MILITARY POLICE FUNCTIONAL AREAS

1-4. With the old battlefield missions, the term "operations" was used extensively and carried too broad of a meaning. To clarify the specific tasks of the MP, the battlefield missions have been redefined into the following five functional areas:

- MMS
- AS
- I/R
- L&O
- PIO

1-5. Each of these MP functions have task areas and tasks that support them. MP functions are the broadest areas for which tasks are placed. Some of these tasks will require groupings that might not be related to the entire function. Therefore, task areas were created to group specific tasks. Specific tasks consist of two types—collective and individual. Individual tasks are further divided into leader and soldier tasks *(Figure 1-1)*. The collective and individual tasks that support the MP task areas are found in the MP mission training plans (MTP) and MP soldier's manuals (SMs).

1-6. MP procedures are the lowest level of detail. They explain the "how to" at the task level. Procedures include the standing methods used by a unit to accomplish tasks, weapon and equipment operating steps, crew drills, and staff action and coordination. They are the building blocks of individual and collective task accomplishment and serve as the foundation of tactics and techniques. Procedures are explained in the unit standing operating procedures (SOPs), MTPs, SMs, and similar publications.

## MANEUVER AND MOBILITY SUPPORT

1-7. The MMS function involves the measures necessary to enhance combat movement and the ability to conduct

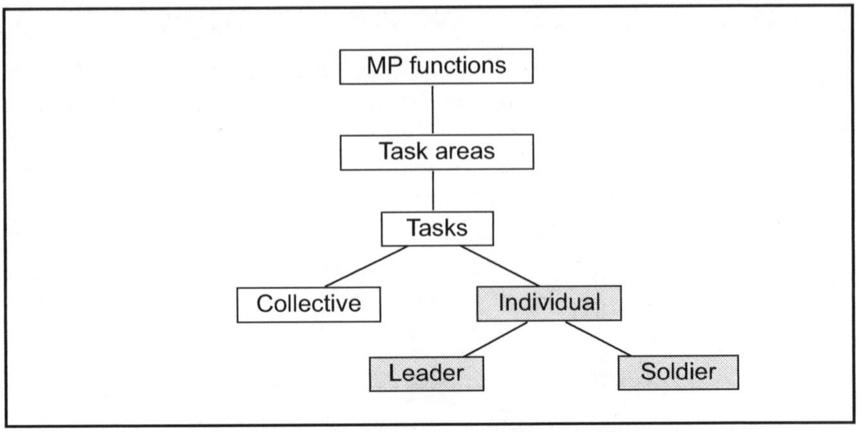


**Figure 1-1. Relationship Between MP Functions and Tasks**

movement of friendly resources in all environments. MP conduct MMS to ensure that the commanders receive personnel, equipment, and supplies when and where they are needed. The task areas that support the function of MMS include—

- MP support for river crossings, breaching, and passage-of-line operations.
- Straggler and dislocated civilian control.
- Route reconnaissance and surveillance.
- Main supply route (MSR) regulation enforcement.

1-8. The security and viability of the operational and tactical lines of communications (LOC) will be critical to continuous sustainment and recovery operations. MP ensure that logistics and supply operations are kept on time and arrive at the right place. Refer to *Chapter 5* for more information about MMS.

## AREA SECURITY

1-9. The AS function consists of those security measures designed to give commanders freedom of maneuver and flexibility to conduct operations. The task areas that support AS include—

- Reconnaissance operations.
- Area damage control (ADC).
- Base and air base defense.
- Response force and tactical combat force (TCF) operations.
- Critical site, asset, and high-risk personnel security.
- Force protection and physical security.
- Antiterrorism.

1-10. MP performing AS contribute to securing and protecting the force and preserving combat power. Refer to *Chapter 6* for more information about AS.

## INTERNMENT AND RESETTLEMENT

1-11. The I/R function consists of those measures necessary to provide shelter, sustain, guard, protect, and account for people (enemy prisoners of war [EPWs] and civilian internees [CIs], US military prisoners, and dislocated civilians [DC]). The task areas that support I/R include—

- EPW and CI handling.
- Populace and resource control.
- US military prisoner confinement.
- DCs control.

1-12. The international community, media, and public perceptions have increased sensitivity to the protection of human rights and the need for absolute accountability of interned, detained personnel, and refugees in military operations. Refer to *Chapter 7* for more information about I/R.

## LAW AND ORDER

1-13. Task areas and tasks that minimize the effects of a criminal threat on friendly forces support the L&O function. MP conduct L&O to remove the conditions and opportunities that promote crime, thereby preventing diversion of military resources and maintaining military discipline. The task areas include—

- Law enforcement.
- Criminal investigations.
- US customs operations.
- Related L&O training.

1-14. Whether patrolling an installation's housing area, conducting counterdrug operations, enhancing security, or investigating war crimes, MP L&O capabilities are invaluable to the commander. Refer to *Chapter 8* for more information about L&O.

## POLICE INTELLIGENCE OPERATIONS

1-15. The PIO function supports, enhances, and contributes to the commander's protection program, situational awareness, and battlefield visualization by portraying relevant threat information that may affect the operational and tactical environment. The task areas that support PIO include—

- Intelligence preparation of the battlefield (IPB).
- Active and passive roles.
- Police assessment process.

1-16. Whether in support of peacetime installation L&O or detecting threat forces in the battle space, the MP force employment provides the commander with substantial information and an intelligence source, particularly where the criminal element is the same as or closely aligned with the opposing forces (OPFOR) and government. Refer to *Chapter 9* for more information about PIO.

# THREAT

1-17. Today, friendly forces encounter a broad range of traditional and nontraditional threats. No single dominating threat will be the undisputed focus of US security policy. Although overt attacks on the US and its strategic interests may be less common, stability operations and support operations will likely increase. Economic development and demographics, as well as the progression of social and cultural movements, will encompass an array of threat forces including nonnation forces (ethnic conflicts and terrorist activities) which challenge traditional nation and state environments. Additionally, nation and state forces (internal security forces and infantry-based and armor-mechanized-based armies) continue to present a global threat. These forces possess varying levels of military and advanced technology capabilities.

1-18. In recent military operations, a nontraditional criminal threat has emerged. The evolving criminal threat operates most often in the rear area, near ports, in built-up areas, and where troop populations are high. This threat is most likely to be detected at border crossings trying to disrupt the relocation efforts of DCs. They may commit crimes against particular ethnic groups or be at checkpoints and roadblocks trying to position weapons, explosives, or personnel in sustainment areas in order to disrupt military operations or kill friendly forces. Such a threat requires commanders to minimize its negative impact on friendly forces, resources, and operations. The MP continue to respond to nonmilitary threats including famine, health epidemics, illegal immigration, illegal drug traffic, and population dislocation.

# MILITARY POLICE PLATOON ORGANIZATION AND LEADERSHIP

1-19. There are two basic MP platoon organizations, corps and division. Corps MP platoons are organized and equipped basically the same. Each division MP platoon supporting a different kind of division (such as heavy, light, airborne, or air assault) is designed under a different table(s) of organization and equipment (TOE).

## CORPS MILITARY POLICE

1-20. Depending on the nature of the operation, corps MP are usually among the first forces deployed to support military operations around the world. They deploy early to areas devastated by natural or man-made disasters to assist disaster relief and damage assessment efforts. They provide security and force protection to friendly forces, critical facilities, and resources as units organize for military operations. In a developing theater, corps MP concentrate mission support to the main effort. Units whose assistance to the main effort is vital normally receive the highest priority for protection. Key facilities, such as traffic choke points, critical tunnels and bridges, and ammunition and fuel storage points may require special protection. As the theater matures, the focus may quickly change to other functions as MP adjust priorities to accommodate the change.

## DIVISION MILITARY POLICE

1-21. Division MP are organized somewhat different depending on the type of division they are supporting. For example, a heavy division has one MP platoon providing direct support (DS) to each maneuver brigade and two MP platoons providing general support (GS) to the division's rear. Both airborne and air assault divisions have four MP platoons providing GS. A light infantry division is supported by three MP platoons.

1-22. In heavy divisions, where highly mobile forces are designed to move quickly over open ground, the overall need for MMS is significant. Division MP are likely to focus on expediting the forward movement of the critical combat resources into the division area. However, the priority could change quickly to removing EPWs from forward areas to freeing maneuver forces from guarding and caring for captives.

1-23. In airborne and air assault divisions, priority of MP support is most often needed for EPW operations and then for MMS to speed the movement of CS vehicles within the airhead.

1-24. For MP supporting any division, certain employment considerations remain constant. MP provide dedicated security for assets deemed critical by the division commander. This includes the division's main command post (CP) where MP operate outside the CP perimeter conducting screening missions designed to detect, disrupt, and delay enemy forces from disrupting the division's primary CP. Another consideration is MP accepting EPWs from capturing troops as far forward as possible.

## SEPARATE BRIGADES

1-25. MP support to a separate brigade is normally provided by a four-squad MP platoon. The platoon is assigned to the brigade headquarters and headquarters company (HHC). The brigade has a separate provost marshal (PM) cell that serves as the command and control ($C^2$) element for the platoon. The brigade PM cell has operational control (OPCON) of all MP assets the same way the division PM has OPCON of the division MP assets. The brigade HHC provides sustainment support for both the PM cell and the MP platoon. The PM advises the separate brigade commander on matters pertaining to MP operations. The platoon leader directs the execution of the platoon's missions based on the priorities set forth by the PM and the supported commander.

1-26. The platoon can perform any of the five functions. The platoon leader may task organize the squads according to mission, enemy, troops, terrain, time available, and civilian considerations (METT-TC), and provide one squad to operate the EPW collecting point, one squad to provide a mobile security screen and occupy observation posts (OPs) around the brigade's CP, and two squads to conduct MMS and AS throughout the brigade's rear area.

## INITIAL BRIGADE COMBAT TEAMS (IBCTs)

1-27. MP support to an IBCT may differ from that of other separate brigades. The IBCT is a preconfigured, ready-to-fight, combined-arms package. It is designed and optimized primarily for employment in small-scale contingencies operations (SSCO) in complex and urban terrain, confronting low-end and mid-range threats. The IBCT participates in war, with augmentation, as a subordinate maneuver component within a division or corps, in a variety of possible roles. It also participates in stability and/or support operations as an initial entry force. Civil unrest or complete turmoil normally characterizes these environments.

1-28. Organically, MP support to the IBCT is a two-person PM planning cell. The PM planning cell is located with the HHC's maneuver support cell and under the direct supervision of the brigade Operations and Training Officer (US Army) (S3). The role of the PM cell is significantly different from that of a traditional division PM or separate brigade PM. The main difference is the lack of organic or habitual MP assets in the IBCT. The absence of organic MP assets makes the job of the PM much more critical. The PM must—

- Understand the organization, capabilities, and limitations of the IBCT.

- Conduct effective liaison with higher HQ PM elements.
- Become an effective planner and anticipator of MP requirements.
- Task organize MP units effectively and efficiently.
- Assume $C^2$ of incoming MP forces or, if operating under a division, relinquish $C^2$ to a division or corps PM (if appropriate for effective employment of MP forces).

1-29. Depending on METT-TC, the IBCT may be augmented by MP elements ranging from a platoon to a battalion. Once the brigade receives MP augmentation, the PM then becomes a staff planner and coordinator for all MP activities.

1-30. Since the IBCT is a divisional brigade, the division PM and the IBCT PM play an important role in developing an optimum MP force package to support the brigade commander's concept of the operation. Despite the brigade's early time lines, the PM must consider and plan for MP augmentation forces as early as possible to free up valuable combat resources. SSCO that result in numerous EPWs, CIs, and refugees will hamper the maneuver force's freedom of movement.

1-31. Initially, MP priority of effort during the offense may be providing MMS for ground combat, CS, and CSS forces and taking control of EPWs and CIs. During the defense, the priority of MP support may shift to conducting AS and counterreconnaissance along the LOC, $C^2$ centers, and CSS bases. MP may also be required to conduct response force operations or become part of the TCF.

1-32. During stability and support operations, MP support may include—

- Order restoration.
- Crowd control.
- AS or force protection.
- DC operations.
- Noncombatant-evacuation operations.

## PLATOON LEADER

1-33. The MP platoon leader is responsible to the company commander for the platoon's combat readiness, training, and discipline and the maintenance of its equipment. To be successful, the platoon leader must demonstrate an ability to lead soldiers and manage an organization, material, and time. He must be able to articulate the capabilities and limitations of the platoon to various non-MP staff sections. In combat, the platoon leader is responsible for accomplishing all the missions assigned to the platoon according to the commander's intent and preserving the platoon's fighting capability.

## PLATOON SERGEANT (PSG)

1-34. The PSG leads elements of the platoon as directed by the platoon leader and assumes command of the platoon in the absence of the platoon leader. He directs the day-to-day activities of the platoon and ensures that the platoon has individual and team training and logistics needed to accomplish its mission. During tactical operations, he may assist in the control of the platoon.

## TEAM AND SQUAD LEADER

1-35. The MP team leader is responsible to the squad leader for individual and team training and team discipline. He is responsible for the tactical employment and control of the team and the maintenance and operation of all vehicles and equipment organic to the team. During combat operations or anytime there is a threat, the team leader quickly assesses the situation, reports to his superiors, and takes appropriate action to protect the team according to the rules of engagement (ROE). A squad leader has the same responsibility for the squad as the team leader has for the team.

## FORCE PROTECTION (FP) MEASURES

1-36. MP leaders at all levels must examine FP requirements and integrate FP measures throughout all the operations. Once higher HQ has established local FP policies, leaders set the example by complying with them. Leaders reduce the soldiers' exposure to hazards by strictly enforcing all the protective postures that may include—

- Traveling with at least two vehicles armed with at least one automatic weapon.
- Hardening of the vehicles.
- Wearing Kevlar® and body armor.
- Not driving off the road or cross-country.
- Placing off-post facilities off limits during nonduty hours.

1-37. MP leaders analyze and compensate for other threats such as disease, weather, crime, complacency, terrorism, morale, safety, and other considerations.

1-38. At the operational level, team and squad leaders establish a safety zone around their teams. The safety zone is the immediate area around the team where threat forces or events could harm the team or inflict casualties. In open terrain, the safety zone may extend out to the maximum effective range of the team's organic-weapon systems. However, when searching vehicles at a checkpoint or conducting crowd control, the safety zone may only be an arm's length from the team.

1-39. Team and squad leaders remain alert to threats that enter the team's safety zone. They must quickly assess any threat to the team and take appropriate action within the ROE to reduce the threat or move the team.

1-40. When required to operate in crowds, maintain eye contact with team members. Establish and maintain a safe distance between the team and the crowd. Never allow the team to become separated or surrounded.

1-41. When patrolling in built-up areas, the gunner scans the upper floors of the buildings and the streets to the vehicle's front, rear, and flanks and immediately reports any suspicious activity. The driver concentrates on the area directly in front of the vehicle looking for any unexploded munitions, scatterable mines, or other road hazards. All team members should stay awake, alert, and ready to react to danger.

## MILITARY POLICE PLATOON MISSION, CAPABILITIES, AND LIMITATIONS

1-42. The platoon has one critical wartime mission which is to provide MP CS to an assigned area of operations (AO). MP CS consists of all five MP functions. The platoon performs its missions primarily mounted, taking full advantage of the high mobility multipurpose wheeled vehicle's (HMMWV's) versatility and the added protection and firepower of the armor security vehicle (ASV).

### CAPABILITIES

1-43. The MP platoon is capable of operating day or night, in various terrain conditions, and under all weather and visibility conditions. Their mode of operation is possible through the deployment and employment of the three-person team throughout the battlefield. However, it is dependent on its parent unit for sustainment support. The platoon has self-protection capabilities such as nuclear, biological, chemical (NBC) detection equipment and a Platoon Early-Warning System (PEWS). The platoon's radio transmission range is increased with an OE-254 antenna. Because of extensive police training and law enforcement missions, the MP are highly skilled in the use of force and the employment of lethal and nonlethal technologies, information-collecting and dissemination, observation and surveillance, and crowd control. The MP platoon has a tremendous combat and noncombat information-collecting capability. This capability is the result of extensive area, zone, and route reconnaissance; daily contact with local nationals; conducting combined police patrols with HN military and civilian police agencies; and conducting field interviews. An MP platoon is capable of covering 500 square kilometers in rolling terrain; however, more severe terrain such as mountains, METT-TC, and mission objectives will affect this capability. For example, consider one mobile MP team per 10 kilometers of route coverage. For area coverage, begin with an estimate of one mobile MP team per 55 square kilometers.

1-44. Unlike most combat arms platoons, which maneuver together in formation, the MP platoon most often operates independently and dispersed over a large area. The platoon conducts combat operations, when required, through the employment of mobile combat systems containing three-man teams operating independently or in concert, and having vehicle crew-served and individual weapons capable of defeating a Level II threat and defending a position against dismounted threats.

1-45. Based on METT-TC, the platoon leader may task organize the platoon for certain missions. Normally, MP are employed as squads; however, individual teams may execute many MP tasks.

### LIMITATIONS

1-46. During combat operations, the platoon is not organized and equipped to fight for extended periods unless it is augmented with indirect fire or close air support (CAS). Although the MP team is a lethal and highly mobile platform, it is not structured or equipped for prolonged autonomous missions. Leaders must use the MP team as a task organizational building block and avoid over tasking based solely on the number of teams available. The platoon has limited antiarmor capability and normally uses antiarmor weapons for self-protection and to break contact.

## PEACETIME TRAINING

1-47. MP units train as they will fight. Peacetime training must replicate battlefield conditions and conform to Army doctrine. Leaders and soldiers must understand standardized doctrinal principles found in applicable manuals to ensure that training is conducted to standard. The following manuals provide the basic foundation for Army training:

- FMs.
- Training circulars (TCs).
- MTPs.
- Drill books.
- SMs.
- Army regulations (ARs)

1-48. *FM 25-100*, *FM 25-101*, and *TC 25-10* provide MP leaders with established training doctrine and assist the leaders in the development and execution of the training programs. These manuals introduce the concept of lane training and define it as a technique for training company, team, and smaller units on a series of selected soldier, leader, and collective tasks using specific terrain.

1-49. Lane training uses multiechelon techniques to maximize the efficient use of limited terrain and control conditions for formal or informal evaluations. Lane training is externally supported, resourced, and evaluated. It enables similar units to simultaneously or sequentially train on mission-related scenarios. Lane training is resource intensive, so commanders must maximize its benefit. Commanders narrow the focus and select only the most critical mission-essential task lists (METLs) items or collective tasks for training. Lane training is especially valuable for conducting specific METL tasks, situational training exercises (STXs), and other training events. It is often associated with training requiring movement over terrain; for example, movement to contact or conducting a route reconnaissance. Lane training can be modified to achieve benefits in L&O scenarios, such as special-reaction team (SRT) incidents, patrol incidents, traffic accidents, and so forth.

1-50. The lane training doctrine outlined in *FM 25-101* and *TC 25-10* can be tailored for small MP units by using the training execution model (TEM). The TEM follows the Army doctrine and training philosophy of hands-on METL training as outlined by *FMs 25-100* and *25-101*. Before the TEM can be implemented, the concept of the operation must be approved, evaluated, and directed from two levels up. For example, a squad leader must receive approval through the chain of command from his company commander to execute the training event; a platoon leader gets approval from the battalion commander and so forth.

1-51. The TEM incorporates the combined-arms training methodology and adjusts it to meet the MP training requirements. The TEM focuses the unit on the time available during the training cycle to train the most critical collective and individual tasks. The TEM consists of an eight-step training methodology that is based on leader certification of the lane expert and an observer/controller (OC) as well as subordinate unit leaders. For more information about TEM refer to *Appendix C*.

# Chapter 2

# Battle Command

This chapter provides the techniques and procedures used by MP leaders at company and platoon level to $C^2$ their organizations.

## OVERVIEW

2-1. Battle command is the art of battle decision making, leading, and motivating soldiers and organizations into action to achieve victory with the least cost to the organization. Commanders must visualize the current and future state of both friendly and enemy forces. The commander positions himself where he can guide and motivate the soldiers and influence the outcome of the missions.

2-2. The company commander is responsible for all that the unit does or fails to do. He cannot delegate this responsibility, and the final decision and responsibility rest with him. He discharges his responsibility through an established chain of command and holds each subordinate leader responsible for the actions of the platoon or the section.

2-3. The commander must be proficient in the tactical employment of the unit. He must know the capabilities and limitations of the soldiers and the equipment. A commander does this through a continuous cycle of planning, executing, and assessing training. Through this training, the commander gets to know the soldiers.

2-4. MP commanders prioritize, assign missions, and allocate resources where they can best support the higher echelon commander's intent. The company commander makes most of the tactical decisions. Technological advances in today's operational environments have reduced the time available for decision making while increasing the possibilities that must be considered.

2-5. Thorough and sound operational planning is the key to successful combat and CS operations. Commanders must identify the opportunities and anticipate and avoid problems. They must analyze their options before making the decisions on which subordinate leaders will base their actions. Commanders balance competing risks and then identify and develop the best course of action (COA).

## MILITARY DECISION-MAKING PROCESS (MDMP)

2-6. The MDMP is a single, established, and proven analytical process used at all the echelons of the commands. This is a seven-step process used when adequate planning time and enough staff support are available *(Table 2-1)*. This process is a detailed, deliberate, sequential, and time-consuming process that helps the commander and his staff examine a battlefield situation and reach logical decisions. The commander uses the entire staff during the process to explore the full range of probable and likely enemy and friendly COAs and to analyze and compare his own organization's capabilities with the enemy's.

**Table 2-1. MDMP**

| Step | Action | Step | Action |
|---|---|---|---|
| 1 | Receipt of the mission | 5 | COA comparison |
| 2 | Mission analysis | 6 | COA approval |
| 3 | COA development | 7 | Orders production |
| 4 | COA analysis | | |

2-7. At company level, the commander normally uses the MDMP in a time-constrained environment without enough staff. A unit can shorten the process if it fully understands the role of each step of the process and the requirements to produce the necessary products. The application of the MDMP at company level and below is called the troop-leading procedures (TLP). *Figure 2-1, page 2-4* shows the relationship between MDMP and TLP.

2-8. MP commanders plan successful operations by anticipating possible future events and planning contingencies. MP leaders enhance both planning and execution of the operations when they—

- Use the military planning and decision-making process.
- Develop short- and long-range goals.
- Identify goals and objectives with a recognizable end-state.
- Coordinate goals and actions internally and externally.
- Base their plans on objective planning factors.
- Review their plans, continuously, in light of the METT-TC and updated information.
- Assign responsibilities and express expectations.
- Identify the options that may develop during an operation.
- Stand ready to accommodate the changes.

## MILITARY PLANNING

2-9. Commanders select and carry out the developed COAs using military planning. Military planning guidelines include—

- Forecasting requirements by analyzing and evaluating facts and trends to predict what may occur.

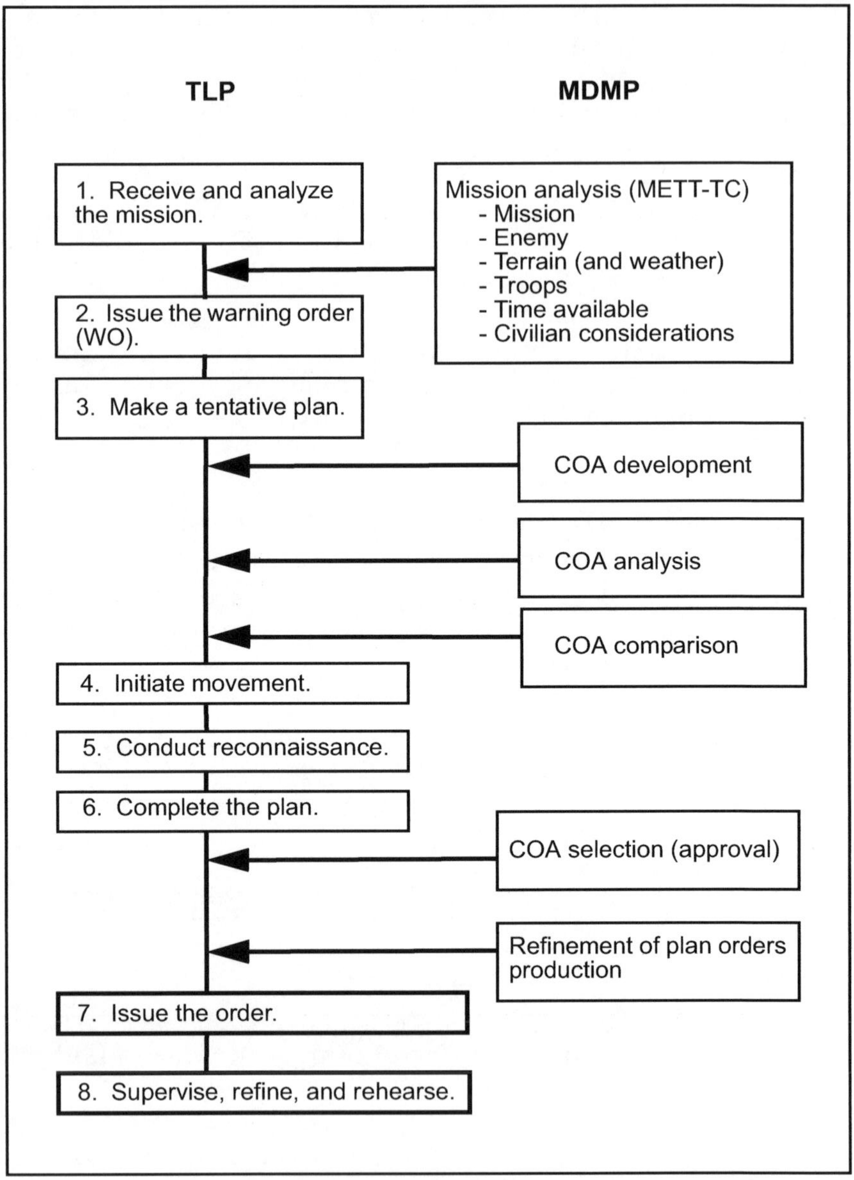


**Figure 2-1. Relationship Between TLP and the MDMP**

- Examining probable requirements and establishing priorities for further preparation.
- Studying implications and interrelationships of probable requirements.
- Analyzing the mission to determine tasks, their complexity, and their relative importance.
- Establishing guidance for further planning that will help keep all the elements focused on the commander's intent.
- Preparing studies and estimates to help formulate the COA and assess its feasibility.
- Selecting the COA, identifying the best course, and retaining other feasible courses for use in contingencies as alternate plans.
- Preparing the plan in detail and conducting rehearsals when time, resources, and security permit.

2-10. Use the following military planning guidelines to answer the three key questions of operational planning:

- What military condition must be produced to achieve the goal?
- What sequence of actions is most likely to produce that condition?
- How should resources be applied to accomplish that sequence of actions?

2-11. If the plan is not implemented immediately, retain it for later use. As events occur or new information becomes available, review and revise the plan accordingly.

## ANALYSIS, FORECASTING, AND RISKS

2-12. Conducting a mission analysis is crucial to planning. The process begins by gathering facts and ascertaining current conditions, such as the—

- Higher-level mission and the commander's intent (one and two levels up).

- Current task organization (two levels down).
- Current unit status (locations, operation capabilities, and activities).
- Logistics situation (refer to *FM 101-5* for the logistic estimate format).

2-13. When facts are not available, the commander will need to develop assumptions. Assumptions must substitute for fact where information is not known. Keep in mind that as time passes between the receipt of a mission and the execution of a plan, facts are increasingly likely to have changed. Develop sound assumptions that can be used in place of facts.

2-14. Analyzing the higher-level mission and the commander's intent will help identify what tasks are required to accomplish the mission. As the mission is analyzed, identify both the *specified* and the *implied* tasks to be performed. *Specified tasks* are those stated in higher HQ orders and plans. *Implied tasks* (like crossing a river or passing through the lines of a unit lying between you and the objective) are not so stated, but must be accomplished to satisfy the overall operation. From among the specified and implied tasks, essential tasks that are crucial to the mission's success must be identified.

2-15. Integral to mission planning is the analysis of mission requirements in terms of time, space, and personnel. If MP are to balance the benefits of detailed planning against the need for immediate action, they must—

- Determine how much time there will be between receiving the mission and the deadline for having completed it.
- Know how long it will take to obtain and process information, make decisions, and issue orders.

- Know how long it will take subordinates to execute the orders, complete the mission, or carry out the operation.

2-16. Because each unit involved in an operation performs its planning based on the plans of the next higher level, allocation of adequate time for subordinate units to plan is a consideration at each level. Publishing SOPs reduces the number of details to be explained. It also promotes understanding and teamwork among commanders, staff, and troops.

2-17. When doing reverse planning, consider the classic allotment of one-third time for planning and two-thirds time for execution *(Table 2-2)*. Move backward from the time of execution to—

- Allocate time to accomplish each phase of an operation.
- Determine how much time is needed to rehearse.
- Determine how much time is needed for developing the plan.

**Table 2-2. Reverse Planning Timetable**

| Time | Action |
|---|---|
| 2230 | Execute the mission. |
| 2130-2215 | Conduct inspection(s). |
| 1845-2130 | Conduct rehearsal(s). |
| 1815-1845 | Issue an operation order (OPORD). |
| 1745-1815 | Complete the plan. |
| 1715-1745 | Conduct the leaders' reconnaissance. |
| 1630-1715 | Issue a tentative plan. |
| 1630 | Issue a WO. |
| 1600 | Receive the mission. |

2-18. When using terrain analysis, consider the layout of the battlefield. Appreciate the opportunities and limitations of the major terrain features, transportation networks, and built-up areas. Fit the operational concept and planning to that environment. Use the IPB to evaluate the area in terms of the military aspects of the terrain. Consider how to exploit the opportunities afforded by weather while minimizing its adverse effects.

2-19. Use current information on the threat to identify known enemy activities and threat capabilities that could affect this and future operations. Attempt to anticipate the enemy's objectives and intentions.

2-20. Consider available assets and determine acceptable levels of risk. At every echelon, MP disperse their assets and prioritize operations to meet the echelon commander's needs within the limits of the resources at hand. MP leaders must concentrate their efforts on key locations and accept risks elsewhere. When possible, recognize and moderate such risks in the choice of operations and in the contingency planning. The five steps to identify, analyze, and reduce risks are listed in *FM 100-14*.

## COURSE OF ACTION IDENTIFICATION, DEVELOPMENT, AND SELECTION

2-21. The commander's selected COA will become the actual plan for accomplishing the mission. To ensure the best plan possible, identify several possible COAs, each significantly different from the others. In identifying COAs, do not overlook a check of the "basics" that include the—

- Commander's intent.
- Essential tasks.
- Effective use of $C^2$.
- Principles of war.
- Ethical considerations.
- Relative force ratio.

2-22. Consider preparing a COA statement (and sketch, if appropriate) for each option. The COA statement is the "how" of the operation. Ensure that it includes the following five elements:

- What – the type of action.
- When – the time the action will begin.
- Where – the assigned area.
- How – the use of available assets.
- Why – the purpose of the operation.

2-23. When developing the COA, analyze the relative combat power. Consider the initial array of the forces and develop the schemes of maneuver. Determine $C^2$ means and maneuver control measures.

2-24. Base doctrinal capabilities and planning on historical planning factors, and then relook the estimates in light of the available assets, the factors of METT-TC, the echelon commander's intent, and the mission's priorities. For example, when planning distribution of mobile assets for route coverage, begin with an estimate of one mobile MP team per 10 kilometers. For area coverage, begin with an estimate of one mobile MP team per 55 square kilometers.

2-25. When dispersing the assets into small combat elements, consider the classic ratios of friendly to enemy forces (3 to 1) to help ensure that the elements can concentrate enough combat power to accomplish the mission. Consider the speed and ease of reassembling the elements if dispersing them to distant sites.

2-26. Attempt to anticipate the enemy's likely moves. Consider ways to obstruct dangerous approaches to the area and avenues leading away from potential landing zones. Plan ways to combine the efforts of different resources, like enhancing the combat power for the base response and counterreconnaissance operations with fire support from field artillery or aviation. Attempt to determine the outcome of the operations by—

- Conserving unit strength through economy of force.
- Using terrain, weather, deception, and operations security (OPSEC) to your advantage.
- Focusing your efforts on enemy vulnerabilities.
- Ensuring unity of effort among subordinates and with your peers.

2-27. When planning for combat operations, whenever possible, develop a COA that avoids an enemy's strength and strikes at his weaknesses. Avoid head-on encounters with an enemy's forces. Seek to gain the element of surprise. When moving, plan to use indirect approaches and flank positions that do not attract immediate attention. Plan for fire support to increase MP combat power. Plan to operate on the enemy's flanks and rear, where direct fire is most effective, psychological shock is the greatest, and the enemy is least prepared to fight. Respond to and implement changes quickly and plan supplementary or alternative control measures to modify the plan as the situation dictates.

2-28. For a combat operation, the COA statement and sketches include the following:

- Allocated forces.
- Unit boundaries.
- Axes of advance.
- Routes for a forward or rearward passage of lines.
- Air axes for the maneuver of attack helicopters.
- Other control measures which may include—
  - Phase lines.
  - Assembly and holding areas.
  - Zones or sectors.
  - Battle positions.
  - Objectives.
  - Obstacles.
  - Routes.

2-29. Assess all of the feasible COAs after developing them. Consider constraints and restrictions on each COA. Weigh the available COA for the AO, for the level of responsibility, the commander's intent, and the mission's priorities. Decide on the best COA. For more information about the MDMP process, refer to *FM 101-5*.

2-30. Once the commander decides on a COA, he announces it in the form of orders that include his intent and concept of the operation. Based on these orders, the platoon leader uses TLP to organize his time during planning and preparation for the mission. Time management is the key. The platoon leader normally uses one-third of the available time to plan, prepare, and issue the order. The squad leaders have the remaining two-thirds of the time to prepare the squads for the mission. Whenever possible, TLP are integrated and accomplished concurrently rather than sequentially. Relationships between TLP and the MDMP are shown in *Figure 2-1, page 2-4*.

## TROOP-LEADING PROCEDURES

2-31. TLP begin when the platoon leader is alerted for a mission and starts again when he receives a change or a new mission. Conducting TLP is an eight-step process *(Table 2-3, page 2-12)*. Steps 3 through 8 may not follow a rigid sequence. Several of the steps may be accomplished concurrently. In CS operations, platoon leaders rarely have enough time to go through each step in detail. However, the procedure must be followed, if only in abbreviated form. This ensures that nothing is left out of the planning and the preparation.

2-32. Risk management is an integral part of TLP and must be integrated into every aspect of mission

Table 2-3. The Eight Steps of TLP

| Step | Action |
|---|---|
| 1 | Receive and analyze the mission. |
| 2 | Issue a WO. |
| 3 | Make a tentative plan. |
| 4 | Initiate movement. |
| 5 | Conduct a reconnaissance. |
| 6 | Complete the plan. |
| 7 | Issue the order. |
| 8 | Supervise, refine, and rehearse. |

planning and execution. Identifying hazards and implementing control measures to mitigate those hazards will minimize operational accidents, thereby preserving combat power. Risk management is not a separate process, but rather infused into the entire TLP. For more information about risk management, refer to *FM 100-14*.

## RECEIVE AND ANALYZE THE MISSION

2-33. TLP begin when the platoon leader is alerted for a mission and start again when he receives a change or a new mission. When the platoon leader receives the mission, he—

- Analyzes the mission.
- Considers the operations underway.
- Identifies the hazards associated with the mission, and considers the aspects of the current and future situations, environment, and known historical problems.
- Considers the time needed to plan and carry out the new mission. For example, will sleep plans be needed to ensure that all the teams are on a similar rest posture?

- Plans the use of available time. The most critical resource may be time, especially during daylight hours.
- Uses reverse planning to make a timetable. The timetable—
  - Identifies what must be done.
  - Works backwards from the time the soldiers have to be ready, allowing them time to do each task.
- Uses no more than one-third of the time for planning. The subordinate leaders need the remaining time to make preparations.

2-34. If time is too short to do the rest of the troop-leading steps in detail, at least do a fast mental review and—

- Make a quick map reconnaissance while sending for the subordinate leaders, depending on the level of the mission.
- Have the minimum control measures needed posted on their maps.
- Give an abbreviated order.
- Cite enemy and friendly situations.
- State the mission of the team, squad, or platoon and the concept of the operation.

2-35. If there is not enough time to do these actions, have the unit move out, then issue a fragmentary order (FRAGO) by radio or at the next scheduled halt. Continue the plan while moving.

## ISSUE A WARNING ORDER

2-36. Issue an oral or written WO to the subordinate leaders as soon as possible. Give enough information for the unit to begin preparing for the mission. If needed, issue several WOs to keep subordinates informed. Refer to *Appendix D* for more information about WOs.

2-37. The unit SOP should detail what actions to take when a WO is received. Such actions may include drawing ammunition, rations, water, and communications gear and checking vehicles and equipment. Keep all the personnel informed of what they are to do and why they are to do it.

## MAKE A TENTATIVE PLAN

2-38. Develop the plan based on the factors of METT-TC (using the OPORD format and the higher HQ order). The order may be specific about the tasks the unit is to do. The time available may be limited and the scheme of maneuver may be dictated. Nevertheless, the leader still must evaluate the mission in terms of METT-TC to see how MP elements can best carry out the commander's order. The leader must—

- Consider each factor and compare the COA to form a base for the plan.
- Include concepts for reconnaissance, coordination with adjacent and/or supporting units, and troop movement.
- Assess the identified hazards. Consider the impact of each hazard in terms of potential loss based on probability and severity.
- Identify control measures that will eliminate the hazards or mitigate them to an acceptable level. Make decisions on acceptable levels of risk based on potential benefits versus cost.
- Issue the plan, when firm, as an order. *Appendix D* contains the OPORD format.

## INITIATE MOVEMENT

2-39. Instruct subordinate leaders to start moving to the assembly area (AA). Allow subordinate leaders enough time for their actions if the element has to move and reorganize for the mission.

## CONDUCT A RECONNAISSANCE

2-40. Ensure that the terrain where the operation is to be conducted is reconnoitered. At a minimum, conduct a map reconnaissance. (A map reconnaissance is the easiest, but least reliable form of reconnaissance. It is a supplement to other types of reconnaissance.) Study the map for terrain features, natural barriers, and other characteristics. Have subordinate leaders help identify key terrain features. Refer to *FM 21-26* for map-reading skills. Follow up with a visual reconnaissance of the area to be used and the terrain over which you will operate. A visual reconnaissance can be done—

- On the ground. A ground reconnaissance is time consuming, but the most reliable type of reconnaissance. MP see terrain features up close and can note problems not easily seen using other reconnaissance methods.
- In the air. If available, air reconnaissance can cover terrain quickly. To do an air reconnaissance, show the pilot a map of the terrain to be reconnoitered. Specify the type of information to be gathered. Have one person in the plane track the patrol's route on a map. At critical points, if the aircraft can land, have part of the patrol dismount to make a ground reconnaissance while the rest of the patrol goes back into the air to provide overwatch security. If the aircraft cannot land, make a visual search for enemy activity or for the required information.

2-41. Use the information from the reconnaissance to verify or to change the plan and to modify the risk-assessment process. Adapt your tactics to the terrain and the abilities of the force. If the reconnaissance cannot be finished due to distance or enemy pressure, make the plan from what has been seen. Give instructions for later actions in general terms and confirm or change while moving over the terrain.

Provide a sketch that will aid with a sand table to help in the OPORD issue, if time permits.

## COMPLETE THE PLAN

2-42. Add details or makes changes to the tentative plan (as a result of the reconnaissance and coordination with nearby and/or supporting agencies) and identify specific tasks for all the subordinate elements.

## ISSUE THE ORDER

2-43. Issue an OPORD or FRAGO ensuring that—

- The soldiers know the plan.
- The instructions are stated clearly and concisely (use the OPORD format). Platoon and squad orders are usually issued orally. However, if time permits, they can be written. When the order is written, delete the service support and command and signal paragraphs if covered by the SOP.
- The soldiers are thoroughly briefed on the hazards associated with the mission and the control measures identified to mitigate the hazards.
- The subordinate leaders back brief the order and spot-check the soldiers.
- When possible, the order is given from the advantage point where soldiers can see the area in which they will operate. This lets the leader point out terrain features on the ground as well as on a map. If this cannot be done, use a terrain model (sand table) or a sketch to help explain the order.

## SUPERVISE, REFINE, AND REHEARSE

2-44. Supervise, refine, and rehearse the preparation to ensure that the soldiers are ready for the mission. This includes—

- Using the feedback received from subordinates.

- Ensuring that every soldier knows the mission and understands the commander's intent for the operation. The unit must be able to carry out the mission in the absence of the leader that developed the plan.
- Continuing to prepare when the troop-leading steps have been completed.
- Having the soldiers rehearse their actions if there is enough time before an operation.

## Rehearsals

2-45. Rehearsals build confidence and improve performance. They allow faults in a plan to surface. If possible, soldiers should rehearse on terrain and under conditions like those at the operation site. Give priority to actions to be taken in the objective area. A rehearsal is especially helpful when operating in reduced visibility.

2-46. Each type of rehearsal reflects an increase in mission realism and a corresponding increase in rehearsal benefit. Each technique increases the realism of the enemy, terrain, team and squad actions, and actual time and distance relationships. Rehearsal techniques fall into the following four categories:

- Back brief.
- Rock drill.
- Walk-through.
- Full-scale.

2-47. Regardless of the rehearsal technique, leaders must demand that soldiers demonstrate the known hazards associated with the mission and understand the control measures identified to counter them. Insist that rehearsals include prescribed control measures (either actions or equipment).

2-48. **Back Brief.** The back brief rehearsal is an event that occurs after an OPORD has been issued. The back brief is the quickest rehearsal technique. Subordinate leaders repeat back to the commander what he expects them to do and why, using a map or a sand table to explain their mission. The subordinate leader identifies all specified and implied tasks, determines their mission-essential tasks, and restates the mission. Items essential to the back brief are the—

- Commander's intent.
- Concept of the operation.
- Scheme of maneuver.
- Time to complete the tasks.

2-49. **Rock Drill.** A rock drill rehearsal is done by acting out the friendly and enemy actions based on the scheme of maneuver and the situation. Subordinate leaders rehearse their actions by moving objects, such as rocks, that represent them or the platoon. In acting out the plan, leaders can talk through their missions, critical tasks, actions, and decisions. All subordinate leaders act out their parts simultaneously so problems and disconnects in synchronization can be more clearly identified.

2-50. **Walk-Through.** A walk-through rehearsal is the acting out of actions that will occur during a mission using the actual vehicles and equipment that will be used to conduct the operation. Participants communicate with the same type of equipment they will use during the operation. During a walk-through, subordinate leaders rehearse—

- Movement techniques, both mounted and dismounted.
- Critical actions.
- Decision making.

2-51. Since the leaders are in a more realistic environment, they rehearse the finer aspects of synchronization, $C^2$, and squad and team actions. Aggressive portrayal of OPFOR is critical in walk-through rehearsals because it increases the rehearsal realism. This type of rehearsal is more difficult to orchestrate. It is the optimum balance between resource constraints and realism. Ensure that a walk-through rehearsal is the minimum goal for all the units.

2-52. **Full-Scale.** During a full-scale rehearsal, participants use real-time mounted and dismounted movements over the actual or similar terrain. It is normally conducted with all teams and squads to be used for the actual mission. At a minimum, one subunit must participate for a full-scale rehearsal. This type of rehearsal is the most resource intensive, but provides the most realistic training environment for the unit. It is often used to rehearse the operation plan (OPLAN) or OPORD when time is not an immediate constraint.

## Inspection

2-53. The last action before an operation is inspecting. Allow ample time for subordinate leaders to correct problems. Inspect soldiers by checking their mental and physical readiness. Inspect their equipment and check for the following:

- Accountability.
- Serviceability.
- Weapons.
- Ammunition.
- Individual uniforms and equipment.
- Mission-essential equipment.
- Water and rations.
- Communications equipment.
- Vehicles.
- Camouflage.

2-54. The equipment used during a mission is based on the unit's SOP, a risk assessment, and special considerations. Ensure that the SOP specifies a combat load and a list of ammunition and equipment usually carried on missions. Base changes from the SOP combat load on the METT-TC.

2-55. *Appendix E* contains a sample precombat-inspection (PCI) checklist. Ensure that the soldiers have everything they need for the mission. Ensure that they—

- Know their duties.
- Have only the equipment needed.
- Are wearing their equipment correctly and securely.

## ORDERS AND REPORTS

2-56. MP leaders translate their thoughts, evaluations, and decisions into understandable reports and orders. Battlefield communication requires standardized, streamlined procedures. Despite personal exhaustion or battle confusion, you must be able to rapidly report information or issue instructions that are simple, clear, and brief.

### ORDERS

2-57. Combat orders are written or oral communications giving details of tactical operations and administration. The three most common types of combat orders at company level and below are—

- WO
- OPORD
- FRAGO

2-58. WOs and OPORDs generally have set formats. This helps ensure that the receiver understands the intent of the message and that all needed information is provided. Standardization helps save time in writing as well as interpreting orders. FRAGOs enhance what has been previously sent out in the OPORD, such as a change in the situation or mission. Refer to *Appendix D* for examples of orders.

**Reports**

2-59. MP report to higher HQ and provide information on which plans, decisions, and orders can be based. The information included in MP reports must be accurate and timely, and complete *negative* information ("There is no enemy at . . . . ") is often as important as positive information. Reports are the main record of operational events. The three broad categories of reports are—

- Administrative.
- Operational.
- Intelligence.

2-60. Commanders may specify report formats in their local tactical SOP. Treat friendly information, including administrative reports, as classified or sensitive in nature to keep information from falling into the enemy's hands. *FM 101-5-2* and *Appendix D* of this manual CONTAIN STANDARD REPORTING FORMATS.

**SOPs**

2-61. SOPs detail how forces will execute unit-specific techniques and procedures that commanders standardize to enhance effectiveness and flexibility. Commanders use the SOP to standardize routine or recurring actions not needing their personal involvement.

## RULES OF ENGAGEMENT AND RULES OF INTERACTION (ROI)

2-62. MP usually are among the first CS forces deployed to troubled areas around the world. As a direct result of a peacetime L&O mission, MP continually train in the prudent use of force, crisis management, and operations requiring restrictive ROE and ROI.

2-63. ROE are the directives established by higher HQ that delineate the circumstances and limitations under which soldiers will initiate and/or continue engagement with belligerent forces. ROE may reflect the law of armed conflict and operational considerations, but are primarily concerned with the restraints on the use of force. ROE are the primary means by which commanders convey legal, political, diplomatic, and military guidance to soldiers. Leaders at every level must train their soldiers carefully and thoroughly concerning ROE and laws that govern armed conflict before deployment. During the conduct of the operation, leaders continue to train soldiers and stress firm, determined, and impartial execution of ROE to preclude inviting challenges from any of the belligerent parties.

2-64. ROI embody those human dimension skills needed to successfully interface with various categories of people. They spell out with whom, under what circumstances, and to what extent soldiers may interact with other forces and the civilian populace. ROI, when applied with good interpersonal communication (IPC) skills, improve the soldier's ability to accomplish the mission while reducing possible hostile confrontations. ROI and IPC, by enhancing the soldier's persuasion, negotiation, and communication skills, also improve his survivability. ROI founded on firm ROE provide the soldier with the tools to address nontraditional threats such as political friction, ideologies, cultural idiosyncrasies, and religious beliefs and rituals. ROI must be regionally and culturally specific. MP leaders must train soldiers on ROE and ROI using tactical vignettes or simulated events.

## SITUATIONAL AWARENESS

2-65. Situational awareness is the ability to maintain a constant, clear mental picture of the tactical situation. This picture includes an understanding of both the friendly and enemy situations and of relevant terrain. It also includes relating events in time to form logical conclusions and make decisions that anticipate events. Since MP platoons normally operate dispersed, it is essential that all MP leaders maintain situational awareness so that they can make quick, sound tactical decisions. Situational awareness also permits MP leaders to anticipate events and relate separate pieces of information to form logical conclusions. One of the critical outcomes of situational awareness on the part of all MP is a reduction in fratricide incidents. Refer to *Appendix F* for fratricide avoidance.

## BATTLEFIELD FRAMEWORK

2-66. The commander will structure the battlefield based on the conditions of METT-TC and his commander's intent. How he does this affects the MP platoon leader's mission planning and his ability to maintain situational awareness. Geographically, available assets, mission, and the AO influence the dispersion of MP assets. The framework of this AO can vary from an area that is dominated by several towns or large cities to an area that includes several bases and base clusters. MP can expect to operate in sustainment areas where there may be clear boundaries and closely tied adjacent units or in a decentralized structure with few secure areas and unit boundaries. Between these extremes are an unlimited number of possible variations. Maintaining situational awareness becomes more difficult as the battlefield becomes less structured. Modern, highly mobile operations with small forces lend themselves to a less rigid framework that challenges the MP ability to maintain an accurate *picture* of the battlefield. MP are the echelon commander's critical link to the battlefield, constantly gathering police, operational, and combat information.

## BATTLEFIELD PICTURE

2-67. To have a clear picture of the battlefield, the MP leader must have virtually perfect knowledge of the friendly situation one level higher. This means the MP platoon leader must know the company situation and the location and mission of the adjacent company and platoons. It is also important that the platoon leader update his subordinate leaders periodically regarding the higher situation. The platoon leader must have a relatively complete knowledge of the terrain, and he must know as much as possible about the enemy. The requirement to maintain a real-time picture of the battlefield one level higher does not relieve the platoon leader of the requirement to understand the situation and the commander's intent two levels higher. The difference is that this understanding of the situation two levels higher does not have to be as specific or in real time.

2-68. Most of the information the platoon leader needs comes in the form of reports over communication channels. Subordinate leaders are required to periodically report their status. If an MP team is operating in an area that does not allow uninterrupted communications, the team leader coordinates with adjacent teams to relay his report. If an MP team does not report in a timely manner, the platoon leader must quickly determine the status of the overdue team.

2-69. If possible, the platoon leader monitors his platoon and company net. How effectively he can accomplish this is, to some degree, experience-dependent; however, there are techniques he can apply to relate the information he is receiving to the map and thereby track the tactical situation.

2-70. The platoon leader's map is the key to maintaining situational awareness. He plots all friendly position reports up to one level higher than his own and plots information from spot reports (SPOTREPs). He uses different colors for friendly and enemy elements to allow quick distinction. To avoid cluttering the map, he places a dot or symbol on the map where the element is located and labels the point with a number. The same number is then written in the map's margin (or beyond the AOs) with the complete SPOTREP or unit identification (ID) next to it. Include the time on this notation. As positions or reports are updated, the old symbol is crossed off and a new one with a corresponding notation is added; it is critical that updates to previous reports be clearly identified as such during transmission. This simple system can greatly increase the ability to track both friendly forces and enemy activity in a particular AO.

## BATTLE SPACE

2-71. As mentioned earlier, an accurate picture of the battlefield provides the platoon leader with important tactical information, including friendly and enemy positions and relevant terrain. In turn, complete understanding of the military significance of this picture requires knowledge of the concept of battle space, the key element in the intellectual process of visualizing the battlefield.

2-72. At the most fundamental level, battle space is the three-dimensional *bubble* or area in which the platoon can acquire enemy forces and influence them with effective fires. This space is defined by the following numerous battlefield factors:

- The locations of the friendly forces, including the platoon's individual teams, OPs, and patrols.
- The effects of the terrain, weather, and movement.
- The ranges of all the available platoon weapons and sensing systems.

2-73. Each squad has its own battle space; the platoon battle space is the sum of the individual squads battle spaces. Platoon battle space is not restricted by boundaries; it can overlap with the battle space of adjacent units. For example, an enemy element that is spotted outside of the platoon's AO can still adversely affect the platoon's mission. Coordination is made with adjacent units to detect and destroy the threat.

2-74. Battle space has applications in all phases of mission planning and execution. During the planning process, it is a critical factor in selection of the routes and tentative positions. Once mission execution begins, the platoon leader's knowledge of the battle space is critical to issuing timely and effective orders as the situation changes.

2-75. The importance of battle space demands that the platoon leader direct most of his battle command effort toward managing and enhancing his space. He must be aware at every moment how battle space is changing as friendly and enemy forces move and terrain and visibility conditions change. He must evaluate how these changes affect his squads.

2-76. As the operation progresses, the platoon leader must take active measures to shape the battle space to his best advantage. One vital step in this process is to eliminate any gaps, or dead space, that exist within the *bubble*. The platoon leader can accomplish this in several ways, including maneuvering teams, repositioning OPs, and deploying patrols or remote sensors.

## COMMAND POST OPERATIONS

2-77. Company- and larger-size elements have a tactical operations center (TOC) and platoons have CPs. The CP is wherever the platoon leader goes. It can be mobile or stationary. No matter the location, there must be communication with and command of the unit and a method for battle tracking. CP activities are a 24-hour operation. The leadership must ensure that there is a plan for continuous operations.

2-78. A mobile CP may be a HMMWV and an ASV. The platoon leader can make decisions on the move while having communication capabilities nearby. A leader's book can hold information on battle rosters, report formats, and a map of the battlefield.

2-79. A stationary CP may be the platoon leader's tent or office. Communications may be telephonic, by messenger, or by radio. Track the status of equipment and personnel by using charts. Use large maps to track the battle. Keep in mind that moving to another location requires taking charts and maps, so everything needs to be mobile.

2-80. Key personnel for CP operations are the platoon leader, the PSG, and the radio/telephone operator (RTO).

## PLATOON LEADER

2-81. The platoon leader is responsible for $C^2$ of his organization and—

- Planning the missions according to the commander's guidance.
- Planning security to include the placement of crew-served weapons.
- Issuing orders and providing work priorities to subordinate leaders.
- Conducting PCI.
- Planning and conducting platoon rehearsals.
- Battle tracking (knowing) on a map the exact location of all the MP teams, to include the position of friendly platoons and known enemy, and a platoon-sector sketch.
- Adapting to new situations, making necessary adjustments, and issuing FRAGOs.

- Controlling the movement of the subordinate units.
- Ensuring that the communication net is established.
- Conducting risk assessment and continuously updating it.
- Reporting to higher HQ using the correct report format.

## PLATOON SERGEANT

2-82. The PSG is responsible for the logistics of the operation and—

- Coordinating and providing all the needed class items.
- Ensuring equipment serviceability and accountability.
- Supervising the maintenance.
- Establishing and supervising a sleep plan.
- Supervising the security plan's execution.
- Supervising the maintenance of work priorities.
- Supervising PCIs.
- Supervising sanitation and hygiene.
- Adjusting, manning, and cross leveling the soldiers, weapons, and equipment.
- Coordinating and supervising morale services.
- Assisting the platoon leader in rehearsals.
- Supervising test firing.
- Assisting with battle tracking.
- Consolidating subordinate units' status reports for the platoon leader.
- Assisting with the reports.
- Providing technical and tactical advice to the platoon leader.
- Maintaining situational awareness.

## RADIO/TELEPHONE OPERATOR

2-83. The RTO should be an experienced MP who is also the platoon leader's driver. The RTO—

- Communicates with higher HQ and separate units.
- Submits the required reports according to orders and SOPs.
- Maintains a record of communications.
- Maintains the radio and communication equipment.
- Assists with battle tracking.

# Chapter 3

# Shoot, Move, and Communicate

The ability of an MP unit to shoot, move, and communicate ensures its ability to detect, disrupt, and defend against the enemy and immeasurably adds to its survivability and maneuverability. MP are structured to be strategically, operationally, and tactically agile to respond to the increased range of worldwide MP requirements.

## SHOOT

3-1. It is important that MP understand their shooting capabilities and limitations. Just as important is the understanding of firing techniques and associated fire distribution, reacting to air and armor attacks, calling for fire, and obtaining various fire support.

### UNDERSTAND FIRE TECHNIQUES

3-2. Fire techniques include fire from or at a moving vehicle, fire distribution and control, and suppressive fire.

#### Fire From or at a Moving Vehicle

3-3. The key to forward maneuver is firing on the enemy. When maneuvering, the fire element—

- Attempts to destroy or suppress the enemy.
- Covers and protects the maneuver element as it advances.
- Moves, when possible, into its firing position undetected. Fire from an unexpected direction

has a greater effect than fire from a known position.

3-4. Firing on the move is less accurate than firing from a halt. However, to halt and fire takes more time and is more dangerous. A stationary vehicle is more likely to be hit than a moving vehicle. The team leader must decide whether to fire while moving or to fire from a short halt. He bases his decision on sound judgment and evaluation of the threat.

3-5. Crew-served weapons engage all targets on the move with free gunfire. To deliver this type of fire, the gunner removes the traversing and elevating (T&E) mechanism from the bottom of the receiver, allowing the gun to move freely in any direction. Accurate firing with crew-served weapons while moving is affected by—

- The terrain.
- The vehicle's speed.
- The team's proficiency.

3-6. When aiming from a moving vehicle or at a moving vehicle, or both, the gunner must lead the target. The speed of the firing vehicle, the time of flight, and the *angle of engagement* affect the amount of lead required. The time of flight is the required time it takes the projectile to move from the firing vehicle to the target. The angle of engagement is the angle found between the centerline of the vehicle and the gun when laid on the target. When a round is fired from the flank of a moving vehicle, the round drifts in the same direction and at the same speed as the vehicle. The longer the flight time and the larger the engagement angle, the greater the drift. Thus, the gunner must apply more lead to the shot. If a lead is required and the gunner is traversing left to keep on target, the gunner must lead left. If the gunner is traversing right to keep on the target, the gunner must lead right. This is true whether the firing vehicle is moving, the target is moving, or both are moving. *Table 3-1* shows the responsibilities of an MP team when firing on the move.

**Table 3-1. Team Responsibilities When Firing While Moving**

| Position | Actions |
|---|---|
| Team leader | • Directs the driver.<br>• Keeps the gunner oriented.<br>• Senses the impact of the rounds–long, short, left, or right of the target.<br>• Identifies additional targets.<br>• Assists the gunner with reloading, if required.<br>• Observes the surrounding terrain. |
| Gunner | • Develops a *feel* for the moving vehicle.<br>• Tracks the position of the target with the MK19 grenade machine gun (GMG) despite the movement of the vehicle.<br>• Remains alert to the sounds of the engine and transmission. These sounds indicate the type of terrain over which the vehicle is traveling and helps the gunner anticipate vehicle movements. |
| Driver | • Tries to maintain a steady gun platform while the gunner engages the targets.<br>• Attempts to time the gear and direction changes so they occur immediately after firing and do not interfere with accuracy.<br>• Informs the gunner of obstacles in the vehicle's path that might affect the gun's accuracy.<br>• Announces "depression," "turn," and the like to warn the gunner of vehicle movements.<br>• Announces, "steady" to let the gunner know when the vehicle is once again on a stable platform. The gunner assumes he has a stable platform unless the driver informs him otherwise. |

## Distribute Fire

3-7. MP leaders must distribute the fires of their organic weapons to destroy or suppress enemy positions. The following are the two methods to distribute fire on a target:

- Point fire. Point fire *(Figure 3-1, page 3-4)* is directed against one target (such as a machine gun position) with all the troops firing at the

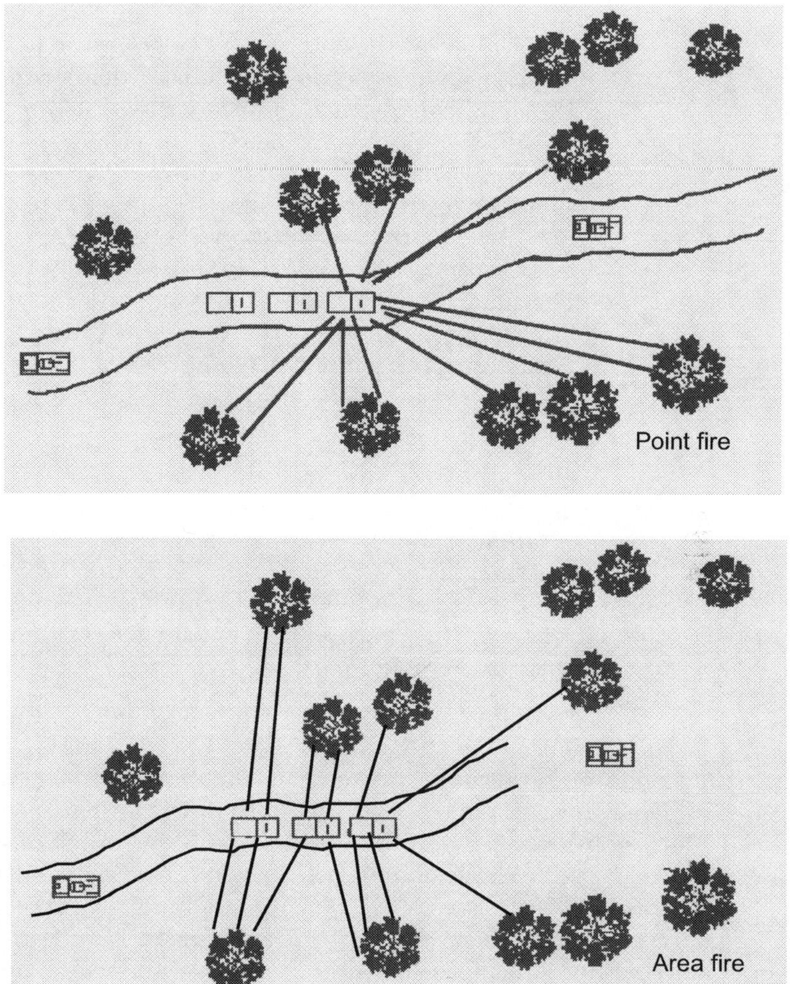


**Figure 3-1. Fire Distribution**

same target. Spreading out the base-of-fire element makes this type of fire particularly effective because the fire is directed from many sources.

- Area fire. Area fire *(Figure 3-1)* permits rapid cover of an entire area with fire from the left to the right and in depth, even if the enemy cannot be seen. This method is used without command and is the quickest and most effective way to bring all parts of a target under fire. Each member in the element is assigned a portion of the target. Fire is placed on likely locations for enemy positions rather than in a general area. If the leader wants fire on a wood line, he may shoot tracers to mark the center of the target. Soldiers to the left of the leader fire to the left of the tracers and soldiers to his right fire to the right of the tracers.

3-8. A rifleman fires his first shot on the part of the target that corresponds to his individual position. If he is left of the leader, he fires to the left of the leader's tracers. He then distributes his remaining shots over the part of the target extending a few meters right and left of his first shot. He covers the part of the target that he can hit without changing position.

3-9. A grenadier fires into the center of the target area of his team. He then distributes his shots over the remaining target area from the center to each side and from front to rear. A machine gunner covers part of the target depending on his position and how much of the target is in range. When possible, he covers the entire target of the team. When placing automatic suppressive fire on the enemy, the tendency is to shoot high. Therefore, he places the first bursts low and works up to the target. The squad leader tells the machine gunners where to shoot by assigning sectors of fire.

3-10. An MK19 gunner engages area targets with traversing and searching fire after the leader designates the width and depth of the target. If one MK19 GMG is being fired, the gunner engages the area target by adjusting his fire on the center of the mass, then traverses and searches to either flank. When he reaches the flank, he reverses direction and traverses and searches in the opposite direction. If two MK19 GMGs are being fired as a pair, the point of the initial lay and adjustment for both guns is on the midpoint of the target. After adjusting the fire on the center of the mass, fire is distributed by applying direction and elevation changes that give the most effective coverage of the target area. Usually, the right gun (number 1) fires on the right half, and the left gun (number 2) fires on the left half. *Appendix G* describes the MK19 qualification and familiarization tables and provides a sample scorecard.

## Control Fire

3-11. Fire control is an essential component of fire distribution. A platoon leader must know what means he will direct the fire element to use when engaging the targets. He will communicate directly or use prearranged signals to identify the location of the target to the other units. He may use sound signals (such as voice, a horn, or a whistle), but must remember that they are only good for short distances and that their reliability and range are reduced by battle noise, weather, terrain, and vegetation. Use a radio to direct the base-of-fire element or adjust fires from reference points or landmarks, because a radio offers immediate voice communication. For example, he may say, "From the burning scout vehicle, northwest 50 meters, machine gun position." If portable radio equipment is not available, he uses prearranged visual signals, such as smoke or flares. A smoke round from a grenade launcher, unless it is being used for some other purpose, and a smoke canister can be used as a signal. Use these items during reduced visibility in addition to aiming stakes, illumination, night-vision devices, infrared chemical lights, and so forth.

### Use Fire Commands

3-12. Leaders use fire commands to direct fire. A fire command has the following six parts:

- Alert. The leader alerts the soldiers to receive further instruction. He alerts the soldiers by name or unit designation, some type of visual or sound signal, personal contact, or any other practical way.
- Direction. The leader tells the soldiers the general direction to the target. In some cases, he pinpoints a target. The following are the three ways the leader can give the direction to the target:
  - Points with his armor rifle.
  - Fires tracer ammunition at a target.
  - Uses either target reference points (TRPs) or easily recognized man-made objects or terrain features. He gives the general direction just before giving the reference points.
- Description. The leader describes the target briefly but accurately and always gives the formation of the enemy soldiers.
- Range. The leader tells the soldiers the range to the target in meters.
- Method of fire. The leader tells the soldiers which weapons to fire, the type and amount of ammunition to fire, and the rate of fire.
- Command to fire. The leader tells the soldiers when to fire by using an oral command or visual signal. When he wants to control the exact moment of fire, he says, "At my command" (then pauses until ready to commence firing). When he wants to start firing on completion of the fire command, he just says, "Fire."

## Use Subsequent Fire Commands

3-13. These commands adjust or change information given in the initial fire command. Only the elements that change are given.

## Terminate Fire

3-14. Fire is terminated by the command or signal for *cease fire, end of mission*.

## Suppress Fires

3-15. When the fire element is in position, it lays a heavy volume of fire on the enemy to suppress them. When the leader senses that the enemy is suppressed, he instructs the fire element to reduce its rate of fire as long as it keeps the enemy suppressed. As the movement element nears its objective, the fire element increases the rate of fire to keep the enemy down. This lets the movement element close enough to assault the enemy before the enemy can react. When the assault begins, or on a signal, the fire element stops firing, shifts its fire to another target, or walks its fire across the objective in front of the movement element, and then shifts or ceases fire.

3-16. Positions for fire elements are located so that movement of the maneuver element does not mask their fires. Fire element positions are often higher and usually to the flank of the maneuver element. The maneuver element neither masks the fire of the fire element nor moves outside the protective umbrella provided by the fire. A platoon or squad can point fire at one target or an area of several targets. In both cases, the leader must control the fire. He must ensure that the fire is directed on the enemy, not on the maneuver element.

## Use Nonlethal Weapons (NLW)

3-17. The Department of Defense (DOD) defines NLW as weapons that are explicitly designed and primarily employed to incapacitate personnel or material while minimizing fatalities, permanent injury to personnel, and undesired damage to property and the environment. Unlike conventional weapons that destroy the targets principally through blast, penetration, and fragmentation, NLW employ means other than gross physical destruction to prevent the target from functioning.

3-18. NLW doctrine and concepts of operation are designed to reinforce deterrence and expand the range of options available to commanders. They enhance the capability of US forces to accomplish the following objectives:

- Discourage, delay, or prevent hostile actions.
- Limit escalation.
- Take military action in situations where use of lethal force is not the preferred option.
- Protect US forces.
- Disable equipment, facilities, and personnel temporarily.

**NOTE: The zero probability of producing fatalities or permanent injuries is not a requirement of NLW. However, while complete avoidance of these effects is not guaranteed or expected, when properly employed, NLW significantly reduce them as compared with physically destroying the same target.**

3-19. When drafting the ROE, it must be clearly articulated and understood that the role of NLW is an additional means of employing force for the particular purpose of limiting the probability of death or serious injury to noncombatants or belligerents. However, the use of deadly force must always remain an inherent right of individuals in instances when they, their fellow soldiers, or personnel in their charge are threatened with death or serious bodily harm. NLW add flexibility to the control of disturbances within the I/R facility and provide an environment where guard forces can permissively engage threatening targets *(Figure 3-2)* with limited risk of noncombatant casualties and collateral damage. Refer to *FM 90-40*.

3-20. The use of lethal force, employed under the standing ROE, will never be denied. At no time will forces be deployed without the ability to defend themselves against a lethal threat, nor will they forego normal training, arming, and equipping for combat. Nonlethal options are a complement to, not a replacement for, lethal force and seek to expand a proactive response across the range of military operations. Refer to *FM 90-40*.

3-21. The decision to use NLW against an adversary during a confrontation is delegated to the lowest possible level, preferably to the platoon or the squad. However, this requires that all personnel, not just the leaders, have a clear understanding of the ROE and the commander's intent. Refer to *FM 90-40*.

3-22. Commanders and public affairs officers must be prepared to address media questions and concerns regarding the use and role of NLW, and they must make it clear that the presence of NLW in no way indicates abandoning the option to employ deadly force in appropriate circumstances.

3-23. **Advantages of Employing Nonlethal Weapons**. NLW provide the commander with the flexibility to influence the situation favorably with reduced risk of noncombatant fatalities and collateral damage.

| NLW | Meters | | | | | | | | | | | | |
|---|---|---|---|---|---|---|---|---|---|---|---|---|---|
| | 0 | 5 | 15 | 20 | 25 | 30 | 35 | 50 | 55 | 65 | 95 | 100 | 120 |
| Modular crowd control munitions (MCCM) | Lethal zone | | | X | X | X | X | X | X | X | X | X | X |
| Stun hand grenade | | | | | X | X | X | X | X | X | X | X | X |
| 12-gauge point | | Lethal zone | Nonlethal zone | | | | X | X | X | X | X | X | X |
| 12-gauge area | | | | | | | X | X | X | X | X | X | X |
| 40-millimeter point | | | Nonlethal zone | | | | | | X | X | X | X | X |
| 66-millimeter sting ball | | | | | | | | Nonlethal zone | | | | | X |
| 66-millimeter flash bang | | | | | | | | | | | | | X |
| 66-millimeter CS grenade | | | | | | | | | | | | X | X |

Figure 3-2. Range of Munitions Contained in a Nonlethal Capability Set

3-24. NLW can be more humane, be consistent with the political and social implications implicit in humanitarian missions, be used during peacekeeping missions, and facilitate post-incident stabilization by reducing internee alienation and collateral damage.

3-25. The force that properly employs nonlcthal options gains advantages over those who rely on lethal options alone, because the degree of provocation required to employ these options is substantially less. This advantage provides a more proactive posture and quicker response as well as a diminished likelihood of having a situation escalate to a point where deadly force is required to resolve a conflict within the I/R facility.

3-26. NLW options are less likely to provoke others and the use of NLW, in fact, may provoke a negative response. However, demonstrated restraint greatly diminishes feelings of anger and remorse when deadly force is required after nonlethal options fail.

3-27. **Military Police Nonlethal Weapons**. I/R facility commanders consider the use of force options discussed in *Chapter 2* and *AR 190-14* when dealing with disruptions within the compound. Facility commanders are encouraged by *AR 190-14* to substitute nonlethal devices for firearms when it is considered adequate for MP to safely perform their duties. Currently, MP have such nonlethal options as riot-control agents chlorobenzul-malononitrile (CS) and oleoresin capsicum (OC), military working dogs (MWD), an MP club, and a riot baton for crowd control. There are other nonlethal devices currently being tested and fielded that will be available to the I/R commander in the future.

3-28. **Nonlethal Training**. Soldiers and their leaders must be trained in the correct employment of NLW available to them. Soldiers and their leaders must understand the limited use of these systems in environments with restrictive ROE. Their training must be continuous at all levels to ensure that NLW are properly employed and that the leaders and the soldiers understand when and how to effectively employ them. They must understand that the incorrect application of NLW can have significant operational and political ramifications. Well-trained MP leaders, providing timely and clear guidance to MP soldiers using NLW, will ensure the mission's successful accomplishment.

3-29. Many NLW have both maximum effective and minimum safety ranges. Individuals struck short of the minimum safety range often suffer severe injuries or death while the effects of most nonlethal devices are greatly mitigated at longer ranges. In order to be effective, engage the threat within the *effective* zone, beyond the minimum safety range, and short of the maximum effective range.

3-30. When training with and planning for the use of NLW consider the following:

- Never apply NLW in a situation where deadly force is appropriate.
- Never apply NLW in a situation that will place troops in undue danger.
- Always cover NLW with deadly force.

3-31. **Nonlethal Tactics**. *FM 90-40* provides an in-depth discussion on the tactics associated with the employment of various NLW available to the commander, such as—

- Riot formations. Riot formations establish riot-control teams with a minimum response time. Because of the physical nature of riot control, individuals in riot control formations should not carry long rifles. Nonlethal attachments should follow closely behind the riot control formation. Lethal coverage should be provided for this entire formation. Refer to *FM 90-40*.

- Designated marksman. During a nonlethal engagement, the use of a designated marksman (DM) provides confidence and safety to those facing a riot. If a lethal threat is presented, the DM who is in an overwatch position and armed with a standard infantry rifle, mounted with a high-powered scope, can scan a crowd and identify agitators and riot leaders for apprehension as well as fire lethal rounds if warranted. Additionally, he is ideally suited for flank security and countersniper operations. Refer to *FM 90-40*.

**React to An Air Attack**

3-32. Passive and active air defense measures are used by MP to protect themselves from enemy air attacks during all missions.

3-33. **Passive Air Defense Measures.** MP employ passive air defense measures that include actions to avoid detection and air attack, and actions to limit damage if they are attacked. MP use active air defense measures to fight back against enemy aircraft. Enemy aircraft will attack and attempt to destroy any target seen. The passive air defense methods that limit enemy detection include—

- Concealment.
- Camouflage.
- Cover.

3-34. Dispersion is another passive air defense method. Its purpose varies from those of concealment, camouflage, and cover. While they are designed to hide MP and their vehicles and equipment from the view of the enemy, dispersion is employed to reduce the effects of an enemy air attack. Dispersion is important when MP are occupying a static position (such as a company TOC or platoon CP) or when they come under air attack. If MP come under air attack, they quickly disperse with their vehicles, move to a concealed position, if possible, and stop. Stationary vehicles are more difficult for the enemy to detect than moving vehicles.

3-35. Early warning (quick recognition of enemy aircraft) is a passive air defense method that affords MP an opportunity to take cover and one that may lead into active air defense measures. The warning may come through communication channels, OP and listening posts (LPs), or from convoy air guards. A whistle, a voice, a radio, or any other method can provide a warning.

3-36. All OP/LPs watch for enemy aircraft as a standard duty. When air sighting, the first person to see an enemy aircraft shouts, "Aircraft," then, "Front (Right, Left, or Rear)." In a convoy, air guards are given sectors of the sky to observe for enemy aircraft. When an enemy aircraft is spotted, the predetermined alarm (such as a horn or hand signal) is given until all vehicles are aware of the situation.

3-37. When an alarm is given, all dismounted troops take cover at once. They go below ground level, if possible. If the aircraft is not firing, MP withhold their fire to avoid disclosing their position and they allow the aircraft to pass. They stay concealed until the all clear is given. The MP leader initiates a size, activity, location, unit, time, and equipment (SALUTE) report (or refer to the SOP) for the sighting of hostile aircraft.

3-38. **Active Air Defense Measures**. Although passive measures are the first line of defense against air an attack, MP must be prepared to engage enemy aircraft. Low-flying hostile aircraft may appear suddenly from behind low hills, trees, or a haze. To gain surprise, they may attack with the sun behind them. Before MP fire at enemy aircraft, they must positively identify the aircraft as hostile. If the aircraft is making a firing run on the patrol, take cover and return fire. However, commanders may restrict active air defense when friendly aircraft are in the area.

3-39. In convoys, drivers alternately pull their vehicles off the road to the right and left, seeking concealment from air observation. Caution must be used if mines are a known threat. If the enemy aircraft is not attacking, the same actions stated earlier are taken. If the aircraft is attacking, MP dismount and seek cover away from the vehicle (the vehicle may be the aircraft's target) and return fire. All personnel remain under cover until the command is given to continue the mission.

3-40. Fire small arms at attacking aircraft during or after the first attack. MP direct fires to saturate the airspace through which the aircraft will fly without trapping the aircraft. For more detailed information refer to *FM 44-8*. When engaging hostile aircraft—

- Fire only on command unless under direct attack (being fired on by aircraft).
- Ensure that the direction of fire does not place rounds on friendly personnel, equipment, or positions.
- Deliver a large volume of fire.
- Lead a slow-moving aircraft and adjust fire by observing the flight of the rounds, especially if tracer rounds become available, using the free-gun technique of fire.
- Aim at the center mass of a grounded or hovering helicopter and a helicopter that is coming directly at your position. Cease-fire when the aircraft passes out of range.

3-41. Slow-moving rotary-winged aircraft that are on the ground, hovering, taking off, or landing are most successfully engaged by the MK19. The MK19 GMG's 40-millimeter ammunition is fired at a relatively slow speed and has a high trajectory at a long distance. The MK19 GMG rounds that do not hit the aircraft will detonate on impact with the ground. The location of friendly elements within range of the weapon must be considered at all times. *Figure 3-3* depicts the rules for selecting aiming points of various aircraft.

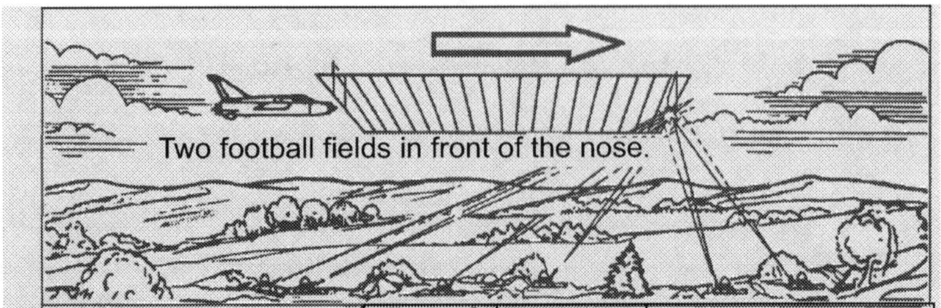


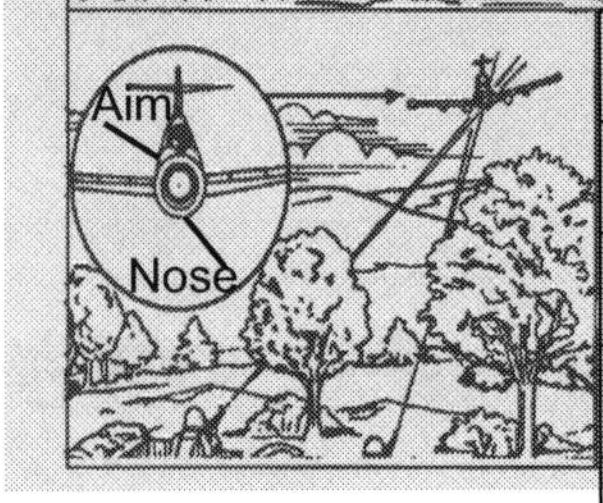


| Type of Aircraft | Course | Aim Point |
|---|---|---|
| Jet | Crossing | Two football fields in front of the nose |
| Jet | Overhead | Two football fields in front of the nose |
| Jet | Directly at you | Slightly above the aircraft's nose |
| Helicopter | Crossing | One-half football field in front of the nose |
| Helicopter | Hovering | Slightly above the helicopter's body |
| Helicopter | Directly at you | Slightly above the helicopter's body |

**Figure 3-3. Rules for Selecting the Aim Point**

## REACT TO ARMOR

3-42. MP maneuver and operate in much of the battle space and can expect to encounter pockets of by-passed enemy armor within the rear area. MP engage enemy armor targets only for self-defense or when total surprise can be achieved. MP place antiarmor weapons on avenues of approach to defend against enemy armor. After initial contact, MP relocate immediately to avoid the fast-moving enemy armor and its firepower.

3-43. The (M136) antitank (AT) 4 provides antiarmor capability for MP teams. The AT4 is primarily employed against light armored vehicles, such as personnel carriers. It has a very limited capability against main battle tanks. The AT4 is issued as ammunition rather than as an individual weapon. It is carried and employed in addition to the basic weapon of MP.

3-44. The most stable firing positions for the AT4 are standing supported, prone, and prone supported. Whenever possible, use a supported position which provides more stability and aids in aiming.

## Engage Armored Vehicles

3-45. The four methods to engage armored vehicles with the AT4 are—

- Single firing.
- Sequence firing.
- Pair firing.
- Volley firing.

3-46. Refer to *Table 3-2* for a description of the four methods. The best methods of engaging armored vehicles are pair and volley. Regardless of the method used, the closer the target, the better the chance for a first-round hit. Aim for the center mass of the target. The most vulnerable spots of armored vehicles are the top and the rear. The sides of the armored vehicles can also be penetrated.

3-47. An armored vehicle without the protection of dismounted infantry is vulnerable to a close attack by well-armed dismounted units. When an armored vehicle is buttoned up, visibility of the crew is restricted. This provides an opportunity for an armor-killer team to approach the vehicle with less risk of detection. The types of vehicles and the methods to engage enemy armored vehicles are—

- Stationary targets. Place the center post at the center of the visible mass for stationary targets. The firer does the same for vehicles that are moving toward or away from him.
- Slow-moving targets. Place the center post on the front leading edge of the vehicle (less than 10 miles per hour [mph]). This method is also applied to oblique moving targets.
- Fast-moving targets. These targets are moving more than 10 mph. If the target is moving to the left, place the right lead post at the center of the mass. If the target is moving toward the right, place the left lead post at the center of the mass.

**Table 3-2. Methods of Engagement with the Light Antiarmor Weapon (LAW)/AT4**

| Method | Gunners | Firing Sequence | Probability of a Hit |
|---|---|---|---|
| Single firing | One | One gunner fires one AT4 at the target. | Low. Use only at ranges up to 200 meters for AT4s. Regardless of the method used, the closer the target the better the chances of a hit. |
| Sequence firing | One | One gunner fires two or more AT4s in turn. He prepares several weapons for firing before engaging the target. He gets the weapon, estimates the sight picture, and shoots the weapon in turn. | Good. If the first round misses, the gunner adjusts the range and the lead of the succeeding rounds until he gets a hit and fires until the target is destroyed. |
| Pair firing | Two or more | Each gunner fires one or more AT4s at a target, one at a time. They prepare several weapons for | Better. Two or more gunners track the target at one time, letting them get a target hit sooner. They can be ready to shoot as soon as an earlier |

**Table 3-2. Methods of Engagement with LAW/AT4 (Continued)**

| Method | Gunners | Firing Sequence | Probability of a Hit |
|---|---|---|---|
| | | firing before engaging the target. | round hits. The first gunner sees a target, identifies it, and states the estimated range and lead to use. For example, the gunner, on spotting a fast-moving scout reconnaissance vehicle, says, "BMP 150 meters; fast target." He then fires at the target. If the first gunner misses, the second gunner gives a revised range and lead. This continues until one gets a hit. Once the correct range or lead has been found, all the gunners fire until the target is destroyed. |
| Volley firing | Two or more | Each gunner fires one or more AT4s on command or on signal until the target is destroyed. They prepare several weapons for firing before engaging the target. | This is the best method of engagement for an AT4 because the gunners shoot more rounds at a target at one time. This method is used only when the range and lead to the target have been determined. Range can be determined by using the map, pacing, or the results of pair firing after a target has been hit. |

## Estimate the Range of the Targets

3-48. A gunner has a better chance of hitting a target with the AT4 if he knows the range to the target. Determining the range is a learned skill and must be mastered by anyone who fires the AT4. Methods of range determination include—

- Using range finders.
- Measuring the distance on a map.

- Pacing.
- Firing the pair and sequence method.
- Using visual range estimation.

3-49. Visual range estimation is the least desirable method of range determination because of its inaccuracy. However, in the offense or in a hasty defense, it may be the only method available.

### Disable Armored Vehicles

3-50. Armored vehicles are hard to destroy when firing at their front. Use the following three ways to disable them:

- Mobility kill. In this disabling method, the vehicle has stopped moving because a track or road wheel has been destroyed, or the vehicle has been hit in the engine compartment. The vehicle can no longer move but can return fire.
- Firepower kill. When a firepower kill has occurred, the main gun cannot return fire because of a hit in the turret, knocking out its capability to fire. The vehicle can still move, thus enabling it to get away.
- Catastrophic kill. In this kill, the vehicle is destroyed. To obtain a catastrophic kill, firers prepare to fire a second or third shot to destroy the vehicle.

## CALL FOR FIRE

3-51. A call for fire is used to obtain fire support from other units. Fire support may be needed in the rear area if the enemy endangers key units or facilities. Fire support may come from mortars, artillery, Army aviation, and US Air Force aircraft. Before a mission, the commander will outline the TRPs and the priority of fires in the OPORD that will affect the call for fire response time. All MP must know how to call for and adjust fire. To call for fire the—

- Leader tells the RTO that a target has been seen.
- RTO initiates the call for fire while the target location is being determined.
- RTO sends the information as it is determined instead of waiting until a complete call for fire has been prepared.

3-52. MP may either go directly to the fire direction center (FDC) of the firing unit for artillery fire support or relay communications to MP leaders, the rear area CP fire support element, or when so directed, a TCF.

### Use an Initial Call for Fire

3-53. Artillery fire support can provide the rear area with on-order fires to assist in countering threat incursions. A standard call-for-fire message is used to obtain artillery or other fire support. Regardless of the method of the target location used, the call for fire consists of six elements transmitted in three parts. There is a break and a read back after each part.

3-54. **First Transmission**. Send elements one and two during the first transmission.

- Element 1 includes the identification of the observer. This element tells the FDC who is calling and clears the net for the remainder of the call.
- Element 2 includes the WO. The type of fire support mission and the method of locating the target are identified in this element. See *Table 3-3* for the types of fire missions.

3-55. **Second Transmission**. Element 3 is sent during this transmission and includes the target's location. A target location may be provided by the grid coordinate

**Table 3-3. Types of Artillery-Fire Missions**

| Mission | Description |
|---|---|
| Adjust fire | This mission is used when the observer is uncertain of the exact location of the target. The observer says, "Adjust fire." |
| Fire for effect | The observer uses this mission when he is certain of the location of the target. He is requesting a first-round fire for effect in this mission to validate the desired effect on the target with little or no adjustment. The observer says, "Fire for effect." |
| Suppress | This mission is used to quickly bring fire only on a preplanned target. A target description is not provided when requesting this mission. It is a simplified call for fire and is sent in one transmission. An example of what an observer may say is, "P25—this is P59—suppress AB2502—over." |
| Immediately suppress | This mission is similar to that of suppression. The difference is a planned target or a target of opportunity is firing at friendly soldiers or aircraft in this mission. The observer may say, "P25—this is P59—immediate suppression AB2503—over." |

(normally six digits), a polar plot, or a shift from a known position method.

3-56. **Third Transmission**. Send elements 4, 5, and 6 during this transmission.

- Element 4 includes the target's description. A brief description of the target is given to the FDC using size, nature, activity, protection (SNAP). SNAP represents the size or shape, the nature or nomenclature, the activity, and the protection or posture of the target.
- Element 5 includes the method of engagement. This element consists of the type of adjustments, danger close, trajectory, ammunition, and distribution. The observer specifies how he wants to attack the target.

- Element 6 includes the methods of fire and control. The observer states who will give the command for fire to begin. If the observer wishes to control the time of firing, he will say, "At my command." If the observer does not say this, the FDC will fire as soon as the element is ready.

3-57. At a minimum, a call for fire must include the first four elements. Untrained observers need to send only the first four elements, and the FDC will decide the methods of engagement, fire, and control. Every MP must know that in order to put indirect fire on a target, the following information must be given to the FDC:

- Who he is.
- Where and what the target is.
- How close the target is to friendly troops.
- Where the target is in relation to his or other known positions.
- The direction from himself to the target. Determine the direction during—
  - Daylight with the mini-eye safe-laser infrared observation set (MELIOS), AN/PVS-6. The lightweight laser rangefinder is capable of determining ranges 50 through 9,995 meters. Refer to *TM 11-5860-202-10*.
  - Hours of darkness with the infrared (IR) illuminator, AN/PEQ-2A. The AN/PEQ-2A is for use with night vision devices (NVD) and can be used as either a handheld illuminator or pointer or can be weapon-mounted. In the weapon-mounted mode, the AN/PEQ-2A can be used to accurately direct fire as well as illuminate and designate targets. This item is fully waterproof and can be taken down to extended depths without risk of leakage.

3-58. Determining the direction to a target is an essential skill for the observer. Direction is an integral part of terrain map association, adjustment of fire, and target location. Use the following paragraphs to manually determine the direction to a target.

- Use a compress. Using an M2 or lensatic compress, the forward observer (FO) can measure the direction. The FO will add or subtract the grid of magnetic (GM) angle to determine the grid direction to send to the FDC.
- Scale from a map. Using a protractor or an observed fire (OF) fan, the FO can scale the direction from a map to an accuracy of 10 mils.
- Measure from a reference point. Using a reference point with a known distance, the FO can measure the angle between the reference point and the target and add or subtract the measured angle to or from the known direction to determine the direction to the target. The angle between the reference point and the target can be measured with binoculars or with the hand measurement technique depicted in *Figure 3-4*.
- Estimate. With a thorough terrain-map analysis, the FO can estimate the direction by visualizing the 8-cardinal directions (north [N], northeast [NE], east [E], southeast [SE], south [S], southwest [SW], west [W], and northwest [NW]).

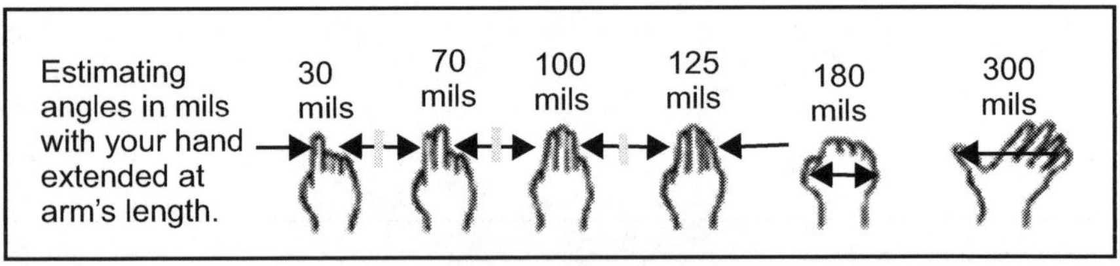


**Figure 3-4. Determine the Direction to a Target**

**NOTE: The observer tries to be as accurate as possible, and the use of mils is preferred. All measured directions sent to the FDC will be rounded to the nearest 10 mils.**

## Identify the Elements of a Call for Fire

3-59. FDC personnel will help in the call for fire and subsequent adjustments by asking leading questions to obtain the information needed. Refer to *FM 6-30* for call for fire. The elements of a call for fire include the following:

- Observer identification. The observer identification tells who you are. Use the call signs from the signal operating instructions (SOI).
- Warning order. The WO alerts the firing units of the—
  - Type of mission. This includes adjusting fire, firing for effect, suppressing fire, and immediately suppressing fire.
  - Size of element to fire. Omission indicates a request for one field artillery (FA) battery. If the fire mission requires an element larger than a FA battery, state the size needed, such as 2 FA batteries (battalions).
  - Method of the target's location. The grid has no announcement. Announce the word "polar" for the polar plot. Shift from a known point by announcing the word "shift" followed immediately by the designation (target number) of the known point.
- Target location. The target location enables the FDC to plot the target.
  - Grid: Two-character, 6-digit grid, such as NA123456.
  - Polar: Direction (grid azimuth) and distance (meters) to the target from the observer's position. Give the difference in elevation if there is a vertical shift of over 35 meters between the observer and the target.
  - Shift from a known point or the direction to the target (grid azimuth). The three types of shifts are the lateral shift (left or right) in

meters, the flange shift (add or drop) in meters, and the vertical shift (up or down) over 35 meters from the known point and target.

- Target description. The target description helps the FDC to select the type and the amount of ammunition. A word picture of the target (for example, the number and type of vehicles or personnel observed).
- Method of engagement. The method of engagement tells the FDC how to attack the target, which includes the following:
  - Type of engagement. Area fire is standard without a request. Request precision fire only to destroy a point target.
  - Trajectory. A low-angle trajectory is standard without a request. An high-angle trajectory is at the request of the observer or when required due to masking terrain.
  - Danger close. Danger close is announced when applicable.
  - Ammunition. Ammunition is the type of projectile, the type of fuse action, and the volume of fire desired in the fire-for-effect phase stated in rounds per howitzer.
  - Distribution. Distribution is the type of sheaf desired and parallel is standard without request.
- Method of fire and control. The method of fire and control tells the FDC how you want to control the delivery and adjustment of the fire.
  - Method of fire. In area fire, the adjustment normally is conducted with one howitzer or with the center gun of a mortar platoon or section. If for any reason the observer

determines that *platoon right (left)* will be more appropriate, he may request it. Adjusting at extreme distances may be easier with two guns firing. The normal interval of time between rounds fired by a platoon or battery right (left) is 5 seconds. If the observer wants some other interval, he may so specify.

- Method of control. Use the following methods of control: *Fire when ready* is standard and no request is required; use *at my command* when weapons fire at the observer's command; use *cannot observe* when fire will not be observed; use *time on target* when rounds land at a specified time; use *continuous illumination* when the FDC will determine when to fire; use *coordinated illumination* when illumination rounds are fired only when the target is engaged; use *cease loading* when missions have two or more rounds in effect, causing the firing unit to stop loading rounds; and use *check firing* when there is a temporary halt in firing.

- Danger close. Include the term *danger close* in the method-of-engagement portion of the call for fire when the target is within 400 meters of any friendly troops or mortars and 600 meters for field artillery. When adjusting naval gunfire, announce "danger close" when the target is located within 750 meters when using naval guns that are 5 inches or smaller. For naval guns larger than 5 inches, announce "danger close" when the target is within 1,000 meters. The creeping method of adjustment will be used exclusively during danger close missions. The FO makes range changes by creeping the rounds to the target using corrections of less than 100 meters.

## Plot a Target Location

3-60. Selecting a targeting method includes giving the directions in mils, degrees, or cardinal points of the compress (N, NE, S, SW, E, SE, W, and NW). Give a deviation of left or right and the distance in meters. Use the following paragraphs when plotting a target location:

3-61. Grid. Determine a 2-character, 6-digit grid for the target. Then, determine a grid direction to the target, and send it after the call for fire and before any subsequent corrections.

3-62. Polar. Determine the grid direction to the target. Determine the distance from the observer to the target. Determine if any significant vertical interval exists.

3-63. Shift. Refer to *Figure 3-5* and determine—

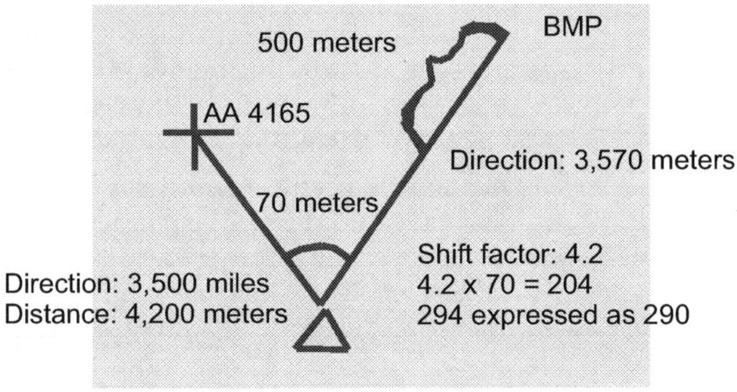


**Figure 3-5. Call for Fire Elements**

- The grid direction to the target.
- The lateral shift to the target from the known point. W = R times m (mil relation formula), when—

$W = R \; x \; m$

*W = width of the lateral shift (the unknown)*

*R = shift factor, the distance to the known point divided by 1,000 and expressed to one decimal place*

*m = measured angle in mils from the known point to the target*

- The range shift from the known point to the target.
- Any significant vertical interval that may exist.

Example of plotting a target location: Complete target location – direction 3,570, right 290, add 500

## Adjust Fire

3-64. The objective in adjusting fire is to move the center of the impact to within 50 meters of the center of the target. The observer accomplishes this by sending the FDC subsequent corrections, which are deviation (lateral) and range corrections. The FDC can talk the observer through the adjustments if they are necessary. The burst is moved to, and kept on, the observer target line in order to get a positive range spotting. The observer target line is the line of sight (an imaginary line) between the observer and the target. When range spotting cannot be determined, the observer makes a request for a lateral correction to place the burst on the observer target line.

3-65. The observer makes range corrections to bracket the target between two successive rounds. The successive bracketing technique is used. After the first definite range spotting is determined, a correction is sent to the FDC to establish a bracket of known distance around the target (such as, one round over the target and one round short of the target). The observer then successively splits this bracket until he is within 50 meters of the target and calls for fire for effect.

3-66. Hasty bracketing is used if the nature of the target dictates that effective fires are needed faster than the above procedures can provide. Hasty bracketing depends on a thorough terrain analysis to give the observer an accurate initial-target location. A bracket is obtained on the first correction in a manner like that used for successive bracketing. Once the initial bracket is established, it is used as a yardstick to find subsequent corrections. The observer sends the FDC the correction to move the rounds to the target and calls for fire for effect. Hasty bracketing improves with observer experience and judgment. Fire for effect consists of one or more rounds from each gun of the unit firing at the target. Dispersion of the guns will cause the rounds to saturate the area with shell fragments. To end a fire mission, the observer states, "End of mission," and reports the results of the fire for effect. The observer may say, "End of mission, three T-62s neutralized, estimate two casualties, over."

## Illuminate the Battlefield

3-67. Battlefield illumination can provide MP with enough light to aid in ground operations at night. Illumination can—

- Mark the targets for CAS.
- Increase visibility in areas of suspected enemy activity.
- Furnish the direction to the patrol activity.

3-68. Illumination is called for and adjusted like other indirect fires except the methods of engagement, fire, and control differ. The observer requests illuminating shells. The methods of fire and control differ in that the adjustment is based on how much visibility is needed in the target area. If the observer calls for—

- "Illumination," the observer gets one round from one gun.
- "Illumination, two guns," the observer gets one round each from two guns. The rounds will burst simultaneously.

- "Illumination, range and lateral spread," the observer gets one round each from four guns. The rounds will burst simultaneously in a diamond pattern.

3-69. The initial request for illumination must include the—

- Date when the illumination is needed, if illumination is preplanned.
- Purpose of the illumination.
- Requested time and duration of the illumination (for example, three minutes at 2150 hours or three minutes on call).
- Grid reference and, if needed, the height of the points or areas to be illuminated.
- Method of controls (any restrictions in the time and the place before and during the operation).

## OBTAIN ARMY AVIATION FIRE SUPPORT

3-70. Army aviation provides the echelon commander with the ability to move combat resources across the battle space with little regard for the terrain's barriers. These units can provide surveillance or screen over a wide area in adverse weather and at night. Attack helicopter units provide the sustainment area with highly maneuverable antiarmor firepower. They are ideally suited for situations in which rapid reaction time is critical.

### Control the Fire

3-71. While en route to a target area, the attack helicopter will contact the caller on the radio. For example, "1L22, this is 1X47, fire team arrives at estimated target area in 4 minutes, over." At this time, a call for fire is transmitted consisting of—

- The target's location and description.

- The proximity of the friendly unit to the target. The words *danger close* must be included when a friendly unit is 600 meters or less from the target. *Danger close* is required because some types of ordnance cannot be used in close proximity to friendly ground forces. When *danger close* is included, MP must mark the unit's location. The method of marking should be one that least reveals the position to the enemy, such as using panels or mirrors.
- The protection of the friendly units (such as good fighting positions, hasty positions, or exposed positions).
- The direction of the friendly unit from the target (cardinal direction).
- Other friendly fire support considerations, including artillery and mortars firing in the area and tactical aircraft (attack direction and altitude).
- Dangers to the flight. Report locations of known or suspected enemy antiaircraft weapons or other dangers to the flight (wires in the target's area, enemy artillery fire impacting in the target's area, or enemy aircraft).

3-72. When the helicopter arrives over the objective, the helicopter's fire team contacts the caller. The caller marks the target and states the method of adjustment. The caller can use the three following methods to mark the target:

- A reference either to a prominent terrain feature that can be identified from the air or to a known point.
- A direction to the target from a reference point, stated in mils or degrees.
- References to friendly fire (such as smoke grenades, tracers, smoke streamers, mortars, artillery, or marking rockets).

## Use the Fire Adjustment Method

3-73. The three methods used to adjust the fire of a fire team are impact-observed, impact sound, and observer target. The preferred methods of adjustment for an attack helicopter in support of a ground force are impact observed and impact sound. Once established, the caller does not change the method of adjustment unless the situation dictates. If the method of adjustment is changed, the caller informs the fire team. When any adjustment is 50 meters or less, the observer transmits the adjustment and calls for fire for effect.

3-74. When using the impact-observed method of adjustment, the observer estimates the direction to the target by using a cardinal heading. He estimates the distance from the point of impact to the target in meters. When the observer cannot see the point of impact, he may use the impact sound method of adjustment. For this method of adjustment, the observer transmits, "Adjust fire. Impact sound. Over." The impact sound method differs from the impact-observed method in that the observer senses by sound, rather than sees, the direction of the impact and makes his corrections accordingly.

3-75. Although the impact-observed method is most preferred for adjusting the fire of the attack helicopters, the observer target method, which is less desirable, may be used. When using the observer target method, the observer must mark his location, possibly compromising his location.

3-76. To use the observer target method, the observer senses the direction, left or right, and the distance, in meters, from the impact to a point on the observer target line. Then, he senses the position of the point on the observer target line relative to the target, long or short, and the distance along the observer target line to the target. The sensing of the observer must be converted to corrections, such as right, left, add, and drop, and transmitted to the fire team. The chance of error for this method is greater than for the other methods. Unobserved rounds are handled the same as for the other methods. An example of an exchange of information between an observer and a fire team using the observer target method follows:

- Observer: "Left, five-zero. Add 100. Fire for effect. Over."
- Fire team: "Roger. Out." (The team commits against the target.)

3-77. When the target is suppressed or destroyed, the following transmission would occur:

- Observer: "End of mission. Target suppressed (destroyed). Over."
- Fire team: "End of mission. Out."

3-78. Reference points are used to visually locate the target. The eyes of the pilot are led to the reference point and from the reference point to the target, sometimes through a series of decreasingly obvious reference points. It is much harder for a pilot to find a target than to keep a target in sight. Any reference point must stand out or contrast with its surroundings.

## Adjust Fire

3-79. Attack helicopter fire allows the pilot to observe the impact and effect of the ordnance on the target. This simplifies the adjustment procedure. However, the observer must still be prepared to adjust direct aerial fire. When adjusting aerial fire—

- Establish a reference point. The point of impact of the first round is the recommended reference point.
- Adjust for target strike. Do not try to bracket the target. The helicopter crew has direct visual contact with the target and needs only specific directions to fix the location.

- Transmit the corrections.

3-80. The following is a typical exchange of information between the observer and the fire team:

- Observer: "Adjust fire. Impact observed. Over."
- Fire team: "Impact observed. Out." (The team fires at the target.)
- Observer: "Northwest, seven-five. Over."
- Fire team: "Roger. Out." (The team fires at the target.)
- Observer: "North, three-zero. Fire for effect. Over."
- Fire team: "Roger. Out" (The team commits against the target.)
- Observer: "End of mission. Target suppressed (destroyed), over."
- Fire team: "End of mission. Over."

3-81. If the observer does not see the impact, the transmission would be—

- Observer: "Unobserved. Over."
- Fire team: "Unobserved. Over." (The team fires at the target.)

3-82. Adjustments continue until the mission is accomplished. *Table 3-4* shows how to direct a pilot to the target.

**Table 3-4. Directing the Pilot to the Target**

| Ways of Directing the Pilot | | How Used |
|---|---|---|
| Ammunition | Smoke rounds | Mortars, artillery, or grenade launchers. Phosphorous is usually the best because smoke clouds blossom quickly and are highly visible. |
| | Ordnance | Ordnance impacting the ground may be an adequate reference point. |

**Table 3-4. Directing the Pilot to the Target (Continued)**

| Ways of Directing the Pilot | | How Used |
|---|---|---|
| | Illumination rounds | Good for guiding the strike aircraft to the target at night, but will not pinpoint small targets if the flares function at the usual height |
| | Trace fire | Used at night. The intersection of the two streams of traces or the impact point of one stream marks the target. |
| Fires | Grass or other | Sometimes used near the target as a night reference |
| Recogniz-able known points | Terrain features or landmarks | If clearly visible from the air, it can help when used with another location method (streams, roads, bridges, tree lines, cultivated areas, prominent hills). |
| | Friendly positions | When clearly recognizable from the air, it may be used day or night for locating close-in targets. |

## OBTAIN AIR FORCE TACTICAL-AIRCRAFT FIRE SUPPORT

3-83. During major enemy incursions in the rear area, fighter aircraft may be available to support ground operations by providing immediate CAS. CAS consists of air attacks against enemy targets that are close to the friendly forces. CAS requires detailed coordination with the maneuver of the ground forces to be effective. The coordination must be responsive, integrated, and controlled. Typical CAS targets are—

- Enemy troop concentrations.
- Fixed positions.
- Armored units of immediate concern to the ground forces.

3-84. CAS missions are flown at the request of the command level. They are planned, directed, and controlled by the Air Force through the Tactical Air Control System.

3-85. Air Force support is directed through a forward air controller who, in turn, talks to the pilots. The controller can be in an aircraft or operating on the ground. In most cases, the controller will come forward to a point where he can see the target. Once the target is in sight, he can adjust the aircraft to the target. If the controller cannot see the target, the observer will have to tell him how it can be identified. The observer must make sure that the controller knows where all the friendly elements close to the target are located.

3-86. If the observer is unable to talk to a forward air controller, he must contact a fire-support team operating in the maneuver area. Fire support teams have the equipment to talk directly to the pilots of the aircraft and are trained observers for CAS.

## Mark Friendly Positions

3-87. Friendly positions are marked during close air strikes if there is no danger of compromise to enemy observers. This may require only a message (such as "All friendly positions are south of the target. Nearest are 500 meters."). As a rule, a mark is usually necessary when friendly troops are within 300 meters of the target. Marking of friendly positions may be overt or covert and include the following:

- Weapons fire. Weapons fire is useful as a signal if it is distinguishable from other types of fire. Tracers are especially useful.
- Smoke. Smoke grenades are the most commonly used overt marker, but they have limitations. Wind may move the smoke away from the source. Red and white smoke show up well with most backgrounds while some colors blend with their background.

- Mirrors. Signal mirrors and panels are probably the best covert ground-to-air devices for attracting attention. When the operator is proficient and the sun is shining, pilots can see a mirror flash for many miles away. Mirror signaling requires practiced training.
- Balloons. Balloons make a good covert marking system for use above a thick forest canopy.
- Flares. Rockets or 40-millimeter flares are good for attracting attention at night. If flares are fired in the direction of the aircraft, they can be mistaken as ground fire.
- Lights. Strobe lights produce a bright pulsing flash that is visible at night from 2 to 5 kilometers. Vehicle lights, such as unshielded red taillights, are visible to a pilot for several kilometers at night. Chemical glow lights may also be used.
- Ground commander's pointer. The pointer is a handheld device that is invisible to the naked eye, but its beam is visible through NVDs. It may be used by ground troops to clearly show air elements the location of friendly elements.

**Select Attack Headings**

3-88. A fighter aircraft is more likely to destroy its target if it attacks along the long axis of the target. Once the fighter aircraft knows where all the friendly units are and where the target is, the forward air controller tells the fighter pilot which attack heading to use. However, if the controller cannot see the target, the observer may have to recommend a direction of approach. The observer must remember that fighters should not attack across friendly positions.

3-89. An attack toward friendly units is undesirable because of ordnance dispersal patterns. An attack from behind and over friendly lines is also undesirable for several reasons. Some fighters dump empty cartridges overboard as they strafe. An empty 20-millimeter case weighs 114 grams and hits the ground at 167 kilometers per hour. An even greater hazard would be an inadvertent bomb release as the pilot repeatedly selects and arms his weapons systems while in the attack pattern.

## MOVE

3-90. Movement by MP teams, squads, or platoons in combat is dependent on the mission, the terrain, and the likelihood of enemy contact. MP apply the fundamentals of movement which include—

- Moving on covered and concealed routes.
- Avoiding likely ambush sites.
- Enforcing camouflage, noise, and light discipline.
- Maintaining all-around security, to include air guards.
- Using formations and movement techniques based on METT-TC.

3-91. In addition to applying the fundamentals of movement, MP leaders ensure that they—

- Maximize the capabilities of HMMWVs and ASVs. This includes considering the speed, mobility, and firepower of the vehicles. Fire and move both vehicles as a weapon system.
- Make contact with the enemy using the smallest force possible. MP move with a small force in the lead with the rest of the unit ready to react. A team leads a squad and a squad leads a platoon. One team leads another when two vehicles are moving. This prevents the whole unit from being pinned down by enemy fire and provides the unit with the flexibility to maneuver.

- Use the terrain. The terrain offers natural cover against enemy fire and conceals them from enemy observation. MP leaders must devote constant attention to protect vehicles and prevent them from skylining. MP make use of all natural cover and concealment when moving or stopped. When MP do stop, they stagger their vehicles on the roadway.
- Control subordinate elements. MP leaders issue clear and complete orders to subordinate elements in order to maintain control. They issue OPORDs, which cite MP actions to be taken on contact and the immediate actions the teams should accomplish. The OPORDs also explains how the MP leader will direct subordinate elements through the use of hand and arm signals, pyrotechnics, and other visual signals.

## USE MOVEMENT TECHNIQUES

3-92. Movement techniques are designed to minimize the exposure of the platoon to enemy fire and to place the platoon in a good position to react to enemy contact. They provide varying degrees of control, security, and flexibility. The selection of their use is based on METT-TC and the likelihood of enemy contact. Their effective use results in the platoon's contact with the enemy with the smallest platoon element.

3-93. MP employ the following three techniques of movement on the battlefield:

- Traveling.
- Traveling overwatch.
- Bounding overwatch.

3-94. While these techniques provide a standard method of movement, the leader must use common sense and sound judgment when employing them as he performs his missions and encounters different situations. The decision of which technique to use is based on terrain considerations and whether enemy contact is not likely, possible, or expected. The techniques are used in both the mounted and dismounted modes of movement.

## Traveling

3-95. An MP leader selects the traveling method of movement when contact with the enemy is not likely and speed is required. This technique allows the lead and trail elements to move together as a unit. It is the fastest but least secure movement technique. Movement is continuous, and interval and dispersion are maintained between the squads as terrain and weather permit. The platoon does not intend to engage in combat, but it is dispersed to prevent destruction in case of unexpected air or ground attack. The distance between the vehicles is based on the factors of METT-TC. This method of movement, with MP mounted, is depicted in *Figure 3-6.*

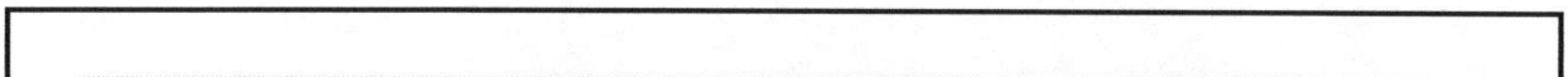

**Figure 3-6. Traveling**

## Traveling Overwatch

3-96. Use the traveling overwatch method of movement when contact with the enemy is possible and speed is desirable. The lead element moves continuously along the best, covered and concealed routes for protection from possible enemy observation and direct fire. The trail element moves at variable speeds, continuously overwatching. It normally maintains contact with the lead element and may stop periodically for better observation. The trail element tries to stay one terrain feature behind the lead element, but close enough to provide immediate suppressive fire and maneuver for support. However, it remains far enough to the rear to avoid contact with the same enemy force that is engaging the lead element. This technique, with MP mounted, is depicted in *Figure 3-7*.

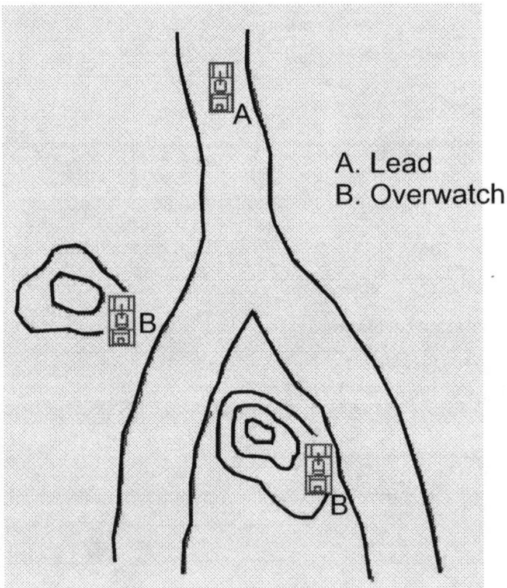


**Figure 3-7. Traveling Overwatch**

### Bounding Overwatch

3-97. When MP expect to make contact with the enemy, they select the bounding overwatch method of movement. It is the slowest, but safest method of movement. In bounding overwatch, the trail element occupies a good, covered and concealed position to overwatch the lead elements. While one element moves, another is always stopped in position to overwatch the bounding element. Although, bounding overwatch is used when enemy contact is expected, use it when time is available regardless of the likelihood of enemy contact. It provides for immediate, direct suppressive fire on an enemy force that engages the bounding element with direct fire. A three-vehicle team uses the V-formation with bounding overwatch. The lead elements advance to a point (first move) where they can support the advance of the overwatch element. On signal, the overwatch element moves forward to a position abreast of the lead elements (second move) and halts. During its move, it is overwatched by both lead elements. The lead elements then move forward, secured by the overwatch. Maximum use is made of folds of the earth and concealment to mask movement from likely enemy positions. MP are mounted in this method of movement as shown in *Figure 3-8.*

## PLATOON LEADER RESPONSIBILITIES

3-98. In the conduct of most tactical missions, the MP platoon moves as separate squads under the $C^2$ of the platoon leader. Regardless of which movement technique is directed, the platoon leader issues the squads an order explaining what each squad will do. This action becomes more critical as the likelihood of enemy contact increases. The platoon leader tells, and if possible, shows the squads—

- The enemy situation as he knows or suspects it to be.

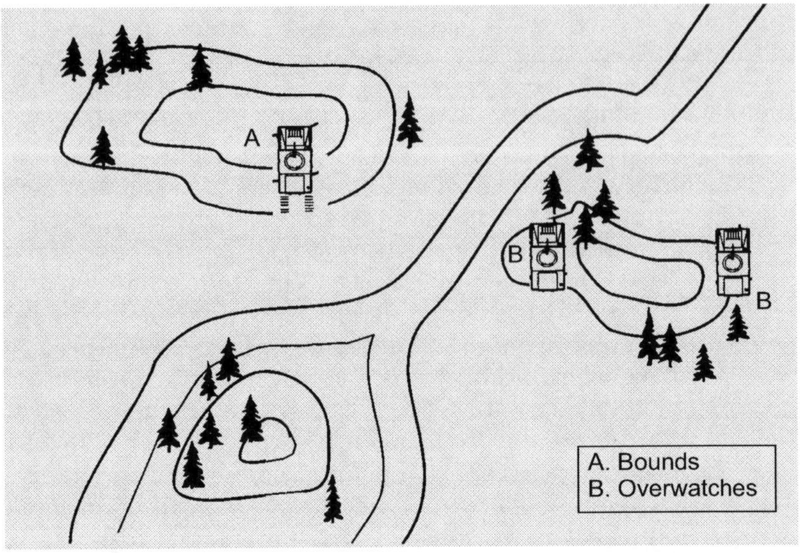


**Figure 3-8. Bounding Overwatch**

- The next overwatch position (objective for the bounding element).
- The route of the bounding element to that position.
- What he wants the squad to do after the bounding element gets to the next position.

## USE EXTENDED AND TEMPORARY HALTS

3-99. When an MP platoon moves as an element, it uses the coil formation for extended halts. This formation provides the platoon with 360-degree observation and fields of fire. The coil is always executed from either the column or staggered column formation. The platoon uses the four-team organization. The lead team assumes the 12 o'clock position (the direction of travel). Teams occupy the 3, 6, and 9 o'clock positions in twos, facing in the appropriate direction. Ensure that there are 50 to 100 meters between the team's vehicles. The interval between the teams will be 100 to 200 meters (terrain dependent). The HQ element occupies the center of the formation. Each platoon must have its own SOP for the formation based on its METL, war plans, and most common organization; it should practice the SOP as a drill to ensure correct execution. *Figure 3-9* depicts an MP platoon in a coil formation.

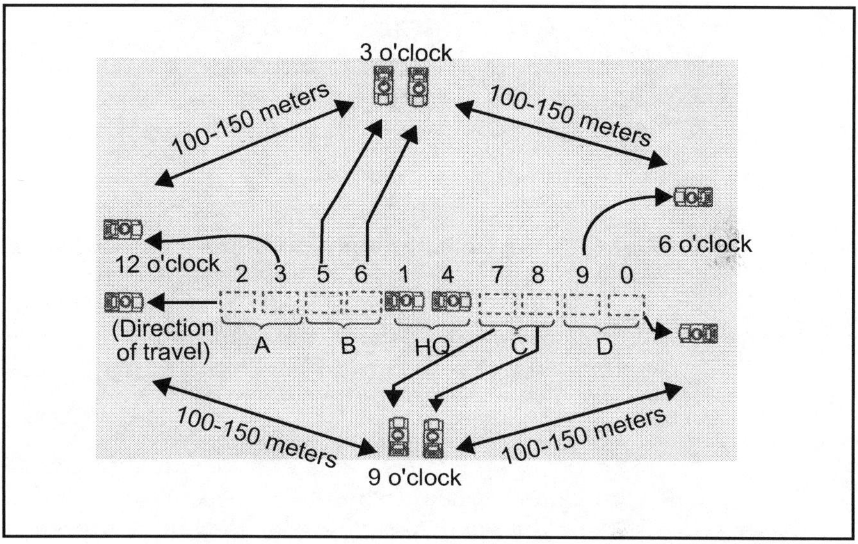


**Figure 3-9. Platoon Coil Formation**

3-100. MP use the herringbone formation for temporary halts from the march column. It provides them with a 360-degree observation and field of fire *(Figure 3-10).*

## USE GRAPHIC CONTROL MEASURES

3-101. MP leaders use graphic control measures to regulate or direct the movement, positions, and fire of the platoon. Control measures—

Figure 3-10. Herringbone Formation

- Are not intended to restrict the exercise of initiative (the function of command). Leaders use control measures to clarify their intent, focus the platoon and squad effort, and ensure synchronization. Each control measure should have a specific purpose that contributes to mission accomplishment. If a control measure fails the purpose test, leaders should not use it.
- Can be drawn on a map, an overlay, a sketch, or a terrain model. Leaders should strive to keep control measures easily identifiable and simple. Graphic control measures may include the AA, the route, the release point (RP) and start point (SP), checkpoints, and so forth. *FM 101-5-1* discusses control measures and provides examples of their use.

## CROSS A DANGER AREA

3-102. Danger areas are specific areas where there may be potential danger because of the increased risk of detection. Plan a way in which the patrol crosses danger areas to reduce the chances of a fight. MP make specific plans for crossing each known danger area and general plans for crossing unexpected danger areas. Patrols must be able to quickly modify these plans to fit the tactical situation. Typical danger areas include the following:

- Curves and blind spots on roads and trails.
- Streams.
- Open areas.
- Hill tops.

3-103. MP move cautiously at danger areas, using the bounding overwatch or variations of it to cross them. The MP leader decides how a patrol will cross danger areas based on the—

- Amount of time available.
- Size of the patrol.
- Size of the danger area.
- Fields of fire into the area.
- Amount of security available.

3-104. To cross a danger area, a patrol must designate nearside and farside rally points, secure the nearside and farside, and cross the danger area.

3-105. A small patrol may cross all at once, in pairs, or one element at a time. A large patrol normally crosses its elements one at a time. The leader positions security teams far enough out on both flanks and to the rear of the crossing point to give warning of the approaching enemy and to overwatch the crossing element. Once flank and rear security are positioned, the team crosses the danger area. The team crosses quickly, reconnoiters, and secures the far side of the danger area. The area on the farside must be large enough for a full patrol employment. When the team leader knows the farside is safe, he signals the rest of the patrol to cross. As each element crosses, it moves to an overwatch position or to the farside rally point until told to continue movement. When the patrol has crossed the danger area, the security teams cross and rejoin the patrol.

3-106. Open areas frequently afford the patrol the opportunity to observe the enemy from long ranges. Conversely, they often require that the patrol be exposed to possible enemy observation and fire for long periods of movement. The leader must make maximum use of the terrain and employ effective observation techniques to avoid exposing the patrol to a well-concealed and camouflaged enemy.

3-107. Before moving across a large open area, the patrol takes a thorough visual scan of the area. This should be done both dismounted and mounted, using all available optics. This scan focuses not only on finding potential enemy positions, but also on locating covered and concealed routes for bounding and a covered and concealed position to move to. If time and terrain permit, use dismounted troops to move to the far side of the open area and secure it. In very large open areas, use of dismounts may not be feasible because of the distances between covered and concealed positions.

3-108. Once the area has been cleared using visual means and/or dismounts, the squad or platoon moves across it. The patrol uses bounding overwatch because of the likelihood of enemy contact. If the open area is very large, the overwatch vehicle remains stationary until the bounding vehicle has moved a distance equal to half the effective range of the weapon system of the overwatching vehicle. When that point is reached, the overwatch vehicle must move out, even if the bounding vehicle has not yet reached a position of cover and concealment.

## MOVE WHILE IN CONTACT

3-109. Maneuver is the technique used for moving while in contact with the enemy. Maneuver involves two actions that occur at the same time. One element moves to a position where it can engage the enemy while another element supports that movement with a base of fire. A patrol maneuvers to move forward, either to close with the enemy or to gain a better position for firing at the enemy. MP can also maneuver to get more information on the position and strength of the enemy. When the position of the enemy is unknown, it may result in an unexpected encounter known as a *chance contact*. MP use maneuver to move away and withdraw safely.

3-110. When maneuver begins, the MP leader most often goes with the base-of-fire element and controls its fire. The base-of-fire element covers the movement element by shooting at the enemy position. The movement element advances within the supporting range of the base-of-fire element, taking a position from which it can fire on the enemy. The movement element then becomes the base-of-fire element, and the former base-of-fire element begins moving. Depending on the distance to the enemy position and the amount of cover and concealment available, the base-of-fire element and the movement element alternate roles as needed to continue moving.

3-111. MP can maneuver mounted, dismounted, or in a combination of both. A fire element using the MK19 GMG will have difficulty moving dismounted. MP move mounted when the terrain protects them from enemy fire, and look for covered and concealed routes.

3-112. When receiving direct fire, the movement element uses maneuver while the base-of-fire element suppresses enemy fire. If the movement element is not receiving direct fire, it uses bounding overwatch or maneuvers internally.

3-113. MP of a dismounted movement element move based on the intensity of the enemy's fire. When facing intense enemy fire with little or no cover, MP may be forced to crawl. They will use the low or high crawl depending on the situation, the need for speed, and the example set by the leader. Although crawling is slow, it reduces exposure to enemy observation and fire. When MP are not moving forward, they place suppressive fires on the enemy. They may need to advance all the way into and through enemy positions by crawling.

3-114. Dismounted MP can use short rushes from one covered position to another when enemy fire allows brief exposure. To do this they should—

- Advance by short rushes to avoid the enemy's fire.
- Try to stay up no more than 3 to 5 seconds. This keeps the enemy from having time to track and engage them.
- Select the next covered position before beginning the rush.
- Rush from cover to cover.
- Not hit the ground just because 3 to 5 seconds are up.

3-115. MP of a mounted maneuver element move based on enemy fire and the terrain. When they move, they should—

- Use the terrain to mask their movement.
- Move quickly between protected positions so that the enemy cannot engage their vehicles.
- Dismount when the terrain no longer provides protection.

## EXECUTE ACTIONS ON CONTACT

3-116. When MP encounter enemy forces, they must quickly execute actions on contact. Whether they remain undetected or are identified by the enemy, MP must first take measures to protect themselves, find out what they are up against, and then decide on a COA. To properly execute actions on contact, MP must take action consistent with the following fundamentals of reconnaissance:

- Remain focused on the reconnaissance objective.
- Report quickly and accurately.
- Maintain contact with the enemy.
- Retain the freedom to maneuver.
- Develop the situation rapidly.

3-117. The platoon leader specifies the actions on contact for the platoon. These specific instructions include the engagement criteria and the desired COA, based on the size and activity of the enemy force encountered. By knowing these details ahead of time, MP can develop the situation more rapidly and arrive at and execute the desired COA. The platoon strives to make contact with the smallest possible element. Visual contact in which the enemy is observed, but MP remain undetected, is the goal. This gives the platoon the greatest possible flexibility to maneuver and develop the situation.

3-118. The steps that make up the actions on contact must be thoroughly trained and rehearsed so that the platoon can react instinctively as a team whenever it encounters enemy forces. The four steps which are executed to allow the platoon to accomplish its mission according to the reconnaissance fundamentals are—

- Deploy and report.
- Evaluate and develop the situation.
- Choose a COA.
- Execute a COA.

## Deploy and Report

3-119. The MP patrol that gains initial visual contact with the enemy deploys to covered terrain that has good observation and fields of fire. The MP patrol in visual contact sends a report using the SALUTE format. If the element in contact is unable to report or cannot report quickly, another squad or team must report. The elements not in contact temporarily halt in covered terrain, monitor the report, and plot the situation on their maps. The platoon or patrol leader immediately determines the COA.

## Evaluate and Develop the Situation

3-120. The patrol concentrates on defining what they are up against. If they have not sent a SPOTREP at this point, they initially focus on getting enough information to send one. If the enemy has not detected them and time is available, the patrol attempts to confirm or determine in detail the enemy's size, composition, activity, orientation, and the locations of the enemy's weapon systems. They search for any additional information regarding the enemy that can help define the situation and update the SPOTREP.

## Choose a Course of Action

3-121. Once the patrol has developed the situation and the platoon or patrol leader has enough information to make a decision, he selects a COA. The COA will be within the capabilities of the patrol and allow the patrol to continue the commander's concept of the operation. The platoon or patrol leader considers various possible COAs, including—

- Breaking contact and bypassing the enemy. This COA may be selected when the enemy sees the patrol before the patrol sees them and comes under—

- Sniper fire. In this situation, the patrol returns fire in the direction of the sniper and conducts the maneuver (fire and movement) to break contact with or destroy the sniper.
- Indirect fire. The patrol quickly gets out of the impact area and does not seek cover, as it may be pinned down by doing so. By continuing to move, the patrol is more difficult to hit. The patrol uses the clock system (described later) to break contact.
- Ambush. In an ambush, a patrol takes immediate action. Personnel in the kill zone return fire immediately and quickly move out of the kill zone. Elements not in the kill zone lay down a base of fire (and smoke if available) to support the withdrawal of the elements in the kill zone. The patrol breaks contact and reorganizes at the last rally point. After or while the elements in the kill zone are being extracted, the leader decides whether to destroy the ambush element or break contact based on the situation and the mission. If no guidance is given, the immediate action of the patrol is geared to breaking contact.

- Maintaining contact to support a hasty attack. This COA is appropriate when the MP patrol discovers enemy elements that the higher HQ wants destroyed. The patrol cannot destroy the enemy because it does not have the combat power (level III threat) or because it has other tasks to perform. In this situation, the patrol maintains contact and continues to develop the situation, focusing on supporting the hasty attack by a TCF. The patrol conducts additional reconnaissance and monitors any changes in the enemy's situation. It focuses on information to support a friendly hasty attack.

- Conducting a hasty attack. If contact is made and the MP patrol and the enemy element become aware of each other at the same time and at such a close range that maneuver is not feasible, the patrol may make an immediate assault. The elements nearest the enemy open fire and shout, "contact, front (right, left, or rear)." The patrol moves swiftly into the assault. It stops the assault if the enemy withdraws and breaks contact. If the enemy fights, the assault is continued until the patrol can break contact, the enemy is destroyed, or the enemy breaks contact. In most cases, the patrol cannot, or should not, mass its combat power to defeat an enemy force. If the patrol concentrates, it risks losing the capability to complete its mission and jeopardizes its ability to conduct subsequent missions. If the patrol attacks an enemy, it should only attack small dismounted formations or lightly-armored or unarmored reconnaissance vehicles. Except in self-defense, patrols should avoid attacking heavily armored vehicles or large formations.
- Establishing a hasty defense. The patrol will establish a hasty defense if it cannot bypass the enemy, all the teams are fixed or suppressed, and the patrol no longer has the ability to break contact by maneuver. Patrols should use a hasty defense when the enemy executes a hasty attack. The patrol maintains contact or fixes the enemy in place until additional combat power arrives or the patrol is ordered to move. If the patrol is required to conduct a hasty defense, the commander then becomes responsible for continuing to develop the situation.

### Execute a Course of Action

3-122. The platoon leader updates his SPOTREP to the commander with any new information and then recommends a COA to the commander. The commander approves or disapproves the recommended COA based on how it will affect the parent unit's mission.

3-123. If the commander has anticipated the enemy situation the platoon is reporting, he will already have addressed the contingency in the OPORD and given guidance to his subordinates on what COA the platoon executes. In such a case, the platoon leader can evaluate the situation, choose a COA consistent with the higher commander's intent or concept, and execute it without further guidance. He keeps the commander informed of what he is doing as he executes the COA.

## BREAK CONTACT

3-124. To break contact with the enemy without disorder, use the clock system. Use this system when the patrol and a larger enemy element see each other at the same time. The patrol must break contact or be destroyed. The direction the patrol moves is always 12 o'clock. When contact is made, the leader shouts a direction and distance to move (such as "4 o'clock, 300 meters" tells the patrol to move in the direction of 4 o'clock for 300 meters). If contact is broken, the patrol rallies at the designated distance and continues with the mission. If contact is not broken, then another direction and distance is given. This action continues until contact is broken. The leader can also use the clock system to shift or direct fire at a certain location.

## CONSOLIDATE AND REORGANIZE

3-125. Once enemy resistance has ceased or the platoon or patrol has broken contact, leaders must quickly take steps to consolidate and prepare to defend against a counterattack. In consolidating on the objective, all-around security is critical because the enemy might counterattack from any direction. The leader must evaluate the terrain thoroughly. Platoons and patrols use the following two techniques when consolidating:

- Clock technique. In using this method, the leader designates either a compass direction or the direction of attack as 12 o'clock. He then uses clock positions to identify the left and right boundaries for squads. The leader positions key weapons along the most likely avenue of approach based on his assessment of the terrain *(Figure 3-11)*.

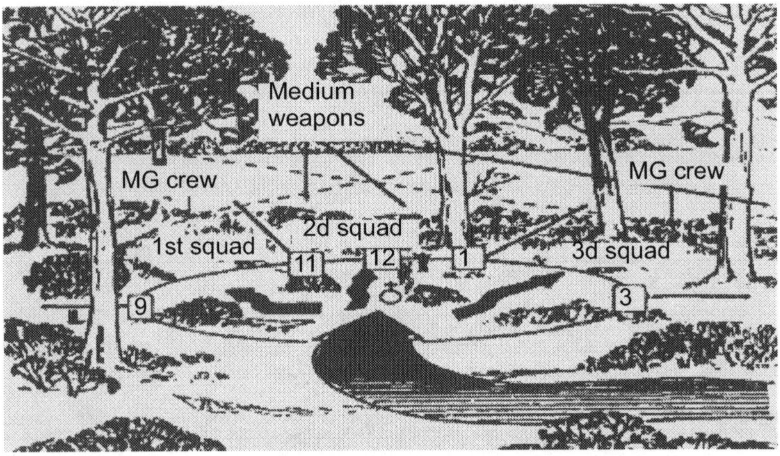


**Figure 3-11. Clock Technique**

- Terrain feature technique. In a similar manner, the leader identifies obvious terrain features *(Figure 3-12, page 3-58)* as the left and right limits for squads. In both techniques, he ensures that squad sectors of fire overlap each other and provide mutual support for adjacent units.

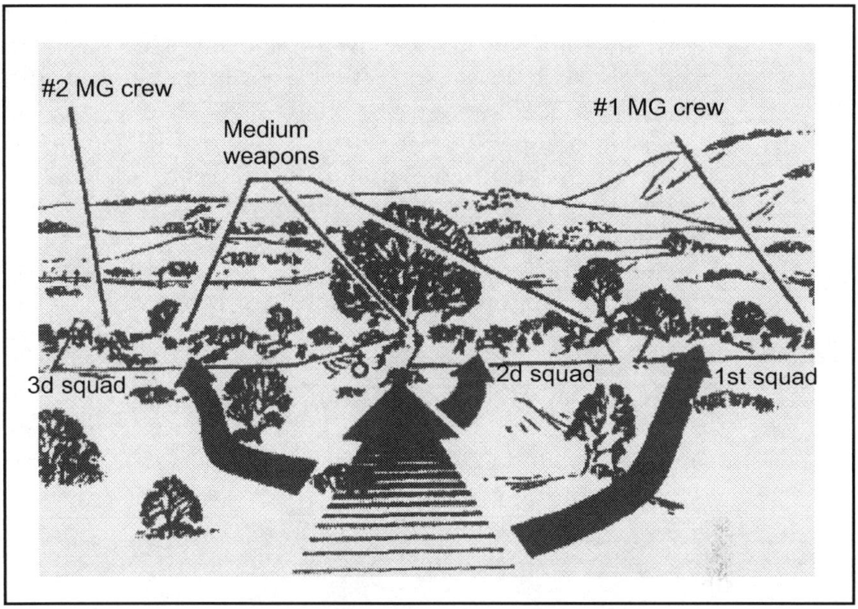


**Figure 3-12. Terrain Feature Technique**

3-126. Once platoons or patrols have consolidated, they begin to reorganize. Platoons reorganize to continue the mission. Reorganization involves the following:

- Reestablishing $C^2$.
- Remanning key weapons and redistributing ammunition and equipment.
- Clearing the objective of casualties and EPWs.
- Assessing and reporting the status of platoon personnel, ammunition, supplies, and essential equipment.

# COMMUNICATE

3-127. On the battlefield, you must be able to communicate. Communication is the means through which battle command is exercised. MP on the battlefield must be able to communicate to maintain $C^2$ of their elements, call for fire or request other support, and respond to orders. The chain of command and succession of command must be known throughout the organization. There must be open lines of communication up, down, and laterally. Situations on the battlefield can change rapidly, and losing the ability to communicate for even a short duration can have a major impact on an operation. Commanders must provide for redundancy in means of communications. MP companies have the capability to place backup communication systems at key locations within an area of responsibility. METT-TC and the battlefield situation usually determine communication means. MP use any combination of the following systems:

- Sight and sound.
- Messenger.
- Wire and radio.

## USE SIGHT-AND-SOUND SIGNALS

3-128. Visual signals are useful for sending prearranged messages over short distances, during radio silence, or when jamming interferes with radio transmissions. Arm-and-hand signals, flashlights, and pyrotechnics may send quick visual signals. Visual signals have some disadvantages, which include the following:

- They are less effective when visibility is limited.
- They may be seen and intercepted or imitated by the enemy.
- They may be masked by terrain features, reducing the chance of a message being received.
- They are easy to misunderstand.

**NOTE: To overcome this last disadvantage, each man in the unit must be able to send, receive, and understand messages using visual signals.**

3-129. Although arm-and-hand and light signals are standard throughout the Army, the meaning of pyrotechnic signals must be set in the $C^2$ portion of the OPORDs and in the SOI. To ensure that a pyrotechnic message was correctly received, confirm the message by some other means as soon as possible. For more information on visual signals, refer to *FM 21-60*.

3-130. Sound signals, like visual signals, work well only for short distances. Simple devices (such as whistles, horns, gongs, and explosives) may be used. Sound signals can be used to—

- Attract attention.
- Transmit prearranged messages.
- Spread alarms.

3-131. A well-known sound signal is the use of metal-on-metal to indicate an NBC hazard or attack. Battlefield noise may blend with or override sound signals, causing confusion and misunderstandings. Sound signals—

- Must be simple to understand.
- May be restricted for security reasons.
- Can be intercepted by the enemy.

## USE MESSENGERS

3-132. Using messengers is the most secure way to communicate long messages and documents. However, it is also the slowest and messengers are vulnerable to enemy action. When using a messenger—

- Put the message in writing.
- Make the text clear, concise, and complete.
- Choose the most expedient transportation on hand.
- Encode the message (using the operational code in the SOI) if there is a risk that the messenger might be captured.

- Send a second messenger by a different route if a backup message is needed.

## USE WIRE AND RADIO COMMUNICATION

3-133. Often, wire communication is more useful than radio. It is hard to jam, and unlike radio, more than one person can talk at one time. It is used most often for communicating with static posts. Wire communications cannot be secured and may be cut by the enemy. When a wire line has to be checked, MP are sent out in pairs. One MP looks for the cut and the other MP provides overwatch security. The enemy can take prisoners by cutting a line and capturing the soldiers who go to repair it.

3-134. Use a radio to communicate with mobile or distant elements. A secure voice radio is best. The enemy can intercept messages on an unsecured radio net. Regardless of the radio type, if the transmission is heard, the enemy may be able to detect the location of the radio or learn what the unit is doing. Keep the transmissions short. Know and use signal security and electronic counter-countermeasures. MP deny the enemy information from friendly telecommunications by following the SOI. To keep transmissions secure, MP—

- Authenticate all transmissions.
- Use only authorized codes.
- Use secure voice transmissions.
- Use encoded messages.

3-135. The enemy is kept from disrupting radio communications, and friendly emitters are protected from enemy detection, location, and identification by—

- Setting radios at low power.
- Placing antennas where terrain blocks the enemy's interception.
- Using remote radios and antennas.
- Using directional antennas.

- Using wire whenever possible.
- Observing listening silence.
- Using short transmissions.
- Using a random transmission schedule.
- Transmitting only when there is a need to do so.

# Chapter 4

# Combat Operations

This chapter provides the MP leader with the information needed to successfully complete a combat mission. When MP platoons conduct the tasks addressed in this chapter, they will mirror the actions of the company.

## PREPARE FOR COMBAT

4-1. Units that are not directly engaged in combat often find it necessary to travel in order to position themselves for combat operations. During these movements, the battalion supports the company and the company supports and monitors the platoons with the movement plan. The move's success depends on the unit's discipline and ability to execute the plan. (Refer to *Appendix E* for precombat inspection checklists.)

### MAINTAIN AND MOVE A COMBAT LOAD

4-2. Combat load is the quantity of supplies, in all classes, that the company must have on hand to sustain operations for a prescribed number of days. The company must be capable of moving the combat load, using organic transportation assets, into combat in a single delivery. To save time, the company combat loads vehicles while the quartering party readies the new site. The combat load ensures that a unit is ready for combat even when it is on the move.

4-3. The principles of combat loading are standard. All equipment, ammunition, and gear are loaded on the vehicles in a logical order and put in predesignated places. Knowing the location of each item allows for quick retrieval during the move. Combat loading also lends to a fast set up at the new site. Like the basic load, the company's combat load is mission-dependent. No single load plan can satisfy all the situations. MP leaders must consider the following:

- METT-TC.
- Vehicle and trailer capacities.
- Weight limits of the unit's vehicles and trailers, being careful not to overload them.
- Whether or not the equipment will fit (cube out). For equipment data, see the applicable technical manual (TM).

4-4. Unit SOPs has load plans tailored for various mission activities. Having a choice of load plans for various deployments reduces the load time. Load plans and diagrams are modified to suit METT-TC and vehicle and trailer capacities. The modifications are shown on the load diagram in the vehicle. HMMWVs may be loaded in many configurations, which include—

- Loading the basic equipment in the mounted standard brackets on the vehicle.
- Modifying and moving the brackets to meet mission or unit requirements.

## PLAN A TACTICAL ROAD MARCH

4-5. The basic considerations in planning any road march are the METT-TC factors (especially the enemy situation and the mission); the march order; and the type, number, and characteristics of vehicles available for the movement. A company conducts a tactical road march for relocating and facilitating rapid movement of the vehicles with a prescribed speed and interval between the vehicles. When preparing for a tactical road march, the company uses the following planning sequence when time permits:

- Prepare and issue the WO as early as possible to allow maximum time for preparation.

- Prepare an estimate of the situation, specifying the organization of the march column.
- Organize and dispatch reconnaissance and quartering parties.
- Prepare detailed movement plans based on the organization of the march column and a review of the available reconnaissance information.
- Use the reconnaissance information to—
  - Choose sites for halts and RPs.
  - Spot problem areas along the route.
  - Select bypasses or alternate routes.
- Select fairly secure locations for halts.
- Choose areas that provide cover and concealment.
- Avoid choosing highly populated areas, curves in the road, or other hard-to-secure areas.
- Plan the timing so that the unit arrives at the SP just before it is scheduled to cross it. The time a unit must cross the SP is provided to the unit. As other units may be planning to use the route, each unit must cross the SP on time.
- Prepare and issue the march order.
- Prepare the overlays and issue them to the vehicle commanders and subordinate leaders. The road march overlay includes, at a minimum, the location of the SP, RP, scheduled halts, and checkpoints at critical points along the route.

## COORDINATE AND DIRECT THE MARCH

4-6. The chain of command controls the column. The march leader—

- Coordinates the road march through the chain of command with the local movement control unit.
- Determines if the convoy needs a movement credit or a clearance to use the given route. If so, the march leader submits a *Department of*

*Defense (DD) Form 1265* through the appropriate movement control center.

- Informs higher HQ and the supported units of the dates and times that the operations will stop at the old site and begin at the new site.
- Tasks subordinate leaders to attend a briefing to discuss unit readiness and load plans and forecast support requirements.
- Submits requests for support based on the forecast developed during the briefing. Requests may include fire support, refueling, vehicle recovery operations, and other support needed to complete the march.
- Issues an OPORD for the movement.
- Requests HQ personnel to prepare a movement table. Refer to *Appendix E* of *FM 55-10* for detailed information on the movement tables.
- Requires unit personnel to analyze the route reconnaissance information looking for likely enemy ambush sites.
- Ensures that a strip map, which may be included as an annex to the OPORD, is prepared. The strip map shows SPs; RPs; route numbers; place names; critical points; directional arrows; distances between the points; scheduled halt locations; and petroleum, oils, and lubricants (POL) refill points. Give copies to the unit drivers.
- Ensures that the drivers and assistant drivers are briefed.
- Ensures that radio communication is kept to a minimum during movement.
- Tasks subordinates to ensure that the road march plan is followed.
- Ensures that safety briefings are conducted and understood.

## CONDUCT THE MARCH

4-7. The commander sets the conditions under which military traffic moves at night. The march leader ensures that personnel are aware of and abide by the set lighting conditions when the company moves at night. Conditions that are more restrictive may be imposed contingent on the threat environment (such as air raids). Lighting conditions may include normal lighting, reduced lighting, or blackout. If the situation warrants, travel by total blackout (use of night-vision goggles [NVGs]) may be prescribed. More often, travel is under partial blackout, using only enough light to see the road and to be seen by other road users. Minimal lighting reduces visibility from the air while it permits drivers to—

- Travel as quickly and safely as possible.
- Apply brakes in time.
- See the side of the road.

### Movement Techniques

4-8. During a tactical road march, the unit leaders (such as the march leader and the platoon sergeant) travel in separate vehicles. This decreases the chance of a unit's primary leaders being lost in one enemy action. The convoy moves en route by close or open column march or by infiltration. In dusty conditions, space the vehicles so that the dust from one vehicle does not blind the driver of the next.

4-9. **Close Column**. In a close column, the elements are close together. Use a close column for marches during limited visibility, blackout conditions, and radio silence. Under these conditions, space the vehicles so that the driver can see the two lights in the blackout marker of the vehicle ahead. Visibility determines the set distance between the vehicles. A close column—

- Reduces the time it takes for the column to pass points on the route.

- Requires fewer guides, escorts, and markers for control than an open column does.
- Enhances movement through congested areas or over poorly marked routes.

4-10. **Open Column**. In an open column, elements are widely spaced as a passive defense measure. Generally, an open column is used during daylight, but may be used at night with infrared lights, blackout lights, or night-vision equipment. A distance of 50 to 100 meters or more may be designated between vehicles depending on METT-TC. Use an open column—

- When enemy contact is likely.
- To enhance security.
- Over dusty roads. Reducing dust is especially important when moving through areas contaminated by radioactive fallout.

4-11. **Infiltration.** Infiltration is the best passive defense against enemy observation and attack, although it may be difficult to control. To move by infiltration, vehicles are dispatched one at a time or in small groups at irregular intervals to keep traffic density low and to prevent undue massing of vehicles. Use infiltration—

- When time and road space allow.
- When maximum security, deception, and dispersion are needed.
- To maintain security during the march. When the unit approaches likely danger areas (such as bridges and tunnels), have one or more teams dismount. The teams check both sides of the road before the vehicles pass. This is critical when only a map reconnaissance was conducted before the move.

## Movement Considerations

4-12. Consider events and control measures that are used during the move in order for the tactical road march to be successful. These measures include the SP and RP, checkpoints, restrictions, speed control, halts, disabled vehicles, and mined areas.

4-13. **Start Point**. A SP gives the vehicles of a march column a common point for starting their movement. When vehicles use more than one route, each route has an SP. The SP is a recognizable place along the route of march (such as a road intersection). It should not be in a defile, on a hill, or at a sharp curve in the road that could cause movement to slow. Ensure that the SP is far enough away from the AAs to allow the vehicles to be organized and move at the prescribed speed when they reach it. Elements of the company reconnoiters the route to the SP to determine the times needed for the serial to arrive at and clear the SP before starting the march.

4-14. **Release Point**. A RP provides all the vehicles of the march column with a common point for reverting to the commander's control. It is a point on the march route that is easy to recognize on the map and on the ground. Guides meet the vehicles as they arrive at the RP and lead them to their new positions, usually in an AA. Multiple routes and cross-country movement from the RP to the assembly areas allow vehicles to disperse rapidly. When leaders select a RP, avoid hills, defiles, and sharp curves that may cause elements to slow or stop on the route. Ensure that vehicles are not required to countermarch or pass through another element to reach their new position.

4-15. **Checkpoints**. Use checkpoints on a route for reference when providing instructions and identifying places where interference with movement might occur or timing may be critical.

4-16. **Restrictions**. Restrictions are points along the march route where the movement may be limited or obstructed during certain time periods (such as bridges, intersections, ferries, or bypasses). The march planner—

- Starts the move early enough to pass such a point before a restriction begins.
- Delays the start of the move to pass a restriction after it has ended.
- Plans to halt the column along the route until the restriction is lifted.

4-17. **Speed Control**. Vehicles in a column of any length may simultaneously encounter different types of routes and obstacles. This causes sections of the column to move at different speeds at the same time, producing an undesirable accordion or whip effect. The movement order specifies the march speed, march rate, and the maximum safe catch-up speed to reduce *column whipping*. The lead vehicle must not exceed the authorized maximum speed of the slowest vehicle in the column. To minimize vehicle congestion on the nearside of an obstacle, vehicle commanders and drivers must be alert and maintain the prescribed minimum following distance. Vehicles should make only gradual speed changes while maintaining their prescribed interval. Vehicle commanders must constantly be aware of the vehicle interval to their front and rear and adjust their speed accordingly.

4-18. **Halts**. Halts are conducted for various reasons. They—

- Allow following traffic to pass.
- Provide time for rest, mess activities, and personal comfort and relief.
- Permit refueling and maintenance of vehicles.
- Provide for maintenance and inspection of equipment.
- Allot time for adjustments in the schedule.

4-19. The time and duration of the halts are usually specified in the movement order or prescribed in the unit SOP. The SOP prescribes actions to take during halts. A short rest halt of 15 minutes is usually taken after the first hour of marching. A 10-minute short halt is taken every 2 hours thereafter. The prescribed march rate includes the time required for short halts. When possible, march elements using the same route should stop at the same time. Route characteristics may make it necessary to halt at a particular point on the route rather than simultaneously at a fixed time.

4-20. Long halts are planned in advance. The length of the halt is added to the total travel time. Locations for long halts are normally selected to allow all the vehicles to clear the road and permit proper dispersion. The unit commander schedules halts for refueling in advance.

4-21. The herringbone formation is used to provide security for the march column during temporary halts. During temporary halts, the MP teams move their vehicles to alternate sides on or off the road in a herringbone pattern that lets vehicles pass down the center of the column. Movement commanders give permission for execution of unscheduled halts. The first priority at any halt is local security. OPs are established and sectors of fire are assigned to each vehicle. These actions should be automatic and part of the unit SOP.

4-22. **Disabled Vehicles**. Disabled vehicles must not obstruct traffic. Their crews must move them off the road and report their status immediately to the PSG. Crews must immediately signal the follow-on vehicles to bypass and continue movement. They then establish security and post guides to direct traffic. If possible, crews repair their vehicles and rejoin the rear of the column just ahead of the trail element. Vehicles that have dropped from the column return to their positions only when the column has halted. The trail party recovers vehicles that cannot be repaired by their crews.

4-23. **Mined Areas**. When a company encounters mined areas, it must remember that the safety of the unit is the most important factor. It bypasses mined areas whenever possible, but considers how the delay will affect the outcome of the mission. Remember to—

- Be cautious. Mines may be used to force an element to take an alternate route into an ambush site.
- Screen the bypass route, if possible, before diverting an element.

**NOTE: Refer to *Appendix H* for countermine operations.**

4-24. Ensure that all efforts are made to bypass mined areas; however, if the element must cross a mined area when engineer assets are not available to breach the minefield, act quickly and cautiously. Mined areas, like other obstacles, are often covered by enemy fire. Before crossing—

- Detonate mines from a protected position.
- Detonate mine trip wires by rigging an object near the trip wire to fall on the wire.
- Use a hand grenade or direct fire to detonate mines.
- Detonate pressure-sensitive mines by rigging an A-frame over the mine and placing a heavy object, attached to a rope, over the mine. Take cover and allow the object to fall on the mine.
- Devise other methods to detonate the detected mines.
- Send a report to the next higher command when mines have been neutralized. Refer *FM 20-32* and *FM 21-75* for information on mines and countering mines.

## ESTABLISH A NEW OPERATIONAL SITE

4-25. MP elements most often will collocate as part of an established base or base cluster. On occasion, MP may need to set up a base of their own. To set up at a new location, whether as part of an established base or base cluster or separately as a company or a platoon base, you must—

- Reconnoiter new sites.
- Pick the most favorable site and its alternate. Choose a site that—
  - Is easily accessible.
  - Can accommodate all the unit's vehicles and equipment.
  - Has a firm, well-drained surface.
  - Has some natural cover and concealment.
  - Is relatively easy to defend.
- Prepare and secure the site.
- Complete the move and establish communication.
- Establish local security to sustain survivability.

## USE A QUARTERING PARTY

4-26. A quartering party is needed whenever a unit relocates. The quartering party's mission is to—

- Reconnoiter the area for enemy presence, booby traps, NBC contamination, and other hazards.
- Establish the dismount point and perimeter security.
- Establish communication.
- Identify the company CP.
- Designate vehicle locations.
- Position guides at the RP to meet the main party.
- Prepare the area for occupation and assist the units with occupation.

4-27. The platoon leader or PSG designates vehicles and personnel from the platoon that will be part of a battalion or company quartering party. The entire platoon may serve as the battalion or company quartering party.

4-28. While the unit loads for deployment, the quartering party moves to and readies the new site. The quartering party's job ends when the last vehicle in the main body arrives at the new site. The size of a quartering party is based on the tactical situation and the amount of work required to prepare the site for occupancy. A quartering party for platoon relocation would be smaller than that of a company. A company quartering party is likely to have personnel from the unit's HQ, each platoon, the maintenance and dining sections, and communications.

4-29. The quartering party leader ensures that the equipment and supplies are available to clear, secure, and set up the new site. A quartering party may need—

- NBC detecting and monitoring equipment.
- Mine detectors.
- Saws or axes to clear the wooded areas.
- White engineer tape.
- Portable route signing material.

4-30. The quartering party leader assigns tasks to the teams based on the size of the quartering party, the work to be done, and METT-TC. He ensures that each team has the equipment needed to complete its tasks (refer to *FM 7-10*) and that they are at the proper mission-oriented protection posture (MOPP) level if they are operating in an NBC environment.

## March Halts

4-31. At march halts, quartering party teams set up local security. If the vehicles can leave the road, the teams form a 360-degree perimeter around the convoy. If the vehicles cannot leave the road, they are parked at an angle so alternate vehicles face opposite sides of the road. Each team is assigned a sector to observe, with sectors overlapping between the vehicles. Each team member has a specific area of responsibility. The troops remain alert and ready to take action on contact with the enemy. All personnel watch for enemy aircraft.

### Site Arrival

4-32. When the quartering party arrives at the site, it clears and then secures the site. One or more teams, after dismounting their vehicles, search the area for mines, booby traps, items of intelligence value, or other signs of enemy presence.

4-33. If nuclear weapons have been used, at least one team using radiation detection, indication, and computation (RADIAC) meters monitors the site for radioactive contaminants. Monitoring for chemical and biological agents must be continuous because it is difficult to detect their first dispersal.

4-34. In urban areas, team members clear buildings that will be used by the unit. Team members clear the structures outside the perimeter if there is a possibility of enemy presence. The priority of buildings to be cleared and the number of teams needed are based on METT-TC. Refer to *FM 90-10-1*.

4-35. When the area is cleared, one or more teams perform the following functions:

- Set up the OPs and the LPs.
- Set up defensive positions on likely enemy avenues of approach. These positions provide early warning and limited protection during the occupation of the new site.
- Prepare the new site for the main body's arrival.

### Company Move

4-36. When setting up a company site, the quartering party—

- Chooses a tentative location for the company CP.

- Sets up the company CP where it can best control the company, be well defended, and have LOC to the subelements.
- Uses buildings (in an urban area) to conceal the CP.
- Considers cover and concealment when choosing the CP location.
- Makes use of natural cover and concealment when possible.
- Uses camouflage screens and man-made cover and concealment where needed.
- Sets up the wire communication net. Marks those areas where other unit elements will be positioned, using signs or materials that cannot be easily seen by the enemy.
- Picks roads and trails that permit an easy flow of traffic.
- Chooses alternate exits and marks them for use as emergency exits.
- Designates parking areas for the heaviest, most awkward vehicles, such as 5-ton trucks.
- Selects a troop area and—
  - Marks the areas where latrines, garbage dumps, and tents will go.
  - Uses ground guides for vehicle movement in areas where troops are sleeping. (For safety, unit personnel should sleep only in the troop area).
  - Chooses a structure (in an urban area) that protects the troops from natural elements and has adequate latrine facilities.
- Locates the following:
  - The food service section inside the perimeter, well away from interior roads to keep dust from contaminating the food. Locate the serving line to take advantage of

cover and concealment. In urban areas, select a building for service and meal activities.

- The latrines away from the bivouac area. Place latrines at least 30 meters down slope from wells or other water sources, and at least 100 meters from the dining facility, downwind and down slope, if possible. In urban areas, use existing latrines if they can serve at least 8 percent of the unit at one time.
- The maintenance section where vehicles can arrive easily from the main road through the site. Ensure that vehicles are able to enter the maintenance tent at one end and exit at the other. Use existing garages for maintenance operations in urban areas.
- The supply section to meet space, roadway access, and drainage needs. In urban areas, use warehouse-type buildings for supply operations.
- The tactical communication section where it has space enough to support the whole operation. Usually it collocates with the maintenance section or the operations section.

4-37. When the main body arrives, the quartering party—

- Maintains security as the main body moves into the site.
- Maintains noise and light discipline.
- Ensures that the vehicles rapidly clear the approach route while maintaining vehicle intervals.
- Dismounts all personnel except drivers at the dismount point.

- Briefs the leader of the main body on the situation and the operational status.

4-38. The leader of the main body informs the higher HQ that the move has been completed. He reports the location coordinates for both the CP and the alternate CP by messenger or other secure means. The leader ensures that the entire party prepares fighting positions and other defensive measures.

### Platoon Relocation

4-39. A quartering party, in advance of the platoon's relocation, has the same considerations scaled to size and need, as one in advance of a company. An MP platoon may collocate with a company HQ or an existing base. However, more often, a platoon's base must be set up where a platoon HQ can best—

- $C^2$ its squads.
- Communicate easily with its squads and higher HQ.
- Link squads, the company CP, and the supported unit.

4-40. The platoon HQ can operate from a static base or vehicles. If the platoon elements are going to operate in one location (as they would for an EPW holding area), the platoon leader sets up a static platoon HQ base. If the platoon elements must operate dispersed over a large area, the platoon leader must remain mobile. In such cases, a platoon leader could elect to set up a temporary platoon base as a rally point to report, resupply, and reorganize the platoon's resources.

4-41. The platoon bases are set up the same. The PSG selects a site that offers good cover and concealment. The site must be defendable and allow the HQ vehicle to be located near the tent. A small tent houses the platoon's HQ. Use a radio set control group to relay remote communication into the tent. An antenna increases the transmission distance and is located based on OPSEC principles. Wire communication is limited to the platoons that can hook into an existing wire net.

## CONDUCT MILITARY POLICE BASE SELF-DEFENSE

4-42. When collocated with a base or a base cluster, the platoon is integrated into that base's or base cluster's self-defense planning and operations. Although bases and base clusters are more prevalent at corps and echelon above corps (EAC), the same principle applies to MP located at the division or brigade support areas. When an MP base is set up on its own, the base is responsible for its own security and protection.

### Collocated

4-43. An MP platoon collocates with a base or a base cluster for logistical support and a means to conduct operations. When collocated, the platoon leader coordinates with the base defense operations center (BDOC) and the base cluster operations center (BCOC) to integrate the platoon's efforts with the base's and base cluster's efforts. The platoon's portion of the base's or base cluster's defense is to help provide early warning of the threat by area security and/or maneuver and mobility support in the area near the base or the base cluster. Because MP resources are austere, the platoon only shares sector efforts on the base's perimeter. MP are used as static posts (such as gates) only under extreme conditions. An MP platoon may be tasked to serve or augment the base cluster's response force. Before the platoon leader accepts this tasking, he consults with the company commander to—

- Ensure that the company commander knows that he may have one less platoon if he is tasked for the TCF or response force for the battalion.
- Allow the company commander to advise the base commander that the MP platoon could

serve as a response force to the base if it is not committed.

4-44. Each base has a BDOC that plans, coordinates, and supervises base defense operations. The BDOC initiates contingency planning that enables the base to—

- Increase the manning posture of the base contingent on the threat.
- Detect and defeat the threat within their capabilities.
- Hold against heavier enemy forces until response forces arrive.
- Maintain control of the fight within the base.
- Support the fire and movement of the response force operating outside the base.

4-45. Each base cluster has a BCOC to monitor base defense plans and establish the base cluster reaction force. The BCOC—

- Provides $C^2$ of the resources for planning, coordinating, and supervising the base cluster's defense.
- Coordinates base defense operations.
- Maintains communication with bases within the cluster as well as MP, BDOCs, and the sustainment area operations center. A great deal of intelligence is provided to a BDOC and BCOC through the rear operations net, which helps in planning the defense.

4-46. The platoon's plans for the interface of MP support into the base's self-defense plans address—

- Cover and concealment of personnel and equipment.
- Signal security.
- Reliable and redundant communication systems at all guard locations (such as land lines, radio

links to the BDOC, and telephone hookups to the center switch).

- Deception.
- Contingency planning.
- Improvement of base defense positions.
- Assistance of the area MP.
- Coordination with the BCOC or rear-area operation center (RAOC), as required.
- OPs/ LPs.
- Noise and light discipline.
- Immediate reaction to enemy threat or attack.
- Rehearsals of defense measures.

4-47. All plans and overlays depicting MP support are forwarded to the BCOC. There they are consolidated and forwarded to the RAOC. (If a base is not part of a base cluster, the base forwards all the plans and overlays directly to the RAOC.)

### Separate Setup

4-48. Set up an MP platoon separately only when there is no other alternative. This is the least desirable means for a platoon to set up. When the platoon sets up as a base separately, it must be able to defend against a wide range of enemy activity. It integrates the defense of its base (including indirect-fire systems, air defense artillery, and tactical aircraft) with the defense efforts of other bases in the sustainment area. Engineers, dismounted troops, armored vehicles, and helicopters contribute to the overall security of the bases. Bases coordinate and synchronize their defense efforts to enhance their strengths and reduce their vulnerabilities. A base's defense priorities include—

- Establishing initial base security.
- Positioning crew-served weapons and troops on assigned sectors of fire.
- Clearing fields of fire and preparing range cards.

- Preparing fighting positions.
- Installing communication.
- Emplacing obstacles and mines.
- Improving primary fighting positions to include overhead cover.
- Preparing alternate and supplementary positions.
- Stockpiling ammunition, food, and water.
- Preparing routes and trenches between positions.
- Developing a counterattack plan.

4-49. Using the IPB process can help the platoon predict threats to base security. Platoons must be aware of the enemy's location, organization, direction of movement, and strength. In the same regard, platoons must have effective OPSEC to deny similar friendly information to the enemy. Platoons can continually improve base defenses by considering what avenues of approach and methods of attack the enemy could use, given the vulnerabilities of the platoon's base. The base defense plan must have overlays depicting weapons positions, sectors of fire, final protective fires, and reaction force contingencies. Plans are updated as often as possible.

4-50. The base's reaction force efforts are coordinated with the designated-area response force. Platoon leaders develop detailed employment plans and exchange as much information as possible with the response force and TCF commander before they are needed. Although the base's reaction force usually would not fight beyond the perimeter of the base, the reaction force must be ready to assist the response force or TCF when it arrives. The following should be considered:

- Command relationships before, during, and after linkup.
- Coordination of fire support before, during, and after linkup.

- Recognition signals and communication procedures to be employed.
- Follow-on operations required.
- Area damage control.

## Setup in a Hide Position (Temporary Defensive Position)

4-51. If the platoon steps down from sustained continuous operations and cannot return to its base or base cluster, it may need to operate briefly from a temporary defensive position. When used properly, a temporary defensive position can enable the platoon to rest, recover, repair damaged equipment, and plan for future operations. It offers concealment with little chance of detection by the enemy. Platoons want to get the best security they can while tasking a minimum of soldiers to provide security. When the platoon leader decides to operate from a temporary defensive position, he notifies the company HQ.

4-52. Locate the position in or near the area of normal operations so that sustained operations can be resumed immediately, on order. METT-TC should be a primary concern. Easily defendable positions are preferred over those that are more difficult to defend. Ensure that the position has more than one exit route and provides communication with the next higher HQ that are enhanced or at least not interfered with by terrain. While built-up, urban areas afford suitable concealment for temporary defensive positions, it is essential that the capability to communicate be assessed before the selection of such a site. Other considerations of a temporary defensive position include—

- Keeping vehicles secure and available in a nearby location.
- Positioning vehicles so that key equipment can be moved or removed without displacement of the entire unit.

- Concealing equipment from the sides and overhead. This prevents detection from aerial observers and some side-looking airborne radar.
- Covering and concealing to reduce security and/or defense requirements.
- Providing enough space between vehicles to allow a vehicle to bypass another vehicle that is inoperable.
- Ensuring that the platoon follows signal security and uses noise and light discipline.
- Setting up fighting positions if the situation calls for them.

## SET UP LOCAL SECURITY

4-53. Self-defense planning and coordination is done as soon as the base is set up. Prior planning and mission analysis are essential elements of a base defense. MP must be able to defend the site even before occupation is complete. When an MP element is located as part of an established base, it helps defend a portion of the larger unit's perimeter. Elements that set up separately usually defend their sites by deploying in a 360-degree perimeter. The techniques and principles of defense are the same for defending a separate squad, platoon, company, or base. To plan a perimeter defense, evaluate the situation. Analyze the terrain in terms of observation and fields of fire, cover and concealment, obstacles, key terrain, and avenues of approach (OCOKA). Defenses are placed where the threat is greatest.

4-54. The platoon leader establishes the CP and the OP. He locates the CP and the OP where he can best see and control the platoon. If this is not possible, he locates it where it can cover the most likely avenue of enemy approach. An alternate CP and OP, operated by the PSG, is placed where it can control the portion of the perimeter that cannot be seen or controlled by the main

CP and OP. The platoon leader then decides what other security measures and means of communication to use.

4-55. Platoon leaders must plan more than cover and concealment to counteract threat infrared, radar, thermal, and other sensors. The platoon leader uses the principles of camouflage and counters the recognition factors that make an object stand out from its background by—

- Locating soldiers, equipment, or structures where they are least discernible. This alone can reduce or eliminate many recognition factors.
- Using any mix of hiding, blending, disrupting, or disguising that conceals visibility.
- Maintaining camouflage discipline continuously.

4-56. When the number of soldiers that will defend a 360-degree perimeter is small, the platoon leader must—

- Vary the size of the defensive sectors.
- Identify alternate fighting positions.
- Retain flexibility of thinking.
- Decide what equipment—
  - Is needed to set up a perimeter defense.
  - Should stay in the vehicles.
  - Must be requisitioned or picked up later.

4-57. The equipment to improve defensive positions includes concertina wire, sandbags and tape (for cover and concealment), trip flares, pyrotechnic devices, mines, or PEWS, and other lethal and nonlethal technologies.

4-58. Platoons must be able to defend during day or night, in reduced visibility, and in a variety of weather conditions. The platoon leader or sergeant ensures that the platoon has the equipment it needs to defend under these conditions. The platoon must know how and when to use the equipment. During reduced visibility—

- Take steps to keep the enemy from observing or surprising the platoon.
- Require OPs and LPs. There should be at least one OP and one LP per squad. OPs and LPs report the enemy's advance and call for illumination and supporting fire. As in a daylight defense, MP manning OPs and LPs withdraw before they become engaged in close combat.
- Use patrols, illumination, PEWSs, and NVDs to help detect the enemy's advance.
- Use trip flares to provide warning and give some illumination. As a rule, do not fire until the targets are visible.
- Use camouflage, movement control, and light and noise discipline.
- Limit radio traffic to essential information.
- Ensure strict fire control to keep from disclosing the fighting positions.
- Ensure that gunners with crew-served and antiarmor weapons use NVDs.
- Provide illumination by using handheld flares or grenade launchers with illuminating rounds. Added light may be provided by fire support.
- Ensure that platoon leaders plan the use of messengers, visual signals, personal contact, or whistles to communicate with the squad leaders. Squad leaders plan to communicate with their team leaders and teams using personal contact or sound and visual signals.

## CONSTRUCT FIGHTING AND SURVIVABILITY POSITIONS

4-59. Fighting positions help protect soldiers and their equipment from the enemy and from the enemy's small-arms fire and fragmentation weapons, while allowing soldiers full weapon system engagement. A fighting position provides cover and concealment from which to engage or defend against the enemy. See *FM 7-8* for individual fighting position construction.

4-60. Fighting positions do not protect against the destructiveness of artillery and other area weapons. Nevertheless, a dug-in fighting position may be the key to survivability. Digging in cannot remove a soldier's vulnerability by itself. It does reduce exposure to the enemy's acquisition, targeting, and engagement systems. Platoons must be able to construct their survivability positions, often without engineer assistance.

4-61. Locate fighting positions for crew-served weapons where gunners can stop dismounted attacks. The sectors of fire must cover infantry avenues of approach and provide the most grazing fire across the platoon or squad front. Overlap the sectors of fire with each other and those of adjacent squads. Prepare fighting positions so that their primary sectors of fire have the guns firing across the front of the unit. Prepare secondary sectors of fire so that the guns fire to the front.

4-62. Usually, one MP team occupies an M249 or MK19 fighting position. One member is the gunner, one is the assistant gunner, and one is the ammunition bearer or rifleman. Each gunner has a primary and a secondary sector of fire. The gunner fires in the secondary sector only on order or when there are no targets in the primary sector. Each gunner uses aiming stakes to set his weapon for a final protective line (FPL) or a principal direction of fire (PDF) within the primary sector. The FPL and PDF are control measures to help defend a position. In an attack, the gunner knows the primary areas. He engages the greatest threat and, on the order of the platoon leader or PSG, fires the FPL.

## PREPARE SECTOR SKETCHES

4-63. After the crew-served weapons are in position, the squad leader positions the remaining MP to protect the gunners and to cover areas not covered by the gunner's. Using the range cards, the squad leader makes a squad sector sketch. He includes a rough sketch of the terrain around the weapon *(Figure 4-1)*. The squad sector sketches are used to plan defense and to control fire. Squad sector sketches show the following:

- The main terrain features in each sector of fire and the ranges to the features.
- Each primary fighting position.
- The primary and secondary sectors of fire for each position.
- MK19 and M249 FPL or PDF.
- The type of weapon at each position.
- The OP and LPs and squad leaders' positions.
- Dead space.
- Mines and obstacles.

4-64. The squad leader checks the range cards and the squad sector sketch for gaps or other flaws in the fire plan. He adjusts the weapons or the sectors as necessary. If the squad leader finds dead space, he takes steps to cover it with mines, grenade-launcher fire, or indirect fire. He then prepares two copies of the squad sector sketch. He keeps one copy and forwards the other copy to the platoon leader who makes a platoon sector sketch *(Figure 4-2, page 4-28)*. The platoon sector sketch shows the following:

- Squad sectors of fire.
- The crew-served and antiarmor weapons positions and sectors of fire, including FPL or PDF for the crew-served weapons and target reference points for the antiarmor weapons.
- Positions of the mines and the obstacles.

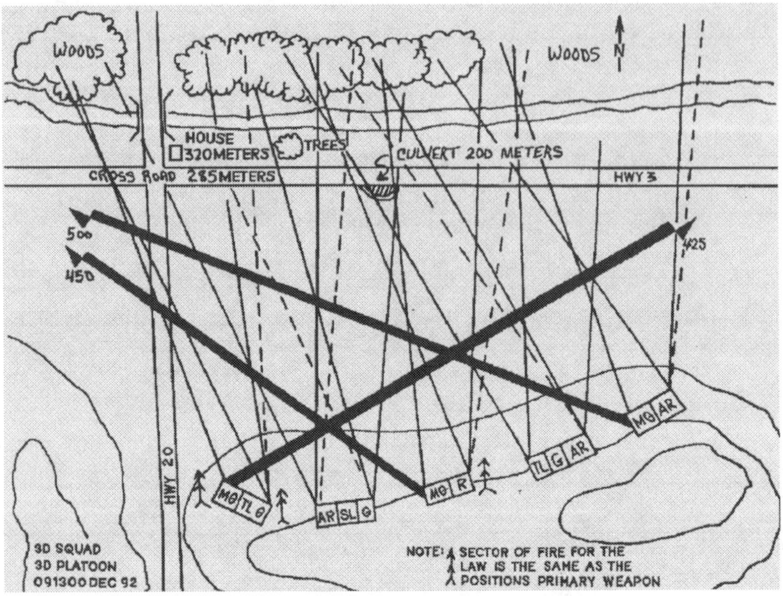


**Figure 4-1. Squad Sector Sketch**

- Indirect fire planned in the platoon's sector of fire.
- The OP and LPs and patrol routes (if any).
- The platoon CP and OP.

4-65. The platoon leader coordinates with the nearby units. He usually coordinates from left to right and from front to rear. The fires of units within the perimeter must be closely coordinated with the platoon's defensive fire plan. Squad leaders coordinate their fire plans with adjacent squads. All positions and units near the platoon are mutually supporting. The platoon leader makes sure gaps between the units are covered by fire, observation, patrols, OPs and LPs, or sensors. The units exchange information on—

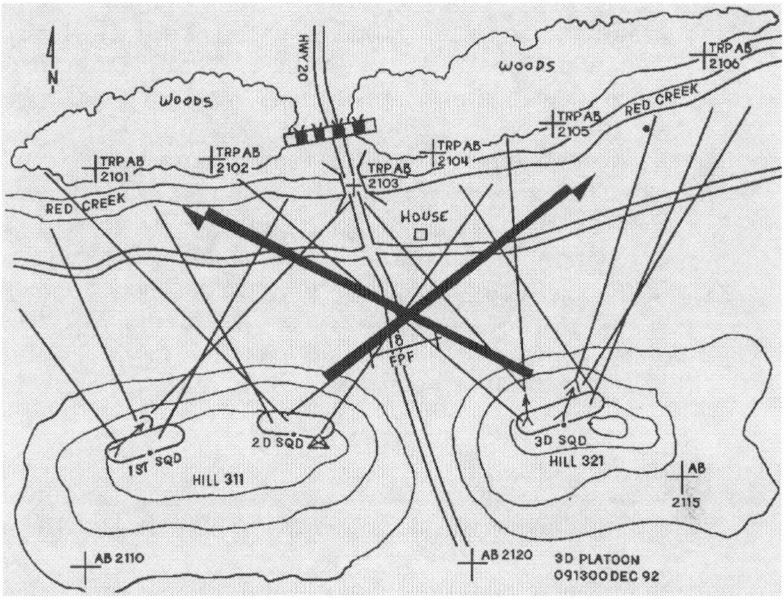


Figure 4-2. Platoon Sector Sketch

- The location of dead space between the elements and how to cover it.
- The locations of primary, alternate, and supplementary positions and sectors of fire for automatic weapons, antiarmor weapons, and subordinate elements.
- The locations of the OPs and LPs.
- The locations and types of obstacles and how they are covered by fire.
- Any patrols to be conducted, giving their size, type, times of departure and return, and routes.

## PREPARE RANGE CARDS

4-66. The FPL for the M249 is the line where an enemy assault is to be checked by interlocking fire from all weapons. Use the M249 on the FPL for grazing fire no more than 1 meter above the ground, about hip high, across the front of the element. Use the MK19 or M203 to cover the dead space. To figure the dead space on the FPL, the gunner watches a person walking down the FPL and marks spaces that cannot be grazed. The gunner records all the dead space data on the range card *(Figure 4-3, page 4-30).* He prepares at least two copies of the range card, keeping one card at the position and giving one copy to the squad leader. Fire on a gunner's FPL is its final protective fire (FPF). FPF is usually used as a last resort to stop an enemy assault. All weapons fire on command, continuously, until the call to stop FPF is given.

4-67. When terrain prevents the use of an FPL, the gunner uses a PDF instead. He directs his fire toward the most threatening avenue of approach that leads to his position. His weapon is positioned to fire directly on this approach rather than across the squad's front.

4-68. Construct fighting positions for the MK19 like M249 fighting positions. However, be aware that it takes added effort to keep the M3 tripod from moving because of the recoil of the MK19. If gunners are using the M249 machine gun, they should use the tripod when firing at an angle and the bipod when firing to the front. When gunners change their fires from the oblique to the front, they must move the machine gun, but leave the tripod in place. If gunners are using the MK19, they position the tripod toward the primary sector of fire. However, because there is no bipod for the MK19, gunners must be prepared to adjust both the weapon and the tripod to the secondary sector, if required. After a crew is positioned and is assigned an FPL or a PDF, the team—

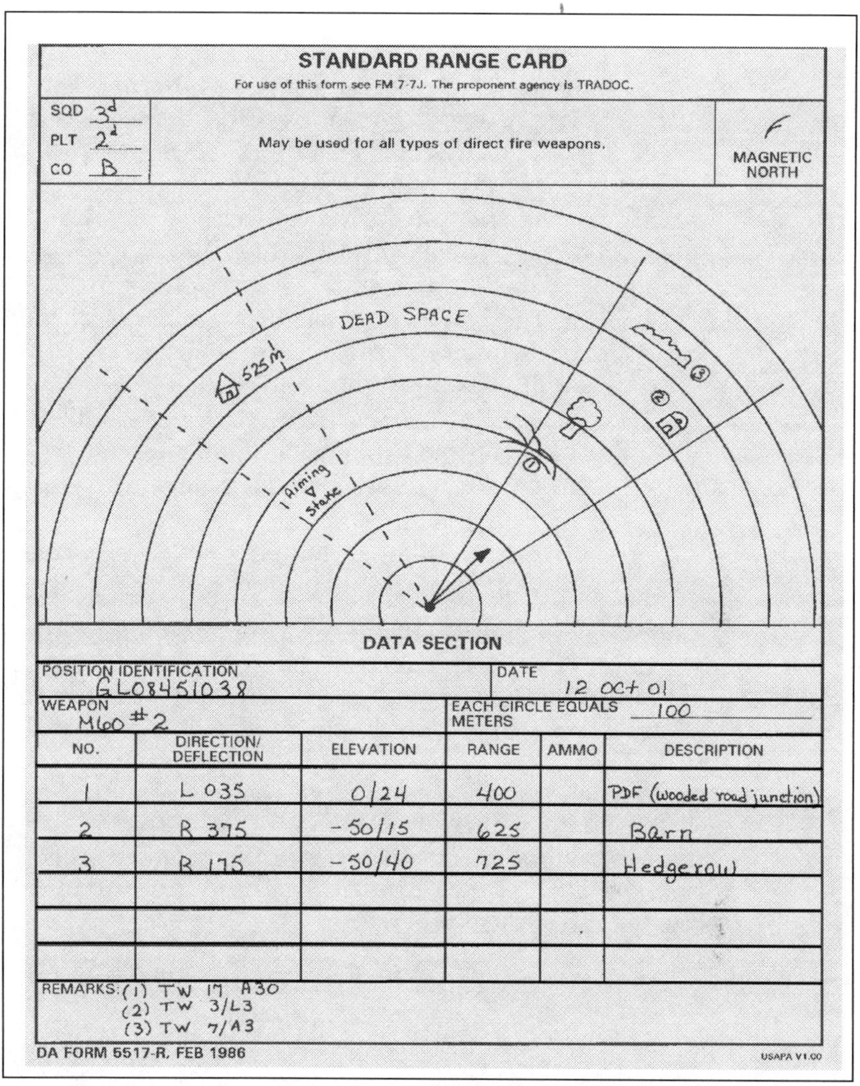

STANDARD RANGE CARD

For use of this form see FM 7-7J. The proponent agency is TRADOC.

SQD 3d
PLT 2d
CO B

May be used for all types of direct fire weapons.

MAGNETIC NORTH



DATA SECTION

POSITION IDENTIFICATION: GL08451038
DATE: 12 oct 01
WEAPON: M60 #2
EACH CIRCLE EQUALS: 100 METERS

| NO. | DIRECTION/ DEFLECTION | ELEVATION | RANGE | AMMO | DESCRIPTION |
|---|---|---|---|---|---|
| 1 | L 035 | 0/24 | 400 | | PDF (wooded road junction) |
| 2 | R 375 | -50/15 | 625 | | Barn |
| 3 | R 175 | -50/40 | 725 | | Hedgerow |
| | | | | | |
| | | | | | |
| | | | | | |

REMARKS: (1) TW 17 A30
(2) TW 3/L3
(3) TW 7/A3

DA FORM 5517-R, FEB 1986 USAPA V1.00

**Figure 4-3. Standard Range Card**

- Marks the tripod's position and the limits of their sectors of fire with aiming stakes.
- Outlines the hole.

- Digs the firing platform first. This lessens their exposure if they have to shoot before the position is completely constructed. Dig the firing platform at a level that allows the gun to traverse the sectors of fire.
- Lowers the gun to reduce the gunner's profile. This also reduces the height of the frontal cover needed.
- Digs the hole deep enough to protect itself and still allow the gunner to shoot in comfort (usually about armpit deep).
- Places the dirt where frontal cover is needed.
- Uses the rest of the dirt to build the flank and rear cover when the frontal cover is high enough and thick enough. Sandbags, wire, hatchets, or saws can be useful for building overhead cover or improving the fighting positions.

4-69. The ammunition bearer digs a one-man fighting position to the flank. He positions himself where he can see and shoot to the front and the oblique. Usually the ammunition bearer is on the same side as the FPL or the PDF. From there he can see and shoot into the machine gun's secondary sector. He can also see the gunner and the assistant gunner. The ammunition bearer connects his position to the machine gun position by a crawl trench. This allows him to provide ammunition or replace one of the gunners.

## SELECT FIGHTING POSITIONS IN BUILT-UP AREAS

4-70. Planning a defense of a platoon on urban terrain is similar to planning a defense in the countryside. Defensive positions must cover likely enemy avenues of approach, be mutually supporting, and provide cover and concealment. Use AT weapons on mounted avenues of approach. Machine guns cover dismounted approaches. AT4s and M203 grenade launchers work well in built-up areas. They are likely to hit enemy armored vehicles on the top or the side where armor is thin.

4-71. The method of defense (such as in-depth or linear) in the two areas is based on the same considerations. Use obstacles to canalize the enemy into kill zones or to deny key terrain. Orders must be very specific. Due to limited resources, use obstacles to channel, divert, or impede movement.

## Position Locations

4-72. Select defensive positions in urban areas based on METT-TC. Often a squad occupies a building, but larger buildings may be defended by a platoon. Select buildings that—

- Are well built. Concrete and steel construction is preferred.
- Have strong floors to keep the structure from collapsing under the weight of debris.
- Have thick walls and floors so that the enemy cannot shoot through roofs and walls to kill defenders.
- Are constructed of nonflammable material. Avoid wood. Strong, fireproof construction provides protection from a nuclear attack as well as conventional firepower.
- Have few glass windows (or break and remove the glass).
- Provide good fields of fire. Buildings located next to vacant lots, alleys, and parks allow better fields of fire than buildings located next to other buildings.
- Allow mutual support between the buildings. No building should be subject to attack without troops in another building being able to provide supporting fire.

4-73. Locate positions so as not to establish a pattern. Avoid obvious firing locations like church steeples (remember the elements of OCOKA) by—

- Placing MK19s in the building where they can cover assigned sectors of fire and FPL.
- Having the squad automatic riflemen and grenadiers cover enemy approach routes to the building.
- Placing most rifle positions at or near ground level to have overhead protection and provide grazing fire on approaches.
- Positioning some MK19 gunners higher to get a longer range. In addition, they can fire into areas that would be dead space for ground-level weapons.
- Positioning AT4s (remember the back blast) so that they can fire down on tracked infantry fighting vehicles and wheeled scout reconnaissance vehicles.

**Building Improvement**

4-74. Change the outside of the building as little as possible, but inside the building—

- Improve the fighting positions to provide overhead and frontal cover. Use firing ports to avoid enemy observation.
- Cut or blow holes between rooms and floors so the soldiers can move quickly by a covered and concealed route to other firing positions in the building.
- Seal off unused basements to prevent enemy entry.
- Barricade doors, halls, and stairs and take down fire escapes to keep the enemy out of the building.
- Reinforce positions with sandbags, solid debris, beds, furniture, and so forth.

- Screen or block windows and other openings. This keeps the enemy from seeing which windows are manned and throwing hand grenades into the building. When firing from the windows or holes in the walls, be sure the muzzle of your weapon does not protrude beyond the wall. This conceals the muzzle flash.
- Remove combustible materials to limit the danger of fire.
- Turn off electricity and gas.
- Stockpile water and dirt to fight fires.
- Wear armored vests, earplugs, and goggles for protection from dust and debris.

## Other Considerations

4-75. Operating in urban terrain can be challenging, so consider the following:

- The employment of weapons is different (shorter ranges).
- The position locations are different, such as LAWs and AT4s go on the upper floors of the buildings (refer to *FM 23-25* for the safety considerations).
- The target acquisition is more difficult (such as, aiming stakes and layered fires are used extensively).
- The CSS will center more on stockpiling materials in positions rather than on traditional resupply methods.
- Controlling indirect fire is more difficult.
- Primary communication must be by messenger, wire, or visual signs rather than radio.
- Avenues of approach are more canalized.
- The three dimensions of the enemy (aboveground, ground level, or below ground)

may be in use. The enemy can easily isolate subordinate units.

- Civilians and fire hazards may be present.

## ESTABLISH AND OPERATE AN OBSERVATION POST/LISTENING POST

4-76. OPs/LPs are selected locations from which to look and listen for enemy activity within an assigned area of observation. The OPs/LPs, the primary means of maintaining surveillance of an assigned avenue or a named area of interest (NAI), are positions from which MP observe the enemy and direct and adjust indirect fires against him. From the OPs/LPs, MP send SALUTE reports *(Figure 4-4, page 4-36)* to the commander when observing enemy activity. Use OPs/LPs for the following:

- On key terrain when the surveillance of a specific area is required.
- To prevent the enemy from a surprise attack on other friendly forces.
- As an early warning security measure in a defensive perimeter.
- For the monitoring of likely enemy avenues of approach, drop zones (DZ), and landing zones (LZ).

4-77. OPs/LPs can be performed either mounted or dismounted. A dismounted OP provides maximum stealth and has the greatest likelihood of remaining undetected by the enemy. The disadvantage of the dismounted OP is the time it takes to remount and move if necessary. If rapid movement or displacement is anticipated, the OP mounts or remains mounted.

4-78. A mounted OP/LP offers the advantages of rapid movement and protection because the enemy can easily detect them; however, it is potentially much less effective than a dismounted OP/LP.

**SIZE**: Give the size of the enemy unit as the number of troops or vehicles seen. Report 10 enemy infantrymen (not an infantry squad). Report three enemy tanks (not an enemy tank platoon).

**ACTIVITY**: Report what the enemy was doing. "They are emplacing antipersonnel mines in the road."

**LOCATION**: Report where the enemy was seen. Report the grid coordinates and the direction the enemy was heading. If a map is not available, relate the location to key terrain, such as the enemy's location is "on the Hahn Road, 300 meters south of the Kell River Bridge."

**UNIT**: An enemy soldier's unit may be hard to determine. Report markings or other distinctive features seen on the vehicles. Some countries have special uniforms and headgear. Some have colored tabs on the uniforms to show the type of unit, or the unit's actions may show its type. The kind of equipment it has may be peculiar to a certain type of unit. For example, a scout reconnaissance vehicle may indicate a reconnaissance unit; an amphibious tracked infantry-fighting vehicle may indicate an airborne unit.

**TIME**: Report the time the enemy was seen, not the time you are reporting.

**EQUIPMENT**: Report all the equipment the enemy is wearing or using. If you do not recognize an item of equipment or a type of vehicle, sketch it. Submit the sketch with the report.

**Figure 4-4. Salute Report**

## SELECT OBSERVATION POST/LISTENING POST SITES

4-79. The platoon leader selects the general location for the platoon's OP/LP after analyzing METT-TC factors. From his analysis, he determines how many OPs and LPs to establish. He decides where they must be positioned to allow long-range observation along the avenues of approach assigned by his commander and to provide depth through the sector. Section and squad leaders select the exact positions for each OP/LP on the ground. The OP/LP must have the following characteristics:

- Covered and concealed routes to and from the OP/LP. Ensure that MP can enter and leave their OP/LP without being seen by the enemy.
- Unobstructed observation of the assigned area or sector. Ideally, the fields of observation of adjacent OPs/LPs overlap to ensure full coverage of the sector.
- Covered and concealed positions that are effective. MP select positions with cover and concealment to reduce their vulnerability on the battlefield. MP may need to pass up a position with favorable observation capability, but with no cover and concealment, to select a position that affords better survivability.
- Located where they will not attract attention. Do not locate OPs/LPs in such locations as a water tower, an isolated grove of trees, or a lone building or tree; these positions draw enemy attention and may be used as enemy artillery TRPs.
- Located where they are not silhouetted. Avoid hilltops and position OPs/LPs further down the slope of the hill or on the side, provided there are covered and concealed routes into and out of the position.

## MAN THE OBSERVATION POST/LISTENING POST

4-80. Ideally, an MP team should man an OP/LP to maintain team integrity. Position OPs/LPs down the slope or on a flank of a hill, if there are covered withdrawal routes. Each of the OP's/LP's fields of observation overlap those of adjacent OPs/LPs. MP may have to selectively clear fields of observation. Ensure that MP are not seen when entering and leaving an OP/LP. Equip OP/LP teams to observe the area, report information, protect themselves, and call for and adjust indirect fire. OPs/LPs on a defensive perimeter need secure communications. Use of field phones or secured radios are usually best. However, messengers can be used. OPs/LPs may use portable radios to supplement wire communication. One MP observes the area while a second MP provides local security and records and reports information. The third MP rests or provides backup security. The team members switch jobs every 20 to 30 minutes because the observer's effectiveness decreases quickly after that time. The observer needs—

- A map of the area.
- A compass.
- Communication equipment (wire and radio).
- Observation devices, such as binoculars, observation telescope, and NVDs.
- An SOI extract.
- The report formats contained in the SOP.
- A radio (this may be the only means of communication from a remote site like a DZ or an LZ).

## POSITION OBSERVATION AND LISTENING POSTS

4-81. Place OPs/LPs either in a linear configuration or in-depth. Linear placement *(Figure 4-5)* allows the platoon to observe the assigned sector from several OP/LP sites, reducing the chance of the enemy entering the sector without being observed. This method works well when the platoon has been assigned a large sector with few avenues of approach or is in desert-type terrain. In-depth OP/LP placement *(Figure 4-6, page 4-40)* allows the platoon to observe the entire sector by placing OP/LP sites where the platoon can observe the most likely avenues of approach in the sector as well as along the sector flanks. This method works well when the platoon is assigned a sector with several avenues of approach or is in heavily wooded terrain. In-depth placement allows for redundancy in observation and better sector coverage.

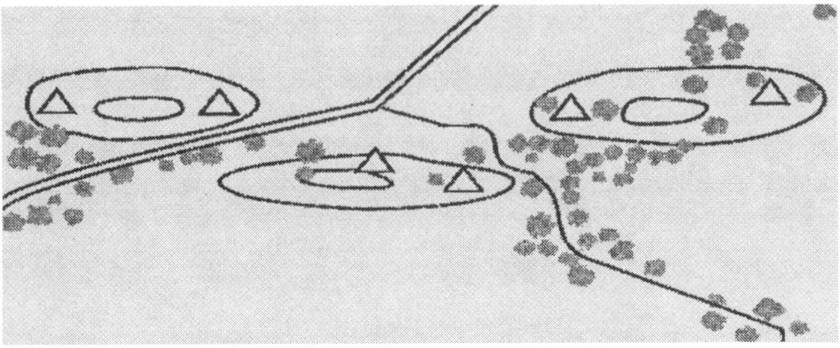

**Figure 4-5. Linear OP/LP Placement**

4-82. OP/LP team emplacement at night depends on METT-TC factors. The platoon leader places OPs/LPs close to the perimeter and within direct fire range of the defensive perimeter for protection. The team leader designates a specific location and primary direction of fire for the crew-served weapon. The OP/LP team builds a hasty-fighting position or a prepared-fighting position depending on METT-TC. The team leader designates a covered and concealed location behind OPs/LPs for the vehicle. The OP/LP team has a covered and concealed withdrawal route to the vehicle from the fighting position. The team camouflages the OP/LP and their vehicle while the gunner clears a field of fire and prepares a range card. The squad leader establishes communication with higher HQ and tells the team when and how to report. He tells them—

- If and when they should fire at the enemy.

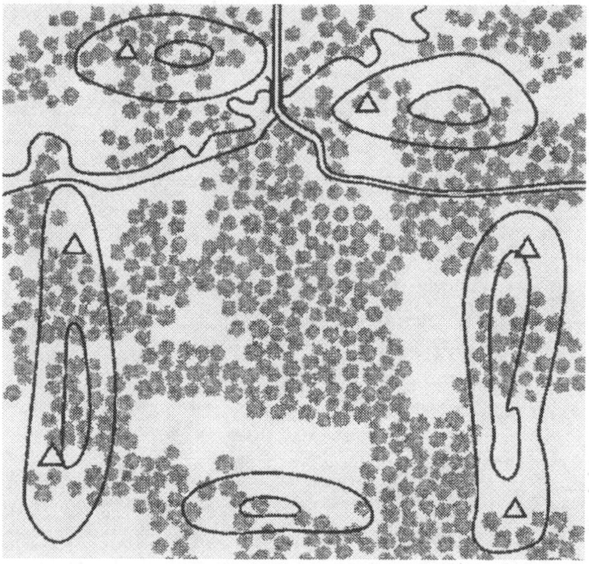

Figure 4-6. In-Depth OP/LP Placement

- How to get back to the squad if they must withdraw.
- What reentry signals to use.
- When they will be replaced, if known.
- To fight or withdraw according to his instructions.
- To be careful not to be drawn away by a small enemy element while the main element attempts to penetrate the perimeter.
- When to pull back or under what conditions they can withdraw without his order.

4-83. The frequency of relief for the OP/LP team depends on the team's physical condition and morale, the weather, the number of troops available, and the next operation. The squad leader carefully plans how each soldier receives rest. When OP/LP team is part of a defensive perimeter, it—

- Ensures that it has rearward cover.
- Builds fighting positions for protection and concealment.
- Uses trip flares, noisemaking devices, and NVDs to detect the enemy.
- Emplaces claymore mines for added protection.
- Coordinates with the perimeter on the reentry procedures to the perimeter from the withdrawal route.

## DEFEND A SITE

4-84. Vigilance is the watchword for local security. When the OPs/LPs detect enemy elements, they notify their superior who calls for indirect fire, if it is available. When the enemy's advance threatens the OP/LP, order the OP/LP to withdraw. As the enemy approaches platoon positions, have the platoon increase its volume of fire. The platoon leader determines if the platoon can destroy the enemy from its assigned positions. If the platoon leader determines that the platoon can destroy the enemy, the platoon continues to fight with the following actions:

- The platoon leader or the FO continues to call for indirect fire as the enemy approaches. The platoon normally begins engaging the enemy at the maximum effective range. It attempts to mass fire and initiate them at the same time to achieve surprise. Long-range fire should disrupt enemy formations, channelize the enemy toward engagement areas, prevent or severely limit the enemy's ability to observe the location of friendly positions, and destroy the enemy as it attempts to breach tactical obstacles.

- The leaders control fire using standard commands, pyrotechnics, and other prearranged signals. The platoon increases the intensity of fire as the enemy closes within range of additional weapons. Squad leaders work to achieve a sustained rate of fire from their positions by having buddy teams fire their weapons so that both are not reloading them at the same time.
- The platoon and squad leaders consider the following when controlling and distributing fires:
  - The enemy's range.
  - The priority of the targets (what to fire at, when to fire, and why).
  - The nearest or most dangerous targets.
  - Shifts to concentrate fires on their own or as directed by higher HQ.
  - The ability of the platoon to engage dismounted enemy with grazing fires, and flank shots against enemy vehicles.
- The platoon leader initiates FPF as the enemy closes on the platoon's perimeter. The following actions occur at the same time:
  - The automatic weapons fire along interlocking PDF or FPLs. Other weapons fire at designated PDF. The M203 grenade launchers engage enemy in dead space or against enemy attempts to breach the protective wire.
  - The platoon continues the fight with claymore mines and hand grenades.
  - The platoon leader requests indirect FPF in support of his positions, if applicable.
  - The platoon continues to defend until the enemy is repelled or until the platoon is ordered to disengage.

4-85. If the platoon leader determines that the platoon can not destroy the enemy, he—

- Reports the situation to the company commander.
- Repositions the platoon to—
  - Continue fires into the platoon sector (engagement area).
  - Shift to alternate or supplementary positions.
  - Reinforce other parts of the company.
  - Counterattack locally to retake lost fighting positions.
  - Withdraw from an untenable position using fire and movement to break contact. (The platoon leader should not move his platoon out of position if it will destroy the integrity of the company's defense.)

**NOTE: In any movement out of a defensive position, the platoon must employ all direct and indirect fire means available to suppress the enemy long enough for the platoon to move.**

4-86. The platoon reorganizes after it has completed the initial battle with the enemy or relocated. The platoon positions key weapons, reestablishes security, provides first aid and prepares wounded soldiers for evacuation, and redistributes ammunition and supplies. The platoon relocates selected weapons to alternate positions if the leaders believe that the enemy may have pinpointed them during the attack and adjusts other positions to maintain mutual support. The platoon also reestablishes communication. It reoccupies and repairs positions and prepares for renewed enemy attack. The platoon repairs damaged obstacles and replaces mines and booby traps. When the platoon reorganizes, it performs the following actions:

- The squad and section leaders provide ammunition, casualty, and equipment (ACE) reports to the PSG. Team leaders provide fuel status. The PSG consolidates the ACE reports, reviews the consolidated ACE report with the platoon leader, and forwards it to the company commander.
- The platoon leader reestablishes the platoon's chain of command.
- The PSG coordinates for resupply and supervises the execution of the casualty and EPW evacuation plan.
- The platoon continues to improve positions. The platoon quickly reestablishes the OP/LP resumes patrolling as directed.

4-87. If the enemy gets through the FPF, repel it by close combat. If the perimeter is penetrated, move teams to block the penetration and cover friendly troops moving to alternate or supplementary positions. Even though the counterattack capability is limited, try to restore the perimeter. When the enemy is repelled—

- Reestablish security.
- Send patrols forward to maintain contact.
- Call for indirect fire on areas where the enemy is likely to regroup.
- Reorganize squads.
- Evacuate seriously wounded MP.
- Redistribute and resupply ammunition.
- Repair positions and continue to improve them.
- Keep the next higher commander informed throughout the conduct of the defense.

## LAY HASTY PROTECTIVE MINEFIELDS

4-88. When possible, lay a hasty protective minefield as part of the unit's defensive perimeter. It can stop, delay, or restrict movement. MP often lay mines to restrict enemy movement near a defensive perimeter or at ambush sites. In the defense, platoons and squads lay hasty protective minefields to supplement weapons, prevent surprise, and give early warning of enemy advance. Hasty minefields must be covered by fire. Ensure that adjacent units are informed of the mine locations.

4-89. Platoons and squads must have permission from higher HQ to install hasty protective minefields. Higher HQ may, however, delegate approval authority to the company commander for emplacement of a hasty protective minefield. Requests for permission go through the normal chain of command.

4-90. If the company is not authorized mines in its basic loads, a special request may be needed. The enemy threat to the rear area requires commanders to issue mines as an additional protective measure. The M18A1 antipersonnel mine (claymore) and the M21 AT mine are the two mines most likely to be available to rear-area units for a hasty protective minefield. Refer to *FM 20-32* and *FM 21-75*.

4-91. MP generally will have claymores available to them, which is mainly a defensive weapon. However, the ways in which the claymore is used is limited only by the imagination. Plan the use of claymore mines to suit METT-TC. Emplace the mines—

- On likely dismounted avenues of approach.
- To cover dead space not covered by FPF of crew-served weapons.
- Outside the hand grenade range, but within the range of small-arms weapons.
- Where they are covered by observation and fire.
- Where back blast will not injure friendly forces.
- Beside buildings or other sturdy structures in urban terrain.

- Strapped to boards (for detonation from around corners).

4-92. MP record the exact location of the mines on *DA Form 1355-1-R (Figure 4-7)* when emplacing the minefield. This enables anyone to recover them. If possible, the unit should recover the mines before the unit relocates, and the same persons who emplaced them should recover them.

# PATROLS

4-93. MP are organized and equipped to conduct mounted operations. The primary offensive weapon system is the MK19. It is designed as a mounted or static position weapon. However, an MP platoon may conduct limited dismounted operations with its other organic weapons. Refer to *FM 7-8* for more information about patrol operations, including organization, planning, and execution. In general, an MP platoon may be required to conduct reconnaissance and combat patrols.

## RECONNAISSANCE PATROLS

4-94. Dismounted reconnaissance patrols are directed by higher HQ and conducted to gather detailed information on the enemy, terrain, specific NAIs, or avenues of approach. When executed as part of a screen or other security mission, a reconnaissance patrol can ensure the security of an OP/LP or the platoon's defensive perimeter. This is also referred to as a security patrol. Refer to Chapter 6 for area and zone reconnaissance.

## COMBAT PATROLS

4-95. An MP platoon may conduct a combat patrol to establish an ambush on a dismounted enemy avenue of approach.

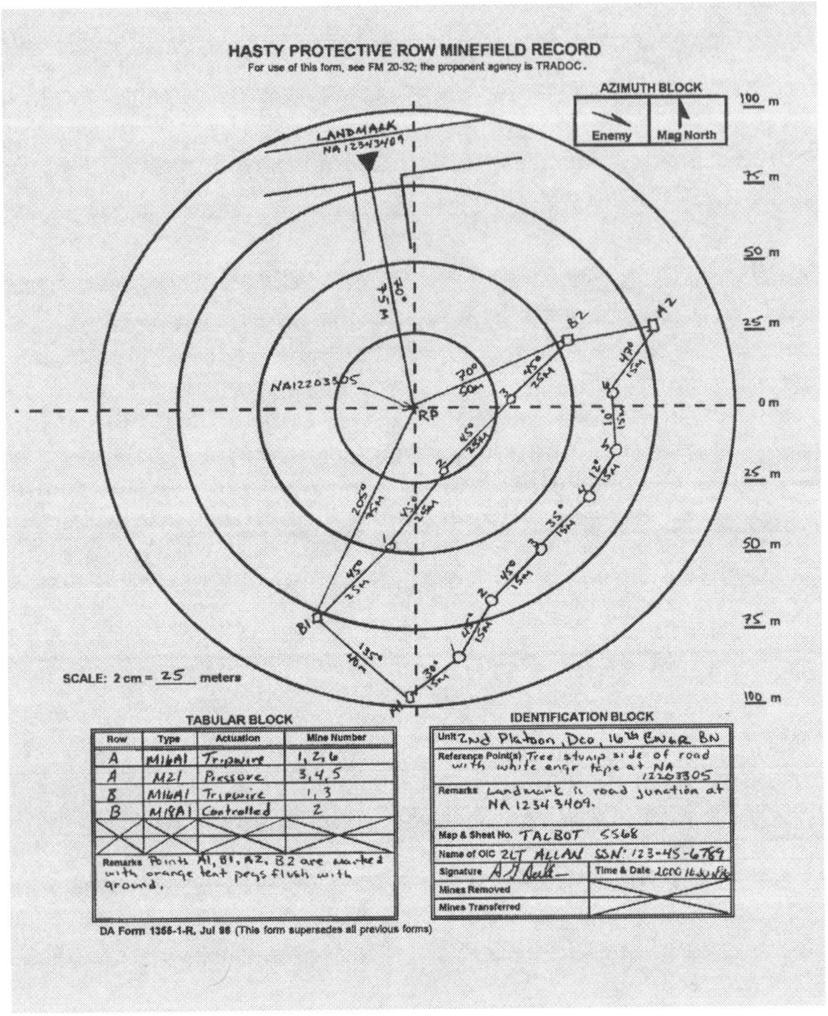
HASTY PROTECTIVE ROW MINEFIELD RECORD

For use of this form, see FM 20-32; the proponent agency is TRADOC.



TABULAR BLOCK

| Row | Type | Actuation | Mine Number |
|---|---|---|---|
| A | M16A1 | Tripwire | 1, 2, 6 |
| A | M21 | Pressure | 3, 4, 5 |
| B | M16A1 | Tripwire | 1, 3 |
| B | M18A1 | Controlled | 2 |

Remarks Points A1, B1, A2, B2 are marked with orange tent pegs flush with ground.

IDENTIFICATION BLOCK

Unit 2nd Platoon, D co, 16th Engr BN

Reference Point(s) Tree stump side of road with white engr tape at NA 12203305

Remarks Landmark is road junction at NA 1234 3409.

Map & Sheet No. TALBOT 5568

Name of OIC 2LT ALLAN SSN: 123-45-6789

Signature A J Allen  Time & Date 2000 16 Jul

Mines Removed

Mines Transferred

DA Form 1355-1-R, Jul 98 (This form supersedes all previous forms)

**Figure 4-7. Sample DA Form 1355-1-R**

## Ambush

4-96. MP elements, normally no smaller than a platoon, use an ambush along suspected enemy routes and elsewhere against Level II threats in the rear area. An ambush enables a small unit with light weapons to harass or destroy a larger, better-equipped unit. An ambush may be a surprise attack from a concealed position on a moving or temporarily halted target or an attack by fire only. A successful ambush requires the following:

- Surprise. The platoon must seize and maintain control of the situation.
- Coordinated fire. The platoon must deliver a large volume of fire into the kill zone, using individual and crew-served weapons, mines, demolitions, and indirect fire (if available), and isolate the kill zone to keep the enemy from escaping or being reinforced.
- Control. Before, during, and after the ambush all elements must be able to communicate effectively with the platoon leader, primarily by using hand and arm signals.

4-97. The platoon leader organizes the platoon into assault, support, and security elements. An ambush is laid on an enemy's expected approach route. The platoon leader selects the site and members are positioned to provide—

- Good visibility of the avenues of approach and the kill zone.
- Good fields of fire into the kill zone.
- Cover and concealment.
- Obstacles between the teams and the kill zone.
- Covered and concealed withdrawal routes.

4-98. A good ambush site restricts the enemy's movement to one flank by natural or man-made obstacles. Natural obstacles include cliffs, steep embankments, swamps, steep grades, sharp curves in the road, narrow trails, streams, and heavily wooded areas. Man-made obstacles can include mines, booby traps, and roadblocks. The ambush is configured to suit the—

- Type of ambush.
- Terrain.
- Troops available.
- Weapons.
- Equipment.
- Ease of control.
- Overall combat situation.

4-99. To conduct an ambush, the platoon leader is positioned so he can best control the ambush elements, normally with crew-served weapons or the AT4, especially if the enemy has armor. The platoon leader—

- Positions the flank security elements.
- Emplaces obstacles and mines.
- Improves fighting positions, if time permits.
- Places a crew-served weapon to cover the left and right limits of the kill zone. These weapons must ensure that once an element is in the kill zone, it cannot leave it laterally.
- Reports to higher HQ when the ambush is in place.

4-100. The platoon leader initiates the ambush with a casualty-producing weapon, such as a claymore mine or a crew-served weapon. He ensures that there is a back-up method in case the primary means fails. The remainder of the platoon opens fire once the ambush has begun.

4-101. Most often, platoons will deploy a squad-size element for an attack on a single kill zone (a point ambush). If the company is deploying a platoon-size force to conduct a number of coordinated, related ambushes (an area ambush), the principles are the same. An area ambush works best where close terrain keeps enemy movement largely limited to trails or roads. For an area ambush—

- Choose one central ambush site around which you can control and organize the outlying ambushes.
- Select outlying ambush sites on the enemy's possible avenues of approach and escape from the central site.
- Set up and maintain communication with all the outlying sites.
- Assign the general locations of the outlying sites to the squad leaders. They will each set and conduct a point ambush.
- Direct the squad leaders to let the enemy pass through the kill zones until the central ambush begins.
- Provide specific instructions to the squad leaders in case the enemy detects an outlying site before the central ambush begins.

## Ambush Types

4-102. There are three types of ambushes. They are the line and L-shaped ambush formations and the antiarmor ambush (refer to *FM 7-8* for more information).

4-103. **Line Ambush Formation**. A line formation (refer to *FM 7-8* for more information) is easy to control and is useful in all levels of visibility. The assault and support elements parallel the long axis of the kill zone to engage the enemy with flanking fire. The target may be so dispersed that it extends beyond the kill zone. Leaders must—

- Position the assault and support elements parallel to the enemy's movement route (such as on a road or trail or at a stream).

- Limit the kill zone to the size area that the ambush can cover with a great volume of fire.
- Place obstacles (such as claymore mines or existing natural obstacles) between the kill zone and the ambush element to prevent counter-ambush actions.
- Leave access lanes through the obstacles so the kill zone can be assaulted (if directed).

4-104. **L-Shaped Ambush Formation**. An L-shaped formation is useful on a straight stretch of a trail, road, or stream. It also works well at a sharp bend in a trail, road, or stream. The assault element is the long leg of an "L," paralleling the kill zone to provide flanking fire. The support element is the short leg, capping the end of the kill zone at a right angle to the assault element. The support element provides enfilade fire to interlock with fire from the other leg.

4-105. **Antiarmor Ambush**. The principles for an antiarmor ambush are the same as for an area or point ambush. MP are likely to encounter bypassed enemy armor in the rear area. The primary antiarmor weapon for MP is an AT4. This is a light antiarmor weapon with limited capability against medium and heavy armor vehicles. However, MP may be required to set up a hasty antiarmor ambush to destroy one or two light enemy armor vehicles. An antiarmor ambush is best performed by a platoon. The platoon leader positions the antiarmor weapons where they can engage the target from the rear, flank, or top. Multiple AT4s are used to ensure destruction. The platoon provides support and security.

# CLEARING TECHNIQUES

## HIGH-INTENSITY VERSUS PRECISION CLEARING TECHNIQUES

4-106. Precision clearing techniques do not replace other techniques currently being used to conduct building and room clearing during high-intensity combat. Specifically, they do not replace the clearing technique in which a fragmentation or concussion grenade is thrown into a room before US forces enter. Use precision room clearing techniques when the tactical situation calls for room-by-room clearing of a relatively intact building in which enemy combatants and noncombatants may be intermixed. They involve increased risk in order to clear a building methodically, rather than using overwhelming firepower to eliminate or neutralize all its inhabitants.

4-107. From a conceptual standpoint, standard high-intensity room-clearing drills can be thought of as a deliberate attack. The task is to seize control of the room, with neutralization of the enemy in the room the purpose. The fragmentation or concussion grenade can be thought of as the preparatory fire used before the assault. As in a deliberate attack against any objective, the assaulting element moves into position using covered and concealed routes. The preparatory fire is initiated when the assaulting element is as close to the objective as it can get without being injured by the enemy. The assault element follows the preparatory fire (fragmentation or concussion grenade) onto the objective as closely as possible. A rapid, violent assault overwhelms and destroys the enemy force and seizes the objective.

4-108. Compared to the deliberate attack represented by high-intensity room-clearing techniques, precision room-clearing techniques are more conceptually akin to a reconnaissance in force or perhaps an infiltration attack. During a reconnaissance in force, the friendly unit seeks to determine the enemy's locations, dispositions, strength, and intentions. Once the enemy is located, the friendly force is fully prepared to engage and destroy it, especially if surprise is achieved. The friendly force retains the options of not employing preparatory fire (fragmentation or concussion grenades) if it is not called for (the enemy is not in the room) or if it is inappropriate (there are noncombatants present also). The attacking unit may choose to create a diversion (use a stun-hand grenade) to momentarily distract the defender while it enters and achieves domination of the objective.

4-109. The determination of which techniques to employ is up to the leader on the scene and is based on his analysis of the existing set of METT-TC conditions. The deliberate attack (high-intensity techniques), with its devastating suppressive and preparatory fire neutralizes everyone in the room and is less dangerous to the assaulting troops. The reconnaissance in force (precision techniques) conserves ammunition, speeds up the clearing process, reduces damage, and minimizes the chance of noncombatant casualties. Unfortunately, even when well executed, it is very stressful and hazardous for friendly troops.

4-110. Certain precision room-clearing techniques, such as methods of squad and fire team movement, the various firing stances, weapon positioning, and reflexive shooting, are useful for all combat in confined areas. Other techniques, such as entering a room without first neutralizing the known enemy occupants by fire or explosives, are appropriate in only some tactical situations.

4-111. Generally, if an alerted enemy force that is determined to resist occupies a room or building, and if most or all noncombatants are clear, employ overwhelming firepower to avoid friendly casualties. In such a situation, use supporting fires, demolitions, and fragmentation grenades to neutralize a space before friendly troops enter.

4-112. In some combat situations, the use of heavy supporting fire and demolitions would cause unacceptable collateral damage or unnecessarily slow the unit's movement. In other situations, often during stability and support operations, enemy combatants are so intermixed with noncombatants that US forces can not in good conscience use all available supporting fire. At such times, room-by-room clearing may be necessary and precision room-clearing techniques are most appropriate.

## PRINCIPLES OF PRECISION ROOM-CLEARING

4-113. Battles that occur at close quarters, such as within a room or hallway, must be planned and executed with care. Units must train, practice, and rehearse precision room-clearing techniques until each fire team and squad operates smoothly. Each unit member must understand the principles of precision room-clearing, such as surprise, speed, and controlled violence of action.

### Surprise

4-114. This is the key to a successful assault at close quarters. The fire team or squad clearing the room must achieve surprise, if only for seconds, by deceiving, distracting, or startling the enemy. Sometimes stun grenades are used to achieve surprise. These are more effective against a nonalert, poorly trained enemy than against alert, well-trained soldiers.

### Speed

4-115. This provides a measure of security to the clearing unit. Speed allows soldiers to use the first few vital seconds provided by surprise to their maximum advantage. In precision room clearing, speed does not mean incautious haste. It can best be described as a careful hurry.

### Controlled Violence of Action

4-116. This eliminates or neutralizes the enemy while giving him the least chance of inflicting friendly casualties. Controlled violence of action is not limited to the application of firepower only. It involves a soldier's mind-set of complete domination. Each of the principles of precision room-clearing has a synergistic relationship to the others. Controlled violence coupled with speed increases surprise. Hence, successful surprise allows increased speed.

## FUNDAMENTALS OF PRECISION ROOM-CLEARING

4-117. The ten fundamentals of precision room-clearing address actions soldiers take while moving along confined corridors to the room to be cleared, while preparing to enter the room, during room entry and target engagement, and after contact. Team members should—

- Move tactically and silently while securing the corridors to the room to be cleared. Carry only the minimum amount of equipment.

**NOTE: Rucksacks and loose items carried by soldiers tire them, slow their pace, and cause noise.**

- Arrive undetected at the entry to the room in the correct order of entrance and be prepared to enter on a single command.
- Enter quickly and dominate the room. Move immediately to positions that allow complete control of the room and provide unobstructed fields of fire.
- Eliminate the entire enemy in the room by fast, accurate, and discriminating fires.

- Gain and maintain immediate control of the situation and all personnel in the room.
- Confirm whether enemy casualties are wounded or dead. Disarm and segregate the wounded. Search all enemy casualties.
- Perform a cursory search of the room. Determine if a detailed search is required.
- Evacuate all wounded and any friendly dead.
- Mark the room as cleared using a simple, clearly identifiable marking according to the unit SOP.
- Maintain security and be prepared to react to more enemy contact at any moment. Do not neglect rear security.

## COMPOSITION OF THE CLEARING TEAM

4-118. Execute precision room-clearing techniques by the standard four-man fire team. Because of the confined spaces typical of building- and room-clearing operations, units larger than squads quickly become unwieldy. When shortages of personnel demand it, conduct room-clearing with two- or three-man teams; four-man teams are preferred. Using fewer personnel greatly increases the combat strain and risks.

## BREACHING

4-119. An integral part of precision room-clearing is the ability to gain access quickly to the rooms to be cleared. Breaching techniques vary based on the type of construction encountered and the types of munitions available to the breaching element. Techniques range from simple mechanical breaching to complex, specialized demolitions.

### Shotgun Ballistic

4-120. A useful method of breaching is the shotgun ballistic breach for forced entry of standard doors. Use a

12-gauge shotgun loaded with buckshot or slugs to breach most standard doors quickly. When done properly, the shotgun breach requires only a few seconds. The two standard techniques of shotgun breaching are the doorknob breach and the hinge breach. When attempting either technique, the gunner approaches the door from an angle, avoiding standing in the area directly in front of the door. While holding the stock of the shotgun in the pocket of his shoulder, the gunner places the muzzle tightly against the door, and aims down at a 45-degree angle.

4-121. **Doorknob Breach**. For the doorknob breach, *(Figure 4-8, page 4-58)* the aim point is a spot halfway between the doorknob and the frame, not at the doorknob itself. The gunner fires two quick shots in the same location, ensuring that the second shot is aimed as carefully as the first. Weak locks may fly apart with the first shot, but the gunner should always fire twice. Some locks that appear to be blown apart have parts still connected that will delay entry. If the lock is not defeated by the second shot, the gunner repeats the procedure.

4-122. **Hinge Breach**. The hinge breach technique is performed much the same as the doorknob breach, except the gunner aims at the hinges. He fires three shots per hinge—the first at the middle, then at the top and bottom *(Figure 4-8)*. He fires all shots from less than an inch away from the hinge. Because the hinges are often hidden from view, the hinge breach is more difficult. Regardless of which technique the gunner uses, immediately after he fires, he kicks the door in or pulls it out. He then pulls the shotgun barrel sharply upward and quickly turns away from the doorway to signal that the breach point has been cleared. This rapid clearing of the doorway allows the following man in the fire team a

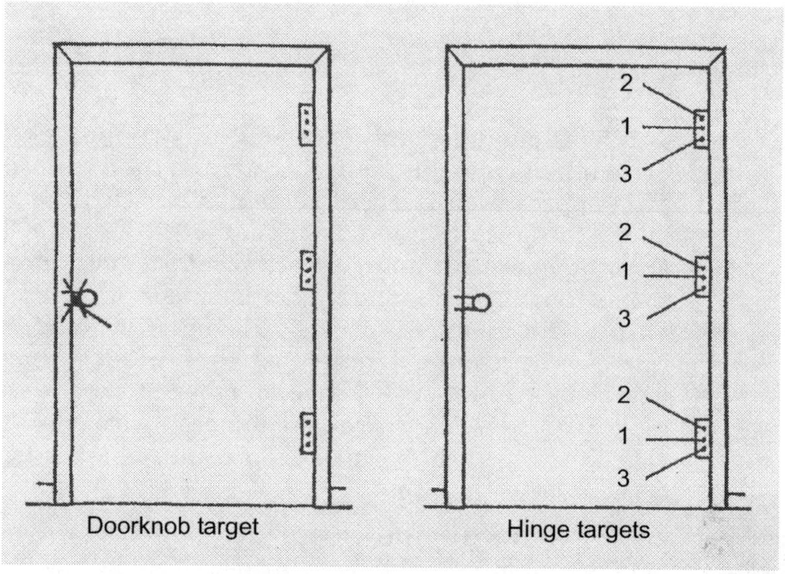


**Figure 4-8. Aim Points for a Shotgun Breach of a Standard Door**

clear shot at any enemy who may be blocking the immediate breach site.

> **WARNING**
> **Do not use small arms (5.56 or 7.62 millimeters) as a ballistic breach on doorknobs and hinges except as a last resort. It is unsafe and could result in death.**

4-123. Demolitions are often needed to defeat more elaborate barriers or to produce a desired effect to aid the initial entry.

## Mechanical Breaching

4-124. Mechanical breaching is planned as a backup to a ballistic or explosive breach. Mechanical breaching is an assumed capability within all units. Taking the time to defeat weak barriers such as doors or windows by means of crowbars, saws, sledgehammers, battering rams, or axes is a decision that must be made based on the conditions of METT-TC.

## Order of Movement

4-125. Clearing-team members must approach the breach point quickly, quietly, and in standard order. This approach preserves the element of surprise and allows for quick entry and domination of the room. The order of movement to the breach point is determined by the method of breach and intended actions at the breach point. The members of the fire team are assigned numbers 1 through 4, with the team leader normally designated number 3. If one member of the clearing team is armed with the squad automatic weapon (SAW) rather than an M16 rifle or carbine, he should be designated as number 4.

4-126. **Ballistic (Shot Gun) Breach**. The order of movement for a shotgun breach has the gunner up front, followed by the number 1 man, number 2 man, and then the number 3 man (team leader). After the door is breached, the gunner moves to the rear of the lineup and assumes the position of the number 4 man.

4-127. **Explosive (Demolition) Breach**. The order of movement for an explosive breach without engineer support is number 1, number 3 (team leader), number 2, and then number 4. The number 1 man provides security at the doorway. The number 3 man (team leader) carries the demolition charge and places it. The number 2 man provides security overhead, and the number 4 man provides rear security. After the demolition charge is placed, the team moves to covered positions and prepares to enter in the standard 1, 2, 3, 4 order.

4-128. **Mechanical Breach**. A suggested order of movement for a mechanical breach is the initial assault team in order, followed by the breach man/element. At the breach point the assault team leader will bring the breach team forward while the assault team provides local security. After the breach is conducted, the breach team moves aside and provides local security as the assault team enters the breach.

## ENTRY CONSIDERATIONS

4-129. The entire team enters the room as quickly and smoothly as possible and clears the doorway immediately. If possible the team moves from a covered or concealed position already in their entry order. Ideally, the team arrives and passes through the entry point without having to stop.

4-130. The door is the focal point of anyone in the room. It is known as the *fatal funnel*, because it focuses attention at the precise point where the individual team members are the most vulnerable. Moving into the room quickly reduces the chance anyone will be hit by enemy fire directed at the doorway.

4-131. On the signal to go, the clearing team moves from covered or concealed positions through the door quickly and takes up positions inside the room that allow it to completely dominate the room and eliminate the threat. Team members stop movement only after they have cleared the door and reached their designated point of domination. The first man's position is deep into the far corner of the room. The depth of his movement is determined by the size of the room, any obstacles in the room such as furniture, and by the number and location of enemy and noncombatants in the room.

4-132. To make precision room-clearing techniques work, each member of the team must know his sector of fire and how his sector overlaps and links with the sectors of the other team members. Team members do not move to the point of domination and then engage their targets. They engage targets as they move to their designated point. However, engagements must not slow movement to their points of domination. Team members may shoot from as short a range as 1 to 2 inches. They engage the most immediate enemy threats first. Examples of immediate threats are enemy personnel who are—

- Armed and prepared to return fire immediately.
- Blocking movement to the position of domination.
- Within arm's reach of a clearing team member.
- Within 3 to 5 feet of the breach point.

## CLEAR A SINGLE ROOM

4-133. Precision military operations on urbanized terrain (MOUT) room clearing is generally accomplished using a four-person team, consisting of three riflemen and one SAW gunner or breacher. Entry does not always require four personnel. If a specific room is to small or there are less than four clearing members due to attrition, the room may be cleared with less. However, never clear a room with less than two soldiers. If the leadership determines that the use of four-soldier teams is impractical for the situation, use the standard three-soldier team filling the numbers 1, 2, and 3 entry positions. The term *point of domination* refers to the two corners of the room assigned to the numbers 1 and 2 soldiers. These points allow the team to gain complete control of the objective. Each team member is assigned a different, but interlocking field of fire. This ensures mutual supporting fires. This type of entry is dynamic in nature. When completed with precision, this technique overwhelms the enemy and allows the team to move on to the next objective very quickly. When a room is cleared, the exiting procedure for leaving the room varies depending on the location of the entry point (either in the middle or at the corner of the room).

> **WARNING**
>
> **The muzzle of the clearing team's weapons should always be pointing wherever that soldier is looking. The clearing team should remain one foot away from all the walls to avoid injury due to bullets traveling down the wall (the path of least resistance).**

## Stack Position

4-134. Once the approach to the room or building is completed, the team stacks *(Figure 4-9)* parallel to the outside wall. The number 1 soldier provides security on the entry point. This point is normally a door in a precision MOUT environment. The number 2 soldier provides security to the front of the team. The number 3 soldier (normally the team leader) is in the high-ready position. The high-ready position consists of the soldier placing the buttstock of the weapon in his shoulder pocket and raising the muzzle of the weapon from a 90-degree angle to a 45-degree angle. The number 4 soldier is in the low-ready position. The low-ready position consists of the soldier placing the buttstock of the weapon in his shoulder pocket and lowering the muzzle of the weapon to a 45-degree angle. The high- and low-ready positions aid in ensuring that the soldiers do not flag their fellow team members with their weapons during entry or clearing. A support element covers the teams from a distance, concentrating on the team's blind spots.

## Initial Entry

4-135. When the situation allows (time, concealment, and so forth), it is important to ensure that each team member is prepared to enter the initial-entry point to the building before the assault. This is achieved while in

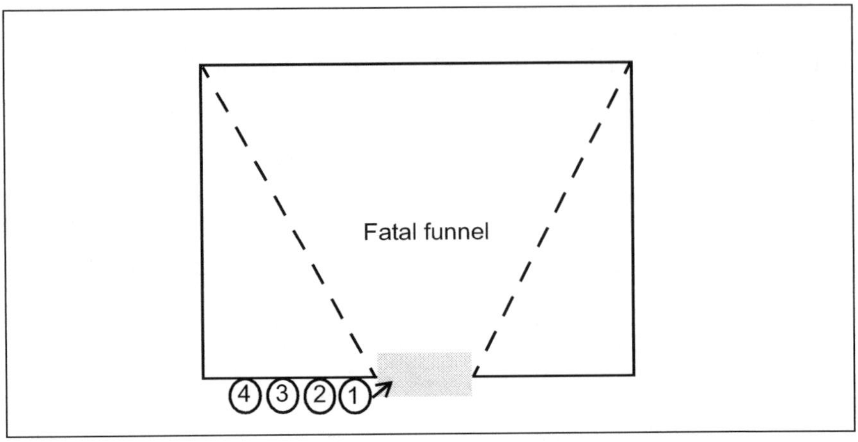


**Figure 4-9. Stack**

the stack position. When ready, the number 1 soldier nods his head while maintaining security on the door. When soldier number 4 is ready, he squeezes the shoulder of soldier number 3 with the nonfiring hand. This action continues up the line. When the signal reaches soldier number 1, he silently counts to three and then enters the building with the rest of the team following. If someone inside opens the door before this sequence is completed, the team will immediately enter the structure.

## Initial-Entry Breach

4-136. *Figure 4-10, page 4-64,* depicts the process of a nonexplosive breach from the standard stack. Soldier number 4 moves up to the side of the door that contains the locking mechanisms. When soldier number 3 gives the signal, soldier number 4 breaches the door (by shotgun, battering-ram, and so forth) and kicks it open. Soldier number 4 then moves back from the door to allow the rest of the team to enter. Soldier number 4 follows them into the room. Refer to *FM 90-10-1* for shotgun breaching techniques.

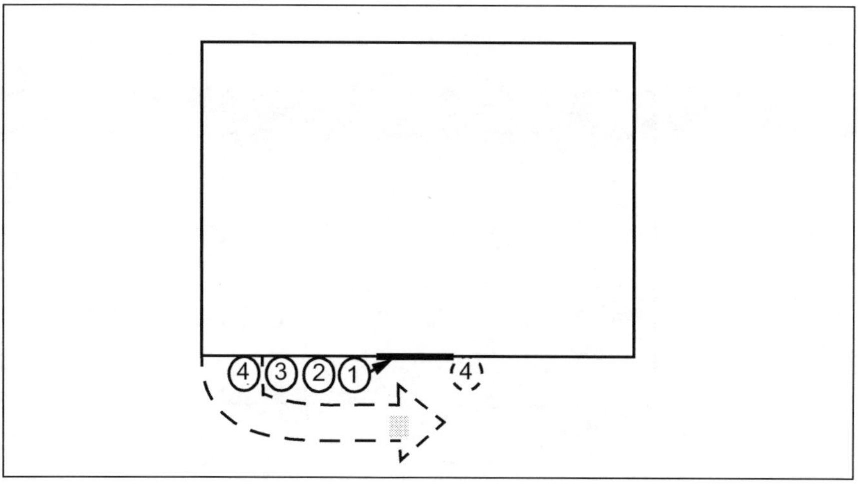


Figure 4-10. Nonexplosive Breach

## Room Entry

4-137. Clear a room in the following manner when the entry point is located in the middle of the room:

- Soldier number 1 *(Figure 4-11)* enters the room using the path of least resistance, (moves across the doorway, into the room, and does not make a turn around the doorframe) clearing the fatal funnel as fast as possible. The term fatal funnel is derived from the natural reaction of the enemy soldiers inside a room (the large end of the funnel) to focus on and fire at the doorway (the spout of the funnel). The rounds are funneled into the doorway from almost any position in the room. An attacker should never move slowly through (or stop in) the fatal funnel. Besides providing a silhouetted target to the enemy, these actions will slow the entry of the rest of the team (reducing supportive fires). The number 1 soldier's first responsibility (in this scenario) is to clear the bottom right corner. This soldier will not be concerned with the middle of the room unless a threat is located to his direct front as he

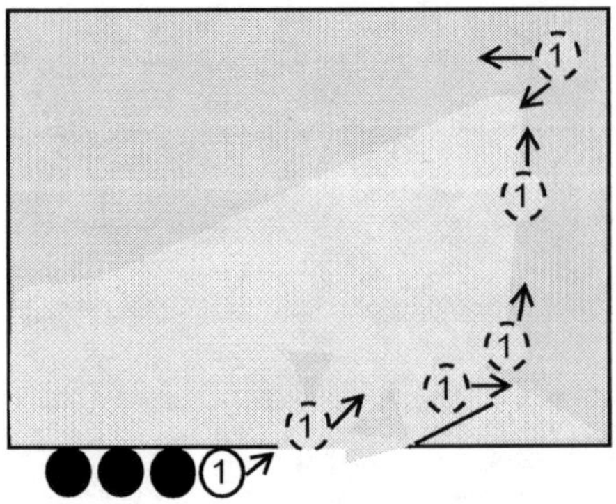


**Figure 4-11. Number 1 Man, Middle-Door Entry**

enters the room. If the number 1 soldier assesses this threat as being capable of hindering his movement to the top right corner of the room (his point of domination), he may engage that threat in a manner that avoids losing momentum. If the door obstructs or partially obstructs the route, soldier number 1 pushes hard through it. This helps to dissipate any loss of momentum and knocks an enemy soldier off balance if he is hiding behind the door. Once the number 1 soldier visually clears the corner, he moves up the wall to his front, clearing it as he goes. The number 1 soldier stops at his point of domination. He then turns and begins a scan from the top left corner to 1 meter off the number 2 soldier's weapon, located in the bottom left corner in this scenario (number 1 has farther to go, soldier number 2 should be in position once number 1 arrives).

- Soldier number 2 *(Figure 4-12, page 4-66)* enters the room immediately after the number 1 soldier enters. Soldier number 2, in this case, executes a hard left turn and moves in the opposite direction of soldier number 1. Soldier number 2's first concern is to clear the bottom left corner. Once this corner is visually cleared, he continues to clear up the left wall. Once soldier number 2 reaches this corner, he has reached the second point of domination. Soldier number 2 stops and scans from the top left corner to 1 meter off soldier number 1's weapon.

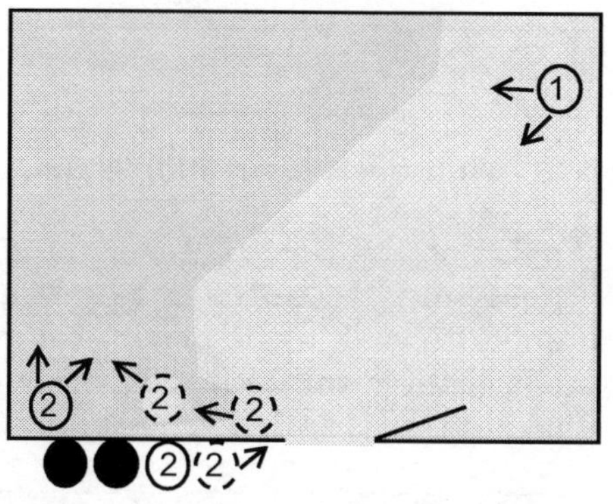


**Figure 4-12. Number 2 Man, Middle-Door Entry**

- Soldier number 3 enters the room at the same time as the number 1 and number 2 soldiers *(Figure 4-13)*. He moves his weapon down from the high-ready position, between soldiers number 1 and number 2 (using his height advantage, if it exists). Soldier number 3's first and primary concern is the middle of the room. If soldier number 3 enters the room quickly, effective supporting fires will be provided for the number 1 and number 2 soldiers as they clear the corners and move to their points of domination. Soldier number 3 begins with a

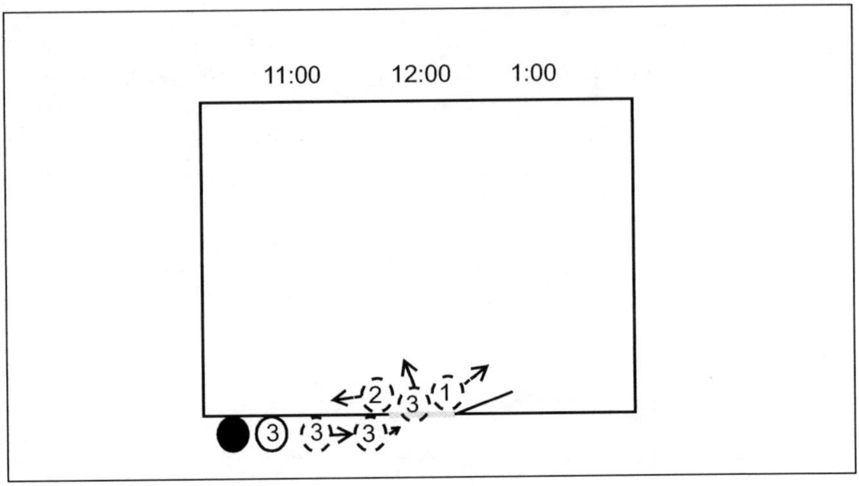


**Figure 4-13. Number 3 Man, Middle-Door Entry**

point of aim (in this scenario) at the 11 o'clock position. He indexes his weapon down the opposite wall, while moving laterally (cross step) out of the fatal funnel, in the opposite direction of soldier number 2 *(Figure 4-14)*. Soldier

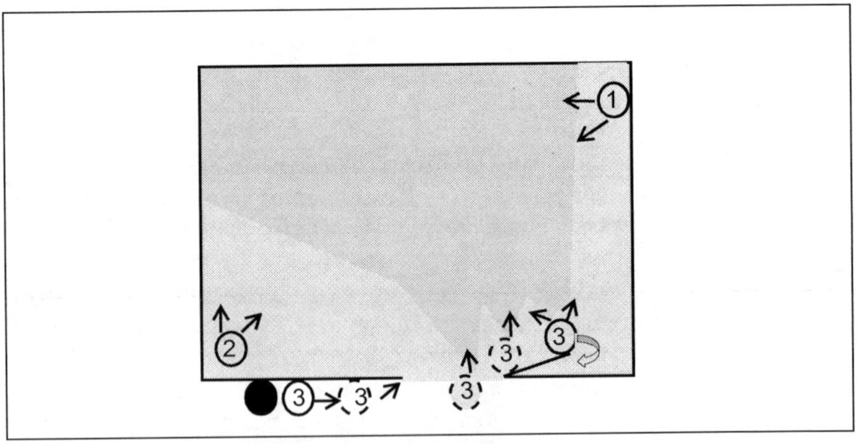


**Figure 4-14. Number 3 Man, Middle-Door Entry**

number 3 stops when out of the fatal funnel and begins to scan from 1 meter off soldier number 1's weapon to 1 meter off soldier number 2's weapon. Once one full scan is completed, soldier number 3 clears behind the door and then returns to scanning the room.

- Soldier number 4 clears in the manner depicted in *Figure 4-15* if the room is large enough for the SAW gunner or breacher to enter. He enters the room directly behind soldier number 3 and clears the fatal funnel as quickly as possible, moving in the opposite direction of soldier number 3. Soldier number 4's primary concern is the middle of the room. He begins with an overlapping point of aim at the 1 o'clock position and indexes his weapon down the far wall as he moves laterally towards the number 2 soldier. Soldier number 4 stops, once clear of the fatal funnel, and continues to scan from 1 meter off the number 2 soldier to 1 meter off the number 1 soldier.

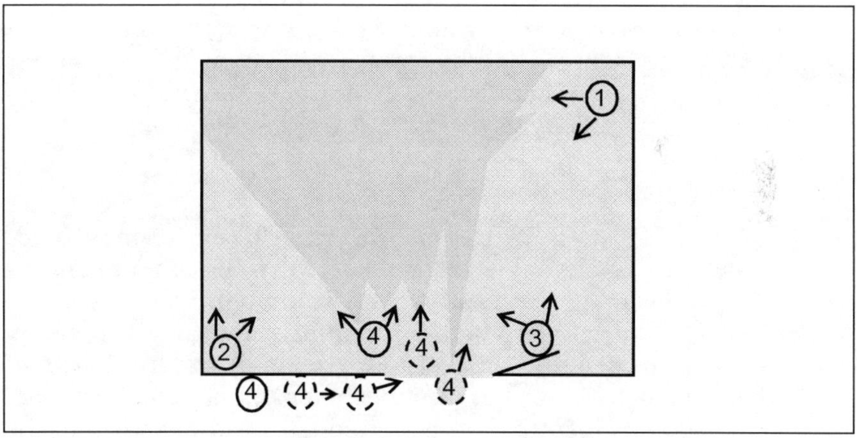


**Figure 4-15. Initial Coverage**

**NOTE: The M249 or shotgun are not precision weapons. The number 4 soldier only engages when a clear shot at a confirmed combatant presents itself. The M249 is also very useful when the situation has degraded and heavier support is required. Leaders should train SAW gunners or breachers to be prudent in the use of their weapons in a precision MOUT environment.**

- *Figure 4-16, page 4-70,* shows the coverage area of each team member once they have reached their positions. The room can be visually cleared very quickly because it has been dominated in depth. This is especially important considering the fact that most rooms contain furniture (hiding positions) and many are irregular shaped. A majority of the room is covered at all times, should an undiscovered or wounded enemy soldier attempt to engage the clearing team. The bottom right corner is the only area not effectively covered by fire at this time. Both the number 1 and number 3 soldiers cleared this area during the initial entry. If the number 1 soldier or number 3 soldier determines that a threat might exist in this corner, it is soldier number 3's responsibility to provide security in this area.

## Communication

4-138. When the room has been dominated, communication is required between the team members in order to clear the room quickly and move on to the next objective. Whether scanning the room or providing security on possible threats (open doors, unidentified persons, or dead or wounded enemy), the team members cannot shift their gaze around the room to ascertain the current situation. To solve this problem, the team leader calls for a status when he determines that no immediate threat exists. This status or situation report (SITREP)

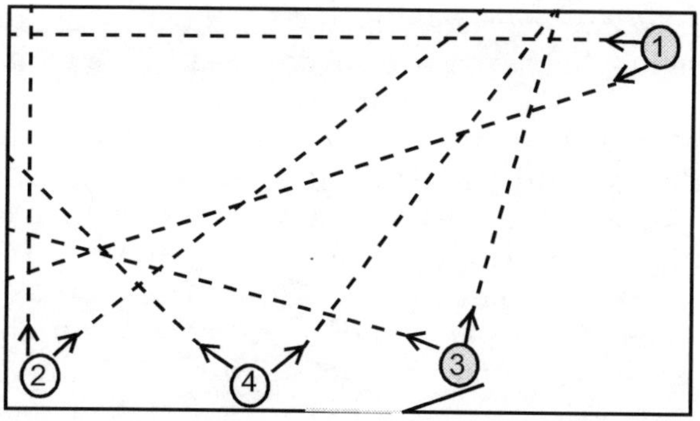


Figure 4-16. Number 1 Man, Center-Door Entry

consists of each clearing team member's current condition and any possible threat in his sector. The number 1 soldier states his situation to the rest of the team. The number 2 soldier follows with his report and so on down the line. The team leader absorbs all this information and then gives directions based on the information. The conversation should be at a level that each team member may hear, but no louder. If soldier number 1 fails to sound off, soldier number 2 states his status. If soldier number 1 is down, the team will know it. If soldier number 1 fails to sound off due to a mental lapse, he may give a SITREP once the other team members have given theirs. An example of this type of communication is as follows:

- Team leader (TL) (soldier number 3 in this scenario): "Status."
- Soldier number 1: "One up!"
- Soldier number 2: "Two, one down to my front!"
- Soldier number 3: "Three up!"
- Soldier number 4: "Four up!"
- TL: "Soldier number 1 secure the subject."

4-139. This scenario is based on a one-room dwelling objective for simplicity. Once the subjects are secured, they are extracted them from the dwelling by a support team and taken to a secure area where their status can be determined and first aid may be rendered, if needed. Meanwhile, the clearing team continues its mission. This would include (dependant on the mission) searching the room for any hidden threats, contraband, or enemy information or equipment. If there are multiple rooms to clear, the TL either directs one of the clearing team members to remain in the room as security (such as "soldier number 1 security") or (if available) calls for a reserve, who is positioned outside of the room, to enter the room and secure the subjects inside it until the building is cleared. The team would then be free to move on to the next room.

## Corner-Door Position

4-140. When an entry point is located in the corner of the room, slight modifications are made to the middle-door entry method. The number 1 soldier *(Figure 4-17, page 4-72)* enters through the path of least resistance and clears the path to the corner to his front and stops. This is soldier number 1's point of domination. Soldier number 1 does not continue on to the next corner, as in the middle-door method, because it would take to long for him to reach this location. The number 2 soldier makes a hard turn around the doorframe and clears the path to the corner diagonally opposite of soldier number 1. This is soldier number 2's point of domination. The number 3 and number 4 soldiers use the far corner of the room as the 12 o'clock reference point *(Figure 4-18, page 4-72)*. All other actions in the room are the same as the middle-door method.

## Rules

4-141. No matter what the room configuration, there are a few rules that should always be adhered to. These rules include the following:

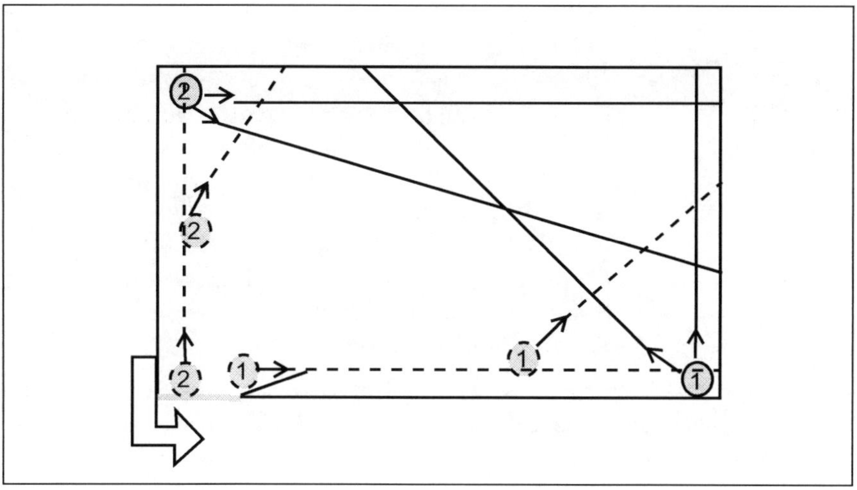


Figure 4-17. Number 1 and 2 Man, Corner-Door Entry

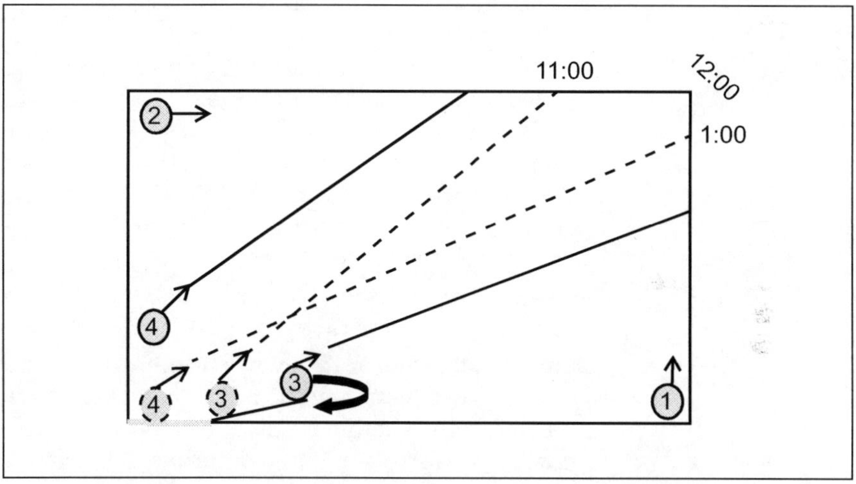


Figure 4-18. Number 3 and 4 Man, Corner-Door Entry

- Know that it does not matter if the person in front of you goes the wrong way when entering the room. Just go the opposite way as the soldier in front of you and it will work out.

- Enter the room as quickly and smoothly as possible and do not waste movements. Remember smooth is fast. The faster each team member picks up its initial point of aim, the more difficult it becomes for the defender. Even a prepared defender can be caught off guard.
- Clear (do not stop) the fatal funnel. The number 3 and number 4 soldiers are especially prone to stopping in the funnel.

**NOTE: Ensure that the doorway is completely cleared before assuming a final position in the room.**

- Stay focused. Never stop scanning your sector for targets unless—
    - You identify a threat in your sector. This threat could be an open door leading to an uncleared room, a person in the room other than one of the team members, an obstacle that cannot be cleared visually from your position, or anything else that you may determine as a threat. If such a threat exists, provide security for the team by covering it with your weapon.
    - You are ordered by the team leader to perform another task.
- Ensure that if you fall down while entering the room, you stay down and do not move. Do not get up until a team member places his hands on you and lifts you up. If a team member arbitrarily gets up, he may become an impediment to the rest of the team's movements. Another possibility is that he would rise into the path of a team member's bullet.
- Rehearse communication. As stated above, speed and momentum will make or break this type of

action. Poor communication techniques will slow your building clearing to a crawl.

- Ensure that each team member knows the procedures for each position. After the first room is cleared in a multiroom objective, a team member may find that he is in a different position in the stack than when the assault started. The team leader should always attempt to avoid being the number 1 soldier when entering a room. This can normally be accomplished, but in some situations it will be unavoidable.
- Ensure that you never enter a room alone. Two personnel are the minimum room-clearing team. If two soldiers are entering a room (either due to the size of the room or attrition), they should assume the number 1 and number 2 soldiers' positions.
- Ensure that you never flag a fellow team member. Flagging is defined as pointing your weapon at or sweeping your weapon across another team member. Muzzle awareness, at all times, is imperative.

## CLEAR A CONNECTED ROOM

4-142. Clearing multiple rooms using connecting doors is preferable to using hallways. This action reduces exposure and compliments speed and momentum.

4-143. *Figure 4-19* shows a team that has just cleared a room using the corner-door technique. The door to the adjacent room is open, providing the number 2 and number 4 soldiers the opportunity to partially clear the second room. If a threat is identified in the adjoining room, it may be engaged from the cleared room. Once this firing has stopped, the SITREP will be restarted if necessary.

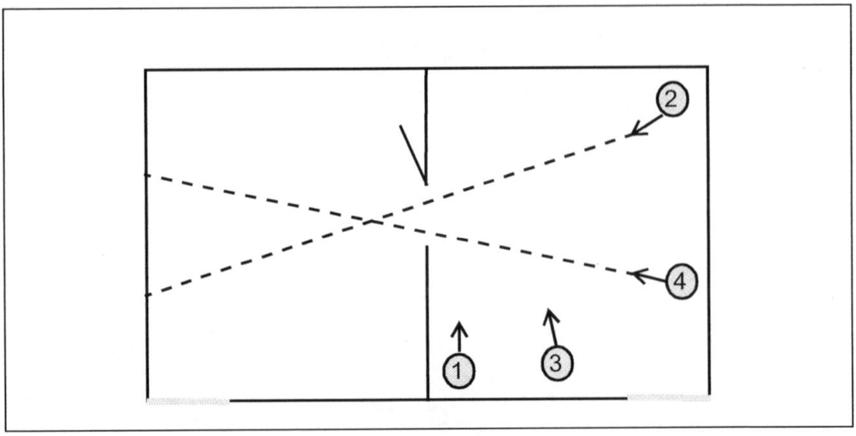


**Figure 4-19. Partial Clearing of a Connected Room**

4-144. *Figure 4-20, page 4-76,* shows how the team stacks in preparation for entry into the next room. When the team leader declares the first room clear, the team will automatically stack on the entrance to the next room. In this example, crossing the doorway would not only place a soldier in the fatal funnel unnecessarily, it would also mask the cover fire of the number 4 soldier. Therefore, soldiers number 1 and 3 stack to the left of the door and soldier number 2 stacks to the right. Soldier number 4 remains in a position to cover a portion of the adjacent room. In this example, soldier number 1 remains the first soldier to enter the room. He awaits a squeeze from soldier number 3. This squeeze acts as a signal to soldier number 1 that at least one other team member is behind him and is ready to enter the room. This form of communication reduces the time spent discussing the order of entry and eliminates the need for a verbal queue (an early warning for enemy soldiers inside the room) to enter the room. If the door had been positioned elsewhere in the room, it could have just as easily been the number 2 or number 3 soldier entering the adjacent room first.

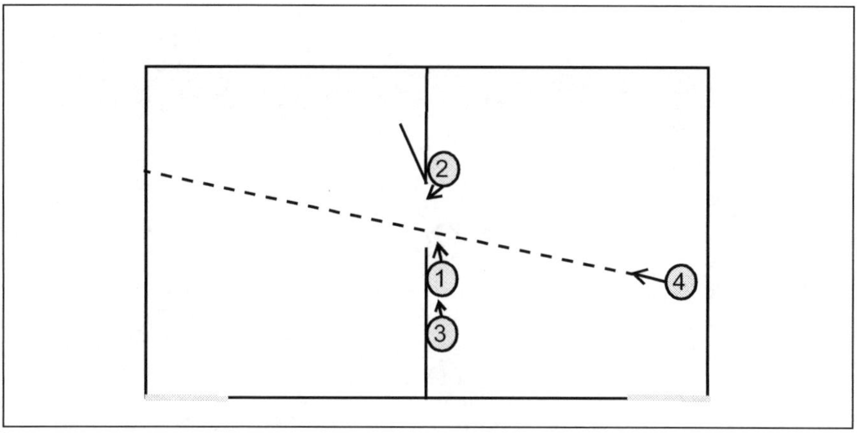


**Figure 4-20. Team Stack**

4-145. *Figure 4-21* depicts the position of the team members while clearing the second room. The soldiers' original numbers are left in place to demonstrate the difference in the positions from the first room. As the team enters the room, soldier number 4 lowers his weapon and moves quickly to the entry point to prevent flagging any team member. He then moves up to the entrance and follows the rest of the team into the next room.

4-146. *Figure 4-22* depicts a closed door to the adjacent room. A closed door is considered locked in all cases when a breacher is present (this is assuming the breacher has the means available to defeat the lock). Once the other team members are stacked, the breacher moves to the door and breaches it on the hand signal of soldier number 3. Soldier number 4 then steps back in order to clear the path for soldier number 2 (the third soldier to enter the room). If no breaching equipment is available, a quick attempt to open the door, from the side may be made. If the door is locked, kick it open. However, this should be the last choice for defeating a locked door. Kicking generally requires more than one

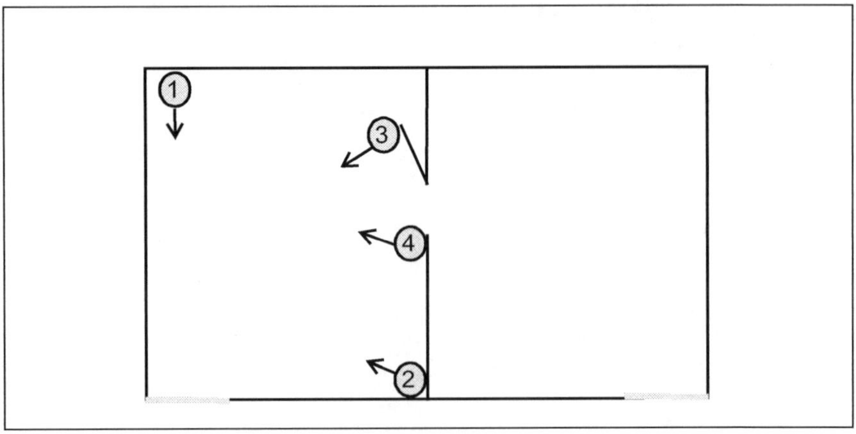


Figure 4-21. Team Members' Positions

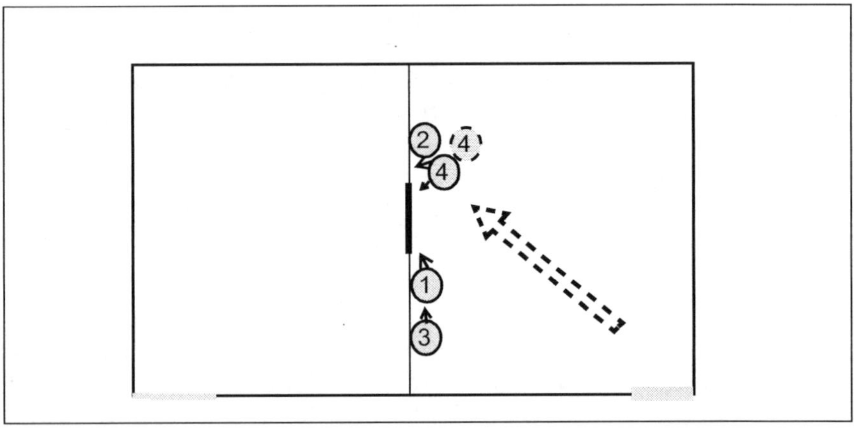


Figure 4-22. Team Positions When the Door to the Adjacent Room is Closed

attempt to defeat the lock. After the first attempt, the breacher's chances of drawing fire through the door are significantly increased.

**NOTE: Do not attempt to open the door by hand. If the door is locked, this action will alert any enemy soldiers inside of the room of your intentions and**

**possibly expose the breacher to fire through the doorway.**

4-147. *Figure 4-23* depicts the team preparing to enter the next room or a hallway. Soldier number 1 moves to a position just inside the doorway. He would have had security on the doorway from his original point of domination. The other team members stack to the left, remaining just inside the doorway. Soldier number 1 is now in the position to partially clear the next room or hallway. The team enters the room the same way as described in the initial coverage.

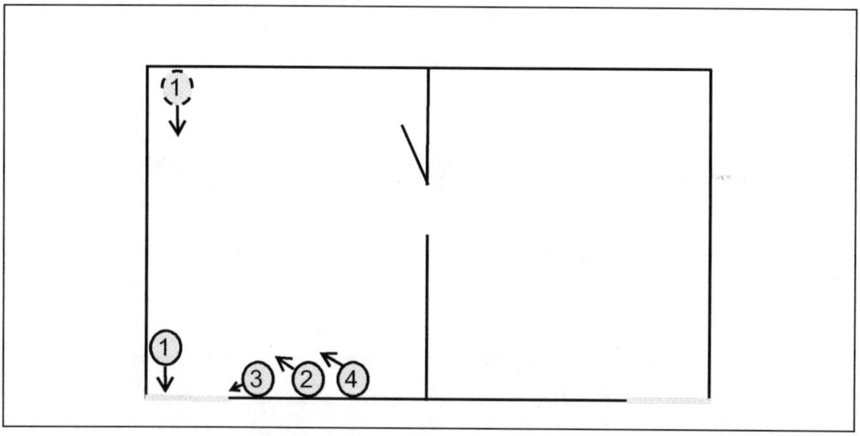


**Figure 4-23. Stack Variation**

## CLEAR A FLOOR

4-148. When clearing a single-floor complex, the principles discussed in single- and connecting-room clearing are used. The only added features are the hallways and an additional squad, when available, to combat attrition. *Figures 4-24* through *4-28* depict the sequence of events in a floor-clearing mission.

4-149. *Figure 4-24* gives an example of a two-squad (four teams, plus HQ element) assault force, preparing

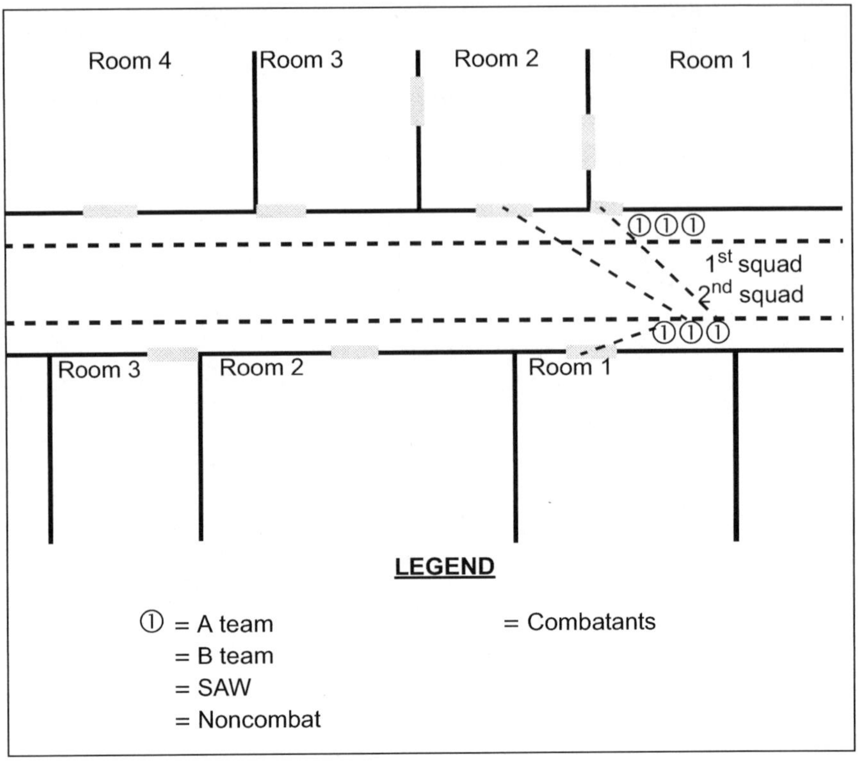


**Figure 4-24. Clearing a Floor**

to clear a floor. The first and second squads are stacked on adjacent opposite walls in the hallway. The first squad prepares to enter or clear the first room on their side. Soldier number 1 (first squad) is continually providing security on the door to his direct front. The second squad halts before exposing itself to the open door in front of the first squad. The second squad's number 1 soldier provides security on the first doorway to his direct front. If no entry point to his direct front exists, soldier number 1 provides security down the hallway. Soldier number 2 (second squad) provides security on the second doorway to the front of the first squad. Soldier number 3 (second squad) provides security on the door to the direct front of the first squad. He lifts that security when the first squad begins to enter that room. The SAW gunners in both teams provide security down the hallway. The HQ RTO (not depicted) provides rear security.

4-150. In *Figure 4-25,* the first squad clears the first room and moves into the second adjoining room. As the second room is cleared, soldier number 3 calls shot and engages an enemy threat in the next room. The B team of the first squad moves into the first room after the A

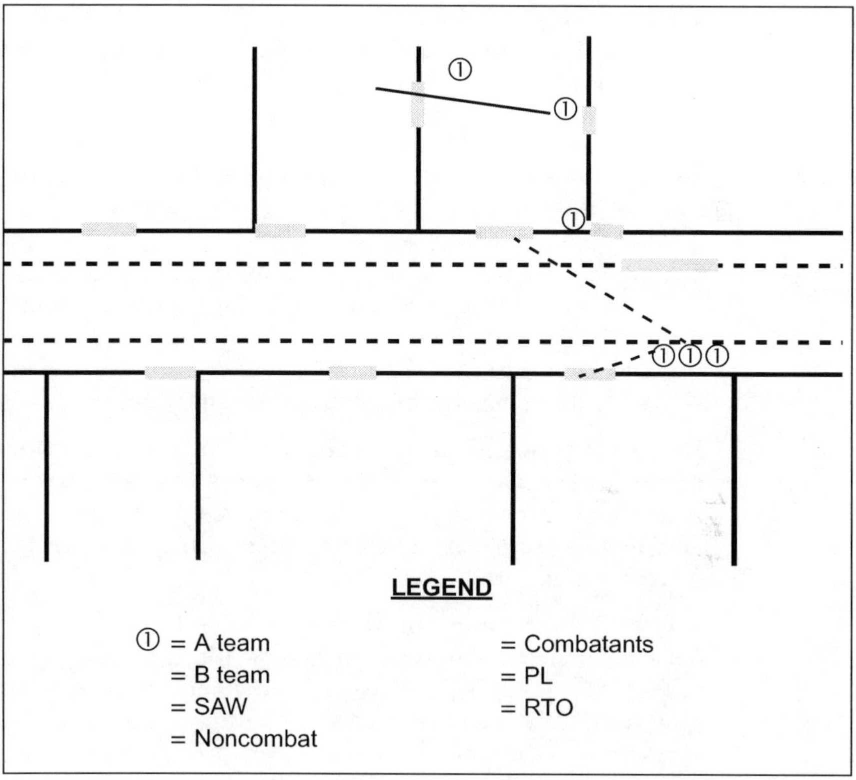


Figure 4-25. Clearing Adjoining Rooms with Hallway Security

team begins to clear the second room. The B team awaits permission from the A team leader to enter the second room. The HQ SAW gunner picks up the hallway security. The second squad has not moved from their original position. A team should never move past an uncleared room. In this case, the second squad will not receive a room clear signal from the first squad until the first squad reaches and clears the third room. This is because entry points interconnect the rooms on this side of the hallway. These entry points should be used in lieu of the hallway, whenever possible.

4-151. In *Figure 4-26, page 4-82,* the first squad has cleared the first three rooms. In this example, one soldier from the B team has been detached to secure the first room, a second has been left in the second room to secure the noncombatant, and a third is securing a noncombatant and a downed enemy soldier in the third room. Soldier number 1 (first squad) remains inside the doorway and provides security on the entry point of the second room to the second squad's front. Soldier number 2 provides security on the third room to second squad's front. The second squad receives the all-clear sign from the first squad (radio, thumbs-up, or hand and arm signals) and moves to their first-room entry point.

**NOTE: If there are not enough personnel available to leave security in cleared rooms (especially in multilevel clearing), disarm and secure all combatants and noncombatants using handcuffs or flex cuffs.**

4-152. If more than one floor is involved, leave a two-person security team on each floor while the building is cleared. In this instance, each room must be systematically recleared once the building is secured. A marking system (such as chalk or chemical lights) must be implemented in order to identify the rooms that have been cleared.

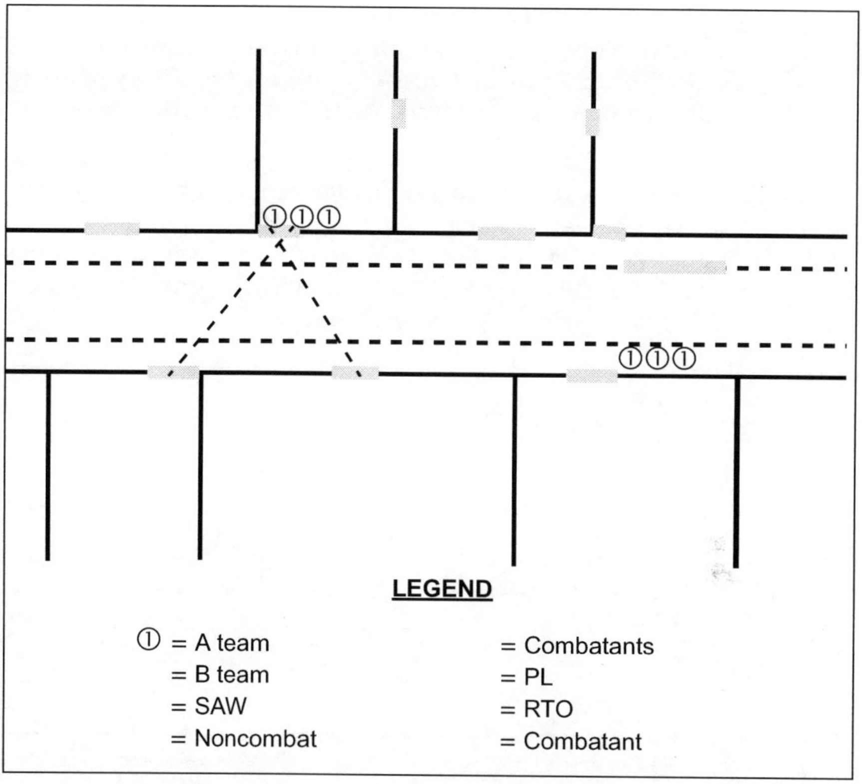


**Figure 4-26. First Squad Clearing a Room with Hallway Security**

4-153. In *Figure 4-27* the second squad clears all three rooms on their side of the hallway. The second squad was forced to clear each room by using the hallway as an access. In this example, the second squad's B team would not follow A team into the room. When the room is cleared, the first member in the B team stacks and (when instructed) enters in order to relieve the A team of security responsibilities. Once this takes place, A team is free to exit the room and move down the hallway to the next room.

**NOTE: Before the team exits the room into the hallway, the lead team member announces that the team is about to exit the room. They should wait for permission to come out into the hallway.**

4-154. Once the second squad has finished clearing their side, the number 1 soldier provides security on the next uncleared room on the right (first squad) side. The first squad receives the all clear signal and then moves up to the next room. Both B team SAW gunners continue to provide security down the hallway.

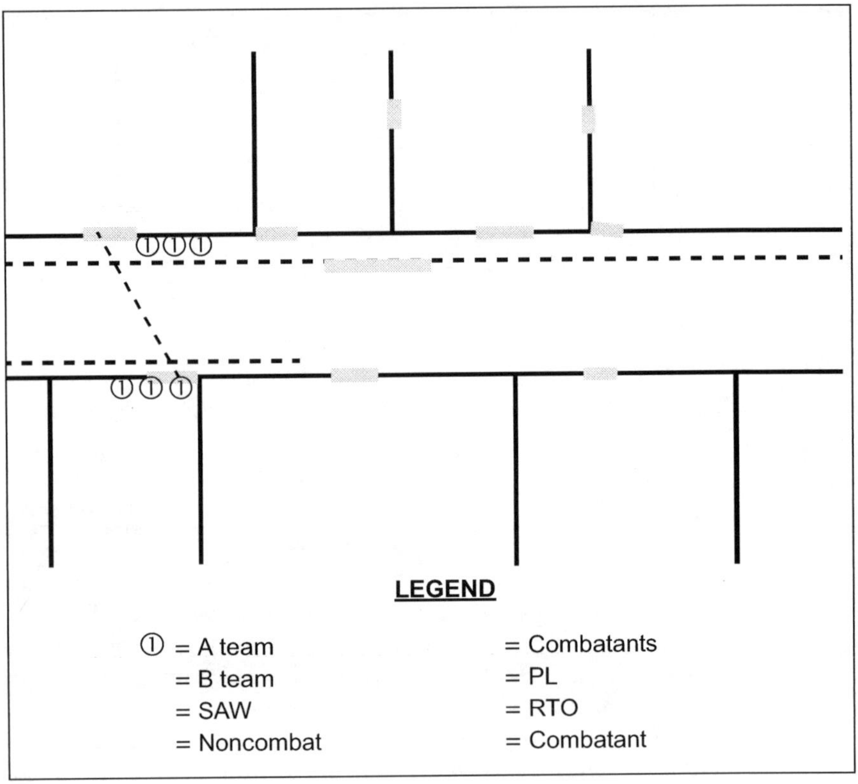


**Figure 4-27. Second Squad Clearing Rooms**

4-155. *Figure 4-28* depicts the events that take place once the building has been cleared. Two SAW gunners are dispatched to each end of the hallway for security. The rooms containing noncombatants or enemy personnel are searched first (mark them to indicate that they have been searched). Once this is completed, it is usually best to remove all of the combatants and noncombatants encountered during the clearing operation (using a small detachment) along an extraction route that is covered by a support element. The support team escorts the noncombatants to a designated holding area to determine their status, and then searches the remaining rooms.

## CLEAR A STAIRWAY

4-156. When multilevel structures are encountered, stairs become an added obstacle that will require maneuver *(Figure 4-29, page 4-86)*. One of the more dangerous stair situations that a team is likely to encounter is a stairway with a turn between floors. Besides the blind spot at the turn, these stairways often have a loft that overlooks the bottom portion of the stairway. If the team is ascending, the number 1 soldier (as always) provides security to his direct front. Soldier number 2 secures the top row of stairs. Soldier number 3 secures the loft area. If a loft area does not exist, soldier number 3 secures the top of the second row of stairs. If the team is descending, each soldier has the same area of responsibility, except that the bottom of the stairway is secured, as opposed to the top.

## CLEAR BLIND SPOTS

4-157. When a blind spot is encountered (such as stairs or a sharp corner), it can be cleared with a technique known as cutting the pie *(Figure 4-30, page 4-87)*. The soldier clears the dead space by slicing it up into pieces and then clearing one piece at a time. By doing this, the soldier exposes only a very small portion of his frame to

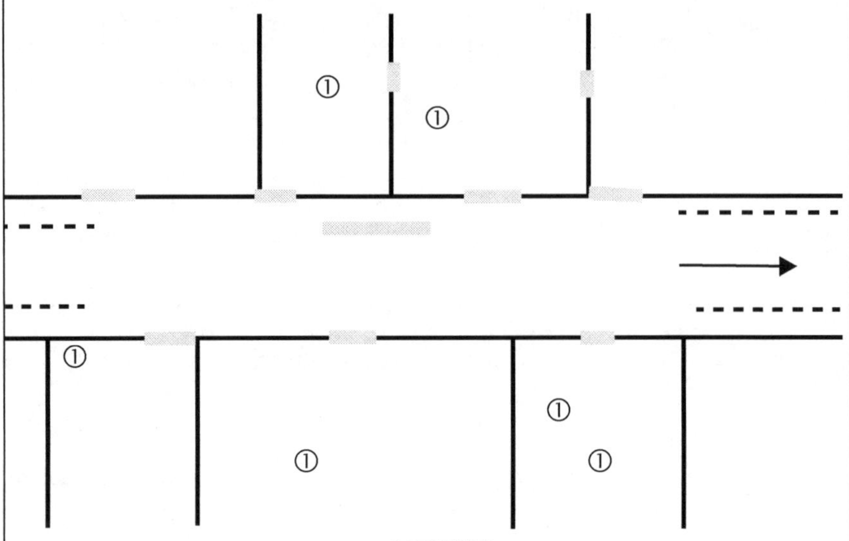


**Figure 4-28. Events After Building is Closed**

any enemy combatants, while placing himself in a position to methodically clear the dead space.

## SECURE PERSONNEL

4-158. The clearing team secures and identifies all personnel found in a room or building that is being cleared or searched. Until identified, (which will not normally occur during the clearance of the building) they approach noncombatants in the same manner as combatants. They do not assume that the frail old man in civilian clothes, cowering in the corner, presents no

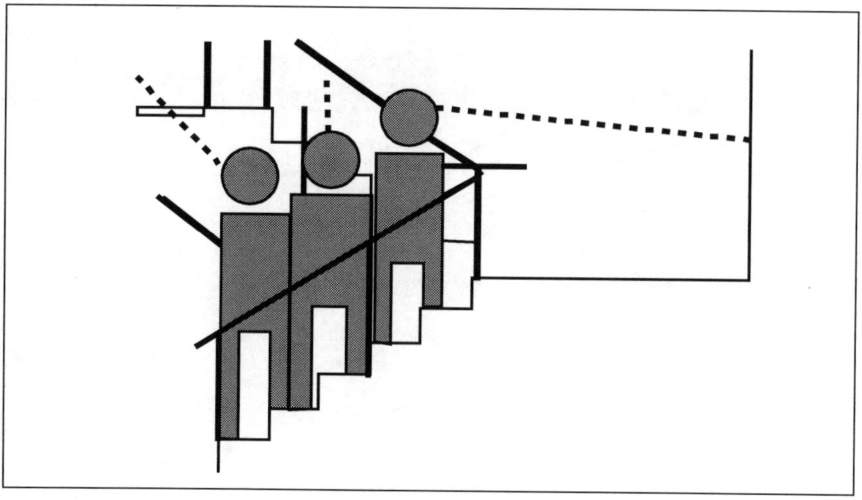

**Figure 4-29. Clearing Stairs**

threat. On contact, the soldier that first indexes his weapon onto the individual orders the individual to his knees, with his hands on his head, and his fingers interlocked. This soldier stays out of arms reach at all times. Do not allow the unidentified individual near any exits. The individuals will remain in that position, with security present, until the building or room has been cleared. When it is time to secure the individual, the soldier that has been providing security does all the talking. In a slow, loud (but calm) voice, the soldier gives short, easy to understand, instructions to the individual, such as "look at me," "sit up straight," and so forth. This talk has a dual purpose. It—

- Gives directions that helps in securing the individual and calms him down.
- Keeps the individuals attention oriented on him and not on the second securing soldier.

4-159. As the first soldier talks to the individual, a second team member unholsters his pistol for close-in work. The second soldier approaches the individual from

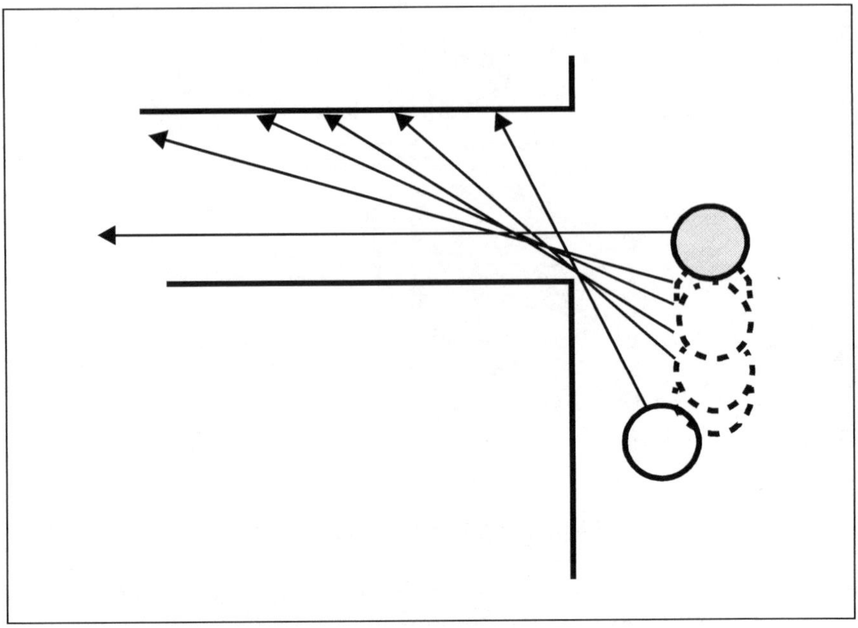

**Figure 4-30. Cutting the Pie**

a blind spot, grasps the middle three interlocked fingers, and places a knee in the middle of the individual's back. Once the individual is under control, the second soldier holsters the pistol and secures the hands of the individual behind his back. If there are numerous individuals to secure and there will be several minutes before they can be exfiltrated from the building, these individuals can be placed on their knees and tilted forward so that their foreheads are against a wall. Their bodies should be at such an angle that their heads support most of the weight. If the prisoner attempts to get up, he will fall to the floor. This position enables a few soldiers to watch several prisoners. If an enemy soldier becomes engaged and the combatant does not appear to be conscious, it must be determined whether the individual is deceased or not. The body should be approached in the same manner as described above. If a kick does not gain a response, a simple eye gouge or sternal rub should identify whether the combatant is indeed dead. It is always best to secure the hands of the individual before applying an eye gouge or sternal rub. Beware of the enemy soldiers *playing opossum*.

## TRAINING FOR MASTERY

4-160. It is obvious that the techniques described above will work best in an environment that has quick-thinking and intelligent soldiers. If the above techniques are practiced using the crawl, walk, and then run method, squads will find themselves catching on very quickly. When the one-room technique is mastered, everything else will fall into place. As teams work together, they will come up with their own systems of signaling, security, and change up in positions. The clearing of a building will be one smooth flow. The platoon leader starts training with single rooms outlined in engineer tape. This system allows for ease in critiquing, and all personnel will be able to observe and learn from the positive and negative performances of each team. The only building layout that cannot be reproduced by tape is a multilevel building (due to the presence of stairwells). Once the platoon is proficient in the one-room technique, the platoon leader moves to connecting rooms outlined in engineer tape. He continues this process until a complete floor in a real building can be cleared smoothly. Once multilevels are covered, he adds simulated breaching, furniture, and combatant or noncombatant scenarios. Once this type of training is completed, an assault team should have the base knowledge required to overcome specific situations that are not covered in this chapter. The platoon leader remembers that this is a perishable skill. As time passes, new platoon members must be integrated into this training and older members must be retrained.

# Chapter 5

# Maneuver and Mobility Support

MMS, formerly known as battlefield circulation control, consists of those measures necessary to enhance combat movement and the ability to conduct movement of friendly resources in all environments. These measures ensure that commanders receive personnel, equipment, and supplies as needed. MMS is conducted across the full spectrum of military operations. The primary focus of MP during MMS is to ensure swift and uninterrupted movement of combat power and logistical support.

## MANEUVER SUPPORT

5-1. Maneuver is the employment of forces on the battlefield in combination with fire (direct or indirect) or fire potential. It is the movement of combat forces to gain a positional advantage, usually to deliver or threaten delivery of direct and indirect fires. MP tasks that support maneuver include—

- MP support to river crossings.
- MP support to breaching operations.
- MP support to a passage of lines.
- Straggler control.
- DC control (refer to *Chapter 7* for more information about DC operations).

### SUPPORT FOR RIVER CROSSINGS

5-2. A river is a significant obstacle that may slow, stop, or impede a unit's ability to maneuver. Units are restricted to moving in column formations along limited routes that come together at crossing sites. Friendly forces are vulnerable while crossing water obstacles. The challenge is to minimize the river's impact on the commander's ability to maneuver. The three types of river crossings include—

- Hasty.
- Deliberate.
- Retrograde.

5-3. MP traffic control is essential to help reduce exposure time and speed units across any obstacle. In addition, effective traffic control contributes to the flexibility of the crossing plan by enabling commanders to change the sequence, the timing, or the site of the crossing units. MP can switch units over different routes or hold them in waiting areas as directed by the tactical commander. This support is vital in reducing congestion, speeding the crossing of any obstacle (not just water), and enabling the maneuver forces to maintain momentum.

## Hasty River Crossing

5-4. A hasty river crossing is a decentralized operation using organic, existing, or expedient crossing means. It is the preferred river crossing method. Conduct a hasty river crossing as a continuation of an attack to ensure little or no loss of momentum by the attacking force. The MP platoon in direct support of a crossing maneuver brigade, may be required to support the crossing without additional support.

## Deliberate River Crossing

5-5. A deliberate river crossing requires planned and augmented MP support. Conduct a deliberate river crossing when a hasty crossing cannot be made successfully, normally when offensive operations must be renewed at the river, and when enemy forces must be cleared from the area. A buildup of firepower and equipment is needed on both entry and exit banks. Normally, MP support from corps is required to augment the division MP company.

### Retrograde Crossing

5-6. Closely plan and control a retrograde crossing. Massed crossing forces could slow momentum or exceed bridge classification limits. Forces moving to the rear may retrograde to defensive positions beyond the water obstacle and may be slowed as they set up to defend the exit bank. MP support retrograde crossings the same as they do deliberate crossings.

### River Crossing Planning

5-7. The crossing force commander plans the river crossing operation. He prepares an OPORD and specifies what support is required. The PM, based on the OPORD, plans MP support for the river crossing. The plan includes how MP assets will be used and what additional resources are needed. The MP commander supporting the operation plans and supervises the mission based on the OPORD and guidance from the PM. The OPORD normally gives OPCON of all units entering the crossing area to the crossing commander.

5-8. The MP leader supporting the crossing site develops a traffic control plan to support the circulation control plan. He must plan for—

- Traffic control posts (TCPs) and temporary route signs at—
  - Major crossroads on the MSR and near crossing sites and lateral boundaries to control traffic from adjacent unit areas that could interfere with division surface movements.
  - Staging areas and engineer regulating points (ERPs) to provide directions and

information, control movement to and from staging areas according to planned times, and relay messages between traffic HQ and the moving unit.

- Holding areas on the entrance bank to direct traffic to crossing sites; on the exit bank, inside the traffic regulating line (TRL), to control movement; and on the exit bank, outside the TRL, to temporarily hold sections of a convoy or a unit until it can reassemble and continue its movement.

- Mobile patrols to operate along primary routes to control traffic, spot problems, guide and escort vehicles, and reroute traffic when necessary.
- Temporary EPW collecting points. Set up the collecting points outside the TRL. Evacuate EPWs through the crossing areas as quickly as possible so their transit does not impede the movement of friendly forces.

5-9. For brigade crossings, the MP leader may collocate with the brigade staff to form a small, temporary traffic control cell located at the brigade main CP or the brigade TOC. The brigade main CP controls the maneuver support force that consists of corps engineers, bridge companies, MP, and chemical units.

## Control Measures

5-10. To ease control of large, fast-moving forces, the river crossing plan usually allots one crossing area for each maneuver brigade. The commander uses control measures to delineate areas of responsibility for subordinates and to ease traffic control. *Figure 5-1* shows the following control measures.

5-11. **Release Line (RL)**. As used in river crossing operations, RLs are used to delineate the crossing area. RLs are located on both the far shore and nearshore and

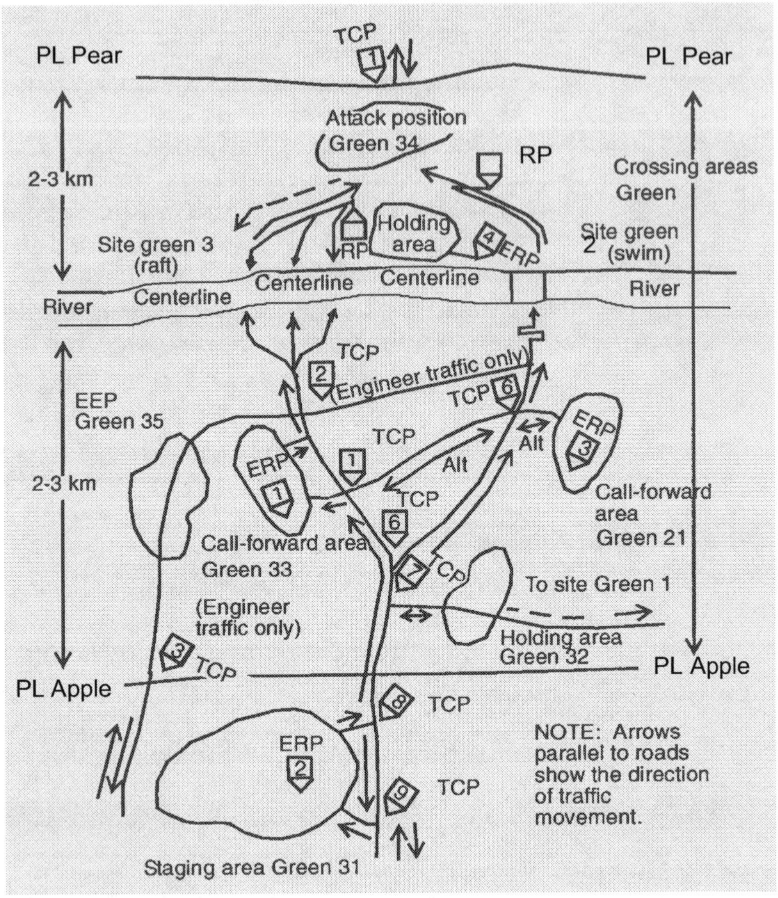


**Figure 5-1. River Crossing Control Measures**

indicate a change in the HQ that is controlling the movement. RLs are normally located within 3 to 4 kilometers of the river and on easily identifiable terrain features, if possible.

5-12. **Crossing Areas**. Crossing areas are controlled-access areas that decrease congestion at the river. This permits swift movement of the forces. Each lead brigade has a crossing area on both sides of the river that is defined by brigade boundaries and the RL. Crossing areas normally extend 3 to 4 kilometers on each side of the river, depending on the terrain and the anticipated battle.

5-13. **Waiting Areas**. Waiting areas are located adjacent to the routes or axes of advance. Commanders use the following waiting areas to conceal vehicles, troops, and equipment while waiting to resume movement or make final crossing preparations:

- Staging areas. These are battalion-size waiting areas outside the crossing area where forces wait to enter the crossing area. The brigade traffic control cell handles the units' movement into the staging areas. The crossing area commander (CAC) controls movement from the staging areas into the crossing areas. MP operate TCPs at the staging areas according to the crossing and traffic circulation plans. They emplace temporary signs along the route from the staging area through the crossing area to guide the convoys. Units make crossing preparations and receive briefings on vehicle speed and spacing in the staging areas. Staging areas—
  - Are located to support the crossing concept.
  - Are far enough back to permit the rerouting of the battalion along other roads or to alternate crossing sites.
  - Are easily accessible from major routes.
  - Have enough area for dispersing a battalion-size unit.
  - Provide concealment.
- Call-forward areas. These areas are company-size waiting areas located within the crossing

area. Engineers use them to organize units into raft loads; crews use them to make final vehicle crossing preparations. The CAC controls movement from the staging area to the call-forward area. The crossing site commander (CSC) directs movement from the call-forward area to the crossing site and on to the far-shore attack position. As a minimum, each CSC operates his own call-forward area. Call-forward are—

- Located to support the crossing plan.
- Company size within the crossing area.
- Easily accessible from routes.
- Planned with a minimum of one per crossing site.
- Collocated with ERPs.
- Used to organize units into raft loads.
- The final preparation areas before going to the crossing site.
- Normally operated by engineers.

- Holding areas. These areas are waiting areas that forces use during traffic interruptions. Units move into these areas when directed by TCP personnel and disperse rather than stay on the roads. Holding areas are battalion size outside of the crossing area and company size within it. Far-shore holding areas are used to organize return traffic. MP operate holding areas according to the crossing and traffic circulation plans and—
  - Are used as call-forward areas for return traffic from the far shore.
  - Are located to support the crossing plan.
  - Are easily accessible from routes.
  - Have enough area for dispersion.
  - Provide cover and concealment.

  - Are defensible.
  - Maximize traffic flow with minimum control.
- Attack positions. The attack positions are the last positions occupied or passed through by the assault echelon or the attacking force before crossing the line of departure. Within the bridgehead, the attack position is the last position before leaving the crossing area or bridgehead line.
  - Assembly areas. These are the areas where forces prepare or regroup for further action.

5-14. **Engineer Equipment Parks (EEPs)**. These are areas located a convenient distance from bridging and rafting sites for assembling, preparing, and storing bridge equipment and material. They are at least 1 kilometer from the river and hold spare equipment and empty bridge trucks that are not required at the crossing sites. EEPs should be located where they do not interfere with the traffic to the crossing sites and where equipment can be concealed and dispersed. Ideally, routes leading from the EEPs to the crossing sites are not the same routes used by units crossing the river.

5-15. **Traffic Control Posts**. In river crossings, TCP personnel assist the crossing-area HQ in traffic control by reporting and regulating the movement of units and convoys. TCP personnel relay messages between the crossing-area HQ and the moving units. The PM identifies locations that need or require TCPs. MP operate TCPs on both banks of the river to control traffic moving toward or away from it. TCPs are operated at major or critical crossroads and road junctions, staging areas, holding areas, and ERP.

5-16. **Engineer Regulating Point**. ERPs are technical checkpoints used to ensure that vehicles do not exceed the capacity of the crossing means. They help maintain traffic flow. Vehicles not allowed to cross are removed so that they do not cause a traffic backup at the actual crossing site. Engineers man the ERPs and report to the CSC. TCPs are collocated with the ERP to ensure that all vehicles clear the call-forward areas. An additional duty of ERP personnel is to give the drivers final instructions on site-specific procedures and other information, such as speed and vehicle intervals. As a minimum, each crossing site requires an ERP at its own call-forward area. If enough engineer assets are available, an ERP may be established at far-shore holding areas to regulate rearward traffic.

**Route Execution**

5-17. MP must be prepared to establish holding areas along movement routes on order. If the road network sustains damage, vehicles will need to be routed into the holding areas until traffic can be restored or rerouted. Refer to *paragraph 5-104* for more information about holding areas.

5-18. MP mobile patrols operate along primary routes, monitoring traffic, spotting problems, and rerouting traffic as necessary and conducting AS around the crossing area. They make frequent checks of temporary signs to prevent the enemy from tampering with them.

5-19. MP may be directed to screen the crossing unit's flanks and rear. The size of such an element is determined by METT-TC. In most environments this mission requires at least a squad. MP conduct screening missions to provide early warning of enemy approach and to provide real-time information, reaction time, and maneuver space for the crossing unit. The squad fights only for self-protection and remains within its capabilities. Refer to *Chapter 6* for more information about screening missions.

5-20. Include at each crossing site a temporary EPW collection point. Initially the collection point will be on the entry bank. Once MP cross as part of the support force, a temporary collection point is established on the exit bank. A division central collection point is established outside of the crossing area. Refer to *Chapter 7* for more information about division forward collection points.

5-21. Rigid control of civilian movement is necessary to preclude congestion on movement routes. The PM coordinates for HN police support to ensure that the civilians who live in the crossing area are kept in place or, if necessary, quickly moved to designated areas away from the river. Normally, civilians are not allowed to cross the river or move along the edge of the river during the river crossing operation. Refer to *Chapter 7* for more information about DC resettlement.

## MILITARY POLICE SUPPORT TO BREACHING OPERATIONS

5-22. Breaching operations are conducted to allow maneuver despite the presence of obstacles. Obstacle breaching is the employment of a combination of tactics and techniques to advance an attacking force to the far side of an obstacle that is covered by fire. Breaching operations begin when friendly forces detect an obstacle and begin to apply the breaching fundamentals. Breaching operations end when the battle handover has occurred between the follow-on forces and a unit conducting the breaching operation.

### Support Planning

5-23. MP support to breaching operations is similar to MP support to river crossing operations. The employment of MP is based on METT-TC, available resources, and the commander's priorities. MP support to breaching operations includes—

- Operating TCPs at the breaching site and along routes leading to or departing from the breaching site.
- Operating holding areas.

- Providing mobile guides to escort the units.

5-24. The platoon leader coordinates with higher HQ and the engineer forces conducting the breach for essential information that includes the—

- Azimuth and distance to the final-approach marker or the 8-digit grid coordinate of the final-approach marker that is entered into the teams Global Positioning System (GPS) receiver.
- Lane marking pattern currently emplaced.
- Type of final-approach marker used.
- Traffic control plan and march order.

5-25. A combined-arms breach is a complex operation and requires precise synchronization. Breaching operations normally require the maximum use of TCPs to assist support, breach, and assault forces to move along various lanes. Refer to *paragraph 5-88* for more information about TCPs. Lanes are marked to safely pass units through the obstacle. The three levels of lane marking are—

- Initial.
- Intermediate.
- Full.

5-26. MP may provide TCPs and guide support to lanes at any level of marking. However, the main effort of MP support may come in later phases of the operation, when larger units (battalion and above) are passed to subsequent objectives, and time permits marking improvements to be made. The increase in traffic and the more diverse forces with different levels of driver experience will increase the need for MP traffic control operations. MP guides are simply mobile MP teams that escort units from one control measure or point to another. Guides and TCPs are essential when there are multiple lanes. *Figure 5-2, page 5-12* shows the flexibility that the combinations of multiple lanes and guides or TCPs provide the commander.

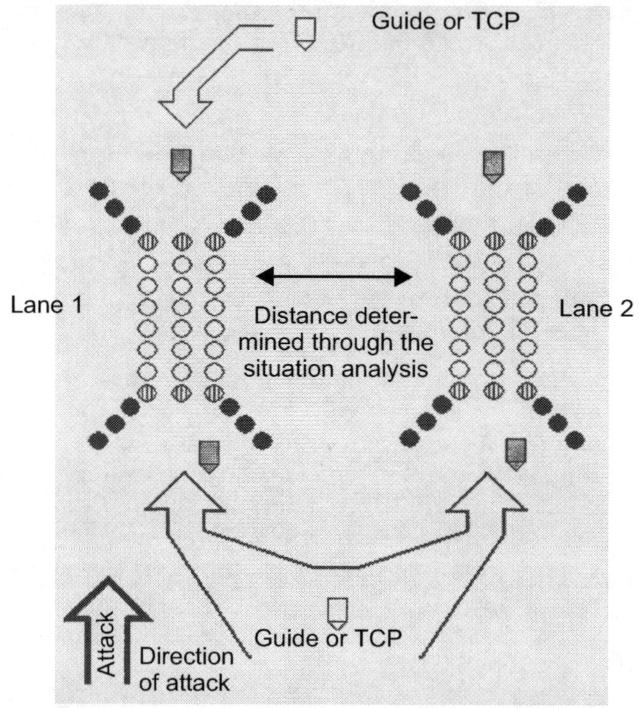


**Figure 5-2. Multiple Lanes (Two-Way Traffic)**

## Movement Execution

5-27. The commander sets the priority of movement based on the situation. MP may concentrate their efforts on assisting the immediate passage of larger combat forces. Or their priority may quickly shift to ground evacuation of casualties or vehicle recovery operations. MP traffic control operations give the commander the ability to make last-minute changes in the traffic flow or lane usage

5-28. MP may be required to establish unit holding areas (battalion and company size) in the event that traffic is disrupted on the lanes due to enemy activity or the need to do maintenance or upgrade a lane. Refer to *paragraph 5-104* for more information about holding areas.

5-29. The commander collocates guides or TCPs at the far recognition marker when he feels the situation requires more positive control.

5-30. Guides and TCPs are briefed on this information and are kept up to date on changes to the traffic control plan and enemy activity in the AO.

5-31. The platoon leader plans for the possible need to establish a forward EPW collection point near the breaching operation. Refer to *Chapter 7* for more information about division forward collection point. He must also plan for an increase in the number of TCPs needed during limited visibility or in restrictive terrain. Refer to *FM 3-34.2* for more information about breaching operations.

## PASSAGE OF LINES SUPPORT

5-32. This area describes how an MP leader is to plan and conduct MP support to the passage of lines. The MP elements described in the following paragraphs are supporting the passing and stationary units. MP conducting a battle handover or passage of lines to a TCF is discussed in *Chapter 6*.

5-33. A passage of lines is a tactical event normally associated with a battle handover. A passage may be designated as a forward or rearward passage of lines. Moving a maneuver unit through the positions of an emplaced unit that is in contact with the enemy is a critical action. It requires detailed coordination; planning; and close, continuous supervision of the movement.

5-34. The main focus of MP support to a passage of lines is normally employing special traffic control measures that include—

- TCPs.
- Temporary route signing.
- Checkpoints and roadblocks.
- Defiles.

5-35. MP may also provide guides to escort the passing unit en route to a release point or AA. Similar to MP support to breaching operations, guides provide the commander a means to change the sequence, timing, or lanes of the passing units.

## Passage of Lines Planning

5-36. MP support the passage of lines operation to assist a maneuver unit in contact with the enemy to maintain movement. Depending on the scope of the operation, a division MP company may not be enough to support a passage of lines operation. METT-TC may necessitate the need for additional corps MP support.

## Control Measures

5-37. When planning control measures for a passage of lines, MP leaders must consider the placement of the following:

- AAs where units prepare for further action.
- The battle handover line (BHL) where the stationary force assumes responsibility for the sector from the covering force.
- The forward edge of the battle area.
- Passage lanes along which the passing units move to avoid stationary units and obstacles.
- Passage points where units will pass through one another. They are located where the commanders want the units to execute the

passage of lines. Designate multiple passage points to help eliminate congestion.

- Contact points (designate an easily identifiable terrain feature) where the units will physically meet.
- SPs where unit elements come under the control of the commander responsible for the movement.
- Phase lines, used in controlling the timing of the operation, are usually recognizable terrain features extending across the zone of action.
- RPs where unit elements revert to their respective commanders and continue moving to their destinations.
- Travel routes from the point of origin to the destination.
- Checkpoints used to coordinate friendly movement. (Checkpoints are not used as reference points for reporting enemy locations.)

### Passage of Lines Execution

5-38. MP support for a passage of lines is conducted at the platoon level. The company monitors the platoon and coordinates with higher HQ. The company operations section is required to conduct detailed coordination with the passing and stationary units that includes—

- Communication requirements.
- Recognition signals.
- Route SP.
- Time of passage.
- Passing lanes.
- Control measures to include TCPs, escort and guide vehicles, temporary route signs, or a combination of these.
- EPW and DC evacuation.

5-39. The platoon leader continuously coordinates with the company operations section to confirm the following:

- The size of the passing unit.
- Locations of AAs.
- Recognition signals.
- The actual time that the passage of lines will commence.

5-40. MP support for a passage of lines begins at the route SP, which serves as a contact point. Recognition signals are displayed at the SP. MP monitor the passing unit's command net during the entire passage. Radio silence is maintained during this time.

5-41. Depending on the situation, MP can support the passing unit with the following:

- TCPs.
- Escort and guide vehicles.
- Temporary route signs.
- A mix of these measures.

5-42. The platoon leader selects the method that best supports the moving unit's passage through the stationary unit. Temporary route signing will decrease the number of TCPs needed, but if routes are not well defined or they cross congested areas, expect to provide TCPs or escort vehicles.

5-43. MP must be prepared to initiate vehicle holding areas at designated locations along movement routes. If the road network sustains damage, vehicles will be routed into a holding area until traffic can be restored or rerouted. (Refer to *paragraph 5-104* for more information about holding areas.)

5-44. The platoon leader plans for the need to establish a temporary EPW collection point, and establishes it near the passage area but out of the view of the operation.

5-45. Strict control of the movement of civilians is necessary to preclude congestion on routes used for the operation. When necessary, MP establish evacuation routes to move DCs to a designated area rear of the staging areas. Local nationals who live in the immediate area will remain in place or be evacuated primarily by HN police from the area.

## STRAGGLER CONTROL

5-46. MP conduct straggler control operations to assist commanders in maintaining combat strength by locating and returning stragglers to their units. MP identify stragglers at the TCPs, checkpoints, roadblocks, defiles, or while patrolling the MSR. For large numbers of stragglers, special collecting points are set up along the MSR.

### Operate Straggler Posts

5-47. When operating a straggler CP, MP teams need to know what units are located or operating in their AO. Most stragglers are soldiers who have become accidentally separated from their command. Stragglers are identified by checking the following:

- Uniforms.
- Unit insignia.
- Bumper markings on the vehicles.
- Identification cards or tags.
- Passes or other authorization documents.

5-48. For each straggler identified, as a minimum, MP record—

- The straggler's name, rank, social security number (SSN), and nationality.
- The straggler's unit.
- The straggler's category ("injured" or "uninjured").
- Whether the straggler is armed or not.
- Where the straggler was coming from and his destination.
- Why and when the straggler left the unit.

- The location where the straggler was sent.

5-49. MP administer first aid to the injured, wounded, or ill stragglers. Seriously ill or injured soldiers are evacuated. Stragglers who have information of immediate tactical value are reported to higher HQ. Soldiers fit for duty who mistakenly became separated are returned to their units or a HQ within their chain of command. The soldiers unit is responsible for any transportation requirements.

5-50. MP treat deliberate stragglers, those who have deserted or are attempting to desert or are absent without leave (AWOL), with caution. These stragglers may resort to violence to avoid military control. MP search, disarm, and detain them. They hold these stragglers until transport and escort can be arranged to take them to their unit, the straggler collecting point, or another facility set by the SOP or the straggler control plan. MP safeguard confiscated property and documents, and dispose of them according to the straggler control plan.

5-51. MP handle stragglers from the HN or other allied forces the same as US stragglers. If the PM has coordinated with other national forces to set up joint straggler posts, allow MP from other national units to handle stragglers from their own forces.

### Operate Straggler Collecting Points

5-52. When large numbers of stragglers exist and TCPs, mounted patrols, and straggler control posts are not able to handle the straggler flow, MP may be tasked to operate a straggler collecting point. MP temporarily hold stragglers at collecting points while they process them for return to their units, placement in medical channels, or placement in other military channels.

5-53. The PM operations section plans the location of a straggler collecting point, placed along a key MSR or at an intersection of the MSR. This allows quicker access to the straggler collecting point to aid in moving stragglers to their appropriate destination.

5-54. Often it is collocated where elements of medical, transportation, and MP units can share efforts to ease the disposition of stragglers. At a straggler collecting point, MP may need food, water, clothing, and shelter for stragglers. If a medical facility is not close by, request extra medical supplies and be prepared to administer first aid.

5-55. Guards separate the injured stragglers from the uninjured. They process each soldier at the collecting point. Guards record the key information on each soldier for a straggler report. They search, segregate, and guard stragglers who refuse to return to their unit. Guards assemble and forward the report to wherever the straggler control plan directs. They hold stragglers at the collecting point until transport arrives.

5-56. Detain stragglers who refuse to return to their unit until their unit provides an escort or until they are transported to a detainment facility.

## MOBILITY SUPPORT

5-57. Mobility is the capability of military forces to move from place to place while retaining the ability to fulfill their primary mission. It includes those activities that enable a force to move personnel and equipment on the battlefield without delays due to terrain or obstacles. MP activities that support mobility include the following:

- Route reconnaissance and surveillance.
- MSR regulation enforcement.
- Special circulation control measures.

## ROUTE RECONNAISSANCE AND SURVEILLANCE

5-58. MP conduct route reconnaissance and surveillance operations to gain detailed information on specific routes to be used as deployment routes, MSRs, or movement corridors. Mobile MP teams record and report the condition of the MSR and other critical roadways, identifying effects of weather on road surfaces, damage to routes, NBC contamination, and the presence of enemy activity. Platoon leaders use this information to develop a road reconnaissance report and a route reconnaissance overlay.

5-59. The platoon leader provides the MP reconnaissance reports and overlays are used to assist the division PM with the development of the division traffic control plan. Movement planners use information gathered from the MP route reconnaissance to update the highway traffic section's (HTS's) traffic circulation control plan and to formulate the highway traffic regulation plan. Refer to *FM 55-10* for more information about traffic circulation control plans and highway traffic regulation plans.

### Plan

5-60. MP leaders plan route reconnaissance operations by examining intelligence reports and maps of the area surrounding the route to be reconnoitered. When time is critical, MP conduct a hasty route reconnaissance to obtain specific information only. The MP leader must clearly understand the following critical tasks to be accomplished:

- Find and report all enemy forces that can influence movement along the route.
- Determine the trafficability of the route.
- Reconnoiter any special areas that could influence movement on the route. These areas may consist of highly populated areas or key

terrain features. Additional teams may be needed to cover these areas based on METT-TC.

- Inspect all bridges on the route.
- Locate fords or crossing sites near all bridges on the route.
- Inspect all overpasses, underpasses, and culverts.
- Locate areas suitable for short halts and holding areas.
- Locate mines, obstacles, and barriers along the route.
- Locate a bypass around built-up areas, obstacles, and contaminated areas.
- Report route information.

5-61. A more detailed route reconnaissance would include additional information concerning the terrain, potential hazards, or obstacles and would include key terrain and built-up areas 2 to 3 kilometers on either side of all MSRs. A route reconnaissance this detailed requires considerably more time and personnel.

**Execute**

5-62. One MP squad can conduct a hasty route reconnaissance of only one route, (about 30 kilometers). MP platoons reconnoiter three routes within the boundaries of the platoon AO, if route reconnaissance is their primary focus.

5-63. The size of the reconnaissance patrol is determined by using METT-TC factors. In most environments route reconnaissance operations are not conducted with less than a squad. The squad leader organizes the squad into security teams and a reconnaissance team. The reconnaissance team records the information, completes a *DA Form 1248*, and prepares a reconnaissance overlay. The other teams provide security.

5-64. MP do not engage the enemy when conducting route reconnaissance except in self-defense or when ordered to do so. They report any visual contact with the enemy and maintain surveillance while gathering as much information as possible. They break visual contact only on order from proper authority.

5-65. Often the main purpose of a reconnaissance is to confirm information already known about a route. Additional information can be obtained from the division PM, the division transportation officer, and from HN police. MP teams talk to convoy commanders, vehicle drivers, local nationals, and highway control regulating teams to gain or verify information about well-traveled roads. Although this information is not as reliable as information gathered from driving a route, its reliability increases when several road users report the same condition.

5-66. Mounted MP patrols continuously collect data at the level of detail required by the PM or commander and report it by the fastest secure means available. The patrols travel the routes within the AO to—

- Identify and locate the recommended route.
- Check the driving time and distances between easily recognized points.
- Look for obstructions and restrictions (bridges, tunnels, steep grades, sharp curves, ferries, snow blockage, defiles, flooding, rock falls, and slides).
- Note the location and type of possible ambush sites on the route.
- Look for terrain where direct fire from the enemy could stop movement on the route.
- Identify natural defense, counterambush, and assembly locations.

- Identify areas where terrain restricts communications.
- Watch for enemy situations that could affect route security or conditions, such as—
  - Enemy elements positioned on key terrain.
  - The enemy emplacing mines and other obstacles.
  - Frequency changes or type of enemy fire in the area.
  - Enemy aerial interdiction.

5-67. To keep from overlooking critical terrain data, the squad leader prepares a checklist of items that may be included on the reconnaissance overlay, such as—

- The route classification formula.
- The identification and location of the reconnoitered route.
- The road distances between the points that are easily recognized both on the ground and on the map.
- The presence and lengths of steep grades (having a slope of 7 percent or greater).
- Curves having radii of less than 45 meters.
- Military load classifications (MLC) and limiting dimensions of bridges. Include suitable bypasses, classifying them as easy, difficult, or impassable.
- Locations and limiting data of fords, ferries, and tunnels. Include suitable classification of bypasses.
- Route restrictions (like underpasses) below minimum standards and, if appropriate, the distances such restrictions extend.
- Areas suitable for short halts, holding areas, or bivouacs that offer easy access to the roadway and adequate dispersion, cover, and

concealment. Include information on the shoulders.

- Rock fall and slide areas that may present a traffic hazard.
- Overhead clearance of less than 4.3 meters.
- Civil or military road numbers or other designations.
- Obstructions to traffic.

5-68. Roads that bisect heavily wooded areas are likely obstacle and ambush sites. Heavily loaded vehicles are vulnerable to ambushes and unable to circumvent obstacles easily. Steep grades and numerous S-turns, where logistical vehicles that are heavily loaded with supplies slow to a crawl, make good ambush points. For more information on route classification, refer to *FM 5-170* and *Appendix I* of this manual.

5-69. If enemy activity is suspected along a route, the squad should—

- Use caution when approaching critical locations.
- Deploy using traveling overwatch or bounding overwatch. Choose movement techniques according to the latest information on suspected enemy activity.
- Avoid danger areas.

5-70. Use caution when approaching a sharp bend or a defile in the road. Such areas are often mined and are ideal sites for an ambush. When necessary, the squad leader has the reconnaissance element conduct a dismounted reconnaissance of these areas while the security element provides overwatch.

5-71. MP check bridges for mines and booby traps. Before crossing a bridge, MP have the security element move to an overwatch position. They have the reconnaissance element dismount and check the bridge and its approaches for mines, booby traps, and demolition charges. If any are found, MP move to a covered and concealed area, report the information, and request engineer support to clear the mines. MP maintain surveillance of the bridge until the mines are cleared. They stop friendly forces and civilians from using the bridge until the engineers have cleared it. When the bridge is cleared, MP have the reconnaissance element gather critical data on the bridge's characteristics and continue the reconnaissance.

5-72. MP reconnoiter key terrain and built-up areas near the route. They move on and off the road to identify enemy activity. The type of terrain dictates whether a reconnaissance is conducted mounted or dismounted. Reconnoitering terrain can be time-consuming. The mission order and the time available for the reconnaissance determine how many and which terrain features are reconnoitered.

## MAIN SUPPLY ROUTE REGULATION ENFORCEMENT

5-73. MP traffic control activities support movement control by enforcing highway regulation plans. Traffic enforcement measures, such as speed control and safety inspection checkpoints, help protect the force and ensure that only authorized traffic uses controlled routes. MP employ special circulation control measures, such as temporary route signing, TCPs, holding areas, defiles, and checkpoint and roadblock operations, to support combat and sustainment operations. Refer to *Chapter 6* for more information about checkpoints and roadblocks.

5-74. Highway regulations are set by the agency having jurisdiction over the road network. MSR regulation measures are stated in the command's highway regulation plan. They also may be in the unit SOPs and command directives. The HTS sets the route classification.

5-75. Mounted MP teams patrol MSRs to monitor traffic and road conditions. They gather information on friendly and enemy activity and assist stranded vehicles and crews. Road condition changes and enemy activity are reported immediately through MP channels.

## SPECIAL CIRCULATION CONTROL MEASURES

5-76. MP limit, control, block, or direct mounted or dismounted forces traveling on the MSR, by employing special circulation control measures. Special circulation control measures include the following:

- Temporary route signing.
- TCPs.
- Holding areas.
- Defiles.

5-77. Many of these measures are used in MP support to breaching operations, river crossings, and passage of lines.

### Temporary Route Signing

5-78. A signed military route system, like the signed US highway system, can enable road users to reach their destinations by following route signs and road markings displayed along the roadside. MP patrols monitor signs on a routine basis, checking specific signs before critical moves. Engineers erect permanent signs, but signs can be damaged, destroyed, or moved by weather, saboteurs, and battle.

**NOTE: For more information about temporary route signing refer to *Appendix I.***

### Traffic Control Post

5-79. TCPs are used to support MMS only when needed. They are used to preclude the interruption of traffic or unit movement along designated routes. TCPs are communication links to units using the MSR. Show on the traffic control plan and the traffic circulation plan the placement of TCPs. MP activities at a TCP include the following:

- Monitoring and assisting traffic authorized to use the MSR.
- Redirecting unauthorized vehicles to the road network they need.
- Providing route security for the MSR at critical locations or intersections.
- Monitoring for NBC contamination.
- Rerouting traffic as needed.
- Gathering information and reporting it.
- Providing information to passing units.
- Assisting stragglers and DCs.

5-80. **Plan**. An MP squad can operate one TCP for an extended period or three TCPs for a short duration. The platoon leader uses METT-TC to analyze the mission and estimate the situation. He decides the appropriate weapons, materials, and equipment needed and considers such factors as the movement of traffic and the degree of control required. If HN police support is needed, the company operations section or the PM arranges for it. The platoon leader uses overlays and the traffic control plan to determine the location for the TCP. The squad leader plans for emergency destruction of documents and equipment in case the TCP is attacked. The squad leader plans for continuous operations by—

- Developing an adequate sleep plan.
- Arranging for maintenance and refueling.
- Arranging for additional rations.
- Constructing fighting positions.
- Camouflaging all vehicles and equipment.

5-81. **Execute**. TCPs are manned at points where two or more MSRs converge or where confusion could affect vehicle movement. They are used to help protect the force at critical locations where civilian or military traffic can cause an accident. Operations in which TCP will be maximized include the following:

- Deliberate river crossings.
- Deliberate breach operations.
- Defile operations.

5-82. When METT-TC requires a TCP to be manned by one MP squad, the squad leader—

- Analyze the terrain location.
- Positions the teams.
- Directs the squad's vehicles to a covered and concealed position near the squad's fighting position. Use camouflage nets, if needed.
- Selects a fighting position from which the squad can cover and secure the TCP.
- Maintains communication.

5-83. Once the squad has occupied the actual TCP site, the squad leader establishes security and provides a grid coordinate to higher HQ. One team in the overwatch covers the TCP while another team watches the flow of traffic from a covered and concealed position near the road. When needed, a member of this team moves to the center of the road to direct the flow of traffic while the other members provide security. If necessary, the squad leader has the third team resting in a covered and concealed position. The squad leader ensures that all three teams communicate by wire (the preferred method) or radio. If neither is available, they use arm and hand signals.

5-84. At a TCP, the main purpose is to ensure smooth and efficient use of the road network according to the traffic circulation plan. The plan contains—

- Military route numbers and directions of travel.
- Light lines and blackout areas.
- Highway regulation points and MP TCP.
- Route control classification.

5-85. Vehicles too wide or heavy for a road will be denied access. MP reroute them to alternate MSRs. No authorization is needed for travel on an open route, but use of a classified route may be restricted to certain units, operations, or types of vehicles.

5-86. All vehicles on the dispatch route will have a current movement credit issued by the HTS. On a supervised route, normally a column of 10 or more vehicles or an individual vehicle of exceptional size or weight will have movement credit from HTS.

5-87. MP stop vehicles or convoys that are not following MSR regulations. They tell the convoy commander why the vehicles are halted. The convoy commander makes immediate corrections. When immediate corrections cannot be made, MP record the key information about the incident and report it through MP channels.

5-88. TCP is used to disseminate information about the AO. MP provide information to authorized personnel only. They support the commander's force protection program by providing friendly forces with current information on route conditions and enemy activity. At the TCP, MP disseminate information on the locations of contaminated areas, supply points, medical facilities, and other units on a need-to-know basis and should not volunteer more information than is needed to avoid creating security and/or intelligence issues.

5-89. MP actively seek information from road users by asking drivers what they have seen of suspected or actual enemy activity along the MSR. When language is not a barrier, MP talk to local civilians and to the HN civilian and military police to gain information about the road network in an AO. MP pass this information up the chain of command so that it can be verified as reliable.

5-90. MP are constantly on watch for enemy aircraft and suspicious activity by the local populace. When such activity is observed they relay SPOTREPs through MP channels. MP use the SALUTE format to report this information.

5-91. When the movement control agency requests it, MP keep track of military movements that pass through a TCP. This helps movement planners keep track of the progress of convoys. This information is compiled into a passing report that includes the—

- TCP location.
- Date.
- Convoy identification (unit or serial number).
- Time the first vehicle passed the TCP.
- Time the last vehicle passed the TCP.
- Number of vehicles in the convoy.

5-92. Usually, TCP passing reports are picked up at the TCP or transmitted by secure radio. The platoon leader compiles the TCP passing reports into one report. He forwards the report through MP channels or as directed by the commander. The report may be written or transmitted. If transmitted, a report is encoded according to the unit SOP. In some instances, the platoon leader may permit a squad leader to bypass the usual report channels and submit a passing report directly to the movement control agency.

## Holding Areas

5-93. MP operate vehicle-holding areas to help regulate the traffic flow. Holding areas can be used as independent measures or with other measures like defiles or checkpoints to support large operations like river crossings or passage of lines. When MP operate holding areas, they direct vehicles, convoys, and troops into and out of the holding areas.

5-94. **Plan**. METT-TC and the size of the holding area determine the number of teams needed to operate it. When one MP squad operates a holding area, the squad leader designates one team to control the entrance to the holding area, another team to control the exit from the holding area, and one team to provide security. He also assigns each squad member a fighting position.

- The general location for a holding area may be designated by the echelon movement control unit, PM, or MP company commander. The exact location is selected by the MP leader with the mission. The holding area's location is noted on the traffic control plan and passed to the echelon movement control unit to keep the traffic circulation plan current. Select a site where—
- Parked vehicles can face the exit so that they can be driven from the area quickly.
- Vehicles can be dispersed.
- There is easy access to and from the roadway.
- The surface of the area is firm enough to hold the weight of the vehicles.
- The area is large enough to allow vehicles to be covered and concealed from air and ground observation.
- The area can be defended.

5-95. **Execute**. MP teams at the entrance and exit to the holding area are positioned in concealed locations. When vehicles approach the holding area, one MP moves to the center of the road and directs the vehicles in. He tells the vehicle driver or convoy commander where to park. He moves back to the concealed location when all vehicles have entered. Vehicles moving in the holding area should be ground-guided by members of the convoy.

5-96. The MP team at the exit operates in a similar manner. MP move from their cover and concealment only when necessary. Depending on the tactical situation, MP teams may use red lens flashlights to direct traffic at night. Colored chemical light sticks are posted to help drivers identify their locations within a holding area if the situation permits.

5-97. Ideally, the MP leader controls the holding area from a position overlooking the entrance and exit. The MP leader receives instructions on when to allow vehicles to pass from higher HQ. When the holding area is to support a river crossing site, the MP leader has a movement schedule to follow. He receives his movement information from the echelon movement control office or the crossing area commander.

5-98. When operating a large holding area, the MP leader may need additional personnel inside the holding area to direct traffic and parking and ensure that the units comply with the flow plan. Large holding areas will have a simple control plan, such as a subdivision system.

5-99. Take the following steps when using a subdivision system:

- Make a map or a sketch of the area showing the road net, trails, and major obstacles.
- Outline the holding area on the map or sketch.
- Divide the area into equal subdivisions and assign a letter or a name to each subdivision. This helps direct units to their section of the holding area.
- Erect signs showing the outline of each area.
- Develop a traffic flow plan and erect directional signs to help users.

5-100. Keep a count of vehicles in the subdivisions by number, size, and unit designation for each vehicle. At night, use chemical light sticks to identify the sections within the holding area and the exit.

5-101. Vehicle holding areas are controlled sites. MP instruct convoy commanders to keep drivers with their vehicles or assign them fighting positions. They do not allow convoy personnel to openly move around within the holding area. They use signs to help control traffic and maintain communications between positions by wire or hand and arm signals.

### Defiles

5-102. Defiles keep traffic moving smoothly despite narrowed passageways. Controls at defiles ensure that traffic moves through the passage, one direction at a time, first from one end and then the other. A defile may be considered a critical site and its security a priority.

5-103. **Plan**. METT-TC and the estimated length of the operation are used to determine the actual size of the element needed to operate a defile. In most environments, a defile will not be operated with less than a squad. An MP platoon may be needed to conduct a large defile operation for extended periods. At a defile *(Figure 5-3)*, MP do the following:

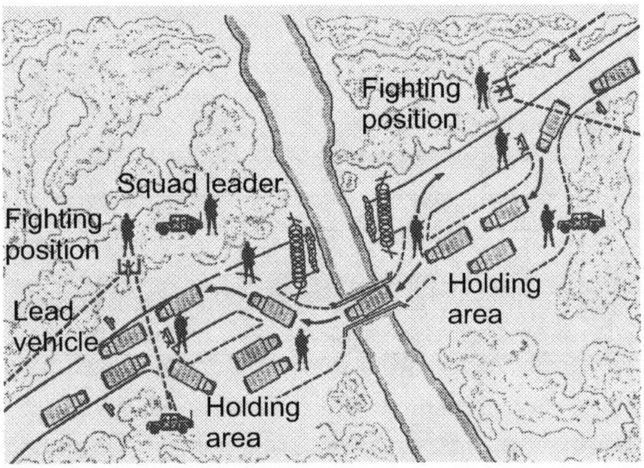


**Figure 5-3. Defile**

- Secure and defend the site.
- Brief drivers about obstructions.

- Control access so vehicles move through quickly.
- Ensure that vehicles enter one at a time.
- Reroute traffic when necessary.

5-104. Terrain or traffic needs may dictate a need for vehicle holding areas and signs or a TCP. Larger defiles require a holding area at each end. The placement of holding areas depends on the site available and the ease of communicating between the sites and the defile.

5-105. **Execute**. Because defiles involve restricted movement, they are an ideal target for the enemy. Security of a defile is a priority. Before a defile is put into operation, an area reconnaissance is conducted to detect the presence of enemy activity in and around the location of the defile. Once the area has been reconnoitered, the squad leader establishes security and reports the exact location of the defile, if not previously known, to higher HQ. He selects the crew-served weapons fighting position, picking key terrain that overlooks the defile. He ensures that the squad's vehicles are covered and concealed. As in most stationary MP operations, communications within the squad is primarily wire and arm and hand signals.

5-106. MP use control measures to keep traffic flowing smoothly. They use the simplest method of control. Plan for the use of two different control measures. This provides a back-up method, if needed. Control measures can include the following:

- Visual signals to tell traffic when to move, such as hand and arm signals, flashlights, or a handheld flags. Use any technique that shows vehicles when to move. Visual signals work best for small defiles where holding areas are not needed.
- Wireless or wire communications to tell teams in holding areas to hold or start traffic through the defile. Link communications directly between holding areas, or route them through the leader

at the defile site. Use wire communications as the main means of communication. Use wireless communications as a backup or when no other means are available.

- A flag to identify the last vehicle moving through a defile. Give the flag to the last driver or attach it to the last vehicle entering the defile. Another MP removes the flag when the vehicle reaches the end of the defile. This serves as a signal for traffic to start in the opposite direction. This is repeated as often as needed.
- An MP rider to indicate the last vehicle of a column. The rider stays in the last vehicle until the column reaches the opposite side. He dismounts and rides back in the last vehicle returning. This technique ensures that all the vehicles clear the defile.
- MP lead and trail vehicles in the front and rear of a column to guide it through the defile. After the column clears the defile, the vehicles guide a column moving in the opposite direction. MP use this method when movement through a defile is complex and requires an escort. The trail vehicle ensures that all the vehicles clear the defile. A single lead or a single trail vehicle can also be used, depending on the number of vehicles and the complexity of the defile.

5-107. To ensure traffic flow is not interrupted by a disabled vehicle, MP plan for a recovery vehicle to stand by at the defile. If a recovery vehicle is not available, they use field expedient measures.

5-108. Regardless of what special control measure is employed, MP leaders coordinate with the PM for the location, duration and special criteria for the execution of the control measure. Once operational, MP monitor its effectiveness and ensure that force protection and security measures are followed and maintained.

# Chapter 6

# Area Security

MP conduct AS operations to protect critical functions, facilities, and forces. They synchronize efforts with base and base cluster defense planners within a specific AO to ensure that support and sustainment operations are not interrupted. The HN, when capable, retains responsibility for security of all areas outside US bases. However, US commanders are always responsible for the defense and security of US forces and bases regardless of HN support. AS is conducted by MP across the full spectrum of army operations to protect the force, impose order, and ensure freedom of movement. MP activities that support AS include reconnaissance operations, ADC, base and air base defense (ABD), response force operations, and critical site asset and high-risk personnel security.

## RECONNAISSANCE OPERATIONS

6-1. MP plan and conduct area and zone reconnaissance, screening and surveillance missions, and counterreconnaissance.

6-2. MP conduct reconnaissance and screening missions to obtain information about the activities and resources of an enemy or potential enemy or to secure data concerning the characteristics of a particular area. MP reconnaissance, screening, and surveillance efforts include area, zone, and route reconnaissance and counterreconnaissance. These missions may be conducted primarily in the rear area, but may occur anywhere sustaining operations are conducted. Refer to *Chapter 5* for more information about route reconnaissance. MP employ NBC detection equipment to determine the absence or presence and extent of NBC contamination. Refer to *Appendix J* for more information about NBC reconnaissance.

## AREA RECONNAISSANCE

6-3. Area reconnaissance is performed to obtain detailed information concerning the terrain or enemy activity within a prescribed area, such as a town, ridgeline, woods, or any terrain critical to the operations. MP conduct area reconnaissance to help guard against unexpected enemy attack in the rear area. Area reconnaissance and surveillance are vital to maintaining AS and contribute to the commander's intelligence collection plan. MP area reconnaissance is a composite of actions. It is initiated from observations and reports gathered over time by MP patrols and information gained through coordination with HN police and other friendly forces. Refer to *FM 7-8*.

6-4. Reconnaissance patrols may differ slightly, depending on the type of reconnaissance to be performed. However, all reconnaissance patrols have a reconnaissance and security team. The size of the patrol is determined by METT-TC. Other considerations to determine the size of the patrol include—

- Size and number of reconnaissance objectives.
- Requirement to secure the objective rally point (ORP) and other points.
- Time allowed for conducting the mission.

6-5. MP leaders plan area reconnaissance based on the IPB and the commander's critical information requirements. Information on enemy activity and likely avenues of approach is coordinated with military intelligence (MI). MP monitor likely enemy avenues of approach and LZ and DZ in critical areas to give early warning of rear-area enemy activity.

6-6. MP area reconnaissance plans include areas near facilities that are designated as critical by the commander, such as—

- NAIs.
- Air bases.
- Bases and base clusters.
- Communications centers.
- Logistic support clusters.
- Key terminals, depots, and bridges.
- Critical terrain features.
- High-value assets.

6-7. When leading an area reconnaissance patrol, in addition to using troop-leading steps and following the general principles for making a reconnaissance, the patrol leader—

- Uses a scheme of maneuver.
- Secures and occupies an ORP.
- Conducts a leader's reconnaissance of the objective area to confirm or change the plan.
- Returns to the ORP, completes the plan, and briefs the soldiers.

6-8. The security elements leave the ORP before the reconnaissance element. The security element leader places security teams at the ORP and on enemy avenues of approach into the objective area. The reconnaissance element conducts the reconnaissance by moving to several vantage points around the objective.

6-9. The reconnaissance element leader may have a small reconnaissance team move to each vantage point instead of having the entire element move as a unit from point to point. This reduces the chances of being spotted.

6-10. After the objective has been reconnoitered for the details outlined in the order, all elements return to the ORP. Teams share their information, consolidate it, and report it, then return to the patrol HQ or continue to the next mission.

## ZONE RECONNAISSANCE

6-11. A zone reconnaissance is a directed effort to obtain detailed information concerning all routes, obstacles (to include chemical or radiological contamination), terrain, and enemy forces within a zone defined by boundaries. A zone reconnaissance is normally assigned when the enemy situation is vague or information concerning cross-country trafficability is desired. Zone reconnaissance techniques include the use of moving elements, stationary teams, or a series of area reconnaissance actions. Refer to *FM 7-8* and *FM 17-98*.

6-12. The four methods used to conduct a zone reconnaissance are—

- Box method.
- Fan method.
- Converging routes method.
- Successive sector method.

### Box Method

6-13. To use the box method *(Figure 6-1)*, the leader sends his reconnaissance and security teams from the ORP along the routes that form a boxed-in area. He sends other teams along routes through the area within the box. All teams meet at a linkup point at the far side of the box from the ORP.

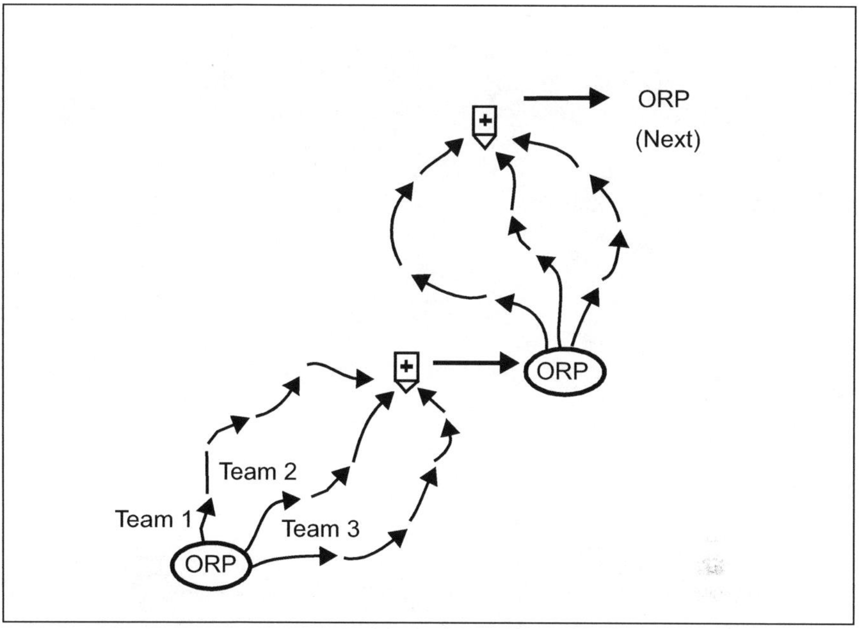


**Figure 6-1. Box Method of Zone Reconnaissance**

## Fan Method

6-14. To use the fan method *(Figure 6-2, page 6-6)*, the platoon leaders selects a series of ORPs throughout the zone. At the first ORP halt and set up security. After confirmation of the patrol's location, the platoon leaders selects reconnaissance routes out from and back to the ORP.

**NOTE: These routes form a fan-shaped pattern around the ORP. They must overlap to ensure that the entire area has been reconnoitered.**

6-15. Once the routes have been selected, send out reconnaissance elements along the routes. Do not send out all the elements at once. The platoon leader keeps a reserve at the ORP. He sends elements out on adjacent

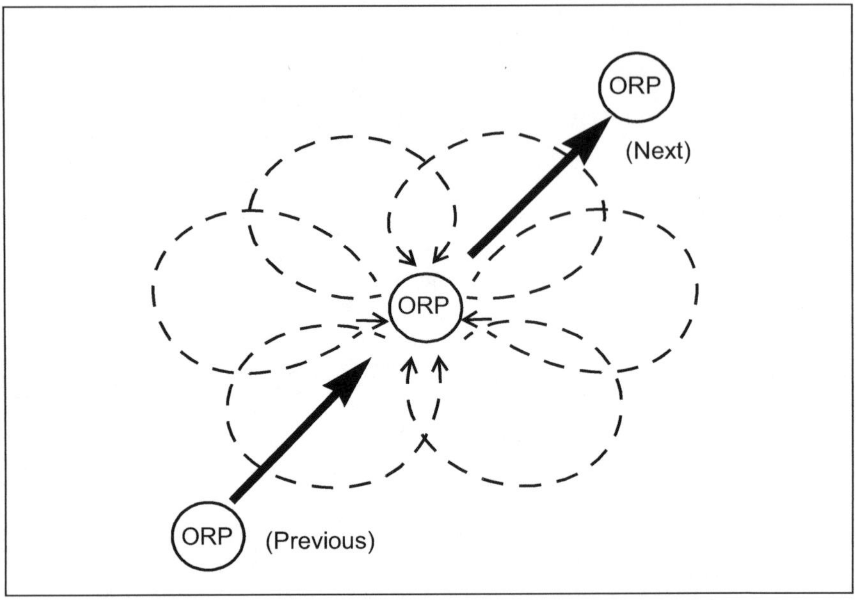


**Figure 6-2. Fan Method of Zone Reconnaissance**

routes to keep from making contact in two different directions.

6-16. After the entire area (fan) has been reconnoitered, report the information then move the patrol to the next ORP. Repeat this action at each successive ORP.

## Converging-Routes Method

6-17. To use the converging-routes method *(Figure 6-3)* (which incorporates the fan method), select an ORP and reconnaissance routes through the zone and the rendezvous point.

**NOTE: The rendezvous point is a place where patrol members link up after the reconnaissance.**

6-18. Halt the patrol at the ORP and set up security. Confirm the patrol's location. Designate a route for each

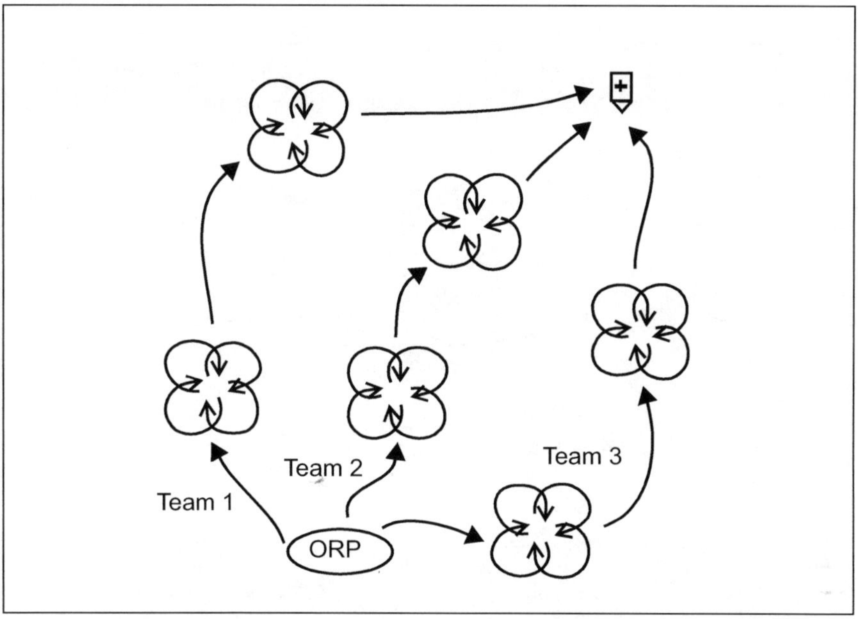


**Figure 6-3. Converging Routes Method of Zone Reconnaissance**

reconnaissance element, a location for the rendezvous, and a linkup time at the rendezvous point. Send a reconnaissance element to reconnoiter each route (usually using the fan method). The leader moves with the center element.

6-19. At linkup, the patrol secures the rendezvous point as it did the ORP. While at the rendezvous point, information gained by each member is exchanged with all the other members. This provides backup to ensure that all information is passed onto higher HQ. The patrol then returns to friendly lines or continues on to another mission.

## Successive-Sector Method

6-20. To use the successive-sector method *(Figure 6-4, page 6-8),* build on the converging-routes method. Select an ORP and a series of reconnaissance routes and rendezvous points. Use the converging-routes method from each ORP to each rendezvous point.

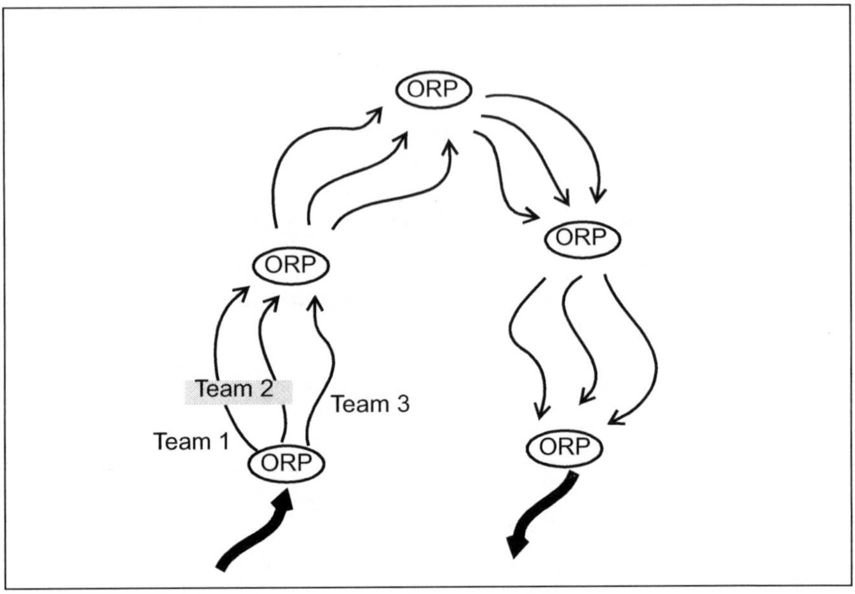


**Figure 6-4. Successive Sector Method of Zone Reconnaissance**

6-21. Each rendezvous point becomes the ORP for the next phase. Designate reconnaissance routes, a linkup time, and the next rendezvous point when the patrol links up at the rendezvous point. Continue this action until the entire zone has been reconnoitered.

6-22. Regardless of the type of method used, report the information through proper MP channels as soon as possible. Commanders rely on fast, accurate reconnaissance information to plan successful operations.

## SCREENING MISSIONS

6-23. Screening missions are defensive in nature and largely accomplished by establishing a series of OPs and conducting patrols to ensure adequate surveillance of the assigned sector. Division cavalry units normally conduct security missions that include a screen for maneuver units during offensive operations.

6-24. MP conduct screening missions for friendly forces in the rear area to provide early warning of enemy approach and to provide real-time information and reaction time for stationary units. In the event of a Level III threat, MP may come under the OPCON of a TCF, which is also referred to as a combined-arms maneuver unit. In this role MP may provide limited security missions, such as a screen to the flank or rear of the main body, with the primary mission of providing early warning and disrupting or destroying enemy reconnaissance vehicles.

6-25. Generally, MP are tasked to observe specific avenues of approach or, more precisely, NAI. The area to observe should be identified in either the reconnaissance and security plan that the platoon leader receives or in the OPORD from higher HQ. If the platoon does not receive an IPB product, the higher OPORD must specifically state where it must focus the screening operation. If the platoon is assigned multiple requirements, the higher HQ must prioritize them.

## SURVEILLANCE

6-26. On order, MP maintain continuous surveillance of all the assigned NAIs or enemy reconnaissance avenues of approach into a particular sector. This is accomplished by setting up a series of OPs. MP may conduct active mounted patrols to extend their observation limits or to cover dead space and the area between OPs. Refer to Chapter 4 for more information about setting up OPs/LPs.

6-27. Once the platoon leader understands what his surveillance requirements are, he task organizes the platoon and any assigned assets to achieve the most effective surveillance of the avenue or NAI.

6-28. Unlike a scout platoon, MP focus on providing early warning of enemy reconnaissance elements rather than gaining and maintaining contact with the enemy's main body or destroying it. During screen missions, it is important to understand that an MP platoon by itself does not have enough assets to both acquire and kill an enemy reconnaissance larger than the engaging element. Generally, other assets will be given the specific mission of killing these forces. If MP are ordered to engage enemy forces, they do so by engaging at the maximum effective range of their organic weapons. If available, the platoon leader also plans for and uses CAS and indirect fire.

6-29. During surveillance the platoon's ability to report is critical. Effective early warning requires detailed planning for uninterrupted communications. The platoon leader considers communication distances and significant terrain features to identify potential wireless communication problems. If problems exist, he requests support from the higher HQ.

## COUNTERRECONNAISSANCE

6-30. MP contribute to the commander's concept of operations by conducting security and reconnaissance missions designed to detect, disrupt, and impede enemy reconnaissance elements. Counterreconnaissance is not a distinct mission; rather, it is a combination of measures taken by friendly forces to reduce the threat's ability to gather information. It contains both active and passive elements and includes combat action to destroy or repel enemy reconnaissance units.

6-31. Counterreconnaissance prevents enemy reconnaissance forces from observing the main body of friendly forces by defeating or blocking the enemy forces. In the execution of counterreconnaissance, MP operate either offensively or defensively using whatever tactics best accomplish the mission. The principal techniques used are—

- A hasty attack.
- An ambush.
- Indirect fire support.

6-32. MP must task organize to defeat enemy reconnaissance forces. Enemy reconnaissance capabilities in any given situation must be compared to the MP unit's capabilities to determine if additional maneuver or CS assets are required.

6-33. Conventional reconnaissance elements are usually squad-size or smaller. However, special-purpose reconnaissance forces can consist of mechanized forces up to company size. In all counterreconnaissance operations, the goal is to acquire, identify, and kill the enemy reconnaissance force after it has penetrated the initial screen line. Defeating such forces usually requires combined-arms forces, but this is dependent on the type, size, and capabilities of the reconnaissance element.

6-34. MP platoons are not organized or equipped to fight for extended periods or to destroy enemy armor vehicles. MP employ AT weapons, such as AT-4s, for defensive purposes (self-protection and breaking contact). However, MP teams are highly skilled at reconnaissance and surveillance and providing early warning of enemy activity.

6-35. A scout platoon acquires and identifies enemy reconnaissance forces along a screen line, which is a control measure usually named as a phase line, and is an established forward of the main body. MP conduct their counterreconnaissance efforts in a similar manner in the rear area or anywhere sustainment operations are taking place.

6-36. In most cases, the scout platoon cannot be expected to acquire, identify, and defeat enemy reconnaissance elements. As a CS asset, MP can assist a scout platoon by locating the enemy reconnaissance element, freeing the scouts or TCF to perform the killing function of counterreconnaissance on larger mechanized enemy reconnaissance elements. MP activities that contribute to counterreconnaissance include—

- Area reconnaissance.
- Zone reconnaissance.
- Route reconnaissance.
- OP operations.
- Physical security and vulnerability risk assessment.
- Critical asset security.
- OPSEC.
- Deception operations.

6-37. Conventional threat reconnaissance elements push far out in front of their combat unit to gain intelligence on their rear area objective. Unconventional threats such as terrorists, criminals, or gangs may try to observe installations, deep-water ports, or other facilities to obtain information. Conventional threat reconnaissance efforts are concentrated on gaining intelligence on the capability of friendly forces. Unconventional threats try to obtain information to plan sabotage or criminal activity, or to simply disrupt the efforts of friendly forces.

6-38. MP leaders plan measures to counter enemy reconnaissance by coordinating with various staff sections and agencies that include—

- MI for information on enemy capabilities, likely rear area targets and objectives, likely enemy reconnaissance avenues of approach, and the commander's critical information requirements (CCIR).

- The PM and criminal investigation division (CID) for HN police information on local gangs, known criminals, and criminal activity.

6-39. To assist commanders with their counterreconnaissance efforts, friendly forces such as MP, CID, engineers, and MI conduct physical security surveys and vulnerability and risk assessments of bases and base clusters, deep-water ports, and air bases. They advise commanders of these facilities on security measures designed to prevent the threat from gaining access to friendly forces and facilities.

6-40. During AS missions, MP conduct security activities around NAI; critical assets, such as communications nodes; and air bases. MP perform area and zone reconnaissance of all terrain that dominates critical facilities. They concentrate their efforts on locating enemy reconnaissance forces. MP deny the enemy the opportunity to observe friendly forces by reporting their location, maintaining surveillance, and assisting in their destruction if required.

6-41. Enemy reconnaissance forces are not likely to use primary reconnaissance avenues of approach to gather information on friendly forces. MP teams are more likely to come in contact with enemy reconnaissance forces operating on trails, rough terrain, and dead space that allows mounted movement. They use the cover of darkness for their operations. MP must make maximum use of NVDs and illumination to help detect their movement. They put the devices on key terrain and along avenues of approach to critical bases, and cover the area with crew-served weapons. Enemy reconnaissance teams are most vulnerable during the day. MP concentrate daytime mounted or dismounted operations on locating their base camp or hide positions.

Once they are discovered, if ordered to do so, MP can lay ambushes on likely routes to destroy them. Refer to *Chapter 4* for more information about ambush patrols.

6-42. Mounted MP patrols use overlapping search techniques to make it difficult for enemy reconnaissance teams to reach their objectives without being exposed. Overlapping searches provide random coverage not easily predictable by simple observation.

6-43. If contact is made, MP may be directed to maintain contact or surveillance until enough force can be assembled to counter the threat. They disrupt their movement and delay them until a large enough force is available to defeat them. If directed to disrupt or delay, they clearly identify the enemy before engaging them; there are generally numerous friendly forces operating in the rear area. MP must consider fratricide preventive measures. Refer to *Appendix F* for more information about fratricide avoidance.

6-44. MP make good use of terrain and maximum use of weapon systems. They use harassing fire from mounted crew-served weapons (MK19/M2), but do not become decisively engaged. They deceive the enemy with heavy harassing fire from crew-served weapons. If available, MP destroy the threat with indirect fire. MP make the threat believe he has encountered a defense. The enemy must not realize these actions are only delaying tactics.

## AREA DAMAGE CONTROL

6-45. ADC is basic to successful rear-area operations. ADC measures are taken before, during, and after hostile actions or natural or man-made disasters to minimize effects and reduce damage. All commanders try to limit the impact of enemy actions and reestablish unit operations as quickly as possible. Commanders at each level plan ADC operations.

6-46. ADC operations integrate the functions of many specialized units. Engineers plan and coordinate ADC operations at the site. Medical teams help sort and treat mass casualties and assist in initial evacuations. Army aviators help evacuate casualties and provide emergency resupply, communications relay operations, area damage assessment, and $C^2$ actions. MP expedite and control battlefield movement into, around, or through damaged or contaminated areas. Signal elements reestablish the signal system. HN civil efforts, like clearing rubble and providing facilities and services, can greatly aid ADC.

6-47. The amount of MP support needed for ADC depends on the extent of the damage, the importance of the affected area, and the effect of the damage on the movement of troops and logistical supplies. Enemy attacks on key military facilities can leave them unprotected. Heavy damage in urban civilian areas can disrupt local government services.

**NOTE: Protecting civilian facilities is a HN responsibility. However, MP preserve law and order in such an area if so doing protects friendly forces and resources.**

6-48. Route and area reconnaissance are key to determining the trafficability of the routes into, out of, and around affected areas; obtaining a damage assessment, and having early warning of the continued presence of the enemy. MP determine the level and extent of NBC contamination and identify the location of critical points affected by damage.

6-49. The enemy's damage to the terrain determines the degree and kind of MP support needed and where to place the priority of effort. Downed trees, urban rubble, damaged or destroyed bridges, cratered roads, and contaminated road networks affect circulation control. If the roads can be traveled, MP provide circulation control operations locally in the affected area. However, damage to an area may be so great that roads must be closed and MSR traffic rerouted.

6-50. MP give priority of movement to ambulances transporting wounded and engineers clearing debris. They maintain law and order in the affected area by employing measures to stop looting and unlawful behavior, according to the ROE and the use of force requirements. They disseminate key information such as the location of first aid stations, emergency shelters, and other emergency facilities.

6-51. Security patrols may be needed around key facilities. MP may need to set up an OPs/LPs to observe sectors of the affected area. The OP/LP teams can watch for enemy agents trying to exploit the effects of a conventional attack through arson, sniper fire on firefighters, or other disruptive acts. The OPs/LPs also watch for theft, pilferage, or arson against military property.

6-52. MP ADC operations may be either part of an ongoing operation or a separate requirement. During ADC, MP—

- Perform route and area reconnaissance in affected areas.
- Evaluate the serviceability of the road network.
- Note and report the development of critical points caused by damage to bridges, tunnels, and the like.
- Monitor the flow of DCs from the damaged area.
- Report and block off affected areas.
- Provide AS for involved critical facilities.
- Reroute battlefield movement to alternate road networks.
- Check for and report NBC hazards and contamination.
- Prevent sabotage, looting, and pilferage in the damaged area.

- Protect property, contain panic, and enforce emergency restrictions.
- Direct persons to first aid stations, emergency shelters, and other emergency operations.
- Post temporary signs to prevent entry into unsafe buildings or redirect activity to temporary locations.
- Help establish populace control in affected areas.
- Operate mounted and dismounted mobile patrols, checkpoints, and roadblocks to—
  - Enforce emergency restrictions on movement into, within, and out of the affected area.
  - Direct DCs.
  - Collect stragglers.
  - Enforce curfews, stand-fast orders, and movement authorizations.

## BASE DEFENSE

6-53. Base defense is the local military measures, both normal and emergency, required to nullify or reduce the effectiveness of enemy attacks on, or sabotage of, a base to ensure that the maximum capacity of its facility is available to friendly forces.

6-54. The base commander organizes and controls the forces assigned to a base to best capitalize on their capabilities. These forces may be joint or single-service forces operating in the joint rear area (JRA). If a base, base cluster, or air base comes under attack, the base commander responds within his capability. Defense fundamentals, as they pertain to the defense of the bases, include the following:

- Understand the enemy. Defenders must be familiar with the capabilities and limitations of the enemy forces, weapons, equipment, and tactics.

- Use situational awareness. Intelligence operations are key to assembling an accurate picture of the battle space.
- Use the defender's advantages. The defender's advantages may permit a numerically inferior force to defeat a much larger one. These advantages include—
  - The ability to fight from cover.
  - A more detailed knowledge of the local terrain and environment.
  - The ability to prepare positions, routes between them, obstacles, and fields of fire in advance.
  - The ability to plan communications, control measures, indirect fires, and logistic support for contingency operations.
  - The ability to deceive enemy forces about friendly defensive capabilities, dispositions, and execution of operations.
- Concentrate at critical times and places. Conduct the defense of a base along interior lines, permitting the timely and secure movement of forces to engage the most critical threats. The commander must mass combat power at points of decision by economizing in some areas, retaining a reserve, and maneuvering to gain local superiority at critical points.
- Conduct counterreconnaissance and counterattacks. Fixed bases have well-established perimeters that usually have limited depth. Counterreconnaissance and counterattack add depth to the battle, outside the perimeter, allowing the base to continue its primary mission with minimal interference.

- Extend counterreconnaissance patrols 3 to 5 kilometers outside the perimeter, based on terrain. This reduces the risk of fratricide, especially at night.
- Coordinate critical defense assets. Synchronization of indirect fires, air defense resources, tactical aircraft, engineers, dismounted troops, armored vehicles, naval surface fire support, and helicopters can produce a combined-arms effect.
- Balance base security with political and legal constraints. This fundamental is especially critical in a low-intensity conflict environment.
- Know the law of war and ROE. Base commanders and their subordinates must comply with ROE. In joint-service operations, reconcile inconsistencies with ROE.

6-55. A base or base cluster has certain inherent capabilities which allows it a degree of defense. Each base has a BDOC that plans, coordinates, and supervises base defense operations. The BDOC initiates contingency planning that enables the base to—

- Increase the manning posture of the base as the threat level increases.
- Detect and defeat the threat within its capabilities.
- Defend against heavier enemy forces until the response force is committed.
- Maintain control of the fight within the base.
- Support the fire and movement of the response force operating outside the base.

6-56. Each base cluster has a BCOC to monitor base defense plans and establish the base cluster response force. The BCOC—

- Provides $C^2$ of resources for planning, coordinating, and supervising the defense of the base cluster.

- Coordinates base defense operations.
- Maintains communications with the BDOC within the cluster, MP, and the supporting RAOC.

6-57. MP leaders plan for the interface of their support into the base plan. The plan requires—

- Cover and concealment of personnel and equipment.
- Secure and redundant communications systems at all guard locations (a landline, radio links to the BDOC, and a telephone hookup to the center switch).
- Deception.
- Contingency planning.
- Improvement of base defense positions.
- Coordination with BCOC or RAOC as required.
- OPs/LPs.
- Noise and light discipline.
- Immediate reaction to enemy threat or attack.
- Rehearsals of defense measures.

6-58. All plans and overlays depicting MP support are forwarded through normal MP channels to the BCOC.

6-59. Any threat to the bases must be dealt with swiftly. Few sustainment units can continue their mission while under attack by even a low-level threat.

6-60. Sustaining operations rely on MP to stay apprised of enemy activity near bases. MP forward early warnings of enemy activity to the operations cell at the rear CP. The rear-operations cell immediately notifies the subordinate RAOC and the base and base clusters. It may also alert the response force.

6-61. If a base or base cluster comes under attack, the base commander responds within his capability. However, interruptions of base sustainment operations must be kept to an absolute minimum. If a threat exceeds a base's ability to defeat it, the base commander requests response force support.

**DANGER**
**Base defense forces must be aware of the disposition of other friendly forces in front of them. Fire discipline between the response force, mobile reserve, and TCF is imperative. Positive control must be maintained at all times.**

## AIR BASE DEFENSE

6-62. The Army is responsible for defending air assets from ground threats outside the boundary of the air base and its area of responsibility (AOR). ABD forces are tasked to defend particularly critical air bases. They may be forces operating in the area specifically on call for ABD or they may be the rear-area response force.

6-63. Air Force security forces (AFSF) are responsible for the internal security of US air bases. HN and US Army forces are primarily responsible for external defense of US air bases. Given the size of an air base and the high priority as a threat target, providing ABD as a dedicated asset requires a force larger than a company. MP do not have the force structure to be dedicated solely to ABD operations. However, an echelon commander could decide to redirect MP mission priorities to ABD and deploy his limited MP assets to serve as part of a larger composite force, such as a TCF. Or an MP response force may be required to respond to threats beyond the base's capabilities.

6-64. MP operating as part of a larger ABD force may provide support that includes —

- Setting up roadblocks and checkpoints to limit access to the base.

- Occupying OPs/LPs and fighting positions on key terrain and likely avenues of approach to the air base.
- Conducting screening missions.
- Conducting combat patrols (security and ambush).
- Conducting DC operations.

6-65. A typical ABD layout consists of a base perimeter that surrounds the runway and the facilities that support air base operations. The tactical perimeter of a base is the outer most ABD tactical AOR. It is where defense forces cover, patrol, or occupy likely avenues of approach to the base and other key terrain. *Figure 6-5* shows a typical ABD layout.

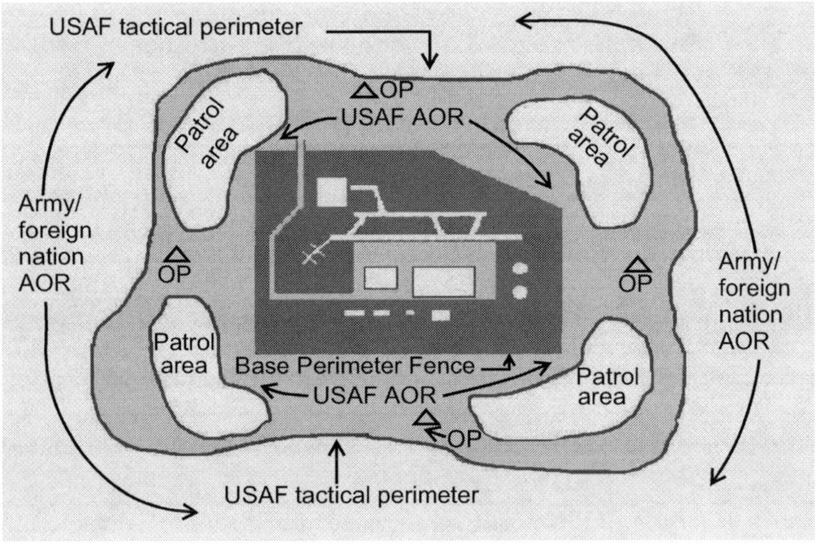


**Figure 6-5. Typical ABD Layout**

6-66. Commanders develop unit SOPs and contingency plans for ABD. SOPs should address $C^2$ and fire support issues that are different for ABD. However, AS plans for ABD provide for increased security patrols and static security measures around an air base the same as they would for other critical facilities.

6-67. The depth of the defense area surrounding the air base depends on METT-TC and other defensive planning factors. MP activities that support ABD are concentrated in the tactical perimeter and on key terrain adjacent to the tactical perimeter.

## FIRE SUPPORT

6-68. Organic AFSF fire assets typically available are the M29 81-millimeter mortar, the MK19, and the M2 .50 caliber machine gun (MG). These fire systems are incorporated into the base defense plan for both the security force (SF) and MP. This ensures compatibility with all ABD measures in the surrounding area and mutual support to other bases when practical.

6-69. The use of mortars and other indirect-fire weapons located in the defense area is coordinated through a 5-man FDC located at the air base BDOC. The AFSF primarily uses its mortars for illumination, suppression of enemy indirect-fire weapons, and suppression of enemy surface-to-air missile (SAM) capability. AFSF weapons controlled by the FDC vary. The number of weapons available depends on the size of the air base.

6-70. CAS is obtained from aircraft at the base. Aircraft may be sent aloft to provide CAS or targets may be assigned to aircraft already airborne. CAS provides suppressive fire or increased detection capabilities. Among the CAS aircraft that may be available, the AC-130 Spectre gunship and the A-10 Thunderbolt II both have night-flying capability. Both have compatible frequency modulated/modulation communications equipment. Both are especially effective for use against targets considered danger close to ABD forces. However, they are very susceptible to SAM.

**DANGER**

**ABD forces must be aware of the disposition of other friendly forces in front of them. Fire discipline between AFSFs response forces, mobile reserves, and the TCF is imperative. Positive control must be maintained at all times.**

## COMMUNICATIONS

6-71. Joint communications for all forces supporting ABD must be compatible. Communications must be secure, have redundancy, and possess the ability to integrate all security forces.

6-72. Effective coordination must be an ongoing process. Call signs, frequencies, and SOI are exchanged between all joint, combined, and coalition forces supporting ABD.

## MILITARY POLICE RESPONSE TO AIR BASE DEFENSE

6-73. Air bases are normally located in the rear area. Threats that are beyond the capability of the base may require the deployment of an MP response force. In general, the response force responds to an air base the same as any other base or base cluster. However, as a joint service operation there are $C^2$ and fire support differences. Company and platoon tactical SOPs for ABD should be developed to help speed the response time.

6-74. The response force leader coordinates with the Air Force defense force commander (DFC) for the current tactical situation and threat. When committed, the MP response force may transition to the operational control of the DFC. This ensures that MP actions do not interfere with air sorties being generated by the air base commander.

6-75. The DFC is in charge of all aspects of ABD operations. He provides guidance on where the response force is needed and informs the response force leader of the tactical situation. The response force leader directs the tactical operations of MP elements, keeping the DCF informed. If a TCF must be called to defeat a Level III threat, all units, Air Force and Army, are under the OPCON of the TCF commander until the threat is defeated.

6-76. For the local ground defense of their base, the Air Force provides a BDOC which operates similar to an Army BDOC. It is responsible for combat intelligence, combat operations, logistics, and personnel administration. It collects the ground combat intelligence within the ABD area.

6-77. The AFSF maintains a mobile reserve force. The mobile reserve's purpose is to mass enough firepower to destroy threat forces within the air base boundary or, at least, delay the threat until a larger force can be assembled. The mobile reserve is usually a mounted force. It is under the direct control of the BDOC. The mobile reserve tries to contain any direct landing by threat forces on the airfield. It responds to penetrations of the defense area by forces that have managed to elude external defense forces.

6-78. If the enemy penetrates deeply into the defense area, the response force leader can request that the BDOC commit the mobile reserve to battle. This may prevent the premature commitment of the TCF. Contact points and AAs are established around the base to assist in integrating the mobile reserve during combat operations outside the ground defense area.

6-79. The response force must keep the enemy from destroying resources on the ground and interrupting or stopping air operations.

## Night Operations

6-80. Air bases are most vulnerable when visibility is limited. In response to the threat, the bulk of both internal and external ABD operations must be dedicated to detecting threat forces during periods of limited visibility. The large gaps between defensive positions facilitate enemy infiltration through external defenses. To narrow these gaps during limited visibility, external defensive efforts may have to be set up closer to the perimeter. External defensive efforts should integrate the extensive use of—

- Night vision equipment.
- Reconnaissance and surveillance.
- Electronic detection devices.
- Tactical deception.
- OPs/LPs.

## Water Obstacles

6-81. Many air bases border water obstacles. Air bases located next to rivers, lakes, or oceans provide unique ABD problems. Threat efforts directed against an air base from the water include infiltration, reconnaissance, and standoff-weapons attacks. As a result, water obstacles that penetrate the defense area may require ABD forces to set up continuous battle positions between the water and any critical air base facility near the water. ABD obstacle plans should include fences, mines, and sensor employment along the portion of the perimeter that borders water. Planned fires are coordinated to defeat watercraft. Floating expedient barricades may have to be assembled. Warning buoys should also be anchored offshore. Underwater approaches provide excellent avenues to air bases. They must be considered during ABD planning. Joint support from the Navy and the Coast Guard may also be needed.

### Urban Terrain

6-82. Air bases bordered by urban areas pose a special challenge for ABD operations. Increased security is a necessity. MP must be aware of underground approaches, such as sewers, that could bypass defensive positions. Threat forces, saboteurs, and special-purpose forces will use these avenues of approach to gain access to the air base.

6-83. Infiltrating groups will be very difficult to detect because they avoid direct confrontation and usually move at night. These teams have the greatest potential for disrupting air operations throughout the rear area.

## CHECKPOINTS AND ROADBLOCKS

6-84. Checkpoints and roadblocks are set up to control the movement of vehicles, personnel, and material and prevent illegal actions or actions that aid the enemy. They are used to prevent trafficking in contraband and stop the movement of known or suspected belligerents. Checkpoint and roadblock operations contribute to the commander's information and intelligence collection process. Additionally, they assist friendly forces in detecting and establishing the behavioral patterns of the local populace.

6-85. When conducted improperly, checkpoints and roadblocks can foster resentment toward US forces. MP minimize the negative impact that checkpoints and roadblocks may have by treating local nationals with dignity and respect. Whenever possible, checkpoints and roadblocks should be conducted with interpreters, HN police, or other HN authorities.

6-86. In most stability and support operations there exists the potential for threat forces to attempt to exercise some level of force against US forces. Checkpoints and roadblocks may become prime targets for threat forces (both traditional and nontraditional).

METT-TC determines the level of self-protection needed at a checkpoint or roadblock. However, MP leaders must always plan force protection measures that allow their forces to quickly implement adequate self-defense when threatened.

6-87. Checkpoints and roadblocks must be designed so that potential threat forces are discouraged from attempting to disrupt its operation or try to inflict casualties. MP use armored vehicles whenever possible to support the dismounted troops operating the checkpoint or roadblock. The MP security element assigned to protect the site should have an antiarmor capability. The vehicle search area should be located in an area that provides blast protection.

### Checkpoints

6-88. Checkpoints are part of a police action. As such, the ROE and the use of deadly force must be clearly understood by all soldiers participating in the operation. MP use the minimum intrusion and imposition necessary to accomplish the mission and protect the force. They allow for a vehicle escape route and plan to destroy a hostile element who uses it. If the checkpoint is completely sealed off, an enemy will have to penetrate it by attempting to run over the barricades. This puts the search team in a position to have to defend itself and fight back.

6-89. MP establish and operate two types of checkpoints—deliberate and hasty.

6-90. **Deliberate Checkpoint.** A deliberate checkpoint is a fixed position set up on a main road in a rural or built-up area. It can be classified as either a *heavy-* or *light-traffic checkpoint*, depending on how much traffic is expected to pass through it. A heavy-traffic deliberate checkpoint normally requires a platoon. An MP squad can only operate a light-traffic checkpoint for a short duration (12 hours or less).

6-91. To operate a heavy-traffic checkpoint, task organize the platoon into—

- A HQ element. The HQ element is responsible for $C^2$ and maintaining communications with subordinate elements and higher HQ.
- A security element. The security element is an MP squad that provides early warning to the search and assault element, watches for and reports suspicious activity, and monitors traffic flow up to and through the checkpoint. The security element should have an antiarmor capability to protect the site from an enemy armor threat.
- A search element. The search element is an MP squad that halts vehicles at the checkpoint, guides them to the designated search point, conducts personnel and vehicle searches, and directs cleared vehicles on through the checkpoint.
- An assault element. The assault element is an MP squad responsible for destroying (consistent with the ROE) any hostile element that forces its way past the search team. The squad leader places his soldiers beyond the search point and emplaced zigzag obstacles and barriers. The soldiers prepare and occupy fortified fighting positions. When confronted by a threatening vehicle, the search element gets out of the way and allows the vehicle to pass. The vehicle passes through the escape lane, and the assault element makes the decision whether to engage or not. If the assault element has to engage, the battle will occur away from the checkpoint. This reduces the possibility of fratricide to friendly forces or injury to innocent civilians.

6-92. In smaller checkpoint operations, a squad can be organized in a similar fashion. MP elements use handheld portable radios, if available, or wire communications. However, much of the needed signals at a checkpoint or roadblock can be easily accomplished using arm and hand signals.

6-93. A deliberate checkpoint is organized into sections *(Figure 6-6)*. The physical layout and detail of preparation depend on the amount of traffic that will pass through it and the duration of its operation. Normally, a deliberate checkpoint will require engineer support to construct obstacles, barriers, escape lanes, and possibly fighting positions. All checkpoints consists of—

- Obstacles or barriers.
- Search areas (personnel and vehicle).
- Security overwatch and fighting positions.
- Holding areas.

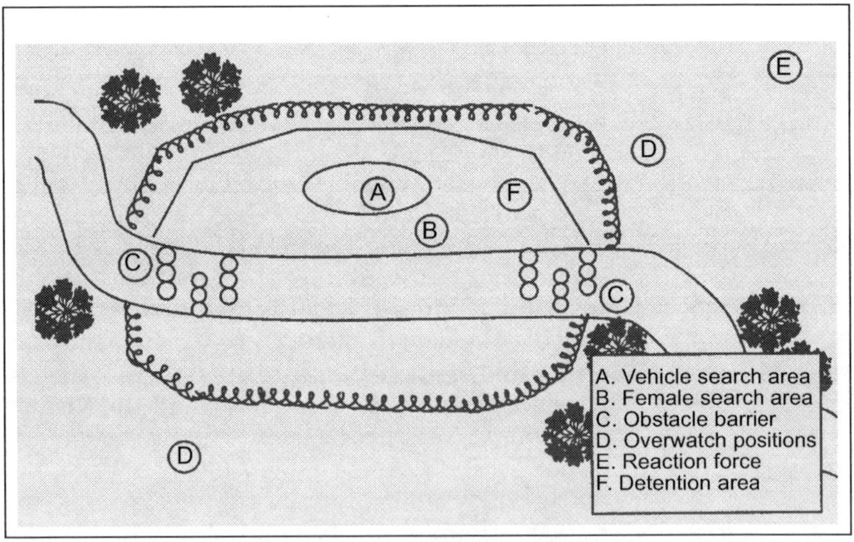


**Figure 6-6. Deliberate Checkpoint**

6-94. **Hasty Checkpoint.** MP set up hasty checkpoints to achieve surprise. They are temporary and should be moved often. The materials used to construct these checkpoints are carried by the platoon. The platoon or squad uses its vehicles, reinforced with concertina wire, as the obstacle. MP may employ tire deflation devices or road spike stripes. These devices are more effective than concertina wire and may be less intrusive in peace operations. They are commercially available or can be locally fabricated.

6-95. MP position the vehicles to partially block the route *(Figure 6-7)*. MP conduct the search in the area

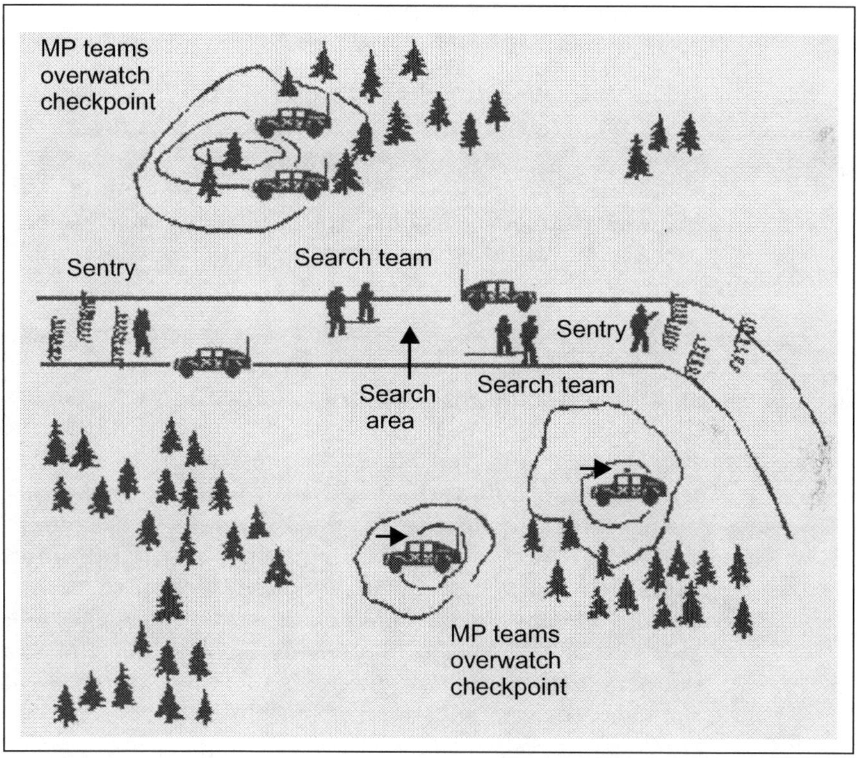


**Figure 6-7. Hasty Checkpoint**

between the vehicles. MP are positioned at each end of the checkpoint. They cover them by mounted or dismounted automatic weapon positions. MP conceal a reaction force (at least one team) nearby to react in case the site is attacked.

6-96. MP establish hasty checkpoints where they cannot be seen by approaching traffic until it is too late to withdraw. Good locations to set up hasty checkpoints include—

- Bridges.
- Defiles.
- Highway intersections.
- The reverse slope of a hill.
- Just beyond a sharp curve

6-97. **Vehicle Searches**. Two members of the search team position themselves at both rear flanks of the vehicle undergoing a search, putting the occupants at a disadvantage. These soldiers maintain eye contact with the occupants once they exit the vehicle and react to any threat attempts by the occupants during the vehicle search.

6-98. The actual search is conducted by two MP armed with pistols only. One MP conducts interior searches and the other performs exterior searches. They instruct the occupants to exit the vehicle during the interior search and instruct the driver to watch the vehicle search. Once the interior search is complete, they escort the driver to the hood of the vehicle and instruct him to open it. Once the engine compartment has been examined, they instruct the driver to open the other outside compartments (tool boxes, gas caps, trunks, and so forth). The driver removes any loose items that are not attached to the vehicle for inspection. Members of the search team rotate positions to allow for mental breaks.

6-99. MP use MWD teams, mirrors, and metal detectors to thoroughly search each vehicle for weapons, explosives, ammunition, and other contraband. Depending on the threat level, the vehicle search area provides blast protection for the surrounding area.

6-100. **Personnel Searches**. MP may be required to conduct personnel searches at the checkpoints. Every attempt should be made for HN authorities to conduct, or at least observe, searches of local nationals. Additionally, MP leaders must plan for same-gender searches. Personnel searches are conducted only when proper authorization has been obtained, usually from higher HQ, according to the ROE, Status of Forces Agreement (SOFA), or HN agreements. This does not preclude MP from searching individuals that pose a threat to US or other friendly forces.

6-101. MP may have to detain local nationals who become belligerent or uncooperative at the checkpoints. The OPORD and the ROE must address the handling of such personnel. In any case, self-protection measures should be planned and implemented according to the orders from higher HQ.

6-102. Searches of local nationals should be performed in a manner that preserves the respect and dignity of the individual. Special consideration must be given to local customs and national cultural differences. In many cultures it is offensive for men to touch or even talk to women in public. Searchers must be polite, considerate, patient, and tactful. MP leaders must make every effort not to unnecessarily offend the local population. Such situations can have a very negative impact on peace operations and can quickly change popular opinion toward US and other friendly forces.

6-103. Each captive is searched for weapons and ammunition, items of intelligence value, and other inappropriate items.

**NOTE: When possible, conduct same gender searches; however, this may not always be possible due to speed and security considerations. Therefore, perform mixed gender searches in a respectful manner using all possible measures to prevent any action that could be interpreted as sexual molestation or assault. The on-site supervisor carefully controls soldiers doing mixed-gender searches to prevent allegations of sexual misconduct.**

6-104. MP conduct individual searches in search teams that consist of the following:

- A searcher. A searcher is the MP that actually conducts the search. He is in the highest-risk position.
- Security. Security includes at least one MP to provide security. He maintains eye contact with the individual being searched.
- An observer. The observer is a MP leader that has supervisory control of the search operation. He also provides early warning for the other members of the team.

6-105. The two most common methods that are used to conduct individual searches are the frisk search and the wall search.

- Frisk search. This method is quick and adequate to detect weapons, evidence, or contraband. However, it is more dangerous because the searcher has less control of the individual being searched.
- Wall search. This method affords more safety for the searcher because the individual is searched in a strained, awkward position. Any upright surface, such as a wall, vehicle, tree, or fence may be used.

6-106. If more control is needed to search an uncooperative individual, the search team places the subject in the kneeling or prone position.

6-107. Strip searches should only be considered when the individual is suspected of carrying documents or other contraband on his person. This extreme search method should be conducted in an enclosed area and by qualified medical personnel when available.

## Additional Checkpoint Considerations

6-108. The effective use of all task organization elements is vital to the success of checkpoint operations is. Roles and responsibilities must be well defined and rehearsed. Additional considerations when conducting checkpoint operations include—

- Preparing and emplacing signs in the local language instructing drivers what to expect and do at the checkpoint (for example, "You are entering a military checkpoint. Prepare to stop your vehicle, and have your identification papers ready for inspection.").
- Determining if it is necessary to apprehend or detain those who see the checkpoint ahead and turn around to avoid it. If it is, HN police are responsible for this mission, if they are available. If they are not available, it may be necessary to position a respond force close to the approach route to block or detain vehicles that try to avoid the operation.
- Clearing and maintaining control of all buildings and terrain that dominate the checkpoint.
- Staying alert for any change of scenery around the checkpoint. A parked car that was not there before, crowds gathering for no apparent reason, or the media waiting for an event are all indicators that something may happen.

- Using artificial illumination for night operations, arrange the lighting to keep those passing through the checkpoint in the light and our forces in the shadows as much as possible.

6-109. If HN personnel are used to assist, commanders ensure that they do not represent a national, ethnic, or religious group of faction feared and hated by the majority of the local population.

6-110. Leaders avoid setting patterns by moving the checkpoint location and changing the method of operation at random.

6-111. MP can gain valuable police, criminal, and combat information while operating checkpoints. They use a checklist to standardize the information collection effort. The following information is included in the checklist:

- The number and type of vehicles stopped. Report identifying markings, license plate number, and any signs displayed on the vehicle.
- The number of passengers in the vehicle. Report the nationality, the ages, and the sex mixes of passengers.
- The type and quantity of cargo.
- The point of origination and destination of the vehicle.
- The stated reason for travel by passengers.
- A description of arms, ammunition, explosives, and sensitive items found and confiscated from the vehicle.
- Possible or actual sightings of weapons, explosives, or threat forces by the passengers.
- The condition of passengers (general health, dress, attitude).
- Anything unusual reported by the passengers.

6-112. When conducting checkpoint operations, MP will need the support of the following:

- Engineers to build obstacles and barriers to channel traffic.
- Linguists familiar with the local language.
- A civil affairs officer or HN police.
- Trained interrogators.

6-113. Sometimes MP are tasked to operate a checkpoint to assist convoys of friendly forces. These checkpoints are set up at the entrance to the controlled route and will have a vehicle holding area to accommodate large convoys. Refer to *Chapter 5* for more information about holding areas.

6-114. At the convoy checkpoints, MP check convoy vehicle movement credits issued by the local movement control unit to ensure that the convoy is moving on the correct route at the correct time. When convoys are ahead of schedule, MP hold them near the checkpoint in a vehicle holding area until it is their scheduled time to pass. They allow convoys that are behind schedule to proceed if route traffic permits and assist drivers who have taken the wrong route.

## Roadblocks

6-115. Roadblocks are set up to stop, slow, or limit movement of vehicles along a route. They are used to limit access to certain areas or roads. A roadblock can be established separate from a checkpoint or used to channel vehicles and personnel to a checkpoint.

6-116. MP use their vehicles with concertina wire, barbed wire, and warning signs as a roadblock *(Figure 6-8, page 6-38)*. If available, armored vehicles make excellent roadblocks. Place a roadblock where unauthorized vehicles or enemy personnel cannot bypass it. If possible, position it so that obstacles like cliffs, swamps, or rivers restrict movement to its flanks or around it. Build man-made obstacles to tie in with and reinforce natural obstacles.

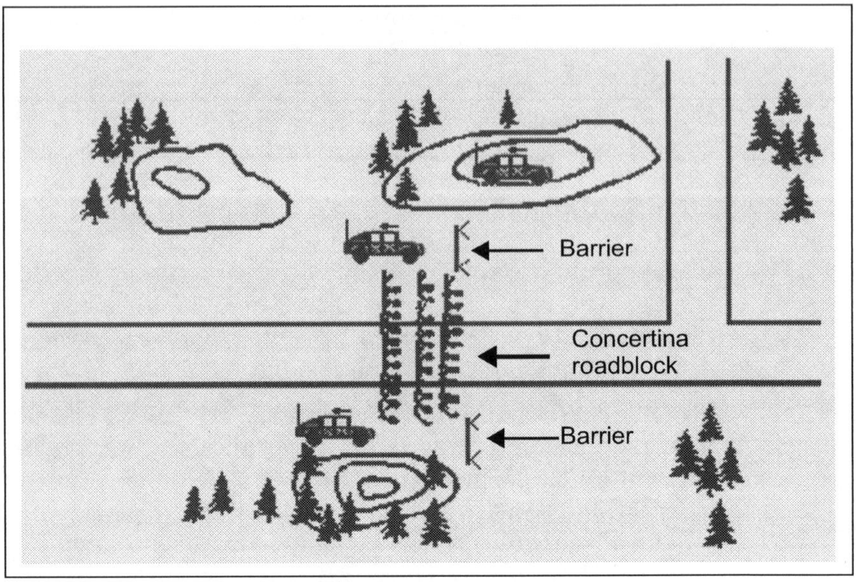


**Figure 6-8. Roadblock**

6-117. When using a roadblock to channel traffic to a checkpoint, place it where drivers cannot see the roadblock until after they have passed all possible turnoffs. When using a roadblock to close off a road, place it at an intersection to let drivers change to another route with little delay.

6-118. Select a defendable site for the roadblock. Cover the roadblock with automatic weapons. Ensure that defensive positions—

- Include a fighting position for the crew-served weapon to provide overwatch for the roadblock.
- Have fields of fire that cover the approaches to the roadblock to keep it from being breached.
- Are not accessible to an attacker and provide cover and concealment.
- Are placed at an intersection to let drivers change to another route with little delay.

- Are placed near an area where drivers can turn their vehicles around easily.

## RESPONSE FORCE OPERATIONS

6-119. MP response forces may respond to threats to a base, a base cluster, or an air base and possibly respond to downed aircraft. MP are normally the commander's designated force to respond to attacks beyond the capabilities of a base or base cluster. The objective of response force operations is to eliminate the threat or make the enemy disengage without requiring the premature commitment of the TCF. MP are organized and equipped to defeat Level II threats. These threats exceed base and base cluster self-defense capabilities but do not require the commitment of a TCF.

6-120. With prior coordination and appropriate supporting fires, MP can defeat the following:

- Small combat units conducting raids, ambushes, and reconnaissance operations.
- Unconventional forces conducting diversionary and sabotage operations.
- Special warfare intelligence collection and operational missions.

6-121. MP response forces are prepared to delay and disrupt Level III threats, protect the flank of a TCF, or allow a base time to establish defenses in greater depth.

### Plan

6-122. METT-TC, the number of MP available, and the commander's intent determine the size of a response force. Normally, an MP response force requires at least a full-strength platoon. However, response force operations conducted in built-up areas or in support of ABD may require a much larger force. MP platoons are not organized and equipped to fight for extended periods. The commander task organizes a platoon for response force operations. A response force should have increased antiarmor capability, integrate available fire support, and have CAS available. The result is a force capable of causing an enemy to break off its attack. The response force can then fix and destroy the threat using close combat techniques and the application of artillery, Army aviation, and CAS.

6-123. MP must know the terrain and be able to use it against the enemy. They mass combat power quickly to destroy the enemy or delay them until a TCF can arrive. The rear CP fire support element, along with the main CP fire support cell, establishes procedures by which MP can call for fire support.

6-124. The response force identifies the location of the bases within their AO, and must know which bases are the most critical and which are the most vulnerable. They include this information in the local IPB. In coordination with the rear CP operations cell and the affected RAOC, they position the response force where they can best—

- Detect enemy incursions (near the DZ, the LZ, and the like).
- Interdict enemy forces enroute to their targets.
- Consolidate quickly in response to threatened key assets in the rear area.

6-125. The response force leader must have the following readily available:

- The base defense status of each base.
- Locations of any obstacles or mines near the base.
- Locations and direction of fire of the crew-served weapons.
- The signal for FPF.
- Locations of TRPs and preplanned fires.
- Methods for contacting the BDOC or the BCOC, to include call signs and frequencies.

- Locations of the OPs/LPs and friendly patrols if employed.
- Call signs and frequencies for supporting artillery and Army aviation units tasked to respond.
- Call signs and frequencies for the TCF, RAOC, and the rear CP.
- Fire support targets that are on the approved list.
- Locations of the nearest medical treatment facility (MTF), NBC decontamination site, and ammunition supply point (ASP).

6-126. Continuous communication with the base and base clusters is the key to knowing how and when an MP response force is needed. Bases and base clusters establish 24-hour communications with the RAOC, if they are located in the corps rear area, or the rear CP, if they are located in the division. This liaison and interface allows timely response and information dissemination.

6-127. The response force will be effective only if it can react swiftly and is thoroughly familiar with the base defense plan. They coordinate all response force actions for a base through its BDOC. (Go through the BCOC for base clusters.) They coordinate the following with BDOC:

- Call signs and frequencies.
- Base defense plans and layouts.
- Positions of critical internal assets, external coordination points, and no-fire areas.
- Indirect-fire support.
- Engineer support, if needed, to help prepare defensive positions or for ADC.

6-128. In coordination with the main CP fire support cell, the operations cell sets procedures by which MP can call for fire support.

## Execute

6-129. Base response force operations rely heavily on MP tactical skills. The choice of action is based on METT-TC and the base commander's tactical assessment. Once a situation develops that requires a response force, there are several options available to disrupt or eliminate the threat. Response force options include the following:

- Call for fire support to make the enemy break contact.
- Attack into the enemy's flank using the base for fire support.
- Use ambushes along likely escape avenues if it is obvious that the attacking force is not strong enough to overrun the base.
- Augment the base's defense forces and provide support from within a base.

6-130. Once the MP company commander receives the mission to direct response force operations, the company's main efforts shift to supporting the response force. The commander conducts a mission analysis and begins the TLPs, culminating in issuance of an OPORD and a FRAGO. Refer to *Chapter 2* for more information about TLPs.

6-131. The company HQ coordinates for support that includes communications, air evacuation, and logistical support (Class I, III, V, and IX). The company operations section coordinates with follow-on forces and other support that includes the following:

- Fire support to include CAS.
- The location of adjacent units.
- The location of threatened friendly forces.
- The location of noncombatants.
- Coordination with the TCF.

6-132. Once the platoon leader receives the mission to conduct response force operations he initiates TLPs. Normally, the PSG assembles the response force at a prearranged AA near the base under attack.

6-133. The base commander gives the response force an objective and a tactical AO. He may designate a small force to assist the response force. This force would be under the OPCON of the response force leader.

6-134. Normally, the response force acts as a maneuver element against the enemy. Once the platoon leader develops a COA he designates the ORP and selects a scheme of maneuver. The response force moves to the ORP, establishes security, and reports its status to higher HQ. Once the response force has established communications with the friendly forces under attack, the platoon leader reconnoiters the area and assesses the situation. Based on the situation and guidance from higher HQ the platoon leader may decide to—

- Request fire support to destroy the enemy.
- Continue to observe for weaknesses in the enemy's flanks.
- Engage and further develop the situation.
- Attack and destroy the enemy.

6-135. If the decision is made to call for fire to destroy or make the enemy break off the attack, the platoon leader follows the procedures set by the main CP fire support cell. When available, MP can call for and adjust indirect fire provided by artillery. Another option is CAS, if there is a tactical air control party or army aviator trained in joint air attack team operations. Refer to *Chapter 3* for more information about call for fire and fire support. Also refer to *FM 6-20*.

6-136. If the response force is ordered to attack and destroy the enemy, the platoon leader immediately prepares the response force for a hasty attack.

6-137. **Plan a Hasty Attack**. A hasty attack is an immediate action taken without extensive preparations, with the resources at hand and, if by so doing, the attacking element can maintain momentum or take advantage of the enemy situation.

6-138. The platoon leader uses fire support to complement the scheme of maneuver. The goal is to kill as many enemies as possible and suppress the rest to keep them from seeing or shooting the maneuver element as it closes on the objective. The attack plan includes the use of all available direct and indirect fire. The response force leader mainly plans and directs the fire of his organic weapons (MG, M203, AT-4, and MK19). The fire element's makeup depends on how much direct-fire support is needed and what support is to be provided. The initial employment of the fire element includes the following:

- Positioning crew-served and M249 weapons on or near the line of departure (LD), when the objective and route to it can be seen and covered by fire.
- Positioning some crew-served and M249 weapons on or near the LD and some with the moving element, when the objective can be seen and covered by fire from the LD, but only a part of the route is visible, or when the entire route is visible from the LD, but the objective is not visible or is out of range from the LD.

6-139. Plan control measures at the objective and initial employment of the fire element. If time is available before moving to the assault position, make a second reconnaissance to see if the plan needs changing.

6-140. **Conduct the Attack.** The platoon leader moves to where he can lead the force and influence the fight. He moves a fire element into the over watch position and has the attacking element move from the LD to the assault position, using covered and concealed positions. If en route to the objective, the patrol should—

- Receive indirect fire. Move quickly out of the impact area to a preset point.
- Meet obstacles, breach, or bypass them. Report their location if the obstacles might affect follow-on units. Take special care when crossing a danger area like a field, roadway, or creek. Refer to *Chapter 3* for more information about moving in contact.
- Meet direct enemy resistance before reaching the objective and return fire. Call for and adjust fire on the enemy.

6-141. The platoon leader coordinates action so that the response force hits the enemy with its full combat power. He maneuvers the force to strike the enemy's flank or rear. The response force engages the hostile element with the appropriate level of force. He uses overpowering force on armed combat forces. While in contact—

- Immediately assume the best available covered positions, when moving, and at the same time, return fire in the direction of the enemy.
- Locate actual or suspected enemy positions and engage them with well-aimed fire.
- Distribute fires evenly over the objective when no individual positions have been identified.
- Report and monitor the situation and recommend COAs.
- Make visual or verbal contact with soldiers on the left or the right. (Platoon members and team leaders make frequent visual contact with the squad leaders. Squad leaders make frequent visual contact with the platoon leader. Relay all arm and hand signals from the platoon chain of command.)

6-142. When resistance is destroyed, the attacking force continues quickly onward to the assault position. They deploy on line just before reaching the assault position. To maintain the momentum, the attacking force passes through the assault position and attack the objective. They halt at the assault position only to allow all the soldiers to come on line to attack at once. The assault position should be as close to the objective as possible to prevent needless exposure to enemy fire.

6-143. As the moving element attacks, the fire element in its overwatch position will cover the attacking element by shooting at the enemy. Before the attacking element moves beyond support range, the fire element moves up to a position from which it can continue its support and shoot at the enemy.

6-144. With the fire element in overwatch, the platoon leader (commander) gives the command or signal to open fire. The fire leader will control the method and the rate of fire. He and the gunners watch the progress of the attacking element and engage targets that threaten it. When crew-served weapons are close together, the fire leader anticipates the masking of their fire and moves the weapons one at a time. When those weapons are separated, each gunner displaces when its fire is masked or when it can no longer support the attacking element. The platoon leader (commander) times the displacement of weapons in a fire element so that the attacking element has continuous fire support.

6-145. Maneuvering begins when the attacking element makes contact with the enemy. As the attacking element maneuvers near the objective, the fire element in overwatch suppresses the enemy with a high rate of fire.

6-146. When the attacking element reaches the objective and closes with the enemy, the fire element walks its fire across the objective in front of the maneuvering element. Then it shifts its fire to supplementary targets (rear of the objective, escape routes) or lifts its fire to keep from endangering the attack element. Soldiers in the element closing with the enemy move singly, by pairs, by teams, by squads, or by a combination of these, using as much cover as possible. As they close, they fire on selected targets to suppress the enemy. Automatic rifles are fired in short bursts across the forces' front. Rifles, M203s, hand grenades, bayonets, and MG are used to overcome pockets of resistance. The advance continues past the objective far enough to shoot at any withdrawing enemy. Firing continues until the enemy is killed or captured or withdraws.

6-147. Once the threat is eliminated, the platoon leader reconsolidates the response force and assembles it at a secure rally point. Refer to *Chapter 3* for more information about consolidating and reorganizing.

6-148. **Conduct an Attack on Urban Terrain.** In military operations in built-up areas, a response force must be able to find, isolate, and evict the enemy concealed in hardened structures. Normally, responding to a threat in a built-up area will call for a larger response force and should be supported by armored fighting vehicles. When attacking and clearing a building in an urban area, try to—

- Determine and isolate the area around the objective. In urban terrain it is often hard to pinpoint the enemy's position.
- Number the area's buildings and use those numbers as reference points to control movement and for clearing operations.
- Clear buildings from the top down if possible.
- Retain a reserve force to have the flexibility essential for urban combat.

6-149. Refer to *FM 90-10-1* for information about attacking on urban terrain.

## Respond to Downed Aircraft

6-150. MP are often required to respond to downed aircraft. This specialized mission requires close coordination with aviation assets. In the event of a downed aircraft, the primary mission is to safeguard the aircraft, render first aid to the crew, evacuate the wounded, and secure the crash site. Normally, this requires MP to be airlifted to the crash site.

6-151. MP leaders plan for downed-aircraft rescue operations as a contingency mission. The primary concern is the safety of personnel onboard the aircraft. The MP leader establishes preplanned lift packages for a mounted or dismounted platoon-size downed-aircraft response force. When organized for downed-aircraft operations, the force must meet the following requirements:

- Possess cross-country capability.
- Be prepared to defend the site against Levels I and II threats.
- Have internal medical support to treat and evacuate injured personnel.

6-152. Engineer support may be necessary to clear lanes to the downed aircraft and medical evacuation (MEDEVAC) LZs.

6-153. Once the response force arrives at the crash site, the platoon leader immediately establishes security and notifies higher HQ of the exact location, the type of aircraft, and the number of wounded or killed personnel. Medical personnel treat the injured and prepare them for evacuation. The platoon leader selects and secures a MEDEVAC LZ. Once all the injured personnel have been evacuated, the response force continues to secure the crash site until relieved.

6-154. Security at the crash site involves MP preventing the handling or disturbance of the wreckage, theft of any parts, or compromising evidence such as gouges or marks on the ground made by the aircraft. Other security measures include—

- Preventing all unauthorized access to the site. This includes media and local authorities.
- Protecting all civil and military property at the site.
- Prohibiting removal of deceased persons until properly identified by military authorities.
- Prohibiting the removal of wreckage.
- Keeping spectators and the media at a reasonable distance from the site for safety.

6-155. All requests for information will be referred to the Public Affairs Office (PAO). MP at the scene will not release information about the crash to the media.

## ENEMY DELAY

6-156. A delay is an operation in which a force under pressure trades space for time by slowing down the enemy's momentum and inflicting maximum damage on the enemy without, in principle, becoming decisively engaged. Normally a delay operation is unlikely to be conducted in a sustainment area. However, bypassed enemy forces could create a situation where a MP response force may be required to delay a threat to buy time for a base to establish defenses in greater depth or to delay an enemy until the arrival of the TCF. MP deliberately conduct a delay only when directed to do so by higher HQ. In that case, higher HQ determines the size of the MP force. In the delay, the destruction of the enemy force is secondary to slowing his advance to gain time.

## PLAN

6-157. The platoon delays either from successive positions or from alternate positions. When delaying from successive positions, the delaying force will fight rearward from one position to another, holding each position for a given time or until it becomes decisively engaged.

6-158. When delaying from alternate positions, the delaying force will fight rearward with the force deployed so elements can provide overwatch and subsequent maneuver. While the forward element is fighting, the rear element prepares to assume the action. The forward element disengages. It passes through or around the rear element to prepare to assume the action from a position in greater depth. Use the following methods to delay the enemy *(Figure 6-9):*

- Delay forward of a specified line or terrain feature if you have—
  - A narrow sector.
  - Cross-compartment terrain.
  - An area restrictive to armor and the enemy can be canalized into a selected area.
  - Terrain dominating mounted avenues of approach.
- Delay from successive positions when you—
  - Have wide sectors or limited observation, fields of fire, and visibility.
  - Have a mobility advantage simultaneously moving one or two elements at a time.
  - Are in dangerous sectors.
  - Lack good defensible terrain in depth.
- Delay in a sector if you have—
  - An area of responsibility that cannot be adequately covered from one battle position.
  - Multiple enemy avenues of approach.

  - Limited fields of fire and observation.
- Delay from alternate positions when —
  - The sector is narrow or there is long-range observation and fields of fire.
  - The enemy is superior in combat power.
  - There is good defensible terrain in depth, and the terrain facilitates movement by bounds.

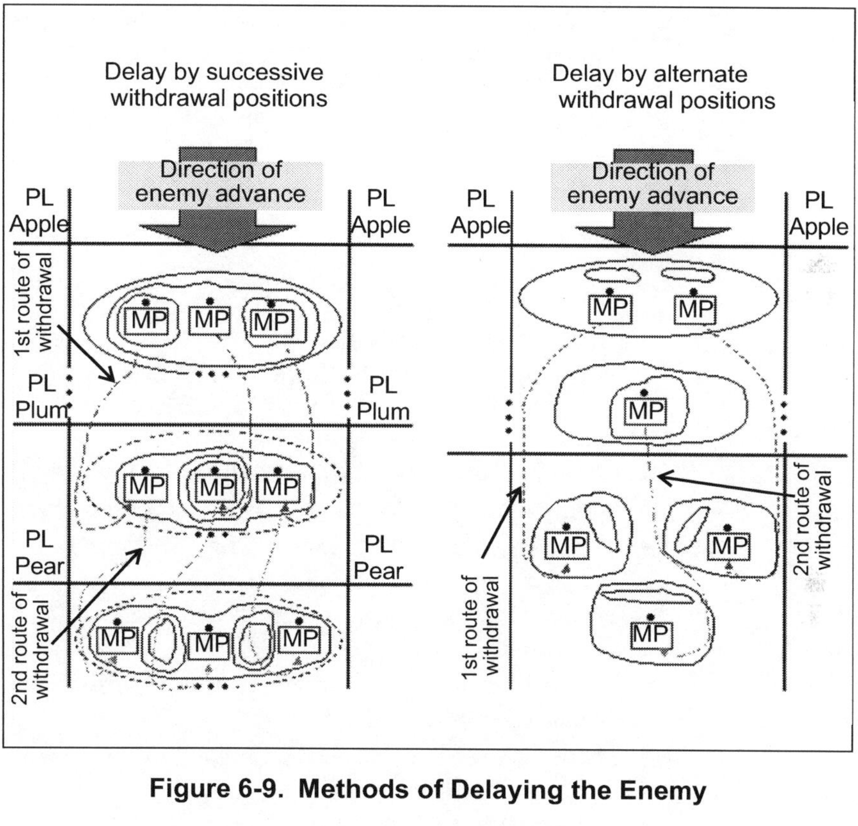


**Figure 6-9. Methods of Delaying the Enemy**

6-159. The platoon leader develops a plan of action using the factors of METT-TC and the commander's concept and intent. He gives priority to—

- Obtaining a detailed knowledge of the terrain.
- Gearing the terrain analysis to the enemy's intent, avenues of approach, and likely positions and targets.
- Gaining information from aggressive area patrol and reconnaissance.
- Having alternate communications, both sound and sight, ready to be used in case radio communications are disrupted.
- Forecasting time. The response force must consider how fast the enemy will close, how long will it take to move off a position, and how long to get to and occupy the next position. In delaying actions, timing can be critical.
- Requesting fire support along the route.

6-160. In the OPORD to subordinate leaders, includes the—

- Concept of conducting the action.
- Initial delay positions.
- Trigger points identifiable on the ground. ("Squad leader, I plan to initiate fires when the enemy reaches the railroad track.")
- PLs and delay lines. Include the length of time to delay on or forward of the given position or PL.
- Coordination and contact points.
- Indirect-fire plan (smoke, high explosives [HEs] munitions, artillery-delivered scatterable mines).
- Responsibility for adjacent-unit coordination (passage and the like).
- Fire control measures, engagement areas, targets, and rate of fire.
- Disengagement criteria. ("Squad leaders, I plan to start disengagement when the enemy has three armor vehicles across the railroad tracks.")

- Plan for moving after disengagement (signals, breakpoint, use of smoke, and the like).
- Subsequent delay positions and routes and the sequence of disengagement.

6-161. Most of the firepower will be oriented toward the enemy. However, flank and rear security must be provided. The platoon leader plans fire support in front of and on both flanks of the positions. He plans fire to cover the disengagement and movement to subsequent positions. He has fire concentrate on breaking up the enemy's advance at long range.

6-162. Higher HQ will resupply, rearm, and refuel the delaying element. On a platoon action, the platoon sergeant is crucial in ensuring that the platoon has the required support to continue a delaying action. The OPORD provides for a central point where the teams or squads can recover.

## EXECUTE

6-163. If directed to delay the enemy, the platoon leader moves to a position where he can best control the action. The platoon leader makes good use of the terrain to mask all movements and provide vantage points for observation and harassing fire.

6-164. The enemy will try to bypass a delaying force. The platoon leader tries to keep the enemy in front of or on an oblique angle to the delaying force's position. He prepares positions as thoroughly as time permits, and reconnoiter routes to subsequent positions.

6-165. When selecting delay positions, look for the following:

- Clear observation and fields of fire onto the avenues of approach.
- Positions offering mutual support and interlocking and flanking fires where possible.
- Restrictions to canalize the enemy.

- Short, covered, concealed, and trafficable routes to rearward positions.
- Positions hidden from enemy observation and fires.
- Defilade positions for HMMWV and ASV.

6-166. If the delaying force has time they—

- Emplace obstacles to slow the enemy while they move to rearward positions.
- Block obvious routes rearward and direct the force to use more hidden routes.
- Continue preparing positions.
- Consider preparing ambush sites along the routes to slow enemy pursuit.
- Select primary and alternate firing positions.
- Have soldiers familiarize themselves with the routes to these positions.
- Place the OPs/LPs, and other security means far enough forward to warn the platoon, particularly if the enemy is using a secondary avenue of approach.

## Control Fire

6-167. Fire control is extremely important. The platoon engages the enemy at maximum range with all weapons systems. They fire on a prearranged signal or event (trigger point). They aim for leaders and $C^2$ vehicles and avoid premature firing. The platoon leader considers not firing until the enemy reaches a preset point on the ground. Then he uses ambush-type fires to quickly destroy lead enemy elements. This also keeps the enemy from learning the friendly force's size and location.

6-168. If the enemy force cannot evade, it will make every effort to envelop and destroy the delaying force. Avoid presenting an exposed flank to the enemy. Use contact points to help ensure flank coordination.

6-169. Each time the delaying force moves back, it will have less time to plan and prepare. Speed, firepower, and maneuver become more important. The platoon leader calls for added fire support or assistance from higher HQ if decisive engagement or envelopment seems likely.

6-170. If additional elements arrive, the platoon leader uses them to augment the element in contact, employing them on the left and right of the element in contact. The platoon leader ensures that as the element in contact moves back, the left and right elements know to also move, but not at the same time. As the element in contact moves back, the flank screens can open harassing fire. Drawing enemy attention to flank elements will allow the element in contact to disengage more easily.

6-171. The platoon leader keeps higher HQ informed of the situation. Forward elements must not become so heavily engaged that they cannot be withdrawn effectively. If contact with higher HQ is lost, the platoon leader uses initiative. He bases each move on METT-TC, the events on the battlefield, and the tactical commander's intent.

## Disengage

6-172. Deciding when to disengage is difficult. If the delaying forces remain too long, they risk decisive engagement or envelopment. If they move too soon, they may give up ground unnecessarily and risk pursuit. The decision of which element to move is based on METT-TC. Move the least heavily engaged element first. To disengage—

- Direct supporting elements to engage enemy forces forward of the disengaging element.
- Concentrate all available fire and mutual support.

- Give priority fire, if available, to the disengaging element to rapidly increase its firepower.
- Begin an orderly movement to successive or alternate positions.
- Use all available fire.
- Place the fire far enough forward to avoid impeding personnel leaving the protected positions.
- Place smoke between the delaying force and the enemy.
- Move into subsequent positions from the rear and send a SITREP to higher HQ.
- Report the delaying force's arrival to higher HQ.
- Tell the elements left in contact to disengage and take their next positions.

6-173. If it becomes necessary, replace key leaders, give crew-served weapons a priority of manning, evacuate casualties, and redistribute ammunition. Refer to *FM 17-95* for more information on defensive tactics used for delays.

## BATTLE HANDOVER TO A TACTICAL COMBAT FORCE

6-174. When the combined efforts of the base and base cluster defense force and the MP response force are unable to jointly defeat an enemy threat, the operation becomes the responsibility of the TCF. Coordination is essential to determine when, where, and how the response force gives up responsibility for the fight and the TCF takes over. The MP response force and the base-defense force will then hand over the fight to the TCF by conducting a battle handover.

6-175. A battle handover *(Figure 6-10)* is a coordinated operation between two units that transfers responsibility for fighting an enemy force from one unit

**ORGANIZATION**

Maintain unit integrity for C2.

**ORDER OF MOVEMENT**

Base the order of movement on the—

- Number of passage points.
- Degree of security required.
- Enemy situation.
- Terrain.
- Prevention of confusion and congestion. Give priorities indicating which units move and when they move.

**COMMUNICATIONS**

Ensure that the following actions occur:

- Both units have SOI compatibility.
- Both units monitor the rear operations net (critical for recognition signals and fire support).
- Both outgoing and incoming unit elements operate on the outgoing unit's command net.
- The outgoing unit maintains routine radio traffic volume while the incoming monitors it.
- When the responsibility of the battle has changed over, the incoming unit operates on its own net only.

**COMMAND AND CONTROL**

Ensure that the—

- Number of passage points can determine the method of C2 used.
- Multiple passage points are set up, dictating decentralized control.
- TCF commander decides how he can best influence the action and positions himself accordingly.
- Response force commander collocates with the command group of the TCF during the passage of lines.
- Time or circumstances when responsibility for the battle is transferred are mutually agreed on before hand by the response force and the TCF commanders.
- Responsibility can be passed by radio communications if need be.

**Figure 6-10. Planning Considerations for a Battle Handover**

**COMMAND AND CONTROL POINTS**

Ensure that the following occurs:

- All units follow the control measures.
- For a response to force passing through a stationary TCF, control measures include the following:
  - AAs. Used by a unit to prepare for an upcoming mission.
  - BHLs. The single, most important control measure in delineating response force and TCF responsibilities. A PL is placed where the TCF assumes responsibility for the battle from the response force.
  - Attack position. The last position occupied or passed through by the TCF before crossing the LD.
  - Passage lanes. Areas on which the response force moves to avoid the stationary TCF.
  - Passage point. A place where the response force will pass through TCF units. It is located where the commander desires subordinate units to physically execute the passage of lines.
  - Time of passage. The time at which the passage is executed by the leading element of the response force to facilitate control through the passage point.
  - Recognition signals. Established signals to prevent the response force from directing fire on the TCF and vice versa.
  - Contact points. Designated, easily identifiable points on the terrain where two friendly forces are required to meet.
  - RPs. Points on the ground where subordinate units are under the control of their respective commanders while en route to their appropriate destinations.
  - Routes. Routes are the prescribed course to be traveled from a specified point of origin to a specific destination.
- For a TCF passing through a delaying response force, control measures include the following:
  - Battle position. A defensive location established by the response force oriented on the most likely enemy avenues of approach.
  - LD and LC. This control measures places the BHL in that control of the battle is assumed by the TCF once its units cross the PL.

**Figure 6-10. Planning Considerations for a Battle Handover (Continued)**

to another in close-in battle. It is designed to sustain continuity of the combined-arms fight and protect the combat potential of both forces involved. Battle handover is usually associated with conducting a passage of lines. MP leaders must establish clear tactical SOPs that allow their forces to quickly establish the necessary coordination to preclude a loss of momentum in the attack. In a battle handover, the MP response force will pass rearward through the TCF lines while handing over the fight or the TCF will move forward through the MP response force lines to engage the enemy.

6-176. However the fight is handed off, coordination between the MP response force commander and TCF commander is critical. Use a BHL (refer to *FM 17-95* and *FM 17-98*) and contact points as control measures to set both forces' responsibilities. Both the TCF commander and the response force leader coordinate the BHL location. The BHL is shown on the operations overlay and in the OPORD as a PL or a boundary. It identifies ground as "owned" by the TCF or by the response force. It sets the location where control of the battle passes from one commander to the other. It is where the TCF can use direct fire and observe indirect fire.

6-177. The handover occurs at the time or event coordinated by the TCF commander. The TCF commander decides the method of handover based on METT-TC. Generally, one of the following methods will be used:

- The response force passes rearward through a stationary TCF.
- The TCF passes forward through a delaying response force.
- The response force sets up a final blocking position, with elements of the TCF conducting an attack to destroy the enemy.

6-178. Coordination for battle handover normally flows from the TCF commander to the response force leader. Coordination normally includes—

- Establishing communications.
- Providing updates on both friendly and enemy situations.
- Coordinating passage.
- Collocating $C^2$.
- Dispatching representatives to contact points.
- Ensuring that there are visual recognition signals for vehicles.
- Ensuring the status of obstacles and routes.
- Coordination of fire support and CSS requirements.

6-179. The most important task for the response force is to maintain contact and continue to engage the enemy. The enemy must continue to see the level of activity that has been established. If the enemy perceives that the handover is occurring, he will attempt to seize the opportunity to destroy the response force.

## PASS A RESPONSE FORCE THROUGH A STATIONARY TACTICAL COMBAT FORCE

6-180. The TCF commander coordinates with the response force leader at a designated contact point. The rear CP operations cell dictates the contact point location and notifies both units.

6-181. The contact point should be located forward (50 to 200 meters) of the BHL. It should be near or on easily identifiable terrain features. At a prescribed time, liaison parties from both forces will meet. Normally, a response force representative is sent to the contact point. At the contact point, the TCF commander or his representative briefs the response force representative. Response force teams may remain in the vicinity of the BHL if passage is imminent. TCF security forces screen along the BHL and monitor the response force command net.

6-182. The actions of the elements on the BHL are critical. At the BHL, the TCF scouts, or others assigned the BHL action, pick up the fight from the passing response force elements. If the response force is in contact, its maneuver elements are bound behind the BHL, preferably to a covered and concealed location. At the same time, the TCF elements on the line relieve the pressure on the response force elements.

6-183. At the passage point, where the response force passes through TCF units, response force teams identify passing response force units. They ensure that the passing vehicles are displaying the correct visual recognition signal. TCF scouts on the BHL notify the forward TCF teams by a prearranged signal that contact has been made and friendly forces are at the BHL. Response force elements continue to delay forward of the BHL. Once behind the BHL and covered from enemy direct fire, the response force should—

- Confirm that the proper visual recognition signal is still displayed on all vehicles.
- Orient weapons systems toward the enemy.
- Move quickly, in the directed formation, through the passage points and along designated routes to the RP.

6-184. Stationary TCF elements overwatch this movement. The only time the TCF fires is when positive enemy identification is made. At the completion of the handover, response forces may either be released by the operations cell to resume other priority missions or tasked to remain in OPCON to the TCF and continue the fight.

## PASS A TACTICAL COMBAT FORCE THROUGH A DELAYING FORCE

6-185. This technique is similar to the response force passing through the stationary TCF. The response force will have more information on the terrain and enemy situation than the TCF.

6-186. The response force leader sets up contact points, passage points, and routes. At the least, he provides guides. The guides meet the TCF at the contact points and lead them along routes to the passage points and/or RPs near the LD and LC.

6-187. The response force maintains normal radio traffic. The TCF and response force operations elements temporarily collocate. The TCF commander assumes control of the battle handover at this time. TCF company teams may maintain listening silence on their TCF command net. The response force guides notify their commander when the TCF begins moving forward from the contact points. As the TCF company teams deploy across the BHL, the response force leader and the TCF commander transfer responsibility for the battle.

6-188. The TCF fire support officer (FSO) may collocate with the response force forward observer (or the individual appointed to act as forward observer for the artillery unit in the AO). Fire missions are approved by the RAOC supporting the response force until the battle handover has occurred. After that, any fire missions for the response force are cleared through the TCF FSO. Response force units provide direct-fire overwatch according to the TCF commander's scheme of maneuver. It is coordinated directly by the TCF commander and the response force leader at collocated command groups.

6-189. Response force direct-fire support depends on the TCF commander's attack plan and the battle handover.

However, the stationary unit lifts and shifts its direct fire as coordinated by the collocated command groups.

## BLOCK FOR A TACTICAL COMBAT FORCE

6-190. When defending from a battle position, the response force positions its elements to achieve the TCF commander's intent. The response force sets up blocking positions to keep the enemy from escaping. It selects battle positions where it can control, maneuver, and concentrate direct fire on the enemy. As blocking positions are set up, the TCF maneuvers to attack the enemy's flank. The TCF commander and the response force commander set up a contact point for coordination and conduct a reconnaissance of this point, time permitting. The TCF commander and response force leader coordinate $C^2$ and exchange the following information:

- When to lift or shift fire.
- What targets must be destroyed to allow the TCF to close on the enemy.
- The time, location, or activity to pass control of the battle.
- The location of the engagement areas.

6-191. The TCF and blocking force coordinate so that the TCF flanks the enemy. The response force suppresses the enemy from the front allowing time for the TCF to maneuver into position.

6-192. The TCF commander specifies tasks for the response force to achieve the TCF scheme of maneuver. The response force normally will have to orient its weapon systems on the enemy avenue of approach using TRPs or engagement areas (EAs). The response force may be tasked to—

- Destroy a certain enemy force from the battle position.

- Control the terrain or block an avenue of approach by holding the battle position against the enemy assault.
- Reorient weapons systems on a secondary avenue of approach from supplemental positions or avenues of escape for the threat.
- Assist in any other task needed to achieve the TCF mission.

6-193. Refer to *FM 17-95* for more information about battle handover.

# CRITICAL SITE, ASSET, AND HIGH-RISK PERSONNEL SECURITY

6-194. MP are often tasked to perform security operations that require integration and prioritization from other MP CS operations. During security operations, MP focus their efforts on measures designed to safeguard resources by conducting special security activities that prevent unauthorized access to installations, equipment, materiel, and documents; and safeguard against espionage, sabotage, damage, and theft. An MP security operation includes protection of CPs, convoys, ASPs, deep-water ports, rail, pipelines, and high-risk personnel.

## COMMAND POST SECURITY

6-195. CPs are priority targets for enemy attack, and their security can be crucial to the success of military operations. CPs at division, corps, and EAC are *designated critical sites and assets*. When CPs relocate, MP provide in-transit security. The amount and type of security provided to an established CP depends on whether the CP is dispersed or massed, large or small. In all cases, the focus of MP security efforts is on providing early warning of threat activity. MP possess mobility and communications assets that make them most valuable in operating AS patrols around the CP. Elements do not share a common defensive perimeter in a dispersed CP. Each cell provides its own local security and access control. MP conduct screening operations to provide early warning of enemy activity. In a massed CP where elements share a common perimeter, MP may contribute to perimeter defense with other elements that are collocated at a base.

### Division Main CP Security

6-196. All Army of Excellence (AOE) divisions have a main, rear, and tactical CP. Division XXI currently has a division main and tactical CP and a security operations center (SOC). The best use of MP assets to secure any division CP is to employ MP teams to conduct AS around the location of the CP. MP provide early warning of enemy activity by conducting area reconnaissance and surveillance (R&S) of likely enemy avenues of approach to the CP. This security measure also provides the CP with standoff protection from direct enemy fire.

### Division Band

6-197. When assigned to augment security operations, the division band should be employed intact, facilitating the commander's ability to quickly return it to its primary mission. The MP commander ensures that the bandmaster and his soldiers know the tactical situation and the defensive plan for the CP when they are assisting in CP security. The band may—

- Assist in perimeter defense of the CP.
- Provide access control on the road to the CP.
- Operate the dismount point for the CP.
- Provide access control to the main CP.

### Corps Command Post Security

6-198. Normally, the corps CPs are echeloned into a tactical CP, a main CP, and a rear CP. The corps main

CP is generally located in the general vicinity of the division rear boundary. Traditionally, the corps main CP is massed and is a large organization with attendant mobility and signature problems. The corps tactical CP is small, mobile, survivable, and normally located well forward in the vicinity of a division main CP.

6-199. MP security is provided to both the corps main and tactical CPs. How MP elements deploy depends on the CP dispersion, the particular situation, and the METT-TC. The MP company provides corps CP security. If the CP is massed, one platoon secures the tactical CP and the remainder of the company secures the main CP. Or, depending on the METT-TC,—

- One squad provides access control within the corps tactical operations center (CTOC).
- Two squads serve as a response force.
- Two squads secure the tactical CP.
- Two squads man dismount points or augment other squads.
- The remaining squads operate the OPs/LPs and the security patrols.

6-200. The main security efforts and emphasis of MP are on early warning and screening. Both internal access and screening force measures provide CP security. These measures are enhanced by—

- Internal guard posts.
- Perimeter access control points.
- LPs/OPs and a security post outside the perimeter.

6-201. When a corps main CP is dispersed in cells, focus MP efforts on screening force measures. Allocate MP based on the location of the function and the METT-TC.

## CONVOY SECURITY

6-202. MP leaders plan and conduct convoy security (specifically convoy escort) using several different methods to secure them as they move through their AO. Primarily, MP secure vehicle movement by maintaining security in the area that a convoy will pass through. MP also conduct route security on a specific route for a designated period during which multiple convoys move unimpeded.

6-203. The main threats to convoys are mines and ambushes. Convoys are vulnerable to sniper fire, long-range fire from artillery, mortar, threat aircraft, and AT-guided missiles. They are also vulnerable to remotely controlled mines and demolitions. Armored vehicles are better suited to provide convoy escort protection. Leaders carefully evaluate the threat before assigning a convoy escort mission to HMMWV-equipped units.

6-204. When MP conduct convoy escort security missions, platoons normally escort the convoy through their AO and pass responsibility for escort to adjacent platoons as the convoy moves from one platoon's AO to another. However, some situations may require a platoon to escort a convoy from its SP to its final destination.

6-205. Whatever method is used to secure the convoy, MP platoons executing the escort mission operate under the control of the convoy commander. The relationship between the MP platoon and the convoy commander provides for unity of command and effort, if combat operations are required during the course of the mission.

### Plan

6-206. The platoon leader plans for security of the convoy in all directions and throughout the length of the convoy. This requires MP teams to be dispersed throughout the convoy formation. When available, locate engineer assets toward the front to respond to obstacles. If a fire-support team (FIST) is attached, locate it near the platoon leader.

6-207. The platoon uses the column formation *(Figure 6-11)* due to its inherent speed and ease of movement.

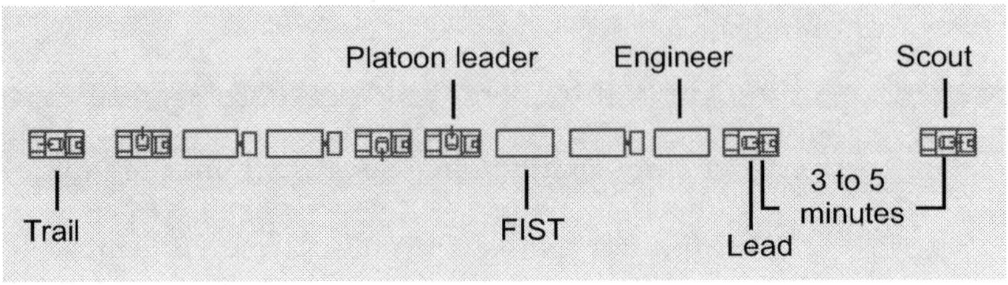


**Figure 6-11. MP Platoon Escorting a Convoy**

6-208. The platoon leader develops a convoy security plan that includes the following:

- HN support.
- The escorted unit's frequencies and call signs.
- The vehicle number and type and the order of the march.
- Checkpoint identification along the route.
- Security for halts and rest stops.
- Enhanced security at critical points such as, bridges, tunnels, key intersections, and other danger areas.
- An alternate route determination.
- Emergency actions to be taken during ambushes, a sniper attack, an air attack, or an artillery attack.
- Ambush reactions.
- Coordination with the response force responsible for the AO.

6-209. The platoon leader assigns selected teams to remain with the convoy to provide close-in security, if attacked, and other teams to maneuver toward the enemy to deliver a high volume of fire. Use a scout vehicle to travel ahead of the convoy (3 to 5 minutes) to alert the convoy commander of enemy activity, obstacles, or anything that might disrupt the convoys movement.

6-210. If a tracked armored vehicle is not available to lead the escort, the platoon leader assigns one team as the lead security element. Hardening of all vehicles used for convoy escort missions must be SOPs.

6-211. The platoon leader coordinates with the convoy commander to ensure that all convoy personnel are briefed and fully understand the convoy security plan. If time permits, the platoon leader conducts a joint rehearsal with the convoy commander and all security elements. At a minimum, the rehearsal should cover reactions to an ambush and air and artillery attacks.

**React to Ambush**

6-212. A well-planned ambush can have a devastating effect on a convoy. The reaction to an ambush must be quick, overwhelming, and decisive. The platoon executes its reaction to an ambush as a drill, taking special care to prevent fratricide. Actions to be taken include the following:

- When the ambush is initiated, security vehicles direct their action toward the enemy *(Figure 6-12, page 6-70)*. Selected teams move to covered positions between the convoy and the enemy to suppress the enemy with the highest possible volume of fire and obscure the enemy's view with smoke. At least one team should remain with the convoy and continue close-in security. The platoon leader sends the appropriate reports to higher HQ.

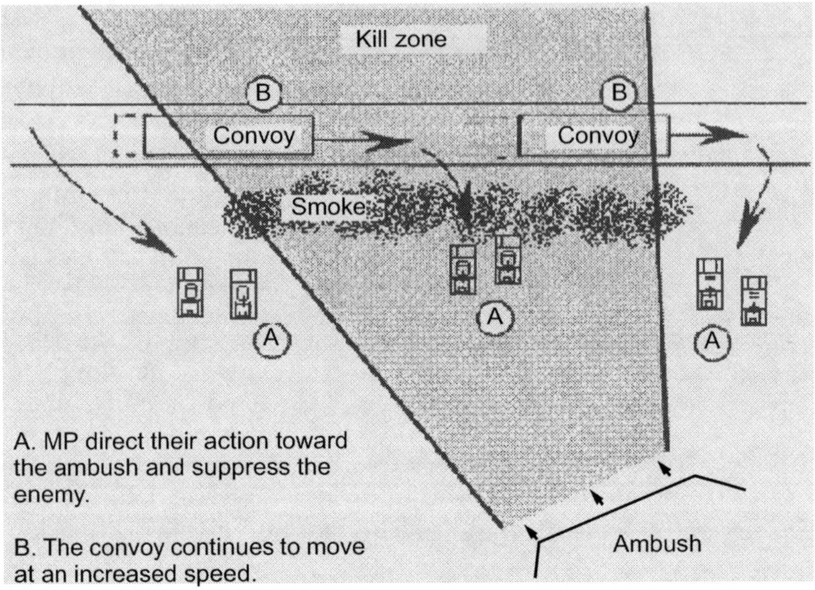


**Figure 6-12. Security Vehicles' Actions During an Ambush**

- The convoy commander continues to control the convoy and maintains radio contact with the security force (MP platoon leader). Vehicles not yet in the kill zone are stopped before they enter. Vehicles in the kill zone are moved out of the zone at the highest possible speed.
- The scout vehicle immediately selects a safe area for the convoy commander to move vehicles. The vehicles use a Light Vehicle Obscuration Smoke System (LVOSS) or other smoke to mask movement if possible. This safe area will be the linkup location when the convoy resumes.
- The convoy vehicles, if armed, return fire only until the security force positions itself between the convoy and the enemy.

- The damaged or disabled vehicles are abandoned and pushed off the route if they are not mission essential.
- The MP platoon leader submits SPOTREPs. Based on his assessment of the situation he can call for fire or CAS, or request the response force.
- Once the convoy is clear of the kill zone, the MP platoon leader chooses one of the following COAs based on the composition of the escort and the strength of the enemy:
  - Continue to suppress the enemy, while response forces move to support.
  - Assault the enemy.
  - Break contact, move out of the kill zone, and calls for indirect fire.

6-213. Normally, MP equipped with HMMWVs move out of the kill zone as soon as the convoy is clear. ASV-equipped security forces may continue to suppress the enemy or execute an assault. The platoon leader decides when to break contact.

6-214. **Mines**. Command-denoted mines are usually to signal the start of an ambush. When ambush fire comes from one side of the road, the enemy will most likely deploy a second, smaller force on the opposite side. They will deploy mines and obstacles between themselves and the convoy to contain US forces. The ambush force often uses mines to protect its flanks. To avoid damage from mines, drivers must—

- Drive on the track of the vehicle in front.
- Avoid driving on the shoulder of the road.
- Avoid running over foreign objects, brush, or grass in the road.
- Watch local traffic and the reactions of people on foot, which may give away locations of mines and booby traps.

6-215. **Artillery Attack**. If a convoy comes under artillery fire, vehicles must not stop. The convoy continues forward as quickly as possible to clear the area.

6-216. **Air Attack**. If the convoy comes under air attack, vehicles are dispersed, and soldiers attempt to find cover for the vehicles. Personnel begin firing all available weapons in an antiaircraft effort.

6-217. **Obstacles**. Obstacles are a major threat to convoys. Obstacles can be used to harass the convoy by delaying it. If the terrain is favorable, it may stop the convoy altogether. In addition, an obstacle or series of obstacles can be used to channel or stop a convoy to set up an ambush. The purpose of a route reconnaissance ahead of a convoy, is to identify obstacles and either breach or find bypasses. The convoy takes actions to reduce or bypass the obstacle. A convoy becomes more vulnerable when it is stopped and its escort force is occupied with tasks required to overcome or bypass the obstacle. For these reasons, security becomes critical, and actions at the obstacle must be accomplished quickly. Security personnel take the following actions at an obstacle:

- Direct (convoy commander) a halt when the lead security vehicle identifies an obstacle.
- Establish dismounted security and provide security throughout all activities directed by the convoy commander (such as conducting a reconnaissance and breaching the obstacle).
- Establish an obstacle overwatch.
- Ensure that convoy vehicles remain on the road.
- Move escort elements to the flanks to provide security.
- Ensure that all convoy vehicles are aware that the enemy may have buried mines in the area, especially on the road flanks.

6-218. **Halts**. When a convoy has to halt for short breaks it assumes a herringbone formation *(Figure 6-13)*. Security forces take up protective positions 360-degrees around the convoy and orient their weapons outward. They remain alert and establish dismounted local security. The convoy pulls into the protected area in the center of the perimeter. Stagger the convoy vehicles off the road, if there is no threat of enemy mines, but position them so they can easily pull back onto the road.

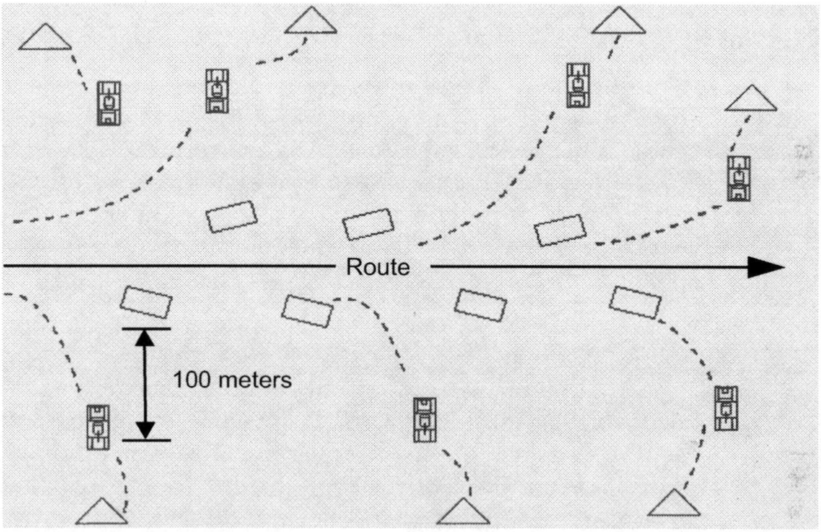


**Figure 6-13. Herringbone Formation**

6-219. When the convoy is ready to resume, the convoy vehicles reestablish the column formation, leaving space for the security vehicles. Once the convoy is in column, the security vehicles join the column, leaving local security dismounted.

6-220. Once all elements are in column, local security personnel mount and the convoy continues.

## CONVENTIONAL AMMUNITION SUPPLY POINT SECURITY

6-221. When an MP platoon has been tasked to provide security for an ASP, the platoon establishes internal and external security. The platoon secures the ASP against enemy or criminal activity so that the ASP is able to maintain normal operations. MP leaders coordinate with ASP personnel to verify communications, distress and duress signals, defense plans, and fratricide prevention.

### Internal Security

6-222. When MP conduct internal security, they establish a dismount point and allow only authorized personnel access to the ASP. Access is granted according to the local SOP and access and entry control rosters. MP conduct perimeter patrols ensuring that they do not set a pattern and maintain communications with platoon HQ.

### External Security

6-223. When tasked to provide external security for an ASP, MP establish an OP/LP for early warning. They establish a TCP at the intersection of the MSR and the ASP access road. The TCP maintains free traffic flow, screens the traffic entering the access road, and controls stragglers and refugees. To further enhance external security, MP employ movement sensors, establish defensive positions, maintain communications with the platoon HQ, and emplace obstacles and physical security measures.

## DEEP-WATER PORT SECURITY

6-224. The very nature of a terminal operation generates some different, and frequently more demanding, aspects of physical security. Ports and harbors are prime targets for enemy and criminal activities. The perimeter areas of these facilities are more vulnerable because of the extensive distance and the exposed beach or pier areas. Terminal areas may include fully developed piers and warehouses or may be an unimproved beach where logistics over the shore (LOTS) or roll-on/roll-off (RORO) operations are conducted.

## Security Measures

6-225. When MP provide security for cargo in a port, the main effort is to provide security from the perimeter of the port outward. Security measures focus on aggressive patrolling and reconnaissance designed to detect, report, or destroy enemy threats before they disrupt port operations. Measures may also include—

- Developing police information and intelligence designed to identify criminal threats operating near the port.
- Controlling traffic in the area surrounding the port.
- Conducting mounted or dismounted patrols, with MWD if available, around the port perimeter.
- Establishing an access control and identification section.
- Providing a response force to react to incidents inside the port's perimeter.
- Providing observation and early warning of threat ground and air attacks.

6-226. When providing security for cargo, MP concentrate their efforts on providing AS in the area used to move cargo from the port to its final destination.

6-227. Inside a port's perimeter, MP security forces limit access to cargoes by—

- Operating random mounted or dismounted patrols (with MWD if available).
- Using the combined patrols as a response force for incidents inside the perimeter.
- Controlling access to the most restricted areas.

6-228. The priority is safeguarding the most critical cargoes waiting to be transferred to land transport. To safeguard stored cargo MP—

- Establish access control procedures.
- Search bundles and packages being taken from the area.
- Examine trip tickets and documentation of cargo vehicles.
- Inspect seals on locked containers.

**Waterside Security**

6-229. If the restricted area is a pier or other maritime environment, access from the water must be controlled as well as from the land. Entry on the landward side of a pier can be limited with fencing, pass control, and aggressive patrolling, but the part of the pier that protrudes over the water is accessible from the sides and below. Methods for securing the pier along its water boundaries include—

- Patrols (both walking on the pier and in small boats).
- Protective lighting.
- Log booms.
- Nets.
- Buoys or floats.
- Anchored or pile-mounted navigational aids and signaling devices.

- Barges.

**DANGER**
**Ports pose special dangers for MP foot patrols. MP leaders must ensure that soldiers are thoroughly briefed on the hazards of patrolling too close to the water's edge.**

6-230. While most of the barriers described above will stop or impede access to facilities from boats or swimmers, nets are among the most effective. Well-marked, partially submerged objects are also effective; however, there may be legal prohibitions against the emplacement of barriers that may constitute a hazard to navigation. The barriers should be emplaced only after coordination with and approval by the appropriate legal and HN authorities. Sometimes it is best to just close off the waterside of a pier. A floating boom will keep small boats out. Suspending a cable or chain link net from the bottom of the boom will deny access underwater.

6-231. MP establish at least two security zones on a facility's waterside: the reaction zone and the keep-out or exclusion zone. MP in these zones notify vessels, craft, and swimmers that they are entering restricted waters and should alter their course. MP may stop and search intruders if necessary. Security zones should be extended at least 1,000 meters from the nearest protected asset; however, in some port areas, a security zone this large is not possible. In such cases, other measures, such as boat patrols, must be increased to mitigate the possibility of attack. The two zones are defined as the—

- Reaction zone. The reaction zone extends from the high water mark to a distance beyond the maximum range of anticipated waterborne threats is the reaction zone. Inside the reaction zone, MP will stop and challenge intruders.

- Keep-out zone. This is the zone closest to the protected assets. The keep-out zone extends from the cargo asset to the maximum range of anticipated threat weapons. MP prevent the entry of all unauthorized craft or vessels into this zone. The tactical response force boat, in this case, may be employed if necessary.

## Cargo Security During the Transfer

6-232. To keep cargo secure during transfer from one transport method to another, MP control the traffic moving in and out of cargo handling areas by—

- Setting up a single access control point.
- Erecting field-expedient barriers and using truck trailers or other large vehicles to constrict the traffic flow if permanent barriers are not in place.
- Limiting entry to mission-essential personnel, vehicles, and equipment, as designated by the port authority.

6-233. If gates are used by other than cargo vehicles, a *turnout* should be available. Cargo vehicles can pull into it while they are being checked. The turnout must be large enough to handle the volume and size of traffic being inspected. A wooden deck or platform at, or slightly higher than, the level of the truck bed hastens checking; it makes it easier to see the cargo. The platform should be as long as the vehicles being inspected. An empty flatbed trailer could be used.

6-234. Cargo is less likely to be diverted when close observation of cargo documentation and container safety is enforced. Containerized cargo is less likely to be stolen or sabotaged if closely watched as they are filled and sealed. Cargo can be pilfered before the seal is applied. An unsealed container can be moved to a stacking area or someone may apply a false seal, break the seal later, remove cargo, and then apply a legitimate seal. At access control points security personnel—

- Inspect inbound and outbound containers. Look for signs of damage or unserviceability.
- Inspect containers for the presence of seals and/or locks and hinges. Check their serviceability.
- Verify that the document's transport number, container number, and seal number match those numbers on the transportation control and movement document (TCMD).

**NOTE: Check seals by handling them, not simply by visually checking them.**

- Allow only containers with valid documents to pass inbound or outbound through the control point.

## TRAIN SECURITY FOR CARGO MOVED BY RAIL

6-235. Military resources moving by rail are often vulnerable to theft, pilferage, and sabotage. Depending on the threat, the cargo's sensitivity, vulnerability, and MP availability, the echelon commander may require MP to protect cargo moving by rail.

6-236. Most train operating crews consist of four or five people–the engineer, a conductor, a fireman, a senior brakeman, and a brakeman or flagman. This crew controls the train. The conductor is the train commander unless a transportation railway service officer is assigned to the train. The train commander is responsible for the train's operation and security. He makes all decisions affecting the train. MP integrate their security efforts with the security efforts of the train's crew. MP watch for and report any discrepancies or interruption to normal procedures at any time during the movement. Information about the movement is usually sent along the movement route by the chief dispatcher through a telephone circuit.

## Safety Considerations

6-237. Many regions use electric trains. These trains use aboveground, high-voltage lines and travel at greater speeds. When these high-speed trains pass parallel to a stopped cargo train and MP security forces are dismounted, the train can create a vacuum strong enough to pull a security force member under the wheels of the passing train. Some of the slower-moving trains are referred to as floaters. They travel at much slower speeds than the express trains, cause less vibration, and are much quieter. These trains may not be heard until they are too close to react. All trains pose a risk to MP working in and around them.

6-238. Track switches are another danger to MP while they are dismounted. Track switches are used to alter the course of a moving train or car onto another rail line. Track switches are remotely controlled and operated. A member of the rail crew activates some switches manually at some sites. The inherent danger to dismounted MP is getting a foot caught or crushed as the track is being switched.

## Security Force

6-239. Normally, an MP squad is an adequate force to secure railway shipments of sensitive freight, but additional MP may be needed for movement of critical cargo. When required, MP conduct railway security to prevent pilferage, detect enemy activity directed against the train and, within their capabilities, defend the train against enemy attack. In addition to MP, the shipper may send specially trained personnel with highly sensitive cargo. The number of MP in a train security force depends on the—

- Sensitivity of the freight.
- Priority of need for the freight.
- Terrain over which the train will pass.
- Length of the train.

- Duration of the trip.
- Degree of enemy threat.

6-240. When planning rail cargo security, MP—

- Review the unit SOP to determine guard force security clearance and weapons qualifications requirements and the use of force and rail security operations training.
- Obtain the time schedule for the rail movement.
- Make a map reconnaissance of the route. If time permits, they conduct an aerial reconnaissance. Bridges and tunnels are especially vulnerable.
- Coordinate with the HN and other agencies, such as the military traffic management command (MTMC) and movement control team (MCT).
- Plan actions at scheduled stops or relief points, deploying MP according to these plans.
- Plot the locations of MP units and other friendly forces along the route and note their radio frequencies and call signs.
- Obtain an intelligence report covering the route to indicate sites where sabotage may occur, or attacks may be expected, and if thefts and pilferage are likely.
- Determine weapons and special-equipment requirements.
- Assess border crossing requirements.
- Develop a security cargo checklist.
- Make a guard force serial-numbered weapons list.

6-241. MP prepare and maintain a record, by car number, of guarded cars in the train. MP may be required to ride in—

- A specific car that requires protection.
- The caboose.

- A special train car provided for the MP to travel in.

**NOTE: If only one security car is used, it should be near the center of the train; if more than one is used, space the cars to provide the best protection for the train.**

## Security Measures

6-242. The shipper is responsible for the security of all carload freight until it is turned over to the transportation railway service (TRS).

6-243. Railway cars are sealed after loading. A seal shows that a car has been inventoried and inspected. The standard method of sealing a railway boxcar door (in addition to padlocks or wires) is by a soft metal strap or cable seal that contains a serial number.

6-244. Maintaining rigid accountability of all the seals is necessary to prevent the undetected replacement of an original seal with another. While sealing does not prevent pilferage, a broken seal is a good indicator that the car and its contents have been tampered with.

6-245. When military vehicles are shipped by railcar, sensitive and high-value items must not be secured in the vehicles. Container express (CONEX) and military-owned demountable containers (MILVANs) are ideal for shipping these and other small items on flatcars since they greatly reduce the chance of pilferage. These containers must be locked and sealed and, if possible, placed door-to-door for additional security.

6-246. Cars containing freight that is likely to be pilfered, high-priority cargo, or special shipments are grouped in the train to permit the most economical use of MP security forces. When flatcars or gondolas are used to transport sensitive or easily pilfered freight, the MP leader positions a security element where it can continuously observe and protect these cars.

6-247. The security force commander conducts a joint inventory and inspection of the seals and locks with the MCT or transportation officer on *DD Form 1911*. The security force commander maintains possession of this document and uses it when transferring custody.

6-248. The MP leader ensures that each seal is checked at every scheduled stop (time permitting) and—

- Reports broken seals immediately to help pinpoint the time and place of possible thefts.
- Dismounts and conducts 360-degree security when the train is stopped.
- Checks for damage to the cars.
- Notes and reports irregularities in procedures, the presence and/or actions of unauthorized personnel, and deficiencies and incidents that occur.
- Conducts a joint inspection of the cars when relieved by another security force while en route. The relief force signs for the train (*DD Form 1911*). The record is kept on the guarded cars.

6-249. Security personnel obtain a receipt for the secured cars when they arrive at their destination. The receipt is attached to the trip report that includes the following:

- The date and time of the SP and the RP.
- Any additional information required by local SOP or command directive.
- Any recommendations for correcting deficiencies or for improving future security on the trains.

6-250. Dock workers unload the cars as soon as possible when arriving at unloading points to reduce the opportunity for loss through pilferage and sabotage. MP assets are normally not available for security of freight in railway yards; however, MP must be prepared to conduct stabling and staging operations.

6-251. Stabling operations are conducted when numerous train shipments are held in a rail yard or other locations, such as a marshalling area, for extended periods pending arrival at the final destination. Stabling operations require continuous security that may include establishing perimeter patrols, access control, and a quick-reaction force to respond to internal and external threats.

6-252. Staging operations are not extended like stabling operations. They occur when parts of a shipment enter a secured or unsecured off-load site. The site may be large enough to accommodate the entire shipment or only several rail cars at a time. Security forces may be split and required to maintain security at two locations. Additional security forces may be required based on the METT-TC. Refer to *FM 55-20* for more information about security of rail cargo.

## PIPELINE SECURITY

6-253. Pipeline systems are widely used in a theater of operations to transport bulk petroleum products or other liquids. Such systems are open to a number of security threats from the point of entry to the point of final delivery. Pipeline systems are composed of storage and dispersing facilities, pump stations, terminals, and extended pipelines. They also include discharging facilities for tankers at ports or other water terminals.

6-254. The type and extent of risk to a pipeline varies with the level of conflict in the AO. In a communications zone (COMMZ), the chief hazard is likely to be pilferage. Pipelines can be tapped by loosening the flange bolts that join sections of pipe or cutting holes in the hose line. The risk rises if gasoline is scarce and expensive on the civilian market. Sabotage is a security hazard during all levels of conflict. It is committed by any method such as simply opening pipe flanges, cutting hose line, or setting fires and causing explosions to destroy portions of the line.

6-255. In areas of greater conflict, the likelihood of sabotage and interdiction increase. Pipeline systems are vulnerable to air attacks, especially at—

- Aboveground sections of the pipeline.
- Pump stations.
- Storage facilities.

6-256. Deploy MP in the best manner to provide coverage to the most vulnerable portions of the pipeline which are at the greatest risk to enemy, terrorist, partisan, and ground attack. MP are best able to provide pipeline security by conducting mounted and dismounted patrols; establishing a LP/OP for early warning; and establishing defensive positions around terminals, pump stations, and aboveground pipeline sections. They patrol isolated areas and remote pumping stations and—

- Detect, report, and respond to attacks on or sabotage of the pipeline.
- Monitor critical parts of the pipeline on a routine but random basis.
- Monitor ground sensors and other intrusion detection devices. These are often used at pump stations and elsewhere along the pipeline to detect and identify threats to the system.
- Check line pressure devices in the pipeline and pumping facilities. These devices monitor the flow and detect breaks in the line, which may indicate pilferage of gasoline or other petroleum products.

6-257. Dedicated MP are rarely enough in number for surveillance of an entire pipeline system. Available supporting forces, in the course of their normal duties, assist in observing and reporting items of intelligence for further investigation. Examples of suspicious activities in the pipeline area might include the unusual presence of commercial tank trucks, the appearance of gasoline drums or cans, or the increased use of motor vehicles in fuel-scarce areas.

## HIGH-RISK PERSONNEL SECURITY

6-258. Security of high-risk personnel is the responsibility of the US Army Criminal Investigation Division Command (USACIDC). However, due to a limited number of CID special agents, in any theater of operation, MP are often tasked to provide close-in security of protected persons.

### Security Considerations

6-259. Normally, the MP company commander designates a platoon leader as the officer in charge (OIC) of the security detail. The platoon leader is given full responsibility for all phases of the security mission, to include the coordination. This allows for a single point of contact for the conduct of the mission. The platoon leader prepares a comprehensive plan to ensure the safety of the designated person. Once the platoon leader receives the designated person's itinerary, he selects an advance team and conducts a reconnaissance of the route and locations the designated person is scheduled to visit.

6-260. Normally, the platoon leader uses organic vehicles and equipment to conduct the mission. If special equipment is required, he makes the necessary coordination through his higher HQ. Personnel selected for the mission are thoroughly briefed on the security plan and their conduct and demeanor during the mission. The security plan should include the following:

- Restrictions on individuals approaching the protected person.

- Evacuation procedures if the protected person is attacked and the threat cannot be immediately reduced.
- Control measures such as designated checkpoints.
- Crowd control measures.
- Alternate routes.
- Emergency MEDEVAC procedures.
- Emergency actions to be taken in the event of an ambush or indirect fire.

6-261. Depending on the threat level, selected members of the security team are armed with a M4, a M16, or a M203. When dismounted with the protected person, they provide standoff security out to the maximum effective range of their weapons. Shotguns should be considered shotguns if there is a threat of attack in a congested area where there is danger of injuring innocent persons if long-range weapons were used. Members of the team that are selected to physically escort the protected person should be armed with sidearms only. This allows them to have both hands free to react to close-in threats.

6-262. The security plan must be flexible to allow for itinerary and weather changes and mechanical failures. Once the platoon leader has finalized the security plan, he assembles the security team and conducts detailed rehearsals.

## Security During the Visit

6-263. Protection may be accomplished by establishing a series of protective cordons around the protected person. The protective measures used must allow the person freedom of movement and must not interfere with the conduct of his duties. Irritating conduct must be avoided at all times.

6-264. MP must remain alert at all times and provide constant security to the protected person. MP provide security by—

- Restricting the circulation of unauthorized personnel around the protected person.
- Remaining a short distance from the protected person.
- Taking positions where they can observe everything and everyone near the protected person.
- Securing the facility of the protected person, but not entering if the protected person does.
- Maintaining constant communications with higher HQ.

6-265. When required to secure a designated person while moving, the mode of transportation must be guarded at all times. Vehicle security normally requires a protected vehicle to transport the designated person, security vehicles to react to threats, and an advance team to travel ahead of the motorcade to provide early warning of threats or hazards before the main body is affected.

6-266. Regardless of the method of transportation, each situation must be evaluated to determine the degree of security that is practical and necessary. Good judgment on the part of the OIC will be necessary in solving the various situations that may exist. Refer to *FM 3-19.30* for more information about the security of designated personnel.

# Chapter 7

# Internment and Resettlement

I/R consist of those measures necessary to guard, protect, and account for people that are captured, detained, confined or evacuated by US forces. In any military operation involving US forces, accountability and the safe and humane treatment of detainees are essential. US policy demands that all persons who are captured, interned, evacuated, or held by US forces are treated humanely. This policy applies from the moment detainees become the responsibility of US forces and continues until the time they are released or repatriated. (Refer to the *Geneva Conventions* and *AR 190-8, AR 190-14, AR 190-47, FM 3-19.40,* and *FM 27-10.*)

## OVERVIEW

7-1. The task areas that support the I/R function are EPW and CI handling, US military prisoner handling, and populace and resource control.

7-2. Captured, detained, and protected persons fall into several different categories that include the following:

- Enemy prisoners of war. EPWs are members of an enemy armed force or militia who must be guarded to prevent escape.
- Civilian internees. CIs are persons who have committed an offense against or poses a threat to friendly forces and must be guarded to prevent escape, but are kept separate from the EPWs.

- Dislocated civilian. DCs are persons that have been removed from their home because of war, disaster, or other reasons. They may be refugees, evacuee, stateless persons, or war victims. DCs are provided sustenance, safety, and humanitarian assistance. They are kept separate from EPWs and CIs. DCs are controlled to prevent interference with military operations and to protect them from combat or to relocate them to safety. DC operations are discussed later in this chapter.
- US military prisoner. US military prisoners are members of the US armed forces being confined, awaiting trial, or waiting transportation to a confinement facility outside the AO. They must be guarded to prevent escape and cannot be confined in immediate association with EPWs and CIs, detainees, or other foreign nationals who are not members of the US armed forces. Refer to *FM 3-19.40* for more information about field confinement of US military prisoners.

7-3. EPWs are more specifically defined in *FM 3-19-40* and the *Geneva Convention Relative to the Treatment of Prisoners of War, August 1949.*

## ENEMY PRISONERS OF WAR AND CIVILIAN INTERNEE

7-4. MP receive EPWs and CIs as far forward as possible to prevent maneuvering units from being burdened with large numbers of prisoners. Prisoners are evacuated from the battle area as quickly as possible. The capturing unit is responsible for guarding prisoners until relieved. They field process captives using the *Five Ss-and-T method (Table 7-1).*

7-5. The capturing unit is usually responsible for delivering the detainees to the collecting point and the

**Table 7-1. Five Ss-and-T Methods**

| Procedure | Description |
|---|---|
| Search | Search each captive for weapons and ammunition, items of intelligence value, and other inappropriate items.<br>**NOTE: When possible, conduct same gender searches; however, this may not always be possible due to speed and security considerations. Therefore, perform mixed gender searches in a respectful manner using all possible measures to prevent any action that could be interpreted as sexual molestation or assault. The on-site supervisor must carefully control soldiers doing mixed gender searches to prevent allegations of sexual misconduct.** |
| Silence | Do not allow the captives to speak or let anyone speak to them. Speak only to captives to give orders. |
| Segregate | Segregate captives by rank, gender, nationally, and status. |
| Speed | Remove the captives from the battlefield as quickly as possible. |
| Safeguard | Safeguard the captives according to the Geneva Convention and US policy. Provide medical care as needed. |
| Tag | Use *DD Form 2745* and include at a minimum the following information:<br>• Date of the capture.<br>• Location of the capture (grid coordinates).<br>• Capturing unit.<br>• Special circumstances of capture (how the EPW was captured, for example, did he resist, did he give up, and so forth). |

nonwalking sick or wounded detainees to the nearest medical-aid station for evacuation through medical channels. Medically evacuated EPWs and CIs must be physically segregated from friendly forces. Detainees are normally turned over to MP at the nearest EPW collecting point or holding area. However, MP must be prepared to go forward to accept EPW from capturing units.

7-6. Traditionally, MP operate collecting points in a division AO and holding areas in a corps or EAC AO. However, collecting points and holding areas should be established wherever they are needed. The evacuation chain normally moves from the division forward or the central collecting point to corps holding area, then to internment facilities. When circumstances permit, such as taking advantage of available transportation, EPW evacuation may bypass one or more stations and deliver the detainees directly to a corps holding area or an internment facility.

7-7. At collecting points and holding areas, MP work closely with MI determining if captives, their equipment, or their weapons have intelligence value. MI interrogation teams conduct interrogations during field processing. Other MI interrogations teams conduct interrogations once EPW have been evacuated to more permanent facilities.

## OPERATE A DIVISION FORWARD COLLECTING POINT

7-8. The number of MP needed to operate a division forward collecting point is based on the number and rate of captives expected and the METT-TC. A division forward collecting point must be mobile and modular and able to set up, expand, and move quickly with little or no notice. The general location of a forward collecting point is given in the brigade OPLAN or OPORD. It often is located near or in the brigade support area (BSA), but should not allow detainees to observe activities in the BSA. The collecting point should be situated close to an MSR. This makes it easier to get supplies, such as water, food, and barrier material from the BSA. Even a moderate number of detainees will put a strain on the equipment and supplies of an MP company. Minor medical treatment may come from the MP company's combat medical section. However, the company's medical resources are very limited and are primarily used to support medical needs within the company.

Units needed to support the division forward collecting point should be specifically tasked in the brigade OPORD. MP leaders operating the division forward collecting point will—

- Conduct a reconnaissance before selecting an exact location for the collecting point.
- Locate the collecting point far enough from the fighting to avoid minor shifts in the main battle area (MBA) (normally 5 to 10 kilometers from the MBA).
- Notify the BSA TOC and the PM operations section of the selected location. The BSA TOC reports the exact location of the collecting point to the brigade TOC. The brigade TOC notifies subordinate units where the collecting point is located so capturing units with detainees can take them there.
- Coordinate with the MI interrogation team if they are to colocate their interrogation site with the division forward collecting point.
- Request transportation, additional medical supplies, and other support through the forward support battalion.
- Ensure that captives do not remain at the division forward collecting point more than 12 hours before being escorted to the division central collecting point.

7-9. A forward collecting point *(Figure 7-1, page 7-6)* should not be set up near local inhabitants. Existing structures like vacant schools, apartments, or warehouses should be used when possible. This reduces construction requirements and minimizes logistical requirements. If existing structures are not used, detainees, except officers, can be tasked to help construct the collecting point. Prisoners may dig or build cover to protect themselves from artillery, mortar, or air attack. There is no set design for a forward collecting

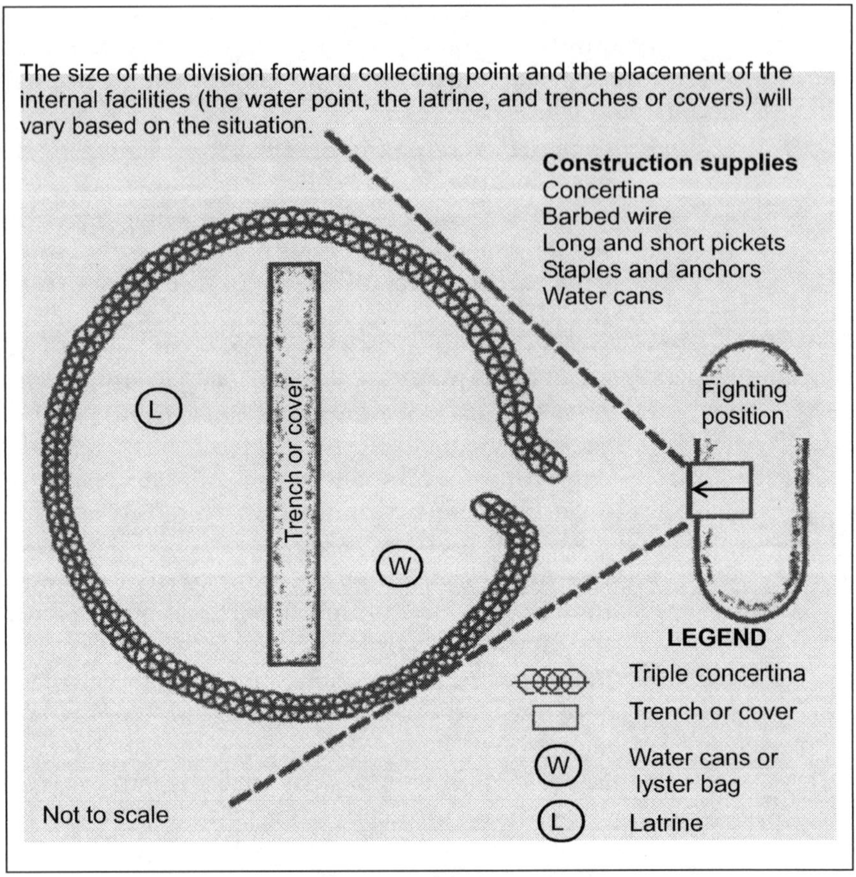


**Figure 7-1. Division Forward Collecting Point**

point. It can be anything from a guarded, roped off area to a secured, existing structure. The collecting point is built to suit the climate, the weather, and the situation. When selecting a collecting point, consider the following:

- The security of the detainees. The perimeters of the enclosure must be clearly defined and understood by the detainees.

- First aid. Injured or ill detainees require the same treatment that would be given to US casualties.
- Food and water. Detainees may have been without food or water for a long time before capture.
- Latrine facilities.
- Field sanitation. If possible, have detainees wash with soap and water to reduce the likelihood of disease.
- Shelter and cover.
- Language barriers. Provide interpreters and/or instructional graphic training aids (GTAs) in the EPW native language to compensate for the language differences.

7-10. MP at collecting points normally receive detainees directly from the capturing troops. MP then process the detainees using the *stress method*. The six principles of stress are search, tag, report, evacuate, segregate, and safeguard.

7-11. **Search**. Search and inspect every EPW and CI and their possessions. Use males to search male prisoners and females to search female prisoners wherever possible unless, in exceptional situations, an individual of the opposite gender must conduct the search. If this is the case, the search of the opposite sex must be performed in a respectful manner using all possible measures to prevent any action that could be interpreted as sexual molestation or assault. Captives may keep the following items found in a search:

- Protective clothing and equipment (such as helmets, protective masks and clothing) for use during evacuation from the combat zone.
- Retained property, such as identification cards or tags, personal property having no intelligence value, clothing, mess equipment (except knives

and forks), badges of rank and nationality, decorations, religious literature, jewelry, and articles that have sentimental value.

- Private rations of the EPW or the CI (in the early stages of captivity).

7-12. Certain items are confiscated from the EPW or the CI and never returned even if the EPW or the CI is released or repatriated. MP confiscate the following items when searching a captive:

- Weapons and ammunition.
- Items of intelligence value (maps, orders, and so forth).
- Other inappropriate items.

7-13. MP will coordinate with the MI interrogation teams to determine which items that have been confiscated are of intelligence value. Personal items, such as diaries, letters from home, and family pictures may be taken by the MI teams for review, but are later returned to the MP for return to the proper owner.

7-14. Currency will only be confiscated on the order of a commissioned officer (*AR 190-8*) and will be receipted for using *DA Form 4137*.

7-15. Impounded articles are items taken from the EPW or the CI during his internment because the articles make escape easier or compromise US security interests. Items normally impounded are cameras, radios, and all currency and negotiable instruments found on the captives. Refer to *AR 190-8* and *Defense Finance and Accounting Service-Indianapolis (DFAS IN) 37-1* for more information about confiscated and impounded property.

7-16. MP prepare a receipt when taking property from a detainee. The MP leader ensures that both the EPW or the CI and the receiving MP sign the receipt (such as *DA Form 4137*). MP consider bundling a detainee's property or placing it in bags to keep each detainee's property intact and separate. They turn in cleared, confiscated property as far forward as possible. MP maintain a strict chain of custody for all items taken from the EPW or the CI. They ensure that a receipt is obtained for any items you release to any other MP or agency. The escorting MP signs for and transports any remaining property that was taken from the EPW or the CI.

7-17. **Tag**. Each EPW or CI is tagged by the capturing troops using *DA Form 2745* as a way of accounting for them. MP check each tag at collecting points and holding areas for—

- The date and time of the capture.
- The capturing unit.
- The place of the capture (grid coordinates).
- The circumstances of the capture (how the EPW was captured).

7-18. The remaining information on the tag will be included as it becomes available. *DD Form 2745* is a perforated, three-part form which has an individual serial number. It is constructed of durable waterproof, tear-resistant material with reinforced eyeholes at the top of Parts A and C. The capturing unit attaches Part A to the captive with wire, string, or another type of durable material. They maintain Part B in their records and attach Part C to the confiscated property so that the owner may be identified later.

7-19. MP at division collecting points will ensure that *DD Form 2745* has been placed on any captive arriving at the collecting point without it. MP may have to direct the capturing units to complete the capture tag before accepting prisoners into the CP. They ensure that the following is done:

- The tag is filled out with the minimum information listed above (also listed on the back of Part C of the form).

- A statement is on the tag if the captive arrived without a tag.
- The captive is instructed not to remove or alter the tag.
- The capture tag's serial number and the captive's name are annotated on a locally developed manifest.

7-20. MP receive detainees from capturing troops using *DD Form 2708* or a similar document. They ensure that the receipt includes the following:

- The capturing unit.
- The time and date the detainee was received.
- The identification of the detainee. (Use the number on the capture tag when the detainee's name, service number, grade, or date of birth is unknown.)
- The name, service number, grade, unit, and signature of the MP who accepts custody of the detainee.
- A statement in the remarks section about the general physical condition of the detainee. For example, *received without wounds, illness, or injury* or *wounded in upper left arm*.

7-21. **Report**. The number of captives at each collecting point is immediately reported through MP channels. This aids in the transportation and security planning process.

7-22. **Evacuate**. Captives are humanely evacuated from the combat zone through appropriate channels as quickly as possible. MP do not delay evacuation to obtain name, rank, service number, or date of birth. When MP evacuate captives, they give them clear, brief instructions in their own language when possible. Military necessity may require a delay in evacuation beyond a reasonable period. When this occurs, MP leaders ensure that there is an adequate supply of food;

potable water; and appropriate clothing, shelter, and medical attention available.

7-23. MP ensure that EPWs or CIs are not be exposed to unnecessary danger and are protected while awaiting evacuation.

7-24. Medical personnel determine if captives with serious wounds or sickness should be kept in the combat zone. Sometimes prompt evacuation would be more dangerous to their survival than retention in the combat zone.

7-25. **Segregate**. The senior officer or noncommissioned officer in charge (NCOIC) having responsibility for custody of the EPWs or CIs will designate how and at what level to segregate them to ensure their security, health, and welfare. EPWs and CIs are segregated into the following categories:

- Officers, noncommissioned officers (NCOs), enlisted, male, and female.
  - Deserters and those that gave up without a fight may be further segregated for their protection.
  - Nationality, ideologies, and recognized ethnic groups are used for further segregation.
- CIs and/or refugees are physically separated from the EPWs and CIs.
- US military prisoners are physically separated from EPWs, CIs, retained persons (RP), other detainees (OD), and refugees.

7-26. MP do not use coercion of any kind to obtain any information from the captives. This includes basic information, such as name, rank, service number, and date of birth, which they are required to provide under the Geneva Conventions. Coercion or inhumane treatment of any EPW, CI, RP is prohibited and is not justified by the stress of combat or with deep provocation. Inhumane treatment is a serious violation of international law and the Uniform Code of Military Justice (UCMJ).

7-27. MP must not speak to captives except to give orders or directions. Captives must not be allowed to talk to or signal each other. This prevents them from plotting ways to counter security and plan escapes. Uncooperative captives may require a gag in certain tactical situations. However, gags should be used for only as long as needed and should not harm the individual.

7-28. **Safeguard**. In order to safeguard captives according to the Geneva Convention and US policy, MP must—

- Provide first aid and medical treatment for any wounded or sick captive. The wounded and sick will be evacuated separately through medical channels using the same assets as those used to medically evacuate US and allied forces.
- Ensure that the detaining power provides their captives with food and water. These supplies must be the same as to that of US and allied forces.
- Provide firm and humane treatment.

7-29. Protecting detainees from attack, preventing their escape, and quickly removing them from the battle area further safeguards them. Detainees should not remain at the division forward collecting point more than 12 hours, if possible. MP from the division central collecting point move forward to escort detainees back to the central collecting points. When detainees are field processed and ready for evacuation, the MP at the division forward collecting point will—

- Report detainee status to the BSA TOC and through MP channels to the PM.

- Request transport, rations, and water for the detainees from the forward support battalion supply officer (US Army) (S4).
- Ensure that the receipts for the detainees are ready for signing by the escort guards.
- Ensure that items taken from detainees for security or intelligence reasons are signed over to the guards taking the detainees to the rear. Ensure that each item is tagged to identify the owner.

## ESTABLISH AND MAINTAIN A DIVISION CENTRAL COLLECTING POINT

7-30. MP in GS are responsible for establishing and maintaining the division central collecting point. They collect detainees from the forward collecting points, then process and secure them until corps MP come forward to evacuate them to the rear. Detainees should be transferred to the corps holding area or directly to an internment facility within 24 hours, if possible. One or more GS MP platoons operate the division central collecting point. The MP platoons are augmented by the division band and/or by the corps MP. Augmentation is based on the number and rate of captives expected.

### Band Augmentation

7-31. When necessary, members of the division, corps, or EAC band augment MP for EPW operations. They guard detainees, operate dismount points, and provide perimeter security. When band members are tasked to augment MP for EPW operations they are OPCON to the MP company for the duration of the mission and released at the earliest opportunity to return to their primary mission.

## Division Central Collecting Point

7-32. A central collecting point *(Figure 7-2)* is larger than a forward collecting point, but the considerations for setting up and operating the collecting points are

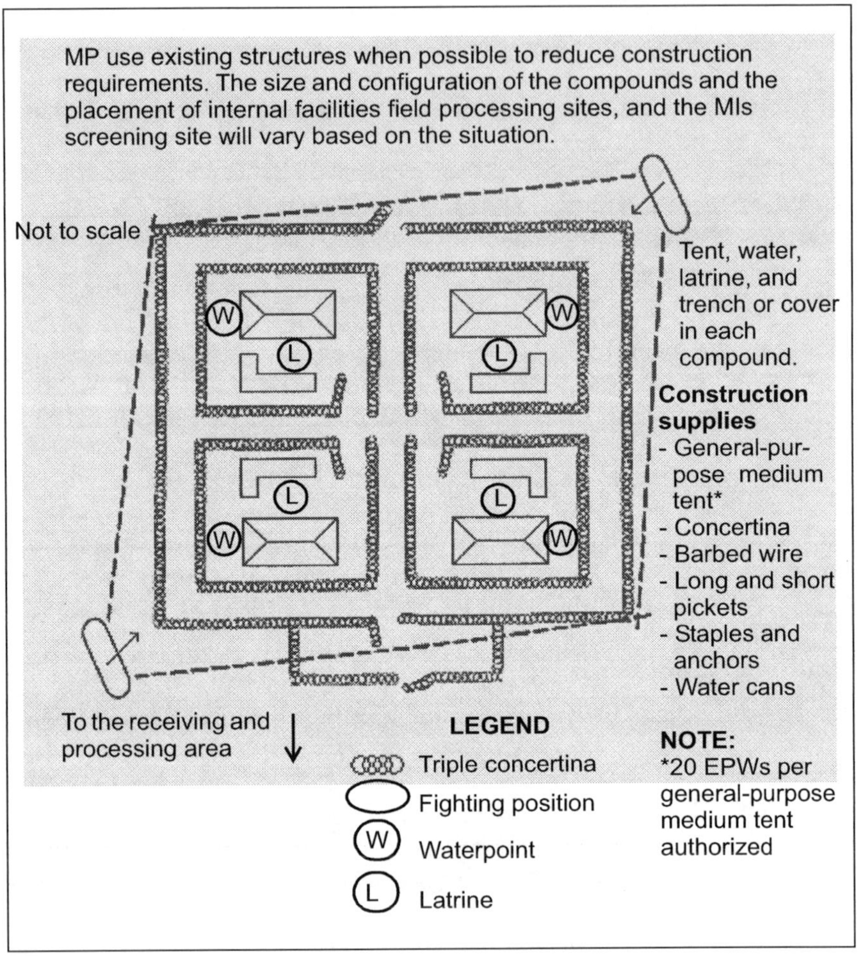


**Figure 7-2. Division Central Collecting Point**

generally the same. The general location of the central collecting point is given in the division OPORD or OPLAN. It is located near the division support area (DSA), preferably close to an MSR. This makes it easier to obtain supplies, transportation, and additional medical support from the DSA. Non-MP units should be specifically tasked in the coordinating instructions of the division OPORD to provide the support needed for the division central collecting point. MP establishing the collecting point should—

- Coordinate with the unit responsible for the area.
- Conduct a reconnaissance before picking the exact location for the collecting point.
- Notify the PM and the operations cell of the division rear CP (through MP channels) of the collecting point location.
- Coordinate with MI for the location of their screening site.
- Use existing structures when possible.
- Request supplies through the division MP company.

## EVACUATE DETAINEES FROM A DIVISION FORWARD COLLECTING POINT TO A DIVISION CENTRAL COLLECTING POINT

7-33. The MP platoon charged with operating the division central collecting point sends MP forward to the division forward collecting point to escort detainees back to the central collecting point. EPWs or CIs must be evacuated from the division forward collecting point as soon as possible, preferably within 12 hours. Before evacuating the detainees, MP checks with MI interrogation teams for any property to be returned to, or evacuated with, the detainees before they are moved.

## PROCESS ENEMY PRISONERS OF WAR AND CIVILIAN INTERNEES FOR EVACUATION

7-34. MP consider the physical status of detainees before evacuating them. Categories for consideration are the sick and wounded EPWs and CIs and the able-bodied EPWs and CIs.

### Sick and Wounded Enemy Prisoners of War and Civilian Internees

7-35. The MP unit's combat medical section will screen detainees and decide if they will be escorted within MP channels or medically evacuated. Generally, the walking wounded go with the other detainees. Litter patients go through medical channels. US forces provide the same medical care for sick or wounded detainees as that given to US and allied soldiers. Sick and wounded EPWs in the combat zone are either treated and returned to the MP for evacuation or stabilized and moved through medical channels to the rear as far and as quickly as possible. If medically evacuated, MP release the detainees to the medical authorities using *DD Form 2708* or other receipt. The corps medical regulating officer (MRO) and the receiving hospital commander coordinate with the Internment Resettlement Information Center (IRIC) to account for detainees in medical channels.

7-36. MP determine when security is required for sick or wounded detainees. Normally, sick or wounded detainees requiring MEDEVAC are not likely to be a security risk. Detainees well enough to present a security risk can be treated by the combat medical section and evacuated through MP channels as soon as possible.

### Able-Bodied Enemy Prisoners of War and Civilian Internees

7-37. Able-bodied detainees are escorted during movement to keep them from escaping. MP planning the movement of detainees consider the following:

- The factors of METT-TC.
- The number of detainees being escorted.
- The condition and morale of the detainees. Fatigued and cooperative detainees may not require as many guards as those who are fresher and more motivated.
- The type of transport to be used. The type of transport may influence the number of guards. A planning consideration is one guard per 5 to 10 detainees. Aircraft is loaded according to the airplane crew's instructions.
- The terrain conditions along the route. Detainees are more likely to attempt escape in close terrain, like dense woods or jungle, and may require more guards than open terrain.
- The level of enemy activity along the route. The more enemy activity in the area, the greater the need for increased security precautions.
- The likelihood or presence of suspected sympathizers and hostile local nationals along the route.
- The scheduled arrival of the transport. Use backhaul transport whenever possible.
- Transportation considerations. Transportation depends on the availability of vehicles delivering cargo in the nearby area.
- The location of MP units or bases and base clusters along the route that could provide assistance during the movement.
- The number and locations of rest stops (based on the type of transportation, distance, and the type of terrain).

7-38. Detainees are evacuated on foot only as a last resort when transport is not available. Transportation for detainees is arranged through the company HQ. At division, the company HQ contacts the local movement control officer.

7-39. Before leaving for the collecting point, the MP in charge of the escort must—

- Conduct a route reconnaissance of the evacuation route.
- Verify the location of the collecting point shortly before departing, since BSAs move often.
- Plan to stop only during daylight and outside towns or installations if possible.
- Designate guards to dismount at halts and supervise the loading of the detainees.
- Segregate detainees by category, if possible.
- Secure the rations and the water. Use captured enemy rations for the detainees, if available. Do not allow the use of utensils or can openers.
- Search detainees and baggage before loading in any transport.
- Use hand irons, leg irons, or special restraining jackets on detainees, if necessary. If hand irons are used, restrain the detainees with their arms in front.
- If prescription drugs are needed, disperse according to the medical officer's instructions.

## EVACUATE ENEMY PRISONERS OF WAR AND CIVILIAN INTERNEES FROM A DIVISION CENTRAL COLLECTING POINT

7-40. In order for MP to conduct successful evacuation of EPWs and CIs, MP brief the escorts and the detainees.

### Brief the Escorts

7-41. MP or other military personnel may perform as guards in evacuating EPWs and CIs. Escort personnel are briefed on evacuation considerations and escape attempts. Considerations include the following:

- Procedures to ensure that the detainees follow instructions and orders. Escorts must be firm, but will not punish detainees who fail to obey.
- Requirements to inspect passenger areas, latrines, and other places that might be accessible to detainees during transport. Escorts should look for the means of escape or items that could be used as weapons and remove the latches from the latrine doors on transports, if possible.
- The necessity to talk to detainees only to give orders and maintain control.
- Emergency actions to secure and safeguard the detainees in case of enemy contact. Members of the escort element must know in advance which of them will control the detainees and which ones will react to the enemy.

### Receive and Brief the Detainees

7-42. The senior MP in the escort element accepts custody of the detainees. Each detainee is tagged and field-processed before being accepted for evacuation. Each detainee is accounted for using *DD Form 2708* or a similar receipt. The senior MP escort ensures that all the detainees (and any equipment) are listed on the custody receipt when custody is accepted. The escort retains a copy of the custody receipt.

7-43. Before moving, the senior MP ensures that the detainees have been briefed on movement discipline in a language understood by them. MP use locally produced GTAs to conduct the briefing. If available, an interpreter should give the instructions to the detainees. They are told—

- The meaning of the word *halt*.
- That the "silence rule" applies at all times (no talking to the guards, no talking to each other).
- The actions to take during an emergency.

## COLLECT ENEMY PRISONERS OF WAR IN OTHER OPERATIONS

7-44. Some offensive operations are executed so rapidly that combat forces completely overwhelm the enemy forces. These operations create special considerations when planning EPW operations.

### River Crossing Operations

7-45. During river crossing operations, an EPW collecting point is established on the nearside of the river far enough to the rear to prevent interference with tactical operations and to afford reasonable protection against hostile fire. The EPWs are evacuated from the bridgehead area as soon as possible to prevent congestion. The movement of EPWs from the farside of the river must be coordinated with traffic control personnel at the crossing sites. Secondary crossing sites must be used if available.

### Amphibious Operations

7-46. During amphibious operations, initially, the shore party or helicopter support teams operate EPW collecting points in the beach support areas or LZs. EPWs are evacuated from the collecting points to designated ships by landing craft, a helicopter, or amphibious vehicles. MP must coordinate with the support force for the handling of EPWs once they have been evacuated from the beachhead. EPWs are retained in the objective area when facilities, supplies, and personnel permit, consistent with the reasonable safety of EPWs from enemy action.

### Airborne Operations

7-47. During an airborne operation, the METT-TC considerations for collecting EPWs include the geographical location of the airhead, the tactical plan, the availability of transportation, and plans for linkup with ground forces. EPWs are evacuated primarily by air, especially during the early stages of the operation. The EPW collecting point should be located near a LZ. Plans should provide for the attachment of MP escort guard units from the area EAC to the airborne force to guard EPW during evacuation.

### Armored Operations

7-48. Armored units are able to quickly penetrate deep into hostile territory, possibly bypassing enemy strong points. They may leave isolated enemy groups, which would hinder the normal evacuation of EPWs. In this circumstance, it may be necessary to hold EPWs in the area of capture until they can be safely evacuated.

### Air Assault Operations

7-49. During an air assault operation, organic military police elements accompany assault elements to the objective areas so that they can collect and evacuate captured EPWs. Collecting points are established as required near heliports or airfields. Arrangements must be made for nondivisional MP to accompany designated incoming or resupply aircraft to guard EPWs during their evacuation from the division.

## UNITED STATES MILITARY PRISONER HANDLING (FIELD DETENTION FACILITIES)

7-50. Use field detention facilities *(Figure 7-3, page 7-22)* to hold US soldiers in custody until they can be tried. Use pretrial confinement only to ensure that an accused appears at trial or when the seriousness of the offense or the threat of violence makes confinement essential. Whenever possible, soldiers awaiting trial remain in their units. Only when they are a hazard to themselves or others are they detained in pretrial confinement under MP control.

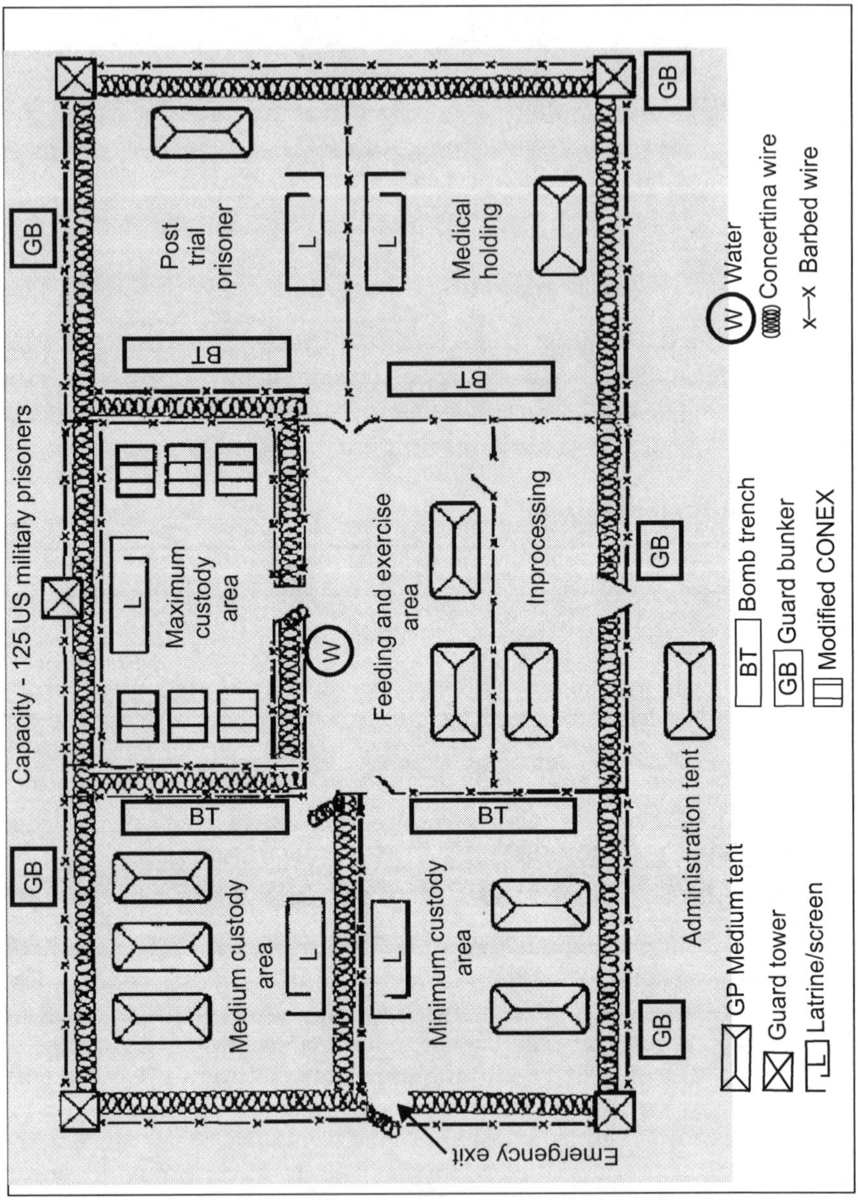


**Figure 7-3. Field Detention Facility**

7-51. Use field detention facilities to hold sentenced prisoners waiting for transfer to a theater's field confinement facility (FCF) or the continental US (CONUS). After trial, move convicted military prisoners, whenever possible, to confinement facilities outside the combat zone.

7-52. Each echelon commander sets procedures and policies for detaining and confining soldiers. Often US military prisoners in a combat zone are placed under the control of an MP unit operating an EPW collecting point. When small numbers of US prisoners are on hand, a squad operating an EPW collecting point can best take responsibility for the security of US prisoners. US military prisoners must be kept physically apart from EPWs. The policy and procedures for the care and treatment of prisoners and the safeguarding of their personal effects remain the same as that set for other Army confinement facilities.

7-53. When prisoners are retained in-theater, separate temporary detention facilities maybe set up in the corps or division areas. US military prisoners should be held in the division rear area for the shortest possible time. At a division facility MP must—

- Safeguard US prisoners.
- Coordinate for their food and medical care.
- Sustain them until they can be evacuated to a corps facility.
- Transfer them to the corps facility as quickly as possible.

7-54. When the situation permits, MP from a detention facility at the corps come forward to pick up the prisoners at the request of the division's detention facility commander. From the corps, the prisoners are evacuated to the theater confinement facility.

7-55. If a temporary detention facility is set up in the corps, it usually is operated by confinement teams from the confinement battalion in a personnel command (PERSCOM). These teams are organized and trained to perform confinement operations. But when corps detention operations are limited to prisoners being evacuated to a confinement battalion in the PERSCOM, elements from a combat support company can operate a temporary facility.

7-56. A field detention facility usually is located near the MP company CP for food, transport, and supply support. MP request construction materials from the engineers to set up and run a facility. Equipment and supplies must include the following:

- Barbed wire (roll and concertina).
- Fence posts.
- Gates and doors.
- Floodlights and spotlights, complete with wiring.
- An emergency generator.
- Mess equipment and equipment for cleaning mess gear.
- Water cans or lyster bags.
- Computers.
- First aid equipment and supplies. Spare clothing and bedding.
- Hand restrains or leg irons.
- Heating equipment (cold climate).
- Field sanitation supplies.

7-57. MP leaders ensure that the facility is large enough to separate prisoners by prisoner status, custody grade, sex, and rank. They locate the facility away from a base's perimeter or any other area of increased risk.

7-58. The size of the facility is based on the number of prisoners being detained. It may be a room or a tent, as long as it provides shelter equal to that offered to other soldiers in the combat zone. The physical criteria for permanent and temporary structures are the same. MP use existing structures if you can. Otherwise, they use tents. Field-expedient facilities must be approved and periodically inspected by a medical corps officer or a designated representative.

7-59. A team or a squad can operate a field detention facility. A team operating a facility may organize so that the team leader controls the operation and relieves the guards. The other team members perform guard duty in alternating 12-hour shifts. When operating a field detention facility, the team—

- Accepts sentenced soldiers on the written order (currently a *DD Form 2707*) of a court martial convening authority (or the authorized representative).
- Ensures that the order states the name, grade, social security number, organization of the prisoner, offense for which convicted, and the sentence.
- Accepts the accused soldiers on the written order (currently a *DD Form 2707*) of the accused's commanding officer.
- Signs a receipt for each prisoner and his property on the correct and current form.

## POPULACE AND RESOURCE CONTROL (PRC)

7-60. PRC operations are the responsibility of the Assistant Chief of Staff, (Civil Affairs) (G5), the Civil Affairs Officer (US Army) (S5), and/or the HN authorities. (Refer to *FM 41-10* for more information about PRC.) PRC is often conducted in stability and support operations where national authority has broken down and the government cannot control the population. MP support PRC by conducting L&O operations designed to restore order and protect the people and property.

7-61. Insurgent organizations often emerge in unstable regions. The aim of such groups is normally to overthrow the established government. The less control the government has, the greater the chance for insurgents to succeed. These organizations try to exploit the population, often through threat and intimidation. When insurgent organizations pose a threat to the population, US forces employ PRC operations that are designed to deny support and assistance to insurgents by controlling the movement of people and goods and restricting access to key facilities. Police activities, such as roadblocks, cordons, curfews, access control, and checkpoints are an important measure in counterinsurgency, but have a high potential for harm if used excessively or incorrectly. MP are specially trained to conduct these operations, as a force focused on security, protection, and assistance. MP are continually trained on the prudent use of force, crisis management, and operations requiring restrictive ROE.

7-62. PRC measures deprive insurgent organizations support and aid in identifying their supporters. Appropriate psychological operations (PSYOP) help make these measures more acceptable to the population by explaining their necessity. The government informs the population that, although its actions may cause inconvenience, the threat posed by the insurgents makes them necessary.

7-63. PRC is often conducted in urban areas. The best use of PRC comes before an organized insurgent movement has the capability for armed conflict. MP intelligence operations support PRC programs. Criminal acts, such as robberies, kidnappings, terrorism, and extortion, may accompany insurgent propaganda or money-raising activities.

7-64. MP employ special control measures to aid populace control that include the following:

- Enforcing curfews.

- Enforcing movement restrictions.
- Verifying travel permits and registration cards.
- Assisting with rescue and evacuation operations.
- Assisting with crowd control.

7-65. MP also employ control measures to aid in resource control. They include—

- Operating roadblocks and checkpoints.
- Inspecting facilities.
- Enforcing local regulations and guidelines.
- Controlling rations.
- Assisting with amnesty programs.

7-66. PRC operations play a vital role in winning support away from the insurgent threat and encouraging support for the goals of the legitimate government.

7-67. MP leaders responsible for supporting PRC must plan for the protection of their forces as these types of operations have a history of turning violent very quickly.

## DISLOCATED CIVILIAN RESETTLEMENT

7-68. International law recognizes the humanitarian practice of providing temporary refuge to anyone, regardless of nationality, who may be in imminent physical danger for the duration of the threat. It is US policy to grant temporary refuge in a foreign country to nationals of that country, or of a third nation, solely for humanitarian reasons when extreme or exceptional circumstances exist.

7-69. Civil affairs (CAs) units are trained to plan, coordinate resources for, and monitor the handling of DCs. (Refer to *FM 41-10* for more information about DC operations.) Whenever possible, resources and control should be arranged with the HN, other governmental agencies, and nongovernmental and private organizations.

7-70. To assist in properly identifying DCs, the definition of the five subdivided categories is provided. These subcategories are defined by legal and political considerations as the following:

- Displaced person. A civilian who is involuntarily outside the national boundary of his country in time of conflict.
- Refugee. A civilian who, because of real or imagined danger, has left home to seek safety.
- Evacuee. A civilian removed from his place or residence by military order.
- Stateless person. A civilian who has been denationalized, whose country of origin cannot be determined, or who cannot establish his right to the nationality claimed.
- War victim. A classification that describes civilians suffering injuries, loss of a family member, or damage to or destruction of his home resulting from war.

7-71. DC operations are a special category of PRC. The goal of DC operations is to minimize civilian interference with military operations, relieve suffering, and protect civilians from combat operations or other threats. When the HN cannot or is unwilling to control DCs, MP may be required to collect, evacuate, and resettle them.

7-72. When directed to conduct DC operations, MP leaders coordinate with CA and the PM to establish a traffic control plan. The plan includes the primary and alternate routes used to move DCs and the location for—

- The TCP.
- Holding areas.
- Roadblocks and checkpoints.

7-73. Temporary route signing is not normally an effective control measure during DC operations. Even when the signs are posted in the local language, DCs usually ignore them.

7-74. MP collect and process DCs in the same manner as EPWs with regard to the difference in their status–they are detained personnel, not prisoners. As such, security of the I/R facility should not give the impression that it is a prison.

7-75. MP collect DCs at assembly points located away from the MSR and areas where combat operations are heaviest. They move DCs along selected routes that have the least military traffic. DCs will normally require frequent rest stops. When possible, holding areas for rest stops should be near a source of water and provide protection from the elements or hostile fire.

7-76. MP provide close-in and standoff security for DCs while en route to a civilian camp or collection point. When required, MP provide external and internal security at a DC camp. Within limits, DCs are allowed freedom of movement once they have been resettled. MP and CA must continue to closely monitor DCs at collection points and camps. Recent operations that involve large numbers of DCs have been known to turn violent very quickly. Such a situation can be avoided when DCs are treated with respect and dignity.

## EVACUATION OPERATIONS

7-77. In addition to EPWs, CIs, and DCs, MP are often required to evacuate other selected individuals or groups from collecting points, holding areas, or areas of hostility. The safe evacuation of noncombatants, US diplomats, or US military prisoners requires close coordination and extensive planning. This type of operation is referred to as noncombatant evacuation operations (NEO). Such missions require contingency plans and unit SOPs that are specifically designed for these special operations. Unit SOPs should cover the evacuation of designated personnel by ground, air, rail, and water. When directed to conduct evacuation operations, MP leaders task organize based on METT-TC and the availability of MP. They perform extensive rehearsals to successfully conduct evacuation operations.

7-78. When a HN can no longer ensure the safety of US civilians in a foreign nation, US military forces may evacuate them. MP are often required to provide security and escort for evacuees from their point of origin to their destination. A NEO is usually a joint operation conducted with other US and HN government agencies and CA.

7-79. An MP platoon is most likely employed for these operations. The platoon leader conducts direct coordination with CA and the higher HQ. The platoon leader begins TLP, issues a WO to the platoon, and requests information that includes the—

- Screening and identification system being used.
- Number of evacuees, their point of origin, and their anticipated direction of movement and destination.
- Location of marshalling areas, collection points, and AAs.

7-80. If time permits, the platoon and squad leaders conduct a reconnaissance of the evacuation route. The purpose of the reconnaissance is to prepare an overlay which is then used to plan control and special security measures at critical areas along the route. If helicopter support is available, platoon and squad leaders conduct an aerial reconnaissance of the evacuation route. An aerial reconnaissance will normally take less time and the observation of the terrain around the route is much better by air than from the ground. At a minimum, a map reconnaissance must be conducted.

7-81. During the reconnaissance, the platoon leader selects possible rest areas and identifies friendly units along the route. He identifies critical areas, such as defiles, bridges, and areas where enemy activity is likely. He marks these areas on the overlay and plans for special security measures at these locations.

7-82. The platoon leader coordinates with the movement control center to ensure that the evacuation route does not interfere with the movement of friendly forces. If available, he coordinates for fire support along the route, normally at critical sites that were identified during the reconnaissance.

7-83. The platoon leader coordinates for MEDEVAC and vehicle recovery. His plan includes emergency reaction to the following:

- An air attack.
- An artillery attack.
- An ambush.
- A riot.

7-84. NEO are normally carried out according to the guidelines established by CA and US policy. MP may be required to screen for authorized personnel to determine who may actually be evacuated. There are generally three groups of personnel. They are—

- Group I. Group I includes US citizens, officials, dependents, tourist, business persons, and non-US family members when the father, mother or wife is a US citizen.
- Group II. Group II includes foreign nationals holding diplomatic papers, visas, or passports who receive Department of state approval.
- Group III. Group III includes all others, to include HN citizens, who do not fit into the first two categories.

7-85. Persons who fall into any of these groups and require immediate medical attention are always evacuated first. MP should give special consideration to the elderly and children.

7-86. MP brief the evacuees on discipline and the actions to take during emergencies. The briefing must be in a language that all the evacuees understand. The CA unit or HN agency should provide an interpreter. The briefing should cover all aspects of the evacuation. This will help calm the evacuees and instill confidence and cooperation.

7-87. The platoon provides security of the evacuees at the marshalling, evacuation, and holding areas and the reception station. The level of protection depends on the level of the threat. Methods of security depend on the type and location of the facilities used. At a minimum, MP must be prepared to provide interior guards for group areas, establish perimeter security, and operate a dismount point to restrict access to the evacuees.

7-88. During movement, the platoon escorts the evacuees by providing close-in security at the lead, middle, and end of the convoy or in front of and behind a single transportation vehicle. MP use a scout vehicle that travels 3 to 5 minutes ahead of the convoy to alert the main body to danger or delays. If a threat tries to disrupt the evacuation operation or destroy the evacuation vehicle, selected MP teams protect the evacuees as other teams counter the threat within the ROE. Throughout the evacuation operation, MP maintain all-around security, protection, and evacuee accountability.

# Chapter 8

# Law and Order

This chapter addresses the L&O function across the full spectrum of military operations. Refer to *Appendix K* for more information about civil disturbance control measures.

## OVERVIEW

8-1. L&O operations consist of those measures necessary to enforce laws, directives, and punitive regulations; conduct criminal investigations; and control populations and resources that ensure commanders the existence of a lawful and orderly environment. MP enforce laws and appropriate commander directives. They maintain liaison and coordinate joint L&O operations with other DOD police organizations; HN military and civilian authorities; multinational police organizations; and US federal, state, and local police agencies. A coordinated law enforcement effort removes the conditions and opportunities that promote crime, thereby preventing diversion of military resources and maintaining military discipline.

8-2. The evolving criminal threat affects military operations and requires commanders to minimize that threat to preclude negative impacts on forces, resources, and operations. The importance of the criminal threat to military operations is a current, as well as future reality. Nationalist ideologies, the instability of a government to effectively govern and control its population, and the breakdown of government infrastructures will foster the linkage between criminal organizations and the government and its armed forces.

8-3. MP provide the capability to train foreign MP or assist in the reorganize of indigenous constabulary forces as part of stability and support post conflict operations. Under the provisions of and exceptions to *Section 660 of the Foreign Assistance Act*, MP provide initial assistance and training to foreign military and civilian police forces or assist in the creation of these forces where national authority has broken down. Additionally, MP forces provides short-term emergency L&O capabilities until the foreign military and civilian forces are functional.

8-4. MP and the USACIDC are the primary collectors of police information and criminal intelligence. They gather information through contacts made with the local populace and from conducting combined and joint patrols with HN military and civilian police agencies. MP and CID conduct field interviews and gather police information from surveillance operations. They investigate serious offenses and maintain a close liaison with the HN or allied civilian and military police agencies. This police information and criminal intelligence is collected, analyzed, and shared with the intelligence community and contributes to the police information assessment process (PIAP).

8-5. The police activities that support L&O operations include the following:

- Criminal investigations.
- Police information collection and dissemination of the information.
- Traffic accident investigations.
- Antiterrorism force protection support.
- Crowd control.
- US customs operations.
- Use of MWDs.

8-6. The enforcement of military laws, orders, and regulations is a command responsibility, as well as an MP responsibility. Each commander is responsible for maintaining order and discipline in the unit. To support commanders, the PM plans the use of MP assets to help enforce military laws, orders, and regulations.

8-7. MP performing L&O operations enhance and extend the tactical commander's $C^2$ by—

- Aiding commanders in maintaining combat strength.
- Helping prevent diversion of military resources.
- Suppressing opportunities for criminal behavior by US and non-US personnel or elements.
- Assisting and protecting military forces.
- Helping ensure the discipline of US forces.
- Assisting intelligence organizations in obtaining a true tactical intelligence picture by providing criminal and operational data and intelligence.

## LAW AND ORDER AUGMENTATION DETACHMENT

8-8. The technical and supervisory expertise to support L&O operations in a mature theater is provided by L&O teams organized under an L&O augmentation detachment. The L&O teams are designed to give the commander the additional flexibility and capability in any environment to conduct a wide range of force protection mission requirements, to include split-based operations. This enables the supported commander to perform needed L&O missions while other MP assets are conducting MMS, AS, or I/R.

8-9. When these teams are not available or have not arrived in theater, L&O missions are prioritized with other MP combat support missions and are usually not full-scale dedicated operations. The theater commander determines when he can afford to dedicate MP assets to L&O operations.

8-10. A complete 45-person L&O augmentation detachment includes a detachment $C^2$ team, an operations team, a desk team, five MP investigation (MPI) teams, five traffic accident investigation (TAI) teams, and two force protection teams. Refer to *Figure 8-1*.

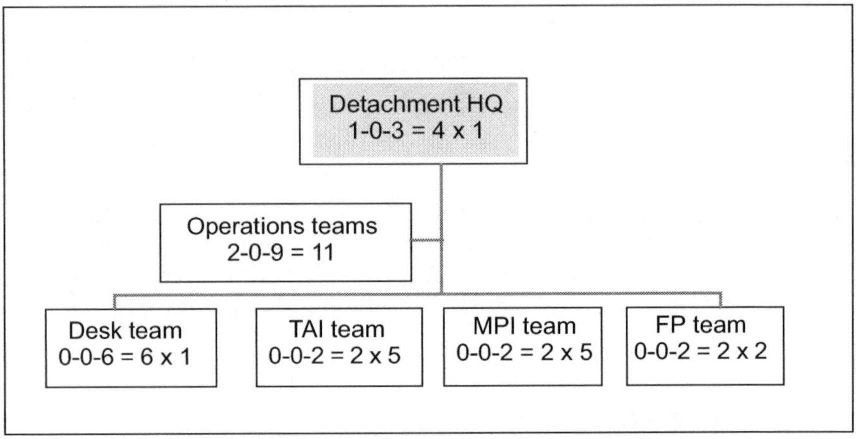


**Figure 8-1. L&O Augmentation Detachment**

8-11. When deployed, detachment and teams are under the overall operational $C^2$ of the PM deployed for the operation. Those teams may be attached to MP battalions and companies, as the situation requires. In war, the L&O augmentation detachment is assigned to an MP brigade and further attached to MP battalions and division MP companies as required. The L&O augmentation detachment consists of the following teams:

- Operations. The operations team provides $C^2$; supervises traffic, force protection, and investigative operations; and provides an evidence custodian to ensure that the chain of custody for evidence is maintained.

- Desk. A desk team establishes and operates the MP desk and coordinates law enforcement patrol activity.
- MPI. MPI teams provide the technical personnel to investigate crimes that do not fall within the investigative purview of the CID and conduct surveillance operations. Each MPI team consists of two MP with additional skill identifier (ASI) V5. These teams have no organic vehicles or communications capabilities.
- TAI. TAI teams provide the technical personnel to conduct TAIs. Based on operational requirements, TAI teams augment PM and MP organizations to ensure unimpeded movement and regulation enforcement on the MSRs. Each TAI team consists of two MP with ASI Q9.
- Force protection. Force protection teams provide technical personnel to assist in the safeguarding of personnel and equipment according to the command policies and field operating procedures. The role of the force protection team is to assist, inspect, and educate. Force protection teams provide expertise for safeguarding personnel and PRC for refugees, displaced persons, and civil disturbance operations. They enhance the unit's ability to ensure that soldiers and units are aware of appropriate personal security safeguards. Additionally, force protection teams assist customs operations teams in the establishment of sterile areas. Each force protection team consists of MP NCOs.

8-12. MP brigade commanders employ L&O augmentation teams based on the METT-TC to support the US forces operating within the brigade AO. While the number of teams is based on the troop population supported, whenever possible the team AO should coincide with the MP battalion and the division AO. The detachment HQ, operations, desk, and force protection teams should be collocated with the supported battalion HQ or division PM. If transportation requirements for these teams exceed the capabilities of the operations team, the supported unit must provide transport.

## COMPANY AND PLATOON LEVEL LAW AND ORDER OPERATIONS

8-13. As previously stated, L&O augmentation teams may not arrive until a theater matures and L&O operations become the priority. Corps and division MP units must be prepared to conduct full-scale L&O missions. When ordered to establish L&O operations, the company commander determines mission requirements that include the following:

- Special equipment (vehicles, communications, and barricades or traffic cones).
- Specialized personnel support (a linguist, HN police, and PSYOPS or civil affairs personnel).
- CCIR.
- Threat assessments from MI and the CID.

8-14. The commander and platoon leaders conduct a map reconnaissance of the AO and determine the platoon areas of responsibility. When the tactical situation allows and adequate structures are available, the commander selects a building that is centrally located within the AO to establish the MP operations center. The facility should, at a minimum, offer areas for the following:

- An MP station desk.
- Offenders processing.
- Good radio transmissions.
- Arms, evidence, and property storage.
- Suspect detainment.

- Latrine facilities.

8-15. The company operations section operates the MP operations center and organizes it into MP sections that include—

- MP operations.
- Administration.
- Force protection.
- An MP desk.
- MP investigations.
- TAI.
- MWDs (if attached or assigned).
- Customs (if attached or assigned).

8-16. MP that are school-trained and have been awarded an ASI should be in charge of the appropriate section.

## MILITARY POLICE OPERATIONS AND ADMINISTRATION OPERATIONS

8-17. The company operations sergeant is in charge of the day-to-day activities of the MP operations center. He instructs platoons regarding patrol patterns and distribution and coordinates for specialized equipment and personnel. He reviews all MP reports and the MP blotter. The center maintains close liaison with HN and allied military and civilian police agencies. The operations sergeant directs and supervises the administration section and provides guidance to other sections.

## FORCE PROTECTION

8-18. The company physical security NCO (ASI H3) coordinates and directs the efforts of force protection for the commander. The force protection section is responsible for reviewing the security measures of all critical facilities and assets within the AO. The force protection NCO reviews points designated as critical by the senior commander and identifies criminal and operational threats. He develops a draft mission-essential or vulnerable area (MEVA) list and submits the list to higher HQ for approval. Once the list is approved, he conducts vulnerability assessments of points on the MEVA list. He applies risk analysis procedures according to *AR 190-51, AR 190-11*, and *DA Pamphlet 190-51* and ensures that appropriate army antiterrorism force protection standards are met according to *AR 525-13*. He reassesses assets for addition to or deletion from the MEVA list or changes in priority. The role of the physical security NCO is to assist, inspect, and advise commanders on force protection measures.

## MILITARY POLICE DESK

8-19. The MP desk coordinates MP patrol activity, documents those activities, and reviews incident reports to ensure accuracy and completeness. The MP desk is also responsible for referring incidents to the appropriate investigative agency for further action.

8-20. The MP desk operates continuously during the conduct of L&O operations. It should be located as near as possible to the center of the L&O operational area, and have communications throughout the AO. When possible, existing facilities should be used for the MP station, but requirements for communications, transportation, and proximity to troops and facilities may rule out existing structures. The MP desk should be located within a reasonable distance of the HN police desk during stability, support, or other operations involving interaction with HN authorities. It may be necessary to establish substations if the area covered or the volume of activity is large. The MP station should have, at a minimum, the following:

- Areas and lighting for preparing reports and other paperwork.

- Communications with the MP elements performing L&O and the command and populace that the L&O operations are supporting.
- Detention areas where persons can be secured by a guard and/or physical constraints, such as a detention cell.
- Latrine facilities.

8-21. MP establish joint service, multiagency, or multinational operations when appropriate for the populace being provided L&O support or as directed. The PM establishes agreements before initiating joint operations to prevent conflicts in jurisdiction, documentation, and communication. Those agreements may be informal at first but should be formalized as soon as the situation permits. Refer to *FM 19-10* for more information about operating a MP desk.

## MILITARY POLICE INVESTIGATIONS

8-22. Crimes occurring within an AO may require an in-depth investigation, depending on the seriousness of the offense and the availability of investigative personnel. Commanders report minor offenses within the unit area to the MP for statistical purposes, but may still investigate the offense themselves.

8-23. MPI investigators handle the most criminal offenses not investigated by CID or the unit commander. MPI normally investigate those offenses cited in *AR 190-30*. The commander ensures that only school trained (ASI V5) personnel are assigned to the MPI section to investigate crimes. Outside continental US (OCONUS) areas, MPI investigate off-post incidents according to SOFA and/or the US and HN agreements.

8-24. CID investigates serious offenses. It has elements in support of all echelons down to division level. CID operates across the full spectrum of army operations, with an emphasis on logistics security (LOGSEC) during wartime operations.

8-25. When MPI has purview to investigate a crime, they use the following investigation measures:

- Take control of the crime scene, ensuring that the crime scene is protected and secure.
- Identify the personnel involved, both suspects and witnesses.
- Identify the type of offense or offenses that have been committed.
- Process the crime scene using notes, sketches, and photographs.
- Collect and secure the physical evidence.

8-26. Once the investigator has processed the crime scene, he pursues leads and conducts investigation activities that include—

- Interviewing victims, suspects, subjects, and witnesses, and obtaining written statements from them.
- Preparing case documents and the required reports.
- Collecting related documents necessary to support the investigation.
- Coordinating with the Staff Judge Advocate (SJA).
- Submitting the final draft report for review by the MPI supervisor.

8-27. The commander appoints an evidence custodian to maintain evidence obtained during criminal investigations. The evidence custodian coordinates the needed laboratory examinations and ensures the proper disposition of evidence. It may be necessary to coordinate with the CID for this support.

8-28. Joint investigative activities, such as drug or black market suppression teams, may involve investigators from CID, MPI, or other US services or agencies. Such operations may require close liaison and cooperation with HN or allied civil or military agencies.

8-29. Refer to *FM 19-20* for more information about criminal investigation techniques and procedures.

## TRAFFIC ACCIDENT AND INVESTIGATION

8-30. MP help reduce nonbattle casualties and the loss of equipment by ensuring that vehicles are operated according to regulations and determining the cause of accidents. Traffic enforcement supports the commander's intent in HNs by reducing the likelihood that military traffic will be seen as a threat to the native population. Thorough investigations of traffic accidents can—

- Identify the personal, environmental, and equipment factors that caused or contributed to the accident.
- Document the facts of the incident for future criminal or civil actions.

8-31. Traffic enforcement measures vary according to the operational and political environment. Those measures may include the following:

- Operator license and dispatch checks.
- Safety inspection checkpoints.
- Vehicle load and route restrictions enforcement.
- Speed control measures.

8-32. Traffic accidents are investigated on the scene whenever the tactical situation permits. MP investigate traffic accidents involving military personnel or vehicles within their AO. When the HN or other authorities have primary responsibility for an accident investigation involving US forces or equipment, MP conduct a concurrent investigation.

8-33. The first MP that responds to a traffic accident is responsible for securing and protecting the accident scene. When necessary, they administer medical attention to the injured persons and implement traffic control measures. MP safeguard classified materials and take the appropriate action to identify and contain hazardous materials at the scene.

8-34. The TAI team takes control of the accident scene when it arrives. Whenever possible a school-trained (ASI Q-9) traffic accident investigator processes the accident scene.

8-35. Once the TAI team receives a briefing from the senior person present, it takes the following steps to investigate the accident:

- Collect physical evidence to include evidence of drug or alcohol involvement.
- Photograph the scene.
- Interview victims and witnesses.

8-36. Once the TAI team has the necessary information, it clears the scene. To do this it ensures that the vehicles and debris are removed and reestablishes traffic flow. The TAI team then prepares a final report and submits it to the proper authority.

8-37. Accidents involving fatalities or life-threatening injuries require a serious incident report (SIR) according to *AR 190-40*.

8-38. Refer to *FM 19-25* for more information about traffic enforcement techniques and conducting TAIs.

## MILITARY WORKING DOGS

8-39. MWD teams further enhance L&O and customs operations with capabilities to detect explosive devices and residue and controlled substances, and detect, track, control, and apprehend personnel.

8-40. Mission support requirements determine the number of functional MWD teams that make up a wartime, mission-oriented MWD team. These mission-oriented MWD teams are assigned to the PM. MWD teams conducting explosive or narcotic detection are unable to provide security for themselves and require security by the supporting unit.

8-41. Explosive, narcotics, and patrol teams each consist of three handlers and three working dogs. This allows each team to provide 24-hour support for a mission that requires one MWD or up to three short-duration missions. There is also a kennel master team, which provides technical supervision and is responsible for establishing kennel operations.

8-42. The explosive or patrol team provides the capability to detect explosive devices or residue in support of personal protection, MOUT, health and welfare, crime scene, and customs operations. When not required for explosive detection, this team functions as a patrol team.

8-43. The narcotic and patrol team provides the capability to detect controlled substances in support of the crime scene, health and welfare, and customs operations. When not required for narcotics detection, the team functions as a patrol team. The patrol team provides the capability to search for, track, detect, and control personnel and augment security in support of personal protection, MOUT, criminal apprehension, force protection, EPW, refugee or displaced person, and crowd control operations.

8-44. Refer to *AR 190-12* for more information about the MWD program, to include standards for selecting and retenting handlers and dogs, training teams, constructing kennels and training areas, maintaining and disposing of dogs, using teams, and using force. Refer to *DA Pamphlet 190-12* for more information about the MWD program.

# UNITED STATES CUSTOMS SUPPORT

8-45. Units (and individuals) returning to the customs territory of the US (CTUS) have the responsibility to ensure that their vehicles and equipment meet US customs, other governmental agencies, and US Department of Agriculture (USDA) standards. Units request and receive customs support to ensure that they comply with all the established requirements. Trained military personnel, normally from the unit deploying to CTUS, conduct military customs inspections in the absence of US customs personnel. Senior military customs inspectors (SMCIs) train and supervise unit personnel on custom's responsibilities and requirements. SMCIs supervise the customs operations. The SMCIs supporting the theater deploy as part of the customs operations team.

## CUSTOMS OPERATIONS TEAM

8-46. The customs operations team supports the commander's requirement to ensure that personnel, equipment, and materiel meet US customs, immigration, USDA, and other federal agency requirements for redeployment back to the CONUS. A complete customs operations team includes an SMCI supervisor team and two 5-member SMCI inspection teams.

8-47. The customs operations team, when deployed, is assigned to the Army service component command (ASCC) PM. It may be under the OPCON of or attached to the theater support command (TSC) PM.

8-48. The SMCI supervisor provides technical supervision, staff planning, and coordination for mission requirements with the Army staffs and agencies, redeploying units, and US federal agencies.

8-49. Two SMCI teams provide the capability to support two ports of embarkation (POEs). Before arrival at the

POE, these teams advise deploying units on requirements and procedures to preclude delays or rescheduling of unit and strategic lift transportation. Team members train and supervise selected redeploying unit personnel as inspectors to augment the SMCI team.

## REDEPLOYMENT OPERATIONS

8-50. *AR 700-93* provides guidance on processing and shipping DOD-sponsor retrograde cargo destined for the CTUS. In order for a unit to redeploy successfully, it must—

- Coordinate for site requirements and standards.
- Establish wash and holding areas.
- Determine the existing USDA preclearance requirements.
- Use the expertise of assigned SMCI.

### Site Requirements

8-51. A site includes those areas necessary to getting the unit's vehicles and equipment ready for the customs inspection and shipment. A suitable site includes the following areas:

- Staging.
- Download.
- Vehicle or equipment wash area.
- Inspection.
- Holding (sterile).

8-52. **Staging Area.** A staging area is an open area large enough to handle the number of vehicles and equipment being processed. Units often underestimate the amount of space needed for the staging area. An accurate count of vehicles and personnel redeploying to the CTUS must be received to accurately determine the staging area requirement.

8-53. **Download Area.** The download area is an area that must have a base of cement, asphalt, or clean or coarse gravel at least 6 inches deep. Vehicles arriving from the staging area download equipment here. The download area breaks out equipment and vehicles for wash area activities. Vehicles are taken to the vehicle wash area and equipment is taken to the equipment wash area.

8-54. **Vehicle and Equipment Wash Area.** A vehicle and equipment wash area or wash site consists of a hardstand area with adequate drainage, water outlets, high-pressure steam cleaners, and ramps. Most wash sites resemble a birdbath using long tubes with holes in them. A vehicle passes by the tubing while it pressure sprays the sides and undercarriage of the vehicle. Unit personnel clean vehicles to free them of dirt, vegetation, insects, and so forth, and check for and clean up fluid spillage. Equipment is cleaned meticulously, ensuring that there is no debris or soil present. When vehicles and equipment are assumed to be clean, they are reloaded and moved to the inspection area.

8-55. **Inspection Area.** The inspection area is where the vehicles and equipment are inspected for cleanliness. If vehicles and equipment are clean, they are moved to sterile holding areas. If they are dirty, they are moved to a spot-wash area and spot-cleaned. Vehicles and equipment that were spot-washed are reinspected, and if clean, go to the sterile holding area.

8-56. **Holding Area.** Cleaned vehicles and equipment are placed in the holding area. Unit personnel guard the holding area to ensure that unit members or other personnel do not contaminate or place restricted or contraband items in vehicles or equipment. The holding area must be free of USDA concerns. Some of these concerns include the following:

- Weeds. The download area must be vegetation free.

- Soil. The download area must be completely free of soil and soil particles.
- Pests. Insects and their debris or excrement are a major concern in the download area. Ensure that insects are kept out of the download area.

**NOTE: SMCIs assist units in identifying and eliminating USDA concerns.**

8-57. Vehicles and equipment in the holding area will be loaded onto rail, ship, or other means of transport for redeployment.

## MULTINATIONAL LAW AND ORDER OPERATIONS

8-58. Multinational police operations may involve joint patrols with allied forces, HN military or civilian police, or a combination of these options. Commanders may also choose to operate certain patrols or stations with MP forces of one nationality, with all police operations answering to a combined, multinational operations cell. Because each nationality has its own statutes, regulations, policies, and traditions, police forces should enforce laws only on persons normally under their own jurisdiction. This reduces confusion about authority and the potential ill feelings that can result when citizens are controlled by other foreign forces. If a situation occurs involving HN or third-nation citizens, the MP should contain or monitor the situation until the appropriate police agency can respond to control the situation. MP may be required to control the situation (consistent with ROE or ROI) if the appropriate agency is not reasonably available or the situation is so severe that a delay would endanger lives.

8-59. *The Foreign Assistance Act of 1961* prohibits the military from training HN civilian police. However, MP may be directed to provide initial assistance and training to foreign military or civilian police forces or assist in the creation of these forces, where the local national authority has broken down. This is an exception to *Section 660 of the Foreign Assistance Act*. MP, when directed, conduct training for foreign military police and assist in reorganizing local constabulary forces during stability and support operations. MP provides short-term emergency L&O capabilities until the foreign military or civilian police forces are functional.

8-60. US military training assistance of foreign military or civilian police normally includes the following:

- Special forces. These detachments teach basic skills needed to perform security duties such as small arms training, unit security tactics, radio procedures, and human rights observation and safeguard.
- Military police. MP teach investigative techniques, law enforcement skills, and confinement operations.
- Civil affairs. CAs coordinate with the HN to effectively integrate training with the established political system. They also provide language interpreter support.
- Psychological operations. PSYOPS personnel work to gain and maintain acceptance of the training effort by the civilian population and encourage their acceptance of the new or reconstituted civilian police force.

8-61. The US ambassador and country team provide guidance for implementing all US Department of Justice (USDOJ) training, including police training. US military forces conduct training under the direction of the civilian authorities, and only until the International Criminal Investigative Training Assistance Program (ICITAP) staff or another agency can assume training. Contractors working for ICITAP, USDOJ, or another international agency may also provide instruction and assistance, and frequently assume those duties from the military as the situation matures. US military involvement in training foreign military or civilian police should not normally exceed 180 days following the end of a conflict.

8-62. MP train HN military or police forces to maintain L&O, using current operational and investigative techniques. They may also act as the agents to equip those forces with the material needed in police operations and train HN personnel on the use and maintenance of that equipment. The training provided may be formal or informal instruction, or consist of on-the-job training with MP working with, and monitoring the performance of, HN police authorities. Classroom instruction is based on training packages developed locally or by higher HQ, or exportable training packages from a service school.

8-63. Once the basics of law enforcement have been learned or reinforced, new or reconstituted police may benefit most from working together with MP in joint operations. ROE and ROI may vary by situation, but some basic guidelines for working with newly constituted or reconstituted police agencies are—

- Allowing the HN authorities to handle situations involving local nationals. US forces assists HN police and monitor the situation. US forces should not appear to be giving orders to the HN agents.
- Ensuring that US forces take care that they do not embarrass or demean HN agents, especially in view of the populace. Soldiers must understand that negative behavior toward their HN counterparts could seriously damage the relationship between US forces and the HN authorities.
- Ensuring that US forces are aware of and sensitive to the cultural traditions and

standards of the HN. What may be acceptable in one culture may be, at best, ill-mannered and possibly, a serious insult or even criminal behavior in another culture.

- Ensuring that US forces appear publicly to be respectful of its agents and leadership and be confident in their abilities. This helps the HN force be effective and respected by the local populace.

8-64. As HN authorities develop professionally and gain the confidence of the HN population, the role of US forces changes. MP shift from being the sole providers of L&O, to being role models and advisors for local authorities, to being monitors of HN police activity.

8-65. Combined police operations reinforce training and provide HN police with the mobility, security, and communications to operate more effectively while conserving critical personnel resources. MP provide support through training and the following, usually combined, operations:

- Law enforcement, security, and criminal information support to the indigenous police force to include L&O operations and administration in a democratic society.
- Patrol and desk.
- Traffic control.
- PRC and civil disturbance.
- MWD.
- Physical security.
- Personnel security.
- Area and route security.
- Counterdrug.
- Antiterrorism.

8-66. In most cases, US forces will continue to exercise jurisdiction over their own forces, especially for offenses of a purely military nature. Local government, once reestablished, may choose to assume jurisdiction in serious or high interest offenses involving US personnel or according to the applicable SOFA or other agreement. MP, in the absence of competent civil authority, establish control over the movement of personnel and supplies and guard critical food supplies and material during production and storage. As the local police and paramilitary forces become effective and assume more security responsibilities, MP elements reduce their activity. They withdraw their support gradually, ensuring that the HN has adequate resources to carry out ongoing programs.

## Chapter 9

# Police Intelligence Operations

The PIO support, enhance, and contribute to the commander's protection program, situational awareness, and battlefield visualization by portraying relevant threat information that may affect the operational and tactical environment. MP and the CID gather police, criminal, or combat threat information during the performance of their functions.

## OVERVIEW

9-1. MP collect and share information during the conduct of their day-to-day operations. This information, which may be police, criminal, or combat, is provided as input to the intelligence collection effort and turned into action or reports. The PIO function ensures that information collected during the conduct of MMS, AS, I/R, and L&O is reported up through the proper channels so that it can be analyzed. MP perform PIO while conducting combat support tasks, such as—

- Checkpoints and roadblocks.
- TCPs.
- Field interviews.
- Criminal investigations.
- Reconnaissance (zone, area, and route).

9-2. Collecting police information during MP activities under the functions of MMS, AS, and I/R may result primarily in combat information, but may lend to police or criminal information. Likewise, the police collection efforts during full-scale L&O operations may result in combat information. MP apply lessons learned from peacetime, conflict, and war environments to the PIO function to gain the most advantage for performing the function. The MP platoon is capable of operating in all functions, day or night, and in various terrain, weather, and visibility conditions. An MP platoon operates independently over large, dispersed areas. MP, along with MI, engineers, and NBC reconnaissance, are key to the IPB (refer to *FM 34-130* for more information about IPB). They are collectors of information in the rear area during sustaining operations or anywhere throughout the battlefield. MP gather police information from contacts that are often very valuable in substantiating or verifying other sources of information. These sources include the following:

- Daily contact with the local populace.
- Combined police patrols with HN military and civilian police agencies.
- Close liaison with local, HN, and multinational police agencies.
- Field interviews.
- Nongovernmental organizations (NGOs).
- Private volunteer organizations (PVOs).

9-3. Refer to *FM 3-19.1* for more information about the processes discussed in this chapter.

## POLICE INFORMATION ASSESSMENT PROCESS

9-4. The PIAP is a tool used to contribute to the PIO function. Information gained through the PIAP may contribute independently or simultaneously to the all-source analysis product (ASAP) and the IPB process. The PIAP, the ASAP, and the IPB enhance and support the commander's force protection program, situational awareness, and battlefield visualization. However, the PIAP independently or collectively—

- Provides the commander with information necessary to improve measures to protect the forces.
- Provides information that clarifies the threat and operational situation.
- Reduces opportunities for threat forces to disrupt military operations and inflict US or friendly casualties.

9-5. The commander and his staff continually monitor the environment at the tactical level consistent with the METT-TC. They apply the military aspects of terrain (OCOKA) as a means of protecting the force. PIO clarify the evolving criminal threat picture for commanders through the PIAP. This helps planners predict threat courses of action against our forces or protected populations.

9-6. MP use PIAP to continuously collect, organize, interpret, and report police and criminal information in support of the IPB. The PIAP consists of 6 steps. Refer to *Table 9-1, pages 9-4 and 9-5*.

## POLICE INFORMATION

9-7. MP gather police information actively or passively. Active collecting efforts result from a direct tasking, and passive collecting efforts result from normal, daily MP operations.

### ACTIVE MODE

9-8. MP perform the PIO function in the active mode when directed by the higher HQ. In this mode, the MP platoon conducts specific missions with the *intent* to actively collect information. Specific MP activities (such as setting up a checkpoint or roadblock) are performed to specifically fulfill a requirement (such as looking for individuals who are in possession of or have knowledge of others in possession of weapons, US military property,

## Table 9-1. Police Information Assessment Process

| Step | Action |
|---|---|
| 1. | Determine the scope of the PIAP by—<br>• Conducting a detailed mission analysis.<br>• Reviewing the mission of the higher HQ and the commander's intent.<br>• Reviewing the CCIR.<br>• Reviewing mission priorities.<br>• Determining the AO and area of interest (AI).<br>• Reviewing the IPB estimates of the higher HQ.<br>• Determining the required information products. |
| 2. | Assemble the working aids and—<br>• Post the applicable maps.<br>• Acquire crime statistics and other related data.<br>• Obtain language aids, such as cultural references and interpreters.<br>• Acquire the necessary automation equipment. |
| 3. | Determine information requirements and coordinate access to data by—<br>• Coordinating with the CID for access to available criminal information developed by CID programs.<br>• Determining police and criminal gaps in the designated geographical area.<br>• Determining the requirement for information collectors to fill police and criminal gaps.<br>• Coordinating with the Intelligence Officer, US Army (S2), the Assistant Chief of Staff, G2 (Intelligence), PSYOP, and other agencies to determine if the information is already available. |

**Table 9-1. Police Information Assessment Process (Continued)**

| | |
|---|---|
| 4. | Recommend and supervise police and criminal collection efforts by—<br>• Coordinating with the S2, the G2, the SJA, the CID, and other applicable police agencies before initiating a collection effort.<br>• Determining which police and criminal information tasks will be assigned to the MP and the CID.<br>• Determining the criteria to satisfy information requirements.<br>• Providing collectors with reporting instructions (such as the reporting frequency and the report format).<br>• Monitoring information collection efforts to prevent duplication of efforts. |
| 5. | Process police and criminal raw data by—<br>• Assembling and assessing the reliability of the data (according to the assessment criteria established above).<br>• Integrating information from the collectors (the MP and CID agents in the field).<br>• Evaluating the data to determine if it meets the requirements.<br>• Developing criminal trends and indicators. |
| 6. | Report and disseminate police and criminal information assessment by—<br>• Recommending MP and CID actions to improve and focus future collection efforts.<br>• Reporting police and criminal information assessments to the S2, the G2, and other MP units.<br>• Reviewing and ensuring that the release of police and criminal information assessments do not violate established guidelines and constraints.<br>**NOTE: Refer to *Appendix E of FM 3-19-1* for a further discussion of PIAP.** |

and so forth). MP gain valuable police, criminal, and combat information while conducting these operations. A standardized checklist enhances the information collection effort and aids in the analysis of the information collected. The checklist indicates a pattern in the behavior of the local nationals. It shows what the local nationals are transporting, to where they are transporting the items, and so forth. The checklist may include—

- The number and types of vehicles stopped. Identifying marks, license plate numbers, and any signs displayed on the vehicles are recorded and reported.
- The number of passengers in the vehicle. The nationality, age, and sex mix of the passengers are recorded and reported.
- The type and quantity of cargo.
- The vehicle's point of origin and destination.
- The stated reason for travel by the passengers.
- The description of arms, ammunition, explosives, and sensitive items found and confiscated from the vehicle.
- The possible or actual sightings of weapons, explosives, or threat forces by the passengers.
- The condition of the passengers.
- The reporting of anything unusual by the passengers.

## PASSIVE MODE

9-9. Every MP conducts the PIO function in the passive mode during their normal day-to-day operations. In the passive mode, PIO are not a stand-alone function and, as such, it cannot be separated from the other MP functions. If while performing MMS, AS, I/R, and L&O, MP receive, observe, or encounter police, criminal, or combat information, they immediately submit a SALUTE, SPOTREP, or other appropriate report to relay information up the chain of command. The information is then integrated into the on-going PIAP and forwarded to the higher echelon S2 and G2 for IPB applications.

# RESPONSIBILITIES

9-10. The responsibilities for the PIO function start with the MP company. The company has the overall responsibility of supervising the collecting and reporting of information to the higher HQ. (Refer to *FM 34-2* for more information about the collection management process). Training scenarios can be developed to enhance the PIO information collecting process of both the active and passive mode. Additionally, the scenarios stress collective training (such as processing police information) and individual training (such as improving interpersonal communication and interview skills).

## PLATOON LEADER

9-11. When the platoon leader is tasked to conduct PIO collecting and reporting, he initiates TLPs for the mission. The platoon leader—

- Coordinates with the higher HQ for the CCIR, police and criminal information requirements (PCIR), and threat estimates. The PCIR allows additional focus on local threats.
- Establishes liaison with civil and MP forces and law enforcement agencies in the AO.
- Coordinates with the local and HN police to determine the existence of organized crime in the AO and the identification of current and emerging criminal leaders and associates.
- Coordinates with the local and HN police and the populace to identify the types of criminal activity (such as smuggling, counterfeiting, narcotics, extortion, and so forth) in the AO.
- Reports information of potential intelligence value by—

- Gathering collected information from squads and teams.
- Consolidating original reports and sketches of potential intelligence value and forwarding them through intelligence channels for analysis and a determination of their intelligence value.

## PLATOON MEMBERS

9-12. The platoon sergeant, squad, and team leaders, down to the junior member of the platoon share in seeking out sources of information by—

- Collecting R&S information that supports the commander's overall PIOR and PCIR collection efforts while on a routine patrol. They accomplish this by—
  - Noting terrain information pertaining to streets, roads, canals, subterranean systems, built-up areas, cities and villages, and the impacts of weather on the terrain.
  - Collecting information on progovernment and antigovernment individuals and groups who might disrupt L&O during protests, strikes, riots, and other spontaneous or organized efforts.
  - Identifying private establishments that may be a target or whose presence or operations contribute to the disruption of L&O (such as gun shops, pawnshops, and liquor stores).
  - Identifying critical infrastructures such as power stations, water works, radio and television stations, telephone and communication facilities, public transportation, and other establishments that may be critical to the sustenance of the community.

- Identifying EPWs, stragglers, and DCs who may have information of potential intelligence value and reporting it to the chain of command. Use a SPOTREP, SITREP, or a SALUTE report or a format directed by the chain of command to report information. SOPs may also be used. Refer to *Appendix D* for report formats

# Appendix A

# Metric Conversion Chart

A-1. Use *Table A-1* to convert from US to metric measurement and metric to US measurement.

**Table A-1. Metric Conversion Chart**

| US Units | Multiplied By | Equals Metric Units |
|---|---|---|
| **Length** | | |
| Feet | 0.30480 | Meters |
| Inches | 2.54000 | Centimeters |
| Inches | 0.02540 | Meters |
| Inches | 25.40010 | Millimeters |
| Miles (statute) | 1.60930 | Kilometers |
| Miles (nautical) | 1.85320 | Kilometers |
| Yards | 0.91400 | Meters |
| **Area** | | |
| Square inches | 6.45160 | Square centimeters |
| Square feet | 0.09290 | Square meters |
| Square yards | 0.83610 | Square meters |
| **Volume** | | |
| Cubic inches | 16.38720 | Cubic centimeters |
| Cubic feet | 0.02830 | Cubic meters |
| Cubic yards | 0.76460 | Cubic meters |
| Gallons | 3.78540 | Liters |
| Fluid ounces | 29.57300 | Milliliters |
| Quarts | 0.94600 | Liters |

## Table A-1. Metric Conversion Chart (Continued)

| US Units | Multiplied By | Equals Metric Units |
|---|---|---|
| **Weight** | | |
| Ounces | 28.34900 | Grams |
| Pounds | 453.59000 | Grams |
| Pounds | 0.45359 | Kilograms |
| Short tons | 0.90700 | Metric tons |
| Long tons | 1.01600 | Metric tons |
| **Metric Units** | **Multiplied By** | **Equals US Units** |
| Centimeters | 0.39370 | Inches |
| Meters per second | 2.23700 | Miles per hour |
| Millimeters | 0.03937 | Inches |
| Kilometers | 0.62137 | Miles (statute) |
| Kilometers | 0.53960 | Miles (nautical) |
| Meters | 3.28080 | Feet |
| Meters | 39.37000 | Inches |
| Meters | 1.09360 | Yards |
| **Area** | | |
| Square centimeters | 0.15500 | Square inches |
| Square meters | 10.76400 | Square feet |
| Square meters | 1.19600 | Square yards |
| **Volume** | | |
| Cubic centimeters | 0.06100 | Cubic inches |
| Cubic meters | 35.31440 | Cubic feet |
| Cubic meters | 1.30790 | Cubic yards |
| Milliliters | 0.03380 | Fluid ounces |
| Liters | 1.05700 | Quarts |
| Liters | 0.26420 | Gallons |
| **Weight** | | |
| Grams | 0.03527 | Ounces |
| Kilograms | 2.20460 | Pounds |
| Metric tons | 1.10200 | Short tons |

## Appendix B

# Media Relations

Military operations are carried out, especially in the initial phases, under the full glare of public scrutiny. The press can distribute reports and pictures faster than the news can be released by the command by using satellites and modern communications technology. Incidents, sometimes embroidered or slanted towards a partisan viewpoint, are screened on the television the same day and in the countries that are parties to the dispute and their allies.

## OVERVIEW

B-1. Journalists fall back on speculation when information is withheld. Such speculation, although usually inaccurate, is often near enough to the truth to be accepted as such by large sections of the public and even by the governments. Belligerents find it advantageous to leak part of a story to the press to build public support for their position. On occasion, such activities can grow into a fully orchestrated press campaign.

## MEDIA INTERACTION

B-2. Establish procedures for media interaction and incorporate them into SOPs. Know what (such as a simple theme which they can tie their responses back to) to discuss with the media and what to refer to the PAO. Ensure that deploying units receive a predeployment briefing from the installation PAO to clarify what may or may not be discussed with the media. Ensure that this action occurs before any exposure to the media. Do not make off-the-record statements in briefings or discussions with the media. Public or media knowledge of any classified activity associated with an operation does not imply or mean that the information is unclassified or may be released or confirmed.

B-3. Ensure that MP are courteous, respond candidly, speak effectively, and continue the mission when interacting with broadcast and print reporters and photographers. This relationship with the media prevents any misgivings. The media will report more accurately and not feel that information is being withheld. The media provides its audiences with real-time information of varying accuracy and completeness.

B-4. The media can be a valuable tool or a lost asset during military operations. It can potentially have a quick and pervasive impact on the plans and operations of commanders, and can change the public's opinions and perceptions about military operations. It can target audiences whose support is crucial to the desired end state of an operation. MP may be the centerpiece for humanitarian assistance or during operations to relieve suffering. A reporter or a photographer may show up at a checkpoint or a TCP where local nationals (belligerents) are refusing to cooperate. Ensure that the media is not only documenting the activities and behavior of the belligerents, but is also recording the military's response to the incident.

# Appendix C

# Training Execution Model

The TEM allows numerous rehearsals and identification of the TTP required for mission accomplishment. It causes each soldier, regardless of position, to understand the unit leader's plan with contingencies and the TTP required. It allows junior leaders and soldiers to see the big picture and understand the tactical environment. Place the company as the OC for the subordinate platoons, and place the platoon leader as the OC for the subordinate squads during internal evaluations.

## OVERVIEW

C-1. The TEM ensures that soldiers are trained to standard, not to time. It is derived from train-up or certification exercises (CERTEXs) before deploying to the National Training Center (NTC).

## TASK IDENTIFICATION

C-2. The TEM takes place after the METL is approved. The approved METL, coupled with an assessment process (such as external or internal evaluations), determines the collective and individual tasks to be covered during training exercises. Once these tasks are identified, implement the following eight-step TEM:

***Step 1.*** Back brief number 1.

***Step 2.*** Back brief number 2.

***Step 3.*** Rock drill.

***Step 4.*** Subunit leaders walk-through.

***Step 5.*** Subunit leaders walk-through with troops.

***Step 6.*** Dry run on the actual lane.

***Step 7.*** Execution.

***Step 8.*** After-action review (AAR).

C-3. The TEM steps complement but do not replace the TLPs. The following outline explains the eight-step model applicable to an MP platoon. The TEM assumes that the platoon's collective task lists are approved at the appropriate level, the OCs are certified and trained, and all resources are available.

## STEP 1: BACK BRIEF NUMBER 1

C-4. During step 1 *(Figure C-1),* the OC issues the mission order to the platoon leader. He briefs the platoon's mission using a sand table, which the platoon built. The platoon leader provides a mission back brief to the OC. The back brief eliminates any confusion or misunderstanding at the very beginning of the mission or plan cycle. The platoon leader's back brief is only with the OC. This allows the interaction between the OC and the platoon leader to take place while leaving the platoon leader's credibility intact. With no subordinate present, the environment to learn and ask questions is established.

C-5. In a nearby AA, leaders train soldiers on critical individual soldier tasks that support the platoon's collective tasks. Here, the PSG checks on the focus and progress of those individual tasks that have been assessed as weak. Additionally, AA actions in step 1 focus on PCIs.

## STEP 2: BACK BRIEF NUMBER 2

C-6. During step 2 *(Figure C-2, page C-4),* the platoon leader returns to the sand table area where the mission order is received from the OC. He briefs the tentative

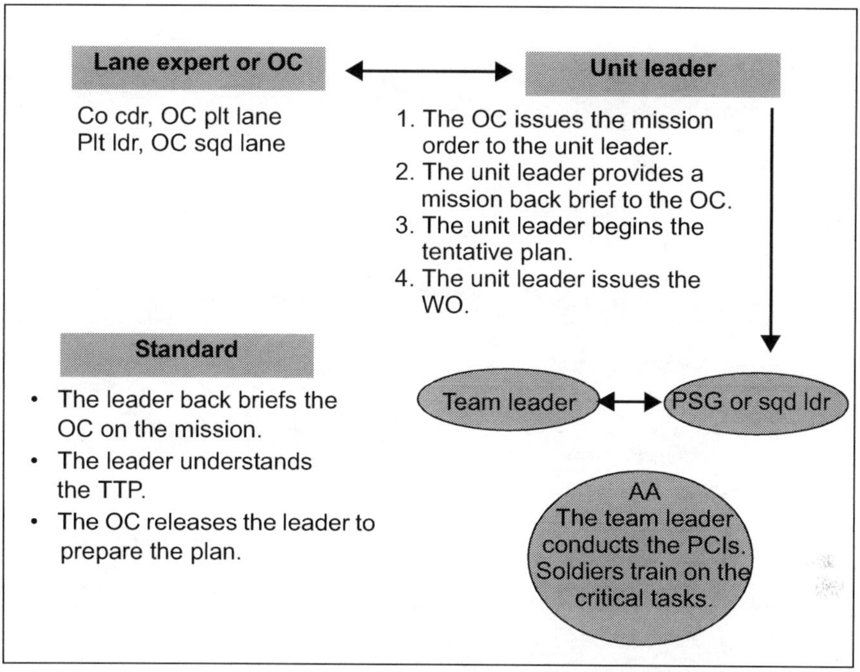


**Figure C-1. Back Brief Number 1**

plan to the OC. Once the OC verifies the MTP standards and pertinent TTPs, the platoon leader briefs squad leaders on the same sand table. The OC remains present to hear the interaction among the platoon's leadership. After the squad leaders are briefed, the platoon leader conducts the leader's reconnaissance. If any modification to the tentative plan is required based on the reconnaissance, the platoon leader briefs the OC. He briefs the squad leaders again if changes are approved. The interaction and planning get more focused after the leader's reconnaissance.

## STEP 3: ROCK DRILL

C-7. During step 3 *(Figure C-3, page C-5)*, the platoon leader and squad leaders conduct a rock drill on a

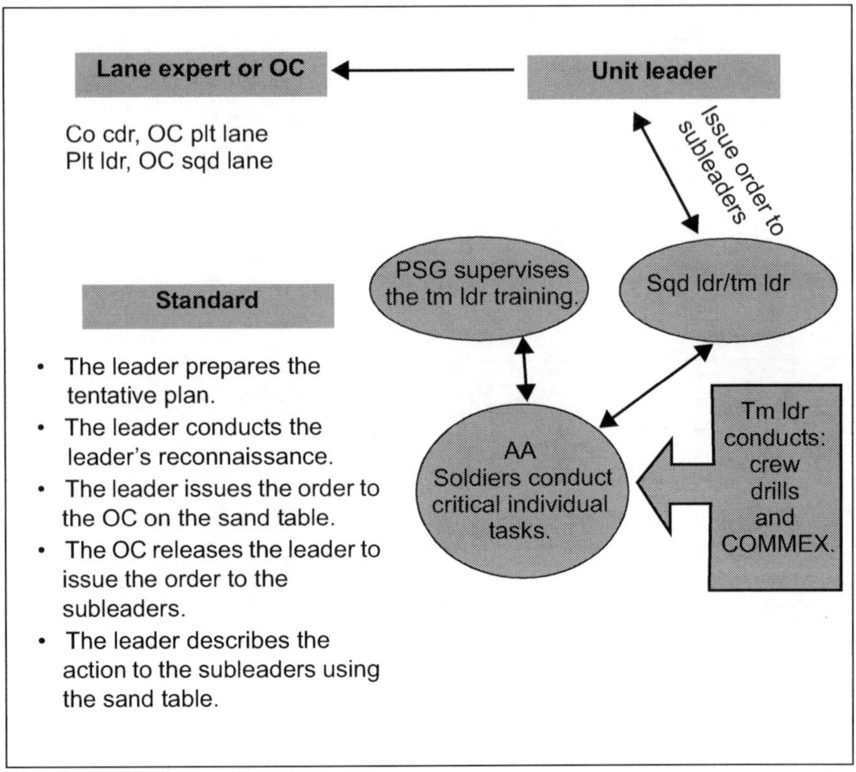


**Figure C-2. Back Brief Number 2**

terrain model that the platoon constructed of the actual training lane. The OC is in an overwatch position. He always assesses the interaction of the leadership and serves as the OPFOR for the rock drill. The platoon leader talks or walks through the plan on the terrain model with all the squad leaders.

C-8. In reaction to the OPFOR, leaders solidify contingencies based on the interaction on the terrain model. The OC's "what if" drills will usually trigger detailed planning. The credibility of the platoon leadership is still intact because the interaction and pertinent questions are not being asked in front of the soldiers. The soldiers are in the AA continuing their individual training and preparing equipment based on the WO received. Locate the PSG in the AA to coach team leaders in their AA activities. Once the OC is sure that the platoon leaders understand the plan, he releases them to conduct a walk-through.

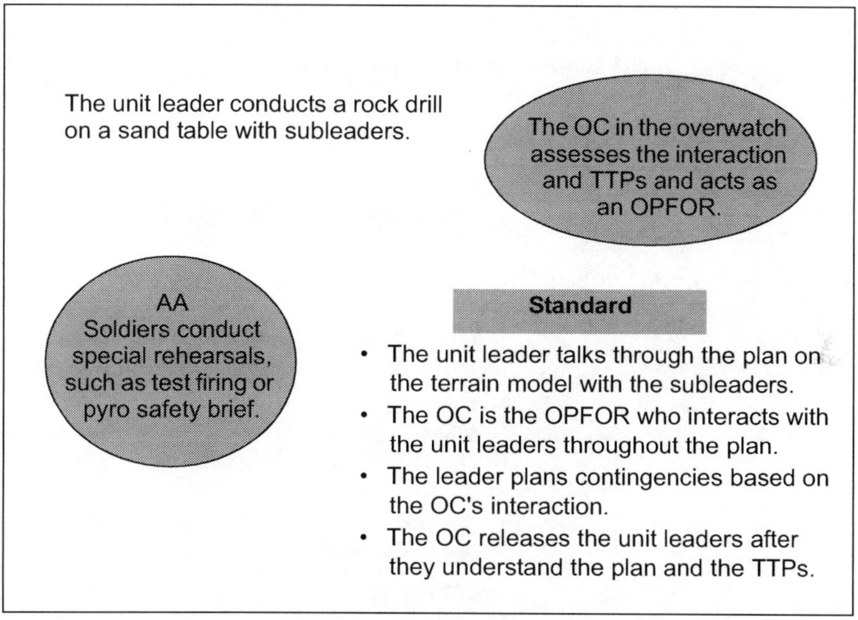


**Figure C-3. Rock Drill**

## STEP 4: SUBUNIT LEADER'S WALK-THROUGH

C-9. During step 4 *(Figure C-4, page C-6)*, the platoon leader displays all terrain features and control measures. During the walk-through, the platoon leadership imitates the movement selected, such as traveling, traveling overwatch, and bounding overwatch. The platoon leadership uses actual radio procedures and call signs according to the platoon's plan. The platoon's leadership demonstrates its grasp of the plan and the mission without troops, showing that credibility is still sound.

C-10. Once the OC is satisfied that the platoon's leadership knows the tasks, he releases them to brief the soldiers on the sand tables previously mentioned. Squad leaders ensure that their team leaders and soldiers understand the mission by showing them on the sand table what their part of the mission is and how it fits into the overall scheme of maneuver.

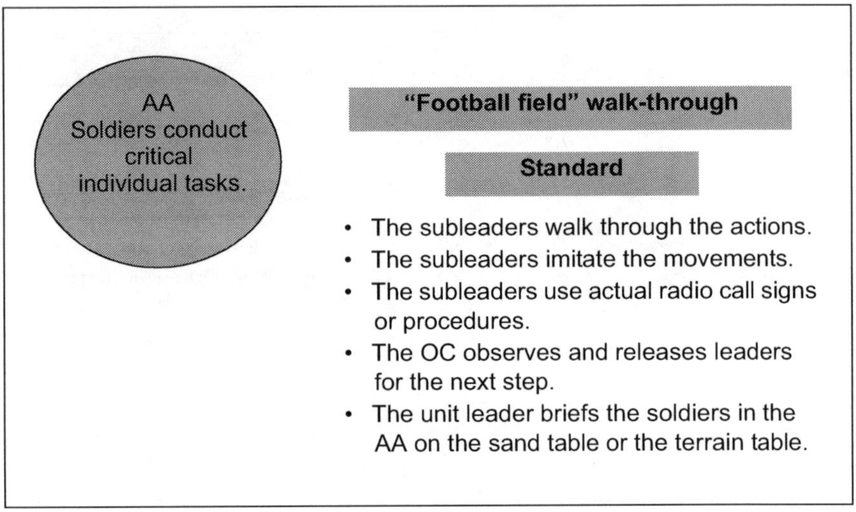


**Figure C-4. Subunit Leader's Walk-Through**

## STEP 5: SUBUNIT LEADER'S WALK-THROUGH WITH THE TROOPS

C-11. During step 5 *(Figure C-5),* the soldiers are assembled and briefed on the mission using sand tables. The MP teams walk and talk through the mission. The platoon walks through its action of the plan; and the soldiers react and dress on their team, squad, and platoon leaders. Shout commands out so that all can hear. Imitate movement as well as radio procedures, call signs, fire commands, calls for suppression, and so forth. The OC who plays the OPFOR and interacts with the platoon through its leaders observes the platoon. Often, while conducting the walk-through with the troops it will be necessary to stop and inform soldiers of any risk hazards and individual responsibilities. The platoon's interaction on the field is a critical component of mission accomplishment. The OC only releases the unit for the next step if he is satisfied that the unit understands the leader's plan and the required TTP.

**Issue warning OPORD**

**Subleaders walk-through**

**All now in final rehearsal**

- The subleader walks through the actions with troops on the sand table.
- The troops act on the leader's commands.
- The subleader imitates movements.
- The subleader uses actual radio call signs or procedures.
- The OC observes and releases the leaders for the next step.

**Figure C-5. Subunit Leader's Walk-Through with the Troops**

## STEP 6: DRY RUN ON THE ACTUAL LANE

C-12. During step 6 *(Figure C-6, page C-8),* the unit follows the leader's plan and conducts a dry run of the mission on the actual training lane. The OC observes to ensure that the platoon follows its plan and conducts the mission to MTP standards. The dry run identifies weak subtasks and allows the platoon to retrain or repeat the dry run. This is a full dress rehearsal and ensures that the platoon is meeting the standard. If the platoon performs satisfactorily on the dry run, the OC releases it to execute at full combat speed.

The unit conducts a dry run of the mission following the leader's plan on the actual lane.

| Actual lane dry run. | **Standard** |
|---|---|
| | • The unit conducts the mission to the MTP standard.<br>• The unit identifies the weak subtasks.<br>• The unit identifies the terrain-induced weakness to the plan.<br>• The unit retrains the weaknesses.<br>• The unit does the lane again.<br>• The OC releases the unit to execute at the full combat speed if satisfied with its performance. |

**Figure C-6. Dry Run on the Actual Lane**

## STEP 7: EXECUTION

C-13. During step 7 *(Figure C-7),* the unit executes the mission at full combat speed against an unrestricted OPFOR that is controlled by the OC. This phase involves force on force, controlled by the OC, with a complete simulation system (such as multiple-integrated laser engagement system [MILES]). The OC ensures that soldiers adhere to MTP standards and TTPs. The OC calls for an end of exercise (ENDEX) when the training lane has run to culmination (such as the mission is accomplished and all the training effects took place). The OC has enough information to assess the platoon's performance and initiate the next step—the AAR.

The unit executes the mission.

| | Standard |
|---|---|
| Actual lane | • The unit conducts the mission to the MTP standards.<br>• The unit has full use of OPFOR and MILES.<br>• The unit uses force on force or is unrestrained.<br>• The OC controls the OPFOR.<br>• The OC calls for an ENDEX. |

**Figure C-7. Execution**

## STEP 8: AFTER-ACTION REVIEW

C-14. During step 8 *(Figure C-8, page C-10),* the unit conducts an AAR of the lane. If possible, it conducts the AAR on the objective site that overlooks the lane. The OC facilitates the AAR. He or the platoon leader states the tasks, condition, and standards for the training objective. The OC ensures that the AAR becomes a review that is driven from the soldier level up toward the top. He elicits the soldiers to identify their actions, both right and wrong, and gets the leaders to do the same. The OC states what was right and wrong with the mission according to the MTP and the appropriate TTPs, always striving to leave the AAR on a positive note to improve soldier and unit morale and performance.

The unit conducts an AAR of the training event.

**Standard**

- The OC facilitates the AAR.
- The unit members state the training objectives (task, condition, standards).
- The unit members have a bottom-up discussion.
- The unit members have the soldiers identify their mistakes.
- The unit members have the leaders identify their mistakes.
- The OC assesses the task according to the MTP.

**NOTE: The assessment dictates whether to go to the next training or redo the training.**

**Figure C-8. AAR**

## Appendix D

# Orders and Reports

Orders are written or oral communications directing action. They are based on plans or the receipt of a new mission. Use reports extensively to provide information to higher, lower, and adjacent commands. Although there are many, this appendix is limited to combat orders and reports. Refer to *FM 101-5* for more information about orders.

## ORDERS

D-1. Combat orders pertain to strategic, operational, or tactical operations and their service support. The combat orders used by MP are WOs, OPORDs, and FRAGOs.

### WARNING ORDERS

D-2. Refer to *FM 101-5* for an example of a WO. WOs provide advance notice of an action so that MP can use available time for preparations, and they—

- Are issued at each level down to the squad.
- Are issued to subordinates in as much detail as possible.
- Are issued as brief, oral, or written messages.
- Are a part of planning the use of available time. WOs should be kept simple.
- Describe the operation and the preparations to be made before the OPORD is issued.

- Include the situation, the mission, special instructions, and the time and place for issuing the OPORD.

## OPERATION ORDERS

D-3. OPORDs coordinate actions to carry out the commander's plan for an operation, and they—

- Explain how leaders at different levels want the operation conducted. To ensure that maneuver and other non-MP units carry out actions or provide a form of support that MP plans and operations depend on, state the required actions or support in the coordinating instructions paragraph of the division and brigade OPORDs.
- Have a great impact on how subordinate leaders employ units and perform missions. For example, the MP platoon leader's latitude to employ the unit could be restricted, based on how the company commander wants the operation conducted.
- May be written, oral, graphic (such as traces and overlays), or a combination of these forms.
- Are usually written when prepared at company level and above.
- Have a standardized system of designating days and hours in relation to an operation or an exercise.
- Follow a prescribed written format, that contains a classification, heading, body, and ending.
- Have a prescribed five-paragraph format for the body of the order, which includes the following:
  - Situation.
  - Mission.
  - Execution.
  - Service support.
  - Command and signal.

D-4. The risk management process is greatly enhanced by incorporating the risk assessment into the OPORD. The commander integrates risk management into the order, rather than treating it as an afterthought. Leaders at all echelons assess the effectiveness of their units by reviewing how well hazards are identified and risk controls are specified in oral and written orders. Refer to *Figure D-1, pages D-5 through D-10,* for an example of an OPORD format. (Refer to *FM 101-5, Appendix H*, for further discussion of an OPORD.)

D-5. FRAGOs issue supplemental instructions to a current OPORD or OPLAN while the operation is in progress, and they–

- Contain missions of immediate concern to subordinate units.
- May be either written or oral.
- Provide brief, specific, and timely information without loss of clarity.
- Have no prescribed format. Prevent confusion OPORD.
- May be issued to change an OPORD that has already been issued. As such, only those items from the original OPORD that have changed are included in the FRAGO, as long as clarity is not sacrificed.

## STANDING OPERATING PROCEDURES

D-6. SOPs prescribe routine methods to be followed in operations, and they–

- Supplement other combat orders.
- Reduce the number, length, and frequency of other orders. Because the SOP is a standing order, do not repeat the information contained therein in other orders unless emphasis is desired.

- Have no prescribed format. However, ensure that the subordinate unit SOPs follows the format of the next higher HQ SOP if possible.
- Prescribe actions of a recurring nature that lend themselves to definite or standardized procedures. Examples include the following items:
  - Troop safety matters
  - Methods of reporting unit locations.
  - Measures for handling captured personnel or equipment
  - Distribution of supplies.
  - Standard communication procedures for exercising $C^2$.
  - Other items that lend themselves to standardization.

**NOTE: These items are generally the constants in what is otherwise a frequently changing set of circumstances.**

## REPORTS

MP contribute to the commander's situational awareness by providing timely information to higher HQ. They collect and report information in the form of administrative, operational, or intelligence reports. These reports ensure that the commander receives continuous current information. *Tables D-1* through *D-3, pages D-11 through D-18,* are examples of reports used by MP when conducting CS operations. Refer to *FM 101-5-2* for a complete listing of the standardized Army report and message formats.

______________________

**(Classification)**

Place the classification at the top and bottom of every page of the OPORD.

**(Change from oral orders, if any.)**

This statement is applicable only if an oral order is issued by the commander. The phrases "No change from oral orders" or "No change from oral orders except paragraph ___ " are necessary here.

**Copy ___ of ___ copies**
**Issuing HQ**
**Place of issue (coordinates)**
**DTG of signature**

Show the place of issue (location of the issuing HQ) on each copy. Show the name of the town or the place in capital letters, the coordinates in parentheses, and the country in capital letters. Both may encode.

The effective time for implementing the plan or order is the same as the DTG, unless coordinating instructions state otherwise. When orders apply to units in different time zones, use time zone Zulu (Z). In OPORDs and service support orders, list the time zone applicable to the operation in the heading of the order following the references. When an order does not specify the actual date and hour for beginning an operation, apply the proper reference designations.

**Message Reference Number**

Message reference numbers are internal control numbers that the unit signal officer issues and assigns to all orders. The unit's SOP normally describes the number's allocation and use. Using the number allows an addressee to acknowledge receiving the message in the clear.

**OPORD** ________ (code name)
(Number)

Orders normally contain a code name and are numbered consecutively within a calendar year.

**Reference(s):** The heading of the order includes a list of maps, charts, datum, or other related documents the unit will need to understand the plan or order. The user does not need to reference the SOP but may refer to it in the body of the order. The user references a map using the map series

**Figure D-1. OPORD Format**

number (and country or geographic area, if required), sheet number, name, edition, and scale, if required. Datum is the mathematical model of the earth used to calculate the coordinates on any map. Different nations use different datums for printing coordinates on their maps. Reference the datum in the marginal information of each map.

**Time Zone Used Throughout the Order:** The time zone used throughout the order (including annexes and appendixes) is the time zone applicable to the operation. Operations across several time zones use Z time.

**Task Organization:** Describe the allocation of forces to support the commander's concept. Show task organization in one of two places: preceeding paragraph one or in an annex, if the task organization is long and complicated.

**OPORD** ________ - __________________________
(Number) (Issuing HQ)

(Place this information at the top of the second and any subsequent pages of the OPORD.)

1. **SITUATION.**

a. **Enemy Forces.** Express this information down to two enemy echelons below yours (for example, battalions address platoons or companies address squads). Describe the enemy's most likely and most dangerous COA. When possible, provide a sketch of the enemy COA in lieu of verbiage (Appendix ___ [sketch] to Annex ___ [title]). Include an assessment of terrorist activities directed against US government interests in the AOs. Reference more sources using the final subparagraph to refer the reader to the documentation.

b. **Friendly Forces.** Include the mission, the commander's intent, and the concept of operations for HQ one and two levels up. Subparagraphs state the missions of the flank units (left, right, front, and rear) and other units whose actions would have a significant bearing on the issuing HQ.

c. **Attachments and Detachments.** State when, and for how long, units are attached or detached to the operation. Do not repeat information already listed under task organization or in a task organization annex. State "See Annex ___ (title)" here if a task organization annex is necessary.

**Figure D-1. OPORD Format (Continued)**

2. **MISSION**. State the mission derived during the planning process. This statement describes the task and purpose of the operation and clearly indicates the action to be taken and its reason. There are no subparagraphs in a mission statement. The mission statement covers on-order missions.

3. **EXECUTION**.

**Intent**: State the commander's intent derived during the planning process. It is a clear, concise statement of what the force must do to succeed with respect to the enemy and the terrain. It provides the link between the mission and the concept of the operation, which provides the basis for subordinate initiative and decentralized mission execution. Always include the desired end state.

a. **Concept of Operations**. The concept of operations may be a single paragraph or divided into two or more subparagraphs. It is based on the COA statement from the decision-making process and, at a minimum, should address close, deep, rear, and security operations as well as designating the main effort. The commander uses this subparagraph when he feels he must supply enough detail to ensure appropriate action by subordinates in the absence of additional communications or further instructions. When an operation involves two or more clearly distinct and separate phases, the concept of operations may be prepared in subparagraphs describing each phase. Ensure that the concept statement is concise and understandable. The concept describes the following:

- The employment of maneuver elements in a scheme of maneuver.
- A plan of fire support or *scheme of fires* supporting the maneuver with fires.
- The integration of other major elements or systems within the operation, including reconnaissance and security elements, intelligence assets, engineer assets, and air defense.

**NOTE: Depending on what the commander considers appropriate, the level of command, and the complexity of any given operation, the following subparagraphs are examples of what may be included within the concept of operations:**

(1) **Maneuver**. State the scheme of maneuver derived during the planning process. Ensure that this paragraph is consistent with the operation overlay. Ensure that this paragraph and the operation overlay is

**Figure D-1. OPORD Format (Continued)**

complementary and adds to the clarity of, rather than duplicating, each other. Do not duplicate information to be incorporated into the unit subparagraphs and coordinating instructions.

(2) **Fires**. State the scheme of fires to support the concept, and include priorities of and restrictions for fire support.

(3) **Reconnaissance and Surveillance**. Specify the reconnaissance and surveillance (R&S) plan and explain how it ties in with the basic concept of operations.

(4) **Engineer**. Clarify the scheme of engineer support to the plan, paying particular attention to the integration of engineer assets and obstacles, including the priorities of effort.

(5) **Air Defense**. State the overall concept of air defense in support of the scheme of maneuver if necessary.

(6) **Information Operations**. State the overall concept of information operations in support of the scheme of maneuver. Refer to the appropriate annexes if necessary.

**NOTE: Units required to accomplish specific tasks for information operations and R&S are specified in the appropriate subparagraphs of 3b (tasks to maneuver [subordinate] units).**

b. **Tasks to Maneuver (Subordinate) Units.** Clearly state the missions or tasks for each maneuver (or subordinate element) unit that reports directly to the HQ issuing the order. List the units in the same sequence as in the task organization, including attachments. Use a separate subparagraph for each subordinate element. State the tasks that are necessary for comprehension, clarity, and emphasis. Place tactical tasks that commonly affect two or more elements in subparagraph 3c (coordinating instructions).

c. **Coordinating Instructions.** List only those instructions applicable to two or more units (subordinate elements) and not routinely covered in unit SOPs. This is always the last subparagraph in paragraph 3.

**NOTE: The following are examples of subparagraphs that are generally included as coordinating instructions. Subparagraphs 1 through 5 below are mandatory.**

**Figure D-1. OPORD Format (Continued)**

(1) Time or condition when an order becomes effective.

(2) CCIR.

(a) Priority intelligence requirements (PIR).

(b) Essential elements of friendly information (EEFI).

(c) Friendly force information requirements (FFIR).

(3) Risk-reduction control measures may include such items as MOPP, operational exposure guidance, vehicle recognition signals, and fratricide prevention.

(4) ROE. When lengthy and complicated, include ROE as a separate annex.

(5) Environmental considerations.

(6) Force protection.

(7) Any additional coordinating instructions.

4. **SERVICE SUPPORT**. Address service support in the areas shown below as needed to clarify the service support concept. Refer to any annexes, where necessary. Subparagraphs can include the following:

a. **Support Concept**. State the concept of logistics support necessary to support the operation. Include the concept for—

- Support HQ or support area locations, including locations of logistical bases if not clearly conveyed in the overlay.
- Commander's support priorities.
- HN support.
- Any significant sustainment risks.
- Unique support requirements in the functional areas of manning, arming, fueling, fixing, moving, and sustaining the soldier and the systems.

b. **Materiel and Services**.

c. **Medical Evacuation and Hospitalization**. At a minimum, include frequencies, call signs, and locations of medical support facilities and units, as well as the plan for casualty evacuation (CASEVAC).

d. **Personnel Support**.

e. **EPW**.

f. **Transportation**.

**Figure D-1. OPORD Format (Continued)**

5. **COMMAND AND SIGNAL**.

a. **Command**. State the map coordinates for the CP locations and potential future locations for each echelon CP applicable to the operation. Identify the chain of command if not clearly addressed in unit SOPs.

b. **Signal**. List signal instructions not specified in unit SOPs; and identify the specific SOI addition in effect, required reports and formats, and the times the reports are submitted. List primary and alternate means of communications, as well as pyrotechnic signals and their meanings.

**Acknowledge**: Include instructions for the acknowledgement of the order by addressees. The word *acknowledge* may suffice or a message reference number. Acknowledgement of an order means that it has been received and understood.

**Commander's last name**
**Rank**

The commander or his authorized representative signs the original copy. If the representative signs the original, add the phrase "For the Commander." The signed copy is the historical copy and remains in the HQ files. It is always designated "Copy 1 of ______ copies."

**OFFICIAL:**
(Authentication) Use only when applicable. If the commander signs the original, no further authentication is required. If the commander does not sign, authorization is required by the signature of the preparing officer or individual and only the last name and rank of the commander appear in the signature block.

**Annexes:** List annexes by letter and title and in the sequence by which they were referenced in the order.

**Distribution:** Furnish distribution copies either for action or for information. List in detail those who are to receive the order. If extremely lengthy, refer to an annex containing the distribution list or to a standard distribution list or SOP.

______________________
**(Classification)**

Place the classification at the top and bottom of every page of the OPORD.

**Figure D-1. OPORD Format (Continued)**

**Table D-1. Administrative Reports**

| Types | Who | What | When | Where | How | Content |
|---|---|---|---|---|---|---|
| Casualty | The unit or the section with casualties | The number of dead, wounded, or sick | Upon experiencing casualties or as required by the SOP or HQ | Higher HQ | By the most secure means and encoded if by unsecure radio | *DA Form 1156* or per SOP |
| Personnel | Submitted at company level and higher (input is needed from the squad or section) | Personnel strength accounting and status | Daily | Higher HQ | *DA Form 5367-R* | *DA Form 5367-R* |
| Periodic logistic | Submitted at company level and higher (input is needed from the squad or the section) | Supported strength and status of critical supplies | As supplies become depleted or as required by the SOP | Higher HQ | By the most secure means and encoded if by unsecure radio | • Logistical situation<br>• Supply<br>• Maintenance<br>• Transportation<br>• Service |

**Table D-1. Administrative Reports (Continued)**

| Type | Who | What | When | Where | How | Content |
|---|---|---|---|---|---|---|
| Journal | Units or sections operating independent of the parent organization | Events about a unit or section during a given or specified period | As events occur over the given or specified period | Maintained locally (provide to higher HQ on request) | *DA Form 1594* | *DA Form 1594*<br>• Item<br>• Time<br>• Incident, message, or order<br>• Action taken<br>• Initials |
| Closing | The unit leader | Notice of change of the location | Upon arrival at the new site | Higher HQ | Secure the radio | • Unit<br>• Date and time<br>• CP location<br>• Vehicles and radios<br>• Personnel and weapons |

**Table D-2. Operational Reports**

| Type | Who | What | When | Where | How | Content |
|---|---|---|---|---|---|---|
| Situation and status | Commander or leader closest to the situation | Tactical situation or status | Immediately after a significant event or as specified | Higher HQ. | By the most secure means and encoded if by unsecured radio | • Enemy<br>• Own situation<br>• CS<br>• General<br>• Commander's evaluation |
| Force tracking (diversion, holding, or passing) | Designated TCP and mounted patrols | Unit movement along the MSR | As requested or designated in the SOP | Local movement control unit | Secure voice or encoded | Division transportation office (DTO) or the Provost Marshal Office (PMO) adopted SOP format |
| Offense | TCP and mounted patrols | Violation of MSR regulations | When tasked to conduct MMS and MSR regulations are in force | Through military channels to the driver's commander | Written report. (refer to *FMs 55-10* and *101-5)* | • Date<br>• Driver's name<br>• Name of vehicle commander<br>• Particulars of the offense observed |

**Table D-2. Operational Reports (Continued)**

| Type | Who | What | When | Where | How | Content |
|---|---|---|---|---|---|---|
| EPW | Platoon, squad, and team operating the collecting points | The number of EPWs collected and evacuated | As required | Higher HQ | By the most secure means and encoded if by unsecured radio | Per SOP |
| Intention of laying a minefield | Platoon, squad, or team leader preparing to lay the minefield | Tactical objectives and characteristics of the minefield | Before emplacing the mines | Higher HQ | Written or by a secure radio (refer to *FM 20-32*) | • Purpose of the minefield<br>• Estimated number and type<br>• Location<br>• Proposed start and completion times |
| Initiation of laying a minefield | Platoon, squad, or team leader of the force laying the minefield | Emplacement of the mine | When emplacement begin | Higher HQ (mandatory) | Written or by a secure radio (refer to *FM 20-32*) | • Time begun<br>• Location and target number |

**Table D-2. Operational Reports (Continued)**

| Type | Who | What | When | Where | How | Content |
|---|---|---|---|---|---|---|
| Completion of laying a minefield | Platoon, squad, or team leader of the force laying the minefield | Completion of the minefield | Upon completion of the minefield | Higher HQ | Written or by a secure radio (refer to *FM 20-32*) | • Field is complete and functional (expedite the report and follow up with a hasty protective minefield report) |

**Table D-3. Intelligence Reports**

| Type | Who | What | When | Where | How | Content |
|---|---|---|---|---|---|---|
| Spot | Unit, section, or individual observing the enemy (all the echelons) | Enemy activity and area information of immediate value | Upon contact or as requested | Higher HQ | By the quickest means and encoded if by unsecured radio | • Reporting unit<br>• Date and time of the event<br>• Location and grid coordinates (encoded)<br>• Event (SALUTE)<br>• Original source<br>• Remarks |

**Table D-3. Intelligence Reports (Continued)**

| Type | Who | What | When | Where | How | Content |
|---|---|---|---|---|---|---|
| Meaconing, intrusion, jamming, interference (MIJI) | Soldier experiencing electronic warfare | Correct report according to the type of interference | As soon as possible after the incident | Through signal channels to the higher HQ. | SOI defines the report and how to prepare it | • Type of report<br>• Frequency or channel affected<br>• Victim designation and call sign<br>• Type of emission or audio characteristics<br>• Coordinates of the affected station |
| NBC 1 | Observer | Observation concerning suspected NBC attacks and resulting hazardous areas | Upon observing a suspected enemy NBC attack | Higher HQ. | Secure radio | • Observer's location<br>• Direction from the observer<br>• Date-time group (DTG) of the attack<br>• Type of burst |

**Table D-3. Intelligence Reports (Continued)**

| Type | Who | What | When | Where | How | Content |
|---|---|---|---|---|---|---|
| NBC 4 | Element in contact with radiation | Radiation dose rates | Upon contact with radiological contamination or as directed | To higher HQ | Secure radio | • Location<br>• Dose rate<br>• DTG of the measurement |
| Patrol | Patrol leader | Pertinent information pertaining to the patrol's activity | Upon completion of the patrol | To higher HQ | Written | • Designation of the patrol<br>• Maps<br>• Terrain<br>• Enemy<br>• Results of encounters with the enemy<br>• Condition of the patrol |
| Enemy minefield location | Soldiers encountering an enemy minefield | Characteristics of the minefield | Upon locating a minefield | To higher HQ | Secure radio (refer to *FM 20-32*) | • Type of minefield<br>• Location and depth<br>• Enemy weapons or surveillance<br>• Routes for bypassing<br>• Coordinates of lane entry and exit<br>• Width of lanes, in meters |

**Table D-3. Intelligence Reports (Continued)**

| Type | Who | What | When | Where | How | Content |
|---|---|---|---|---|---|---|
| | | | | | | • Map sheet designation<br>• Date and time information collected<br>• Coordinates of minefield boundaries<br>• Estimated time to clear the minefield<br>• Estimated material and equipment required to clear the minefield<br>• Other, such as the type of the mine, new mines, or booby traps |
| SALUTE | Observer | Enemy activity; for example, convoy, patrol vehicles, and aircraft | Upon encountering the enemy's activity | Higher HQ | By the most secure means and encoded if by unsecure radio (refer to *FM 101-5*) | S - Size<br>A - Activity<br>L - Location<br>U - Unit<br>T - Time<br>E - Equipment |

# Appendix E

# Precombat Inspections

Combat missions start with PCIs. These PCIs are an essential part of every mission. The objective of PCIs is to confirm the combat readiness of the unit. Subordinate leaders conduct detailed checks, allowing the unit leader to conduct PCIs in an atmosphere of total preparedness. After the PCI, the unit is prepared to move on the receipt of orders.

## MODIFIED PRECOMBAT INSPECTION CHECKLIST

E-1. A modified PCI checklist *(Figure E-1, page E-2)* ensures that the team has the required equipment to conduct its mission when the unit is short on time. The modified PCI allows the team leader to focus on preparing the team for the mission, while the platoon leader issues the mission to the squad leader. Use the checklist as a recovery checklist when the team returns from the mission. Immediate, proper recovery ensures that the team is ready for the next mission. This checklist is only a guide and can be modified based on METT-TC and the unit SOP.

| Steps | Actions |
|---|---|
| 1 | The squad leader and the team leader receive the WO. |
| 2 | The team leader prepares the squad for movement, while the squad leader is receiving the mission from the platoon leader. |
| 3 | The team leader follows the same checklist, while the squad leader debriefs the platoon leader if recovering from a mission. |
| 4 | The team leader inspects the communications equipment and ammunition in HMMWVs. He—<br>• Conducts a communications check with the platoon CP.<br>• Ensures that proper frequencies are set and that call signs are available.<br>• Ensures that there is a Single-Channel, Ground-to-Air Radio System (SINCGARS), a manpack, accessories, and a battery case.<br>• Ensures that there are batteries for the radios (automated net control device [ANCD], precision lightweight global positioning system receiver [PLGR], OE-254).<br>• Ensures that the Class V basic load is accounted for. |
| 5 | The driver conducts preoperation checks and services on the vehicle. He ensures that—<br>• The correct paperwork is completed, such as the dispatch, folder, and signature. Ensures that the correct TM is on hand.<br>• He has a valid driver's license.<br>• Preventive-maintenance checks and services (PMCS) are performed to a -10 standard.<br>• A current *DA Form 2404* is on hand.<br>• The fire extinguisher is sealed, tagged, and dated.<br>• The engine's coolant, oil, transmission fluid, and washer fluid are at the correct level.<br>• The battery level is correct and that cables are secured.<br>• Lights are operational.<br>• Fuel and water cans are filled and that there are meals, ready-to-eat (MREs) available.<br>• There are no exhaust leaks.<br>• Basic-issue item (BII), on-vehicle material (OVM), pioneer tools, locks, and chains are present on the vehicle.<br>• The load plan is followed and that the vehicle's interior is clean. |

**Figure E-1. Modified PCI Checklist**

| Steps | Actions |
|---|---|
| 6 | The gunner conducts preoperation checks and services on the crew-served weapon and NBC equipment. He ensures that—<br>• A TM for the equipment (-10) is available.<br>• The weapon is oiled, cleaned, and functional.<br>• There is a spare barrel and BII.<br>• There is a T&E mechanism.<br>• There is a wrench.<br>• There is a tripod mount with an adapter.<br>• There are weapon cleaning kits present.<br>• There are night vision devices (NVDs) present.<br>• All the NBC equipment is present and serviceable. |

**Figure E-1. Modified PCI Checklist (Continued)**

# PRECOMBAT INSPECTION CHECKLIST

E-2. This PCI checklist is extensive and time-consuming. Use it when the team has a lot of downtime and has a secure environment in which to work. Once combat operations begin, the secure environment and the downtime will be lost. *Figure E-2, page E-4,* is a draft checklist, which may be modified to meet the needs of the unit.

<table>
<tr><th>Steps</th><th>Actions</th></tr>
<tr><td>1</td><td>Soldiers. Soldiers should ensure that they—<br><br>• Have a seasonal uniform. The uniform should include the following:<br>▪ Load-carrying equipment (LCE) that is complete.<br>▪ First aid packet.<br>▪ Canteens that are full and have a M17 drinking cap.<br>▪ M40 with a hood, M258A1 kit, M13 kit, and optical inserts (if required).<br>▪ Kevlar with a camouflage cover and band.<br>▪ Individual weapon with 6 magazines.<br>▪ ID tags around the neck and ID cards.<br>▪ Ear plugs.<br>• Have been briefed on the current situation and the mission.<br>• Possess a current driver's license.<br>• Have been briefed on hot- and cold- weather injuries.<br><br>Leaders. Leaders should conduct a safety briefing, to include—<br><br>• Checking for hazards, such as ticks, chiggers, and spiders, and taking preventive first aid measures.<br>• Checking for snake hazards and ensuring that there is first aid.<br>• Checking for poisonous plants, such as oak, ivy, and sumac.<br>• Checking for hot- and cold- weather injuries and preventing them.<br>• Ensuring that correct driving practices are followed, to include—<br>▪ Speed in relation to the weather.<br>▪ Ground guides.<br>▪ Seat belts.<br>▪ Blackout driving.<br>▪ Gunner positioning (goggles).<br>• Ensuring that equipment is set up, such as camouflage nets or tents.<br>• Ensuring that there are fire-prevention vehicles and tents.<br>• Taking appropriate actions during severe weather.<br>• Using pyrotechnics.<br>• Using the buddy system for safety and during the mission.</td></tr>
</table>

**Figure E-2. Extensive PCI Checklist**

| Steps | Actions |
| --- | --- |
| | Vehicle commander, squad leader, or team leader. The vehicle commander, squad leader, or team leader ensures that—<br>• There is a map with a current overlay.<br>• There is a current SOI.<br>• Includes the following in the leader packet:<br>▪ A unit tactical standing operating procedure (TACSOP).<br>▪ Map markers.<br>▪ Overlay sheets.<br>▪ A notebook and a pen or a pencil.<br>▪ A route reconnaissance GTA card.<br>▪ A bridge classification GTA card.<br>▪ A call for-fire GTA card.<br>▪ A rigging GTA card.<br>▪ An NBC GTA card.<br>▪ An *EPW GTA card (GTA 19-7-1).*<br>▪ Flex cuffs for EPWs.<br>▪ *A DD Form 1408.*<br>▪ *A DA Form 4137.*<br>▪ *A DA Form 3975.*<br>▪ *A DD Form 2708.*<br>▪ *A DA Form 3881.*<br>▪ *A DA Form 2823.*<br>▪ A compass and a wrist watch.<br>▪ Batteries for all the equipment.<br>▪ Binoculars.<br>▪ NVGs.<br>▪ A sensitive item serial number list (radio, weapons, NVGs, and so forth). |
| 2 | Vehicles.<br>• Tie down all the items according to the load plan, and ensure that—<br>▪ There is space left for ammunition.<br>▪ The TA-50 is loaded.<br>▪ The MOPP chemical protective overgarment (CPOG) is stored for ready access.<br>▪ The gas tank is topped off.<br>▪ There are POL package products and weapon oil.<br>▪ Water cans are full. |

**Figure E-2. Extensive PCI Checklist (Continued)**

| Steps | Actions |
|---|---|
| | ▪ Spare fuel cans are full and tied down.<br>▪ MRE rations are inventoried and stored.<br>▪ There are weapon-cleaning kits.<br>▪ The OVM is stored and serviceable.<br>▪ There are chock blocks.<br>▪ That a complete first aid kit is available.<br>▪ A TM for the equipment (-10) on hand.<br>▪ A vehicle dispatch packet and a *DA Form 2404* are available.<br>▪ There are tools and tool kits.<br>▪ There are tire chains.<br>▪ The spare tire is properly stowed.<br>• Perform a premovement inspection and ensure that—<br>▪ There is a current TM (-10) on hand.<br>▪ There are no fuel leaks.<br>▪ The fire extinguisher is sealed, tagged, and dated.<br>▪ The engine coolant level is correct.<br>▪ The oil level is correct.<br>▪ The transmission fluid level is correct.<br>▪ The battery level is correct and the cables are tight.<br>▪ The air filter is clean and secure.<br>▪ The lights are operational.<br>▪ There are no exhaust leaks.<br>▪ The fuel filter is drained.<br>▪ Warning triangles are on hand.<br>▪ The canvas is attached and tied down.<br>▪ The gauges are functional.<br>▪ The tire pressure is correct.<br>▪ The vehicle service is current.<br>▪ The interior of the vehicle is clean and orderly. |
| 3 | Crew-served weapons. Ensure that—<br>• All mounting equipment is present.<br>• The TM for the weapon (-10) is available.<br>• There is a spare barrel and BII (such as gloves, a T&E mechanism, a wrench, a blank adapter, a tripod mount with an adapter, and a cleaning kit).<br>• The NVDs are operational.<br>• The tripod is clean and serviceable.<br>• The weapons are mounted with pins. |

**Figure E-2. Extensive PCI Checklist (Continued)**

| Steps | Actions |
|---|---|
| 4 | Communications equipment. Ensure that—<br>• Radios (vehicles or manpacks)—<br>▪ Are operational.<br>▪ Are secure.<br>▪ Have the proper frequency setting.<br>▪ Have a matching unit operational/set.<br>▪ Have an antenna with the tip protector tied down.<br>▪ Have connectors that are clean and serviceable.<br>▪ Have batteries available.<br>▪ Have the TM for the equipment (-10) is on hand.<br>▪ Have cleaning materials available (such as cotton swabs, alcohol, steel wool, and so forth).<br>• The field telephone (TA 312) is operational.<br>• The digital nonsecure voice terminal (DNVT) with wire is operational (communication).<br>• The appropriate phone listing is available.<br>• Radio antennas (OE 254s) are complete.<br>• Communication test sets, including tools, are on hand.<br>• Frequencies, call signs, challenges, and passwords are disseminated.<br>• Each team has one SINCGARS manpack for quick-dismount operations. |
| 5 | NBC equipment.<br>• Ensure that protective clothing is complete and serviceable.<br>• Ensure that the M40 is complete, such as—<br>▪ M13 deconnaissance kit.<br>▪ M258A1 deconnaissance kit.<br>▪ M256 detector kit.<br>▪ M8 detector paper.<br>▪ M9 detector paper.<br>▪ Antifogging kit.<br>▪ Hood.<br>▪ Optical inserts, if needed.<br>▪ Waterproof bag. |

**Figure E-2. Extensive PCI Checklist (Continued)**

| Steps | Actions |
|---|---|
| | • Ensure that the mask and hood are serviceable.<br>• Implement the proper MOPP level.<br>• Brief the threat condition.<br>• Ensure that nerve agent antidotes are available.<br>• Issue and mount M11 deconnaissance apparatus, to include the following:<br>▪ Two DS2 cans per vehicle.<br>▪ An M8 alarm.<br>▪ An M229 refill kit for each M8.<br>• Ensure that NBC defense equipment is available, such as—<br>▪ An M256 agent detector kit.<br>▪ An M 174 RADIAC meter.<br>▪ Two sets of batteries for the IM 174.<br>▪ An IM 93/147 RADIAC meter (dosimeter).<br>▪ A PP 157A RADIAC charger 1/IM 93/147.<br>▪ NBC contamination signs (marking kits). |

**Figure E-2. Extensive PCI Checklist (Continued)**

## Appendix F

# Fratricide Avoidance

Fratricide is defined as the unforeseen or unintentional death, injury, or damage to friendly personnel or equipment. It is committed during the employment of friendly weapons and munitions with the intent to kill the enemy or destroy his equipment or facilities. Fratricide prevention is the commander's responsibility, and all leaders assist him across all operating systems to accomplish this mission.

## FRATRICIDE

F-1. Friendly forces operating in the sustainment area are particularly vulnerable to fratricide due to the high concentration of soldiers. Most MP missions are conducted in the sustainment area in support of these friendly forces. During MP operations, it is critical that MP know where other friendly elements are operating. With this knowledge, they anticipate dangerous conditions and take steps to either avoid or mitigate them. The platoon leader is always aware of changes and developments in the situation that may place friendly forces in danger from MP long-range weapon systems.

F-2. It is important that other friendly forces operating in the AO know the location of MP to prevent them from becoming victims of fratricide. MP leaders constantly report the location of their forces to higher HQ. Higher HQ disseminates this information to other friendly forces. When the platoon leader perceives potential for fratricide, he quickly responds to prevent it.

## FRATRICIDE EFFECTS

F-3. Fratricide results in unacceptable losses and increases the risk of mission failure; it usually affects the unit's ability to survive and function. Units experiencing fratricide suffer the following consequences:

- Loss of confidence in the unit's leadership.
- Increased self-doubt among the leaders.
- Hesitancy to employ deadly force.
- Oversupervision of units.
- Hesitancy to conduct night operations.
- Loss of aggressiveness.
- Loss of initiative.
- Disrupted operations.
- General degradation of unit cohesiveness, morale, and combat power.

## FRATRICIDE CAUSES

F-4. The following paragraphs discuss the primary causes of fratricide. MP leaders must identify any of the factors that may affect their soldiers and strive to eliminate or correct them.

### TARGET IDENTIFICATION

F-5. The primary cause of fratricide is the lack of positive target identification. MP must positively identify targets before engaging.

### LAND NAVIGATION FAILURES

F-6. MP may stray into unknown locations, report wrong locations, and become disoriented. In this situation, they may unexpectedly encounter an errant unit, that may fire their weapons at the friendly force.

## REPORTING AND COMMUNICATIONS FAILURES

F-7. Units at all levels may fail to generate timely, accurate, and complete reports as locations and tactical situations change. This distorts the tactical picture available at each level and can lead to erroneous clearance of supporting fire.

## WEAPON ERRORS

F-8. Lapses in individual discipline can result in fratricide. These incidents include charge errors, accidental discharges, mistakes with explosives and hand grenades, and incorrect gun data use.

## OPERATION HAZARDS

F-9. A variety of explosive devices and materiel may create danger on the battlefield, such as UXO, unmarked or unrecorded minefields (including scatterable mines), and booby traps. Failure to mark, record, remove, or otherwise anticipate these threats will lead to casualties.

# FRATRICIDE RISK ASSESSMENT

F-10. *Table F-1, page F-4,* shows an example of a work sheet for evaluating fratricide risk in the context of mission requirements. The work sheet lists six mission accomplishment factors that affect the fratricide risk, along with related considerations for each factor. Platoon leaders assess the potential risk in each area (low, medium, or high), and assign a point value to each (one point for low risk, two for medium risk, three for high risk), and add the point values to calculate the overall fratricide assessment score.

F-11. They use the resulting score only as a guide. The platoon leader's final assessment is based on observable risk factors, such as those listed on the work sheet, and on his feel for the intangible factors affecting the operation. Note that descriptive terms are listed only in the low- and high-risk columns of the work sheet. The assessment of each factor determines whether the risk matches one of these extremes or lies somewhere between them as a medium risk.

**Table F-1. Fratricide Risk Assessment Work Sheet**

| Factors Affecting Fratricide | Potential Risk Categories (With Variable Conditions and Point Values) | | |
|---|---|---|---|
| | **Low Risk (1 point)** | **Medium Risk (2 points)** | **High Risk (3 points)** |
| **Understanding of the Plan** | | | |
| Commander's intent | Clear | | Vague |
| Complexity | Simple | | Complex |
| Enemy situation | Known | | Unknown |
| Friendly situation | Clear | | Unclear |
| ROE/ROI | Clear | | Unclear |
| **Environmental Factors** | | | |
| Intervisibility | Favorable | | Unfavorable |
| Obscuration | Clear | | Obscured |
| Battle tempo | Slow | | Fast |
| Positive target ID | 100 percent | | None |
| **Control Measures** | | | |
| Command relationships | Organic | | Joint/ combined |
| Audio communications | Loud/clear | | Jammed |
| Visual communications | Easily seen | | Obscured |
| Graphics | Standard | | Not understood |
| SOP | Standard | | Not used |
| Liaison personnel | Proficient | | Not trained |
| Location/navigation | Sure | | Unsure |

**Table F-1. Fratricide Risk Assessment Work Sheet (Continued)**

| Factors Affecting Fratricide | Potential Risk Categories (With Variable Conditions and Point Values) | | |
|---|---|---|---|
| | **Low Risk (1 point)** | **Medium Risk (2 points)** | **High Risk (3 points)** |
| **Equipment (Compared to US Equipment)** | | | |
| Friendly | Similar | ⟷ | Different |
| Enemy | Different | | Similar |
| **Training** | | | |
| Individual proficiency | MOS qualified | ⟷ | Untrained |
| Unit proficiency | Trained | | Untrained |
| Rehearsals | Realistic | | None |
| Habitual relationship | Yes | | No |
| Endurance | Alert | | Fatigued |
| **Planning Time (Based on the 1/3–2/3 Rule)** | | | |
| Higher HQ | Adequate | ⟷ | Inadequate |
| Own unit | Adequate | | Inadequate |
| Subordinate elements | Adequate | | Inadequate |
| Overall risk assessment (by total point value) | 26-46 points | 42-62 points | 58-78 points |
| **NOTE: Point values alone may not accurately reflect fratricide risk. The commander tailors the assessment to the unit's requirements. Refer to *FM 100-14*.** | | | |

# PREVENTIVE MEASURES

F-12. These measures provide the platoon leader with a guide to actions he can take to reduce and prevent fratricide risk. These guidelines are not directive in nature, nor are they intended to restrict initiative. All MP leaders must learn to apply them as appropriate based on the specific situation and METT-TC factors.

Platoon leaders must educate their soldiers on the following reduction and prevention of fratricide principles:

- Identifying and assessing the potential fratricide risks in the estimate of the situation. Express the risks in the OPORD and applicable FRAGOs.
- Maintaining situational awareness. Focusing on such areas as current intelligence, unit locations and/or dispositions, denial areas (minefields and scatterable mines; contaminated areas, such as improved conventional munitions [ICM] and NBC), SITREPs, and METT-TC factors.
- Ensuring positive target identification. Review vehicle and weapons ID cards. Become familiar with the characteristics of potential friendly and enemy vehicles, including their silhouettes and thermal signatures.
- Establishing a command climate that emphasizes fratricide prevention. Enforce fratricide prevention measures, placing special emphasis on the use of doctrinally sound TTP. Ensure supervision in the execution of orders and in the performance of all tasks and missions to standard.

F-13. Additional considerations for fratricide reduction and prevention include the following:

- Reporting MP patrol locations accurately. The TOC and CP track the patrols' locations in relationship to friendly forces.
- Recognizing the signs of battlefield stress. Maintain unit cohesion by taking quick, effective actions to alleviate the battlefield stress.
- Conducting individual, leader, and collective (unit) training covering fratricide awareness.
- Giving complete and concise mission orders.

- Simplifying mission orders by using SOPs that are consistent with the doctrine. Periodically review and update SOPs as needed.
- Striving to provide maximum planning time for leaders and subordinates.
- Planning and establishing effective communications.
- Ensuring that ROE are clear and understood.
- Conducting rehearsals when the situation allows.
- Being in the right place at the right time. If an MP patrol becomes lost or misoriented, leaders must know how to contact higher HQ immediately for instructions and assistance.
- Including a discussion of fratricide incidents in all AARs.

## FRIENDLY FIRE INCIDENTS

F-14. MP patrol's could become involved in a friendly fire incident in one of several ways: as the victim of the fire, as the firing element, or as an observer observing an attack of one friendly element on another. If a patrol encounters such a situation, it should follow the procedures for the various situations listed below:

### FALL VICTIM TO FRIENDLY FIRES

F-15. When falling victim to friendly fire, react to contact until recognizing friendly fire, then—

- Cease fire.
- Take immediate actions to protect soldiers and vehicles.
- Use a visual recognition signal directing the firing unit to cease fire.
- Report the following on the next higher unit net:

- Announce that friendly fire has been received.
- Request medical assistance as needed.
- Give the location and direction of the firing vehicles.
- Warn the higher unit not to return fire if the firing unit has been identified as friendly.

## ENGAGE FRIENDLY FORCES

F-16. When engaging friendly forces, cease fire and report the following on the next higher unit net:

- The ID of the engaged friendly force (if the unit is unidentified, report the number and type of vehicles).
- The location of the incident.
- The direction and distance to the engaged force.
- The type of fire.
- The target effects.

## OBSERVE A FRIENDLY FIRE INCIDENT

F-17. When observing a friendly fire incident, seek cover for protection, then—

- Use a visual recognition signal directing the firing unit to cease fire.
- Report the following on the next higher unit net:
  - The ID of the engaged friendly force (if the unit is unidentified, then report the number and the type of vehicles).
  - The location of the incident.
  - The direction and distance to the victim and the firing unit.
  - The type of fire.
  - The target effects.

- Provide assistance as needed (when safe to do so).

## LEADER RESPONSIBILITIES

F-18. In all situations involving the risk of fratricide and friendly fire, leaders must be prepared to take immediate actions to prevent casualties and equipment damage or destruction. Recommended actions in fratricide situations include—

- Identifying the incident and ordering the parties involved to cease fire.
- Conducting an in-stride risk assessment.
- Identifying and implementing the controls to prevent the incident from recurring.

# Appendix G

# MK19 Qualification Tables

The tables in this appendix are designed to enable the MK19 gunner and assistant gunner to qualify with the weapon.

**NOTE: These tables are not for ASV qualification. The ASV tables are under development.**

## PRIMARY GUNNER, MK19 QUALIFICATION AND ZERO/PRACTICE TABLES

G-1. *Table G-1, page G-2,* is a zero/practice table for the MK19 primary gunner.

**Table G-1. Primary Gunner Zero/Practice Table**

| Task | Condition | Standard | Ammunition | Time |
|---|---|---|---|---|
| Mount the MK19 on a vehicle or a M3 tripod | Given a mounted MK19, an M3 tripod, an M64 gun mount, a T&E mechanism, an organic carrier, and a selected firing position | The gun crew must be able to mount the MK19 on a vehicle or a tripod. | 0 | 2 minutes |
| Zero the MK19 using a 400-meter vehicle target | Given a mounted MK19, a 400-meter vehicle target, a selected firing position, and 4 40-millimeter training practice (TP) rounds | The gunner must impact on the target with at least one of the 4 40-millimeter TP rounds. | 4 | No time limit |
| Engage a stationary vehicle target at 1,100 meters | Given a mounted MK19, a vehicle target at 1,100 meters, a selected firing position, and 4 40-millimeter TP rounds | The gunner must impact on the vehicle target with at least 1 of the 4 rounds. | 4 | No time limit |
| Engage a stationary vehicle target at 1,500 meters | Given a mounted MK19, a vehicle target at 1,500 meters, a selected firing position, and 4 40-millimeter TP rounds | The gunner must impact on the vehicle target with at least 1 of the 4 rounds. | 4 | No time limit |

### Table G-1. Primary Gunner Zero/Practice Table (Continued)

| Task | Condition | Standard | Ammunition | Time |
|---|---|---|---|---|
| Engage a stationary vehicle target at 600 meters | Given a mounted MK19, a vehicle target at 600 meters, a selected firing position, and 4 40-millimeter TP rounds | The gunner must impact on the vehicle target with at least 1 of the 4 rounds. | 4 | No time limit |
| Dismount the MK19 from the vehicle or the tripod | Given a mounted MK19 and a selected firing position | The gunner must take the MK19 out of operation from the carrier or tripod. | 0 | 1 minute |

## PRIMARY GUNNER, INDIVIDUAL MK19 QUALIFICATION (DAY), DISMOUNTED

G-2. *Table G-2* is the MK19 qualification table for the primary gunner, dismounted. The commander may decide to fire using this table (mounted) based on the METT-TC.

**Table G-2. Primary Gunner (Day), Dismounted**

| Task | Condition | Standard | Ammunition | Time |
|---|---|---|---|---|
| Mount the MK19 on an M3 tripod or a vehicle | Given a mounted MK19, an M3 tripod, an M64 gun mount, a T&E mechanism, an organic carrier, and a selected firing position | The gun crew must be able to mount the MK19 on a vehicle or a tripod. | 0 | 120 seconds |
| Engage a stationary vehicle target at 800 meters | Given a mounted MK19, a selected firing position, 6 40-millimeter TP rounds, and a vehicle target at 800 meters | The gunner must impact on the vehicle target with at least 2 of the 6 rounds. | 6 | 90 seconds |
| Engage a stationary vehicle target at 600 meters | Given a mounted MK19, a selected firing position, 4 40-millimeter TP rounds, and a vehicle target at 600 meters | The gunner must impact on the vehicle target with at least 1 of the 4 rounds. | 4 | 90 seconds |
| Engage a stationary vehicle target at 1,100 meters | Given a mounted MK19, a vehicle target at 1,100 meters, a selected firing position, and 6 40-millimeter TP rounds | The gunner must impact on the vehicle target with at least 2 of the 6 rounds. | 6 | 120 seconds |

**Table G-2. Primary Gunner (Day), Dismounted (Continued)**

| Task | Condition | Standard | Ammunition | Time |
|---|---|---|---|---|
| Engage a stationary vehicle target at 1,500 meters | Given a mounted MK19 vehicle target at 1,500 meters, a selected firing position, and 8 40-millimeter TP rounds | The gunner must impact on the vehicle target with at least 2 of the 8 rounds. | 8 | 180 seconds |
| Dismount the MK19 from the vehicle or the tripod | Given a mounted MK19 and a selected firing position | The gunner must take the MK19 out of operation from the carrier or tripod. | 0 | 60 seconds |

## PRIMARY GUNNER, MK19 QUALIFICATION, MOUNTED

G-3. *Table G-3, page G-6,* is the primary gunner's individual MK19 qualification table, mounted.

**Table G-3. Primary Gunner, Mounted**

| Task | Condition | Standard | Ammunition | Time |
|---|---|---|---|---|
| Mount the MK19 on a vehicle or a M3 tripod | Given a mounted MK19, an organic carrier, an M3 tripod, an M64 gun mount, a T&E mechanism, and a selected firing position | The gun crew must be able to mount the MK19 on a vehicle or a tripod. | 0 | 120 seconds |
| Engage a stationary vehicle target at 400 meters | Given a mounted MK19, a selected firing position, 4 40-millimeter TP rounds, and a vehicle target at 400 meters | The gunner must impact on the vehicle target with at least 2 of the 4 rounds. | 4 | 90 seconds |
| Engage a stationary vehicle target at 1,100 meters | Given a mounted MK19, a selected firing position, 8 40-millimeter TP rounds and a vehicle target at 1,100 meters | The gunner must impact on the vehicle target with at least 2 of the 8 rounds. | 8 | 120 seconds |

**Table G-3. Primary Gunner, Mounted (Continued)**

| Task | Condition | Standard | Ammunition | Time |
|---|---|---|---|---|
| Engage troops at 600 meters | Given a mounted MK19, a selected firing position, 4 40-millimeter TP rounds, and a troop target at 600 meters | The gunner must impact on the target with at least 1 of the 4 rounds | 4 | 60 seconds |
| Engage a stationary vehicle target at 1,500 meters | Given a mounted MK19, a selected firing position, 6 40-millimeter TP rounds, and a vehicle target at 1,500 meters | The gunner must impact on the vehicle target with at least 1 of the 6 rounds | 6 | 120 seconds |
| Engage a stationary vehicle target at 1,100 meters and troops in the open at 600 meters | Given a mounted MK19, a selected firing position, 10 40-millimeter TP rounds, a vehicle target at 1,100 meters, and a troop target at 600 meters | The gunner must impact on the vehicle and troop targets with at least 2 rounds per target | 10 | 180 seconds |
| Dismount the MK19 from a vehicle or a tripod. | Given a mounted MK19 and a selected firing position | The gunner must take the MK19 out of operation from the carrier or tripod | 0 | 60 seconds |

## PRIMARY GUNNER, MK19 QUALIFICATION (NIGHT)

G-4. *Table G-4* is the primary gunner's MK19 qualification table for night firing.

**Table G-4. Primary Gunner (Night)**

| Task | Condition | Standard | Ammunition | Time |
|---|---|---|---|---|
| Mount the AN/TVS-5 to the MK19 | Given a complete AN/TVS-5, an MK19, an M3, and a T&E mechanism or organic carrier | The gunner must mount the AN/TVS-5 on the MK19. | 0 | 120 seconds |
| Use a zeroed MK19 with an AN/TVS-5 mounted for observation and engage a 400-meter vehicle target | Given a complete AN/TVS-5, a mounted MK19, a selected firing position, 6 40-millimeter TP rounds, and a vehicle target at 400 meters. | The gunner must impact the target with at least 2 of the 6 rounds. | 6 | 60 seconds |

**Table G-4. Primary Gunner (Night) (Continued)**

| Task | Condition | Standard | Ammunition | Time |
|---|---|---|---|---|
| Engage a stationary vehicle target at 600 meters | Given a mounted MK19, a selected firing position, 8 40-millimeter TP rounds, and a vehicle target at 600 meters | The gunner must impact the vehicle target with at least 4 of the 8 rounds. | 8 | 120 seconds |
| Engage troops in the open at 800 meters | Given a mounted MK19, a selected firing position, 8 40-millimeter TP rounds, and troops in the open at 800 meters | The gunner must impact the troop target with at least 4 of the 8 rounds. | 8 | 120 seconds |
| Dismount the AN/TVS-5 from the MK19 | Given an MK19 and an AN/TVS-5 | The gun crew must take the AN/TVS-5 out of operation. | 0 | 120 seconds |

# ASSISTANT GUNNER, MK19 FIRING TABLE, MOUNTED

G-5. *Table G-5* is for familiarization only.

**Table G-5. Assistant Gunner, Mounted**

| Task | Condition | Standard | Ammunition | Time |
|---|---|---|---|---|
| Mount the MK19 on a vehicle | Given a zeroed, mounted, MK19; a T&E mechanism; an MK64 gun mount; an organic carrier; and a selected firing position | The gun crew must be able to mount the MK19 on a vehicle or a tripod | 0 | 120 seconds |
| Engage a stationary vehicle target at 400 meters | Given a mounted MK19, a vehicle target at 400 meters, a selected firing position, and 4 40-millimeter TP rounds | The gunner must impact the vehicle target with at least 2 of the 4 rounds | 4 | 90 seconds |
| Engage troops in the open at 600 meters | Given a mounted MK19, a troop target at 600 meters, a selected firing position, and 4 40-millimeter TP rounds | The gunner must impact the troop target with at least 1 of the 4 rounds. | 4 | 60 seconds |

**Table G-5. Assistant Gunner, Mounted (Continued)**

| Task | Condition | Standard | Ammunition | Time |
|---|---|---|---|---|
| Engage a stationary vehicle target at 1,100 meters | Given a mounted MK19, a vehicle target at 1,100 meters, a selected firing position, and 8 40-millimeter TP rounds | The gunner must impact the vehicle target with at least 2 of the 8 rounds. | 8 | 120 seconds |
| Engage a stationary vehicle target at 1,100 meters and troops at 600 meters | Given a mounted MK19, a vehicle target at 1,100 meters, troops in the open at 600 meters, a selected firing position, and 10 40-millimeter TP rounds | The gunner must impact the vehicle and troop targets with at least 2 rounds per target. | 10 | 180 seconds |
| Dismount the MK19 from a vehicle or a tripod | Given a mounted MK19 and a selected firing position | The gunner must take the MK19 out of operation from the carrier or tripod. | 0 | NA |

## ASSISTANT GUNNER, MK19 FIRING TABLE, DISMOUNTED

G-6. *Table G-6* is a dismounted firing table. The commander may decide to fire using *Table G-5* based on the METT-TC.

**Table G-6. Assistant Gunner, Dismounted**

| Task | Condition | Standard | Ammunition | Time |
|---|---|---|---|---|
| Mount the MK19 on a vehicle or an M3 tripod | Given a mounted MK19, an M3 tripod, an M64 gun mount, a T&E mechanism, an organic carrier, and a selected firing position | The gun crew must be able to mount the MK19 on a vehicle or a tripod. | 0 | 120 seconds |
| Engage a stationary vehicle target at 800 meters | Given a mounted MK19, a vehicle target at 800 meters, a selected firing position, and 6 40-millimeter TP rounds | The gunner must impact the vehicle target with at least 2 of the 6 rounds. | 6 | 60 seconds |
| Engage a stationary vehicle target at 600 meters | Given a mounted MK19, a vehicle target at 600 meters, a selected firing position, and 4 40-millimeter TP rounds | The gunner must impact the vehicle target with at least 1 of the 4 rounds. | 4 | 90 seconds |

**Table G-6. Assistant Gunner, Dismounted (Continued)**

| Task | Condition | Standard | Ammunition | Time |
|---|---|---|---|---|
| Engage a stationary vehicle target at 1,100 meters | Given a mounted MK19, a vehicle target at 1,100 meters, a selected firing position, and 6 40-millimeter TP rounds | The gunner must impact the vehicle target with at least 2 of the 6 rounds. | 6 | 90 seconds |
| Engage a stationary vehicle target at 1,500 meters | Given a mounted MK19, a vehicle target at 1,100 meters, a selected firing position, and 6 40-millimeter TP rounds | The gunner must impact the vehicle target with at least 2 of the 6 rounds. | 6 | 120 seconds |
| Engage a stationary vehicle target at 1,500 meters | Given a mounted MK19 vehicle target at 1,500 meters, a selected firing position, and 8 40-millimeter TP rounds | The gunner must impact the vehicle target with at least 2 of the 8 rounds. | 8 | 120 seconds |
| Dismount the MK19 from a vehicle or a tripod | Given a mounted MK19 and a selected firing position | The gunner must take the MK19 out of operation from the carrier or tripod. | 0 | 60 seconds |

## ASSISTANT GUNNER, MK19 FIRING TABLE (NIGHT)

G-7. *Table G-7* is for night firing familiarization only.

**Table G-7. Assistant Gunner (Night)**

| Task | Condition | Standard | Ammunition | Time |
|---|---|---|---|---|
| Mount the AN/TVS-5 to the MK19 | Given a mounted AN/TVS 5 and a T&E mechanism | The gunner must mount the AN/TVS-5 on an MK19. | 0 | 120 seconds |
| Engage a stationary vehicle target at 600 meter | Given a mounted MK19, a vehicle target at 600 meters, a selected firing position, and 6 40-millimeter TP rounds | The gunner must impact on the vehicle target with at least 1 of the 6 rounds. | 6 | 120 seconds |
| Engage troops in the open at 800 meters | Given a mounted MK19, troops in the open at 800 meters, a selected firing position, and 6 40-millimeter TP rounds | The gunner must impact on the vehicle target with at least 1 of the 6 rounds. | 6 | 120 seconds |
| Dismount the AN/TVS-5 from the MK19 | Given an MK19 and an AN/TVS-5 | The gun crew must take the AN/TVS-5 out of operation within 2 minutes. | 0 | 120 seconds |

# MK19 SCORECARD

G-8. *Figure G-1* is the scorecard for primary and assistant gunner qualification.

**MK19 Qualification Scorecard**

| 1. Name (Last, First, Middle Initial) | 2. SSN | 3.Grade | 4. Unit | 5. Date |
|---|---|---|---|---|
| | | | | |

**Mounted Table**

| Rounds | Range (Meters) | Target | Time (Seconds) | Required Hits | Hits |
|---|---|---|---|---|---|
| 4 | 400 | Vehicle | 90 | 2 | |
| 8 | 1,100 | Vehicle | 120 | 2 | |
| 4 | 600 | Troops | 60 | 1 | |
| 6 | 1,500 | Vehicle | 120 | 1 | |
| 10 | 1,100/600 | Vehicle/ troops | 180 | 2/2 | |
| | | | | Total hits | |

**Rating:**
Distinguished = 5 of 5 qualified engagements.
Qualified = 4 of 5 qualified engagements.
Unqualified = 3 and below.

**Dismounted Table**

| Rounds | Range (Meters) | Target | Time (Seconds) | Required Hits | Hits |
|---|---|---|---|---|---|
| 6 | 800 | Vehicle | 90 | 2 | |
| 4 | 600 | Vehicle | 90 | 1 | |
| 6 | 1,100 | Vehicle | 120 | 2 | |
| 8 | 1,500 | Vehicle | 180 | 2 | |
| | | | | Total hits | |

**Rating:**
Distinguished = 4 of 4 qualified engagements.
Qualified = 3 of 4 qualified engagements.
Unqualified = 2 and below.

**Night-Firing Table**

| Rounds | Range (Meters) | Target | Time (Seconds) | Required Hits | Hits |
|---|---|---|---|---|---|
| 6 | 400 | Vehicle | 60 | 2 | |
| 8 | 600 | Vehicle | 120 | 4 | |
| 8 | 800 | Troops | 120 | 4 | |
| | | | | Total hits | |

**Rating:**
Distinguished = 3 of 3 qualified engagements.
Qualified = 2 of 3 qualified engagements.
Unqualified = 1 and below.

**Standards**

All standards achieved within table = distinguished.
Missed 1 standard within table = qualified.
two or more missed standards = unqualified.

**Overall Rating:**
Three distinguished tables = distinguished.
Two distinguished tables = qualified.
One distinguished table = unqualified.

**Figure G-1. MK19 Qualification Tables**

# Appendix H

# Countermine Operations

Military operations are often conducted in areas where warring factions have left unrecorded mines and minefields scattered across the landscape. In these environments, emphasize mine awareness at all levels of command and plan countermine operations into every military operation.

## OVERVIEW

H-1. Undertake countermine operations to breach or clear a minefield. All the tasks fall under breaching or clearing operations and include detecting, reporting, reducing, proofing, and marking.

H-2. MP efforts in countermine operations are in detecting, reporting, and marking. They conduct countermine operations in conjunction with their other missions. Countermine operations include–

- Denying infiltrators, insurgents, and terrorist groups the opportunity to use mines.
- Using mine detection equipment.
- Detecting mines visually or by probing.
- Reporting the location of suspected mines and UXO.
- Marking the location of these devices.

H-3. MP patrols look for suspicious persons along MSRs; watch approaches to critical points like bridges or defiles; and maintain surveillance of MSRs, key terrain, and critical facilities. Check with the HN police and local nationals for information on unusual activity in a particular AO. Mines are usually emplaced at night. Use night vision and early warning devices to maintain surveillance and detect enemy activity.

H-4. Mine and UXO awareness involve soldier and leader skills. Soldier skills are a mix of individual and collective tasks that are required for an element to maintain its combat effectiveness in and around a mined environment. Soldier skills involve individual and collective tasks that are required for basic survival in a mined environment. They include minefield indicators, probing techniques, mine detector operations, extraction drills, survival rules, casualty treatment, and evacuation drills. The soldier's basic mine awareness skills are critical to his and the unit's survival. Leader skills involve planning missions, assessing situations, and tracking and disseminating mine information. Soldiers must be proficient in all mine awareness skills to effectively operate in a mined environment (refer to *FM 20-32*).

## DETECT

H-5. Detection is the actual confirmation and location of the mines and may be accomplished through reconnaissance or unintentionally (such as a vehicle running into a mine). Use mine detection in conjunction with intelligence-gathering operations, minefield bypass reconnaissance, and breaching and clearing operations. Use the following three methods to detect mines and UXO:

- Visual.
- Physical (probing).
- Electronic (mine detector).

### VISUAL INDICATORS

H-6. Mine and UXO indicators are part of all combat operations. Understanding and recognizing mine indicators could determine whether or not a soldier becomes a casualty. The following may indicate the presence of mines and UXO:

- Trip wires.
- Signs of road repair (such as new fill or paving, road patches, ditching, or culvert work).
- Signs, usually understood only by local populace, placed on trees, posts, or stakes. Threat forces mark their minefields to protect their own forces.
- Dead animals.
- Vehicles that are damaged.
- Disturbances in previous tire tracks or tracks that stop unexplainably.
- Wires leading away from the side of the road. They may be firing wires that are partially buried.

H-7. Check for odd features in the ground or patterns that are not present in nature. Plant growth may wilt or change color, rain may wash away some of the cover, the cover may sink or crack around the edges, or the material covering the mines may look like mounds of dirt. Civilians may know where mines or booby traps are located in the residential area. Civilians staying away from certain places or out of certain buildings are good indications of the presence of mines or booby traps. Question civilians to determine the exact location of these devices.

H-8. Pieces of wood or other debris on a road may be indicators of pressure or pressure-release firing devices. These devices may be on the surface or partially buried. The enemy uses mines that are fired by command, so search road shoulders and areas close to the objects.

## PHYSICAL INDICATORS

H-9. Physical detection (probing) is very time-consuming. Use it primarily for clearing operations, self-extraction, and covert breaching operations. Detection of mines by visual or electronic methods should be confirmed by probing. Use the following procedures and techniques when probing for mines:

- Roll up your sleeves, and remove any jewelry to increase sensitivity. Wear a Kevlar helmet, with the chin strap buckled, and a protective fragmentation vest.
- Stay close to the ground, and move in a prone position to reduce the effects of an accidental blast. When moving into a prone position–
  - Squat down without touching your knees to the ground.
  - Scan forward up to 2 meters and to the sides up to 3 meters for mine indicators.
  - Probe the area around your feet and as far forward as possible.
  - Kneel on the ground after the area is found to be clear. Continue probing forward until you are in a prone position.
- Use sight and touch to detect trip wires, fuses, and pressure prongs.
- Use a slender, nonmetallic object as a probe, and—
  - Probe every 5 centimeters across a 1-meter front.
  - Push the probe gently into the ground at an angle that is less than 45 degrees.
  - Apply just enough pressure on the probe to sink it slowly into the ground.
  - Check the probe for resistance. If the probe encounters resistance and does not go into the ground freely, carefully pick the soil

away with the tip of the probe and remove the loose dirt by hand. Take care to prevent functioning the mine.

- Stop probing when you touch a solid object, and use two fingers from each hand to carefully remove the surrounding soil and identify the object. If the object is a mine, remove enough soil to show the mine type and mark its location.

**DANGER**
**Do not attempt to remove or disarm the mine.**

H-10. Probing is extremely stressful and tedious. The senior leader sets a limit to the time a prober can actually probe in the minefield. To determine a reasonable time, the leader considers the METT-TC factors, weather conditions, the threat level, the unit's stress level, and the prober's fatigue level and state of mind. As a rule, 20 to 30 minutes is the maximum amount of time that an individual can probe effectively.

**DANGER**
**Use extreme caution when probing. If the probe is pushed straight down, its tip may detonate a pressure fuse.**

## ELECTRONIC INDICATORS

H-11. Electronic detection is effective for locating mines, but this method is time-consuming and exposes personnel to enemy fire. Confirm suspected mines by probing.

H-12. The AN/PSS-12 mine detector can only detect metal, but most mines have metal components in their design. The detector locates and identifies plastic or wooden mines by a slight metallic signature. Refer to *TM 5-6665-298-10* for more information about the employment and operation procedures for the AN/PSS-12. The detector is handheld and identifies suspected mines by an audio signal in the headphones.

H-13. As in probing, take consideration for the maximum amount of time an individual can operate the detector. The leader considers the METT-TC factors, weather conditions, the threat level, the unit's stress level, and the individual's fatigue level and state of mind. As a rule, 20 to 30 minutes is the maximum amount of time an individual can use the detector effectively.

## REPORT

H-14. Intelligence concerning enemy mines and UXO is reported by the fastest means available. Report sightings of these devices using a SPOTREP format. SPOTREPs originate from patrols that have been sent on specific reconnaissance missions or from MP patrols that discover mine information in the course of their normal route operations.

## MARK

H-15. When mines and UXO are detected, mark the location to prevent friendly follow-on forces and local nationals from accidentally encountering them. Mark the mines and UXO with standard North Atlantic Treaty Organization (NATO) markers (a red triangle with white letters). Use concertina wire to construct a perimeter around the mined area and place markers on the fencing about waist-high.

## Appendix I

# Route Classification and Signing System

The military route classification system helps in planning and executing battlefield movements. Military engineers develop route classifications using a route classification formula. This formula consists of a series of numbers and letters that express, in standard sequence, the route width, the route type, the lowest MLC, overhead clearance, obstructions to the traffic flow, and special conditions on a given route. Findings are based on information extracted from route reconnaissance reports.

## OVERVIEW

I-1. The first step in completing a route reconnaissance overlay is to understand what information must be included on it. As a minimum, include the following information on the overlay:

- The route classification formula.
- The name, rank, and social security number of the person preparing the overlay.
- The unit conducting the reconnaissance.
- The DTG that the overlay was conducted.
- The map name, edition, and scale.
- Any remarks necessary to ensure the complete understanding of the information contained on the overlay.

I-2. This appendix focuses on the route classification system and the methods necessary to determine the classification of a road. Refer to *FM 5-170* for more information about route classification and reconnaissance.

## ROUTE CLASSIFICATION FORMULA

I-3. The route classification is derived from the information gathered during the route reconnaissance. The formula is recorded on the route reconnaissance overlay *(Figure I-1)* and consists of the following:

- The route width (in meters).
- The route type (based on the ability to withstand weather).
- The lowest MLC.
- The lowest overhead clearance (in meters).
- Any obstructions (OB) to traffic flow, if applicable.
- Any special conditions, such as snow blockage (T) or flooding (W).

I-4. The following are examples of route classification formulas:

- 6.1/Z/40/$\infty$. A fair-weather route (Z) with a minimum traveled way of 6.1 meters and an MLC of 40. Overhead clearance is unlimited ($\infty$), and there are no obstructions to traffic flow. This route accommodates both wheeled and tracked, single-flow traffic without obstruction.
- 7/Y/50/4.6 (OB) (W). A limited, all-weather route (Y) with a minimum traveled way of 7 meters, an MLC of 50, an overhead clearance of 4.6 meters, and an obstruction. This route width is not suitable for double-flow traffic (wheeled or tracked). This route is subject to regular, recurrent flooding.

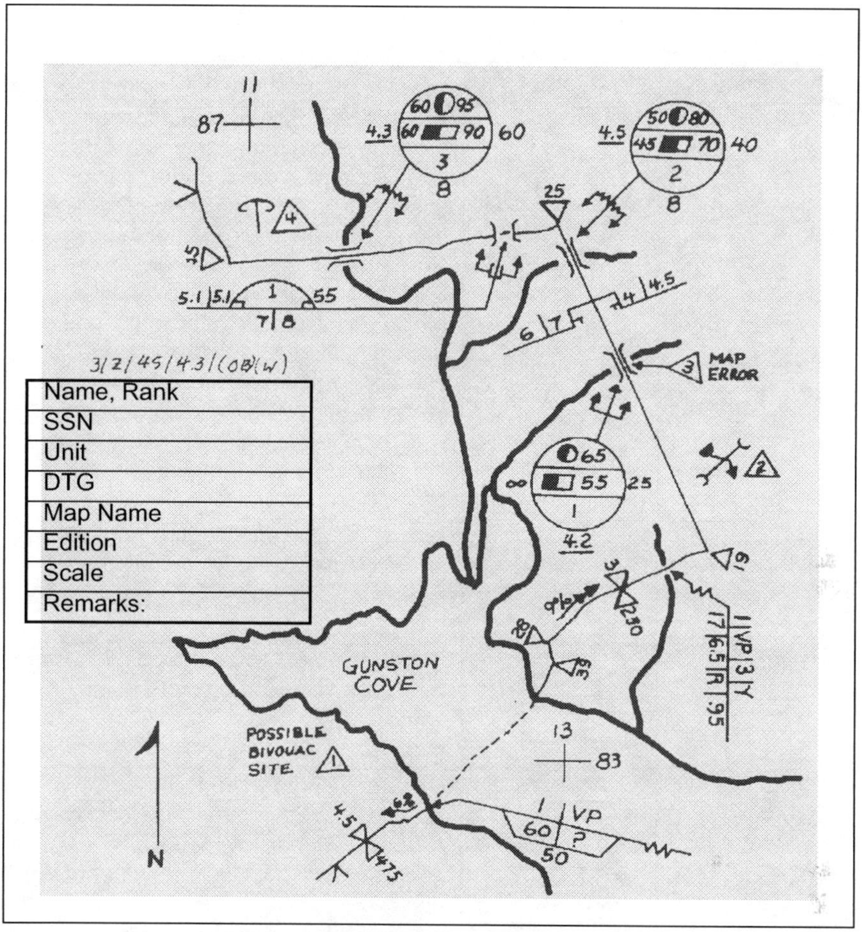


**Figure I-1. Route Reconnaissance Overlay**

I-5. *Table I-1, page I-4,* gives route reconnaissance symbols used on a route reconnaissance overlay.

## Table I-1. Route Reconnaissance Symbols

| Explanation | Symbol | Remarks |
|---|---|---|
| 1. Abbreviated bridge symbol | 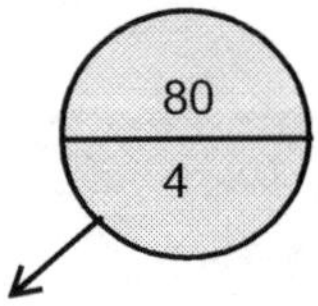 | Use this symbol only when the map scale does not permit the use of the full NATO bridge symbol. If this symbol is used, submit DA Form 1249. Draw an arrow to the map location of the bridge. Show the bridge's serial number in the lower portion of the symbol and the MLC for single-flow traffic in the upper portion. If there are separate load classifications for tracked or wheeled vehicles, show the lesser classification. Underline the classification number if the width or overhead clearance is below the minimum standard. |
| 2. Axial route | 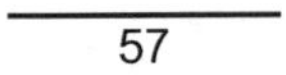 | Use a solid line and identify the route by an odd number. |
| 3. Bypass difficult | 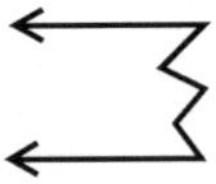 | Use this symbol when the obstacle can be crossed in the immediate vicinity, but some work is necessary to improve the bypass. |
| 4. Bypass easy | 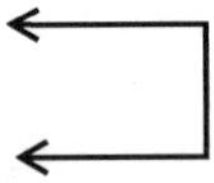 | Use this symbol when the obstacle can be crossed in the immediate vicinity by a US 2 1/2-ton truck (or NATO equivalent) without work to improve the bypass. |
| 5. Bypass impossible | 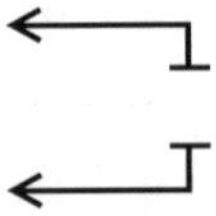 | Use this symbol when the obstacle can be crossed only by repairing or constructing a feature or by detouring around the obstacle. |
| 6. Civil or military route designation |  | Write the designation in parentheses along the route. |

## Table I-1. Route Reconnaissance Symbols (Continued)

| Explanation | Symbol | Remarks |
|---|---|---|
| 7.Concealment |  | Show roads lined with trees by a single line of circles for deciduous trees and a single line of inverted Vs for evergreen trees. Show woods bordering a road by several rows of circles for deciduous trees and several rows of inverted Vs for evergreen trees. |
| 8. Critical points |  | Number, in order, and describe critical points on *DA Form 1711-R*. Use critical points to show the features not adequately covered by other symbols on an overlay. |
| 9. Damage or destruction | 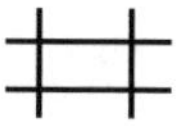 | Damage or destruction that prevents movement along the route. |
| 10. Ferry<br>*Ferry type*<br>P–pedestrian<br>V–vehicular | 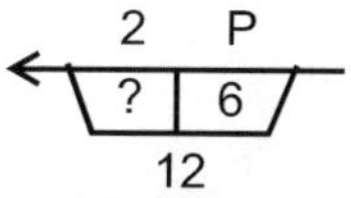<br>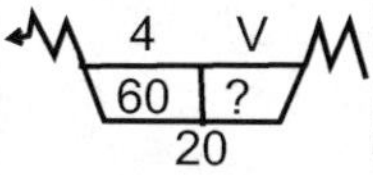 | Draw an arrow to the ferry location. The data above the symbols shows, in order, the left approach, the ferry's serial number, the ferry type, and right approach. The data inside the symbol shows, from left to right, the MLC and the dead weight capacity in tons. The number below the symbol shows the turnaround time, in minutes. A question mark indicates unknown information. Show the different approaches by a zigzag line, and show easy approaches by a straight line. |

## Table I-1. Route Reconnaissance Symbols (Continued)

| Explanation | Symbol | Remarks |
|---|---|---|
| 11. Ford<br>*Ford type*<br>P–pedestrian<br>V–vehicular | 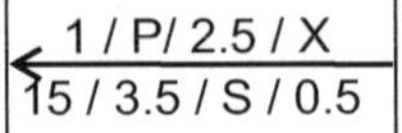 <br>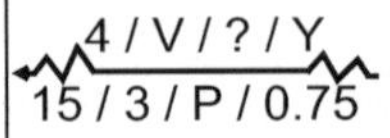  | Draw an arrow to the ford location. The data above the line shows, in order, the left-bank approach, the ford serial number, the ford type, the stream velocity (in meters per second), seasonal limitations, and the right-bank approach. A zigzag line corresponding in position to the shore where the approach is located represents difficult approaches. Straight lines identify an easy approach. The left and right banks are determined by looking downstream. The data below the line shows, in order, the length, the width, the bottom type, and the depth. All measurements are in meters.<br><br>Seasonal limiting factors: X – none; Y – significant; ? – unknown information.<br><br>Bottom type: M – mud; C – clay, S – sand; G – gravel; R – rock; P – artificial paving. |
| 12. Full NATO bridge symbol | 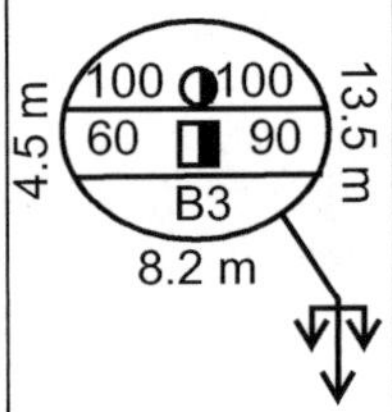  | Indicate wheeled vehicles in the upper third of the symbol with the two-way wheeled classification on the left and the one-way wheeled classification on the right. Show tracked vehicles in the center third of the symbol with the two-way tracked classification on the left and the one-way tracked classification on the right. Place the bridge serial number in the lower third of the symbol. Draw an arrow to the bridge location and show the bypass conditions on the arrow shaft. Place the traveled-way width below the symbol, the overhead clearance to the left of the symbol, and the overall length to the right of the symbol. |

## Table I-1. Route Reconnaissance Symbols (Continued)

| Explanation | Symbol | Remarks |
|---|---|---|
| 13. Grades | 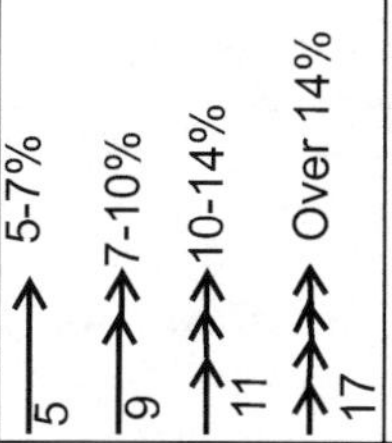  | Show the actual percent of grade to the right of the symbol. Any grade of 7 percent or more is an obstruction. Include the obstruction in the route classification formula. Arrows should point uphill. The length of the arrow represents the length of the grade if the map scale permits. |
| 14. Lateral route | 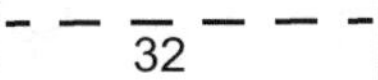  | Use a broken line, and identify the route by an even number. |
| 15. Limits of sector |  | Show the beginning and ending of a reconnoitered section of a route or road with this symbol. |
| 16. MSR |   | Label the route as MSR and assign a code name. |
| 17. Obstacle<br>a. Proposed block<br>b. Prepared but passable<br>c. Completed block | 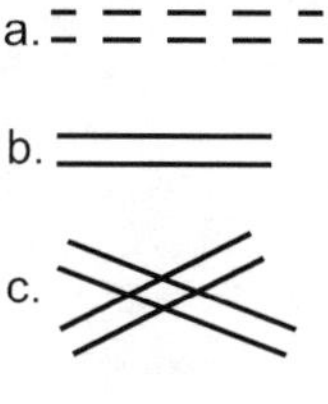  | Place the center of the symbol over the location of the blocked part of the route. Use parallel broken lines for a proposed block, parallel lines for a prepared but passable block, and crossed lines for a completed block. |
| 18. Clearance unlimited |  | Use this symbol when overhead clearance is unlimited. |
| 19. Parking area |  | Designate parking areas. |
| 20. Railroad crossing | 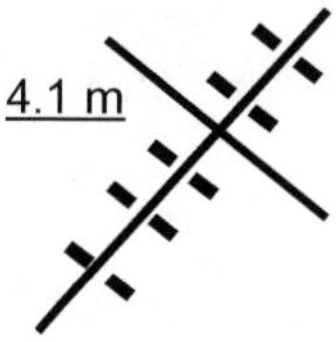  | Use this symbol to show a level crossing where passing trains would interrupt the traffic flow. If there is a power line present, show its height, in meters, from the ground. Underline the overhead clearance if it is less than 4.3 meters. |

## Table I-1. Route Reconnaissance Symbols (Continued)

<table>
<tr><th>Explanation</th><th>Symbol</th><th>Remarks</th></tr>
<tr><td>21. Railway bridge symbol</td><td>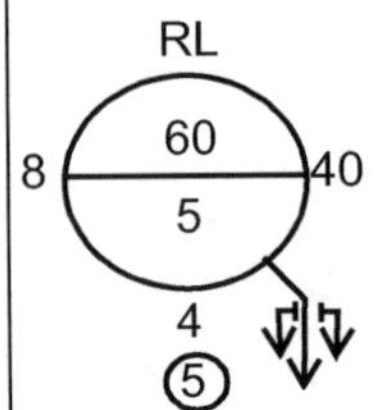

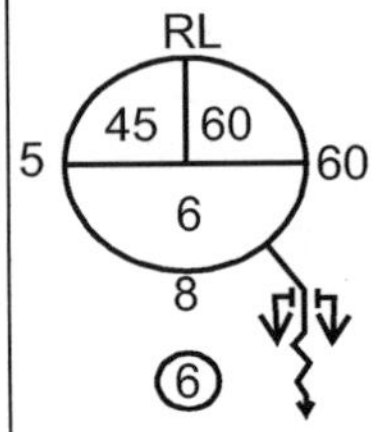

RL = railroad bridge</td><td>Place RL above the symbol to indicate a railway bridge. At the left of the symbol, show the overhead clearance. Show the overall length of the bridge on the right of the symbol. Indicate the traveled-way width below the symbol, and underline it if it is below standard for the classification. Inside the symbol, show the bridge classification in the upper half. If the class is different for single- and double-flow traffic, show single flow on the left and double flow on the right. Place the railway bridge serial number in the lower half of the symbol. Draw an arrow to the bridge location. On the arrow shaft, indicate the ease of adapting the bridge for road vehicle use. A zigzag line means it would be difficult to adapt, and a straight line means it would be easy to adapt. Place the bypass symbol on the arrow shaft to indicate the bypass conditions.</td></tr>
<tr><td>22. Route classification formula</td><td>6 / Z / 30 / 4 / (OB)</td><td>Express the formula in the order of the route width, the route type, MLC, minimum clearance, obstructions (if present), and special conditions. All measurements are in meters.<br>Route types: X – all weather; Y – limited all-weather route; Z – fair-weather route. Special conditions: (T) – regular snow blockage; (W) - regular flooding.</td></tr>
<tr><td>23.Series of sharp curves</td><td>
</td><td>Write the number of curves and the radius of the sharpest curve of the series to the outside of the triangle.</td></tr>
</table>

## Table I-1. Route Reconnaissance Symbols (Continued)

| Explanation | Symbol | Remarks |
|---|---|---|
| 24. Sharp curve | 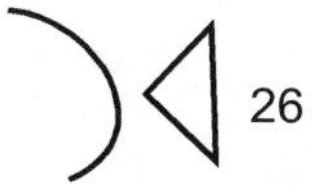  | Point the vertex of the triangle to the curve location and indicate the radius of the curve, in meters, outside the triangle. A curve of 45 meters or less must be reported on the overlay, and a curve of 25 meters or less is an obstacle. |
| 25. Traffic control HQ | 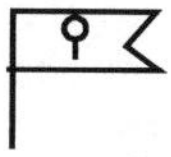 | None |
| 26. Traffic control post |  | Manned traffic control post. |
| 27. Tunnel | 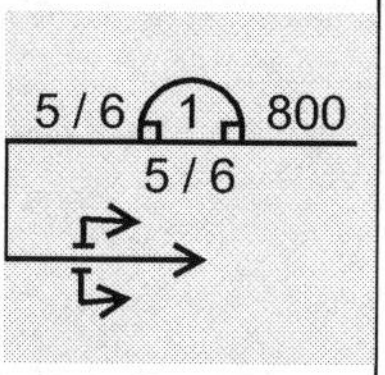  | Draw an arrow to the tunnel location. Place the bypass condition symbol on the arrow. Show the minimum and maximum overhead clearances to the left of the symbol, the tunnel serial number inside the symbol, and the total tunnel length to the right of the symbol. Below the symbol, show the traveled-way width. If sidewalks are present, follow with a slash and the total traveled way, including sidewalks. Underline the traveled way if the road entering the tunnel is wider than the traveled way of the tunnel. Use a question mark to show unknown information. |

## Table I-1. Route Reconnaissance Symbols (Continued)

| Explanation | Symbol | Remarks |
|---|---|---|
| 28. Turnout; the symbol may be amplified as follows: a. Wheeled vehicle b. Tracked vehicle c. A length of road exceeding 1 kilometer | 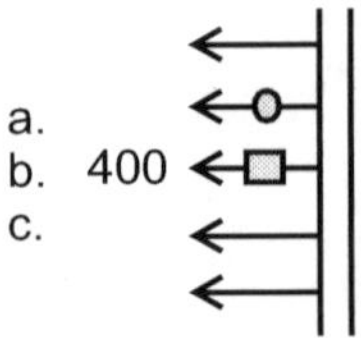  | Use this symbol to show the possibility of driving off the road. Draw the arrow in the direction of the turnout (right or left of the road). For wheeled vehicles, draw a small circle on the shaft of the arrow. For tracked vehicles, draw a small square on the shaft of the arrow and place the length of the turnout, in meters, at the tip of the arrow. When the turnout is longer than 1 kilometer, use double arrows. |
| 29. Underpass constrictions (arched or rectangular) | 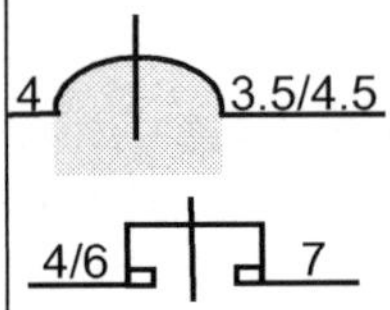  | Draw the symbol over the road. Place the width of the traveled way, in meters, to the left of the symbol. If sidewalks are present, follow the traveled-way width with a slash and the total width, including sidewalks. Underline the traveled-way width if the road entering the underpass is wider than the underpass traveled way. Show the overhead clearance, in meters, to the right of the symbol. Show both minimum and maximum overhead clearances, if different. |
| 30. Unknown information | ? | Use a question mark with almost any symbol when information required for the symbol is unknown or doubtful. |
| 31. Width constriction | 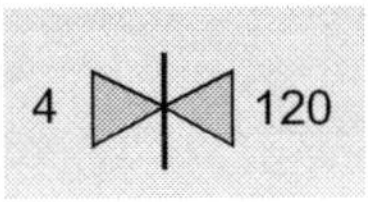  | The minimum width is to the left of the symbol. The length of the minimum width is to the right of the symbol. |

## ROUTE WIDTHS

I-6. The route width is the narrowest width of a traveled way on a route *(Figure I-2)* and is expressed in meters. It may be the width of a bridge, a tunnel, an underpass, or any other constriction that limits the traveled-way width. The establishment of the traveled-way width will help determine the number of lanes of traffic that the route may accommodate.

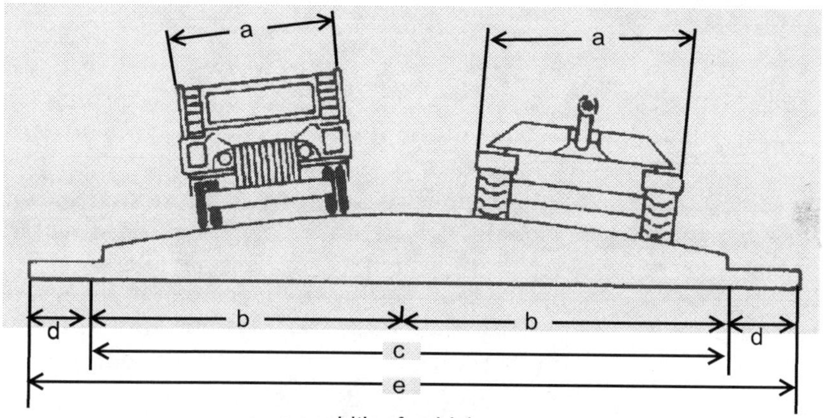


a = width of vehicle
b = width of lane
c = width of traveled way
d = width of hard shoulder
e = width of grading

**Figure I-2. Road Width**

I-7. Based on the number of lanes of traffic, categorize the route as either—

- Limited access. Limited access permits passage of isolated vehicles of the appropriate width in one direction only.

- Single lane. A single lane permits use in only one direction at any one time. Passing or movement in the opposite direction is impossible.
- Single flow. Single flow permits the passage of a column of vehicles and allows isolated vehicles to pass or travel in the opposite direction at predetermined points only.
- Double flow. Double flow permits two columns of vehicles to proceed simultaneously.

## ROUTE TYPES

I-8. Route types are designated by their ability to withstand the effects of weather, and are determined by the worst section of the route. The three types of routes are—

- Type X. Type X is an all-weather route that, with reasonable maintenance, is passable throughout the year to its maximum volume of traffic. This type of route is normally formed of roads having waterproof surfaces, is only minimally affected by precipitation or temperature changes, and is never closed from the effects of weather.
- Type Y. Type Y is a limited, all-weather route that, with reasonable maintenance, is passable throughout the year, but at times having a volume of traffic considerably less than maximum capacity. This type of route is normally formed of roads that do not have waterproof surfaces and are affected by weather or temperature changes. This type of route is closed for short periods (up to one day at a time) by adverse weather conditions during which heavy use of the road would probably lead to complete collapse.
- Type Z. Type Z is a fair-weather route that may become impassable in adverse weather. It may

be closed for long periods, and may require major maintenance or construction efforts to reopen it.

## MILITARY LOAD CLASSIFICATION

I-9. A route MLC is a class number representing the safe load-carrying capacity and indicating the maximum vehicle class that can be accepted under normal conditions. Usually, the lowest bridge MLC (regardless of the vehicle type or conditions of traffic flow) determines the route MLC. If there is not a bridge on the route, the worst section of road determines the route overall classification.

I-10. Whenever possible, the basic military road network is composed of average routes and includes a number of heavy traffic routes and a few very heavy traffic routes. Individual routes are grouped and identified in broad categories, such as—

- Average traffic routes are Class 50.
- Heavy traffic routes are Class 80.
- Very heavy traffic routes are Class 120.

## OVERHEAD CLEARANCE

I-11. Overhead clearance is the vertical distance between the road's surface and any overhead obstacle (power lines, overpasses, tunnels, and so forth) that denies the use of the route or road to all vehicles or loads that exceed this height. This distance is represented in the route classification formula, in meters. Use the infinity symbol (∞) in the formula for unlimited overhead clearance. Any point along the route where the minimum overhead clearance is less than 4.3 meters is considered an obstruction.

## ROUTE OBSTRUCTIONS

I-12. Route obstructions are conditions that restrict the type, amount, or speed of the traffic flow. They are indicated in the route classification formula as OB. If an obstruction is encountered and reported, depict its exact nature on the route classification overlay. Reportable obstructions include the following:

- Overhead obstructions, such as tunnels, underpasses, power lines, and so forth, with a clearance of less than 4.3 meters.
- Reductions in traveled-way widths that are below the standard minimums prescribed for the type of traffic flow *(Table I-2)*. This includes reductions caused by bridges, tunnels, craters, and lanes through mined areas, protruding buildings, or rubble.
- Slopes (gradients) of 7 percent or greater.
- Curves with a radius of 25 meters or less. Curves with a radius of 25.1 to 45 meters are not considered obstructions; however, report them on the reconnaissance overlay.
- Ferries.
- Fords.

**Table I-2. Traffic Flow**

| Type of Vehicles | Limited Access (In Meters) | Single Lane (In Meters) | Single Flow (In Meters) | Double Flow (In Meters) |
|---|---|---|---|---|
| Wheeled | At least 3.5 | 3.5 to 5.5 | 5.5 to 7.3 | Over 7.3 |
| Tracked or combination vehicles | At least 4.0 | 4.0 to 6.0 | 6.0 to 8.0 | Over 8.0 |

### SPECIAL CONDITIONS (SNOW BLOCKAGE AND FLOODING)

I-13. Conditions, such as snow and water, are not normally classified as obstructions, except where these conditions are regular, recurrent, and serious. In cases where snow accumulation is excessive and is either blocking traffic or has the potential to block traffic, the symbol following the route classification formula is T. Similarly, where flooding poses the same problem, the symbol following the formula is W.

## CALCULATIONS

I-14. Make calculations for the various curves, slopes, and fords that may affect vehicle movement along the route.

### CURVES

I-15. The speeds that vehicles move along the route and security planning measures are affected by sharp curves in the roadway. As previously stated, curves with a radius of 25 meters and less are obstructions to traffic and are indicated by OB in the route classification formula and identified on *DA Form 1248*. There are several ways to measure the radius of a curve, but the two easiest are the tape measure and formula methods.

#### Tape Measure Method

I-16. The quickest way to estimate the radius of a sharp curve is by using a tape measure to find the radius *(Figure I-3, page I-16)*. Imagine the outer edge of the curve as the outer edge of a circle. Find (estimate) the center of this imaginary circle; then measure the radius using a tape measure. Start from the center of the circle and measure to the outside edge of the curve. The length of the tape measure from the center of the imaginary circle to its outer edge is the curve radius. This method is practical for curves having a radius up to 15 meters and located on relatively flat ground.

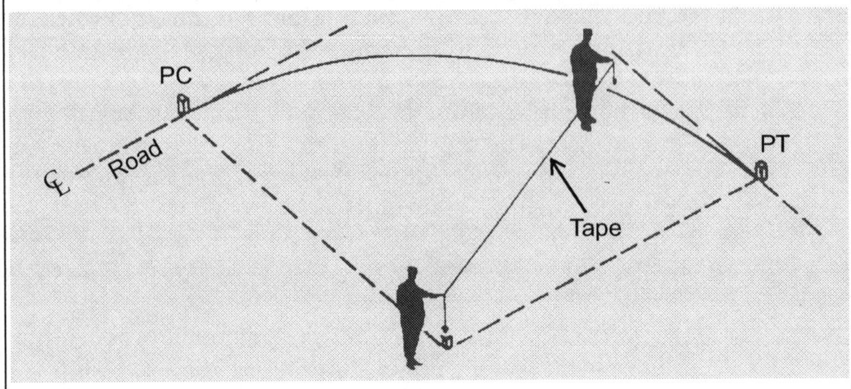


**Figure I-3. Find the Radius with a Tape Measure**

## Formula Method

I-17. The other, and often more practical, method of determining a curve radius *(Figure I-4)* is based on the following formula:

$$R = (C^2/8m) + (M/2)$$

*where—*

*R = radius of the curve*

*C = distance from the centerline of the road to the centerline of the road at the outer extremities (beginning and end) of the curve*

*M = perpendicular distance from the center of the tape to the centerline of the road*

I-18. When conditions warrant, set M at 2 meters from the centerline, then measure C at 2 meters from the centerline. Use this method when there is a time limitation or because natural or man-made restrictions prevent proper measurements.

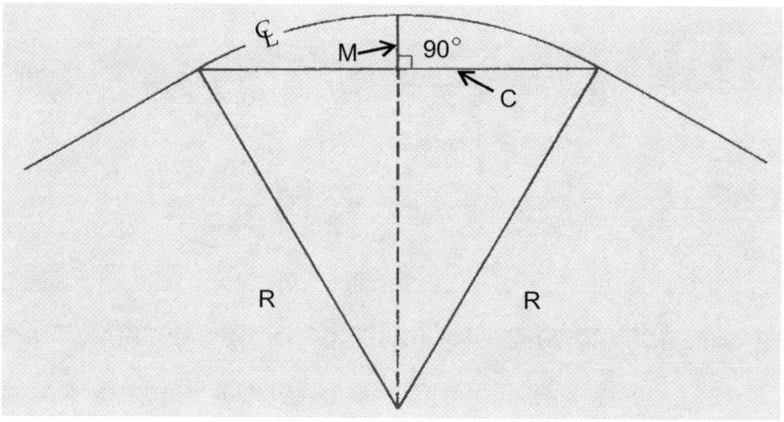


**Figure I-4. Determine a Curve Radius**

I-19. If C is 15 meters and M is fixed at 2 meters, the formula becomes:

$R = (15^2/16) + 2/2$

I-20. The result of this calculation would be an obstruction to traffic flow, and OB would be placed in the route classification formula.

## SLOPES

I-21. The rise and fall of the ground is known as the slope or gradient (grade). Roadways with a slope of 7 percent or greater affect the movement speed along the route and are considered an obstruction. The percent of slope is the ratio of the change in elevation (the vertical distance to the horizontal ground distance) multiplied by 100 *(Figure I-5, page I-18)*. It is important to express the vertical and horizontal distance in the same unit of measure. Although not an obstruction, report all hills with a slope greater than 5 percent on the reconnaissance overlay.

I-22. There are numerous methods and instruments available for determining the percent of slope, but the

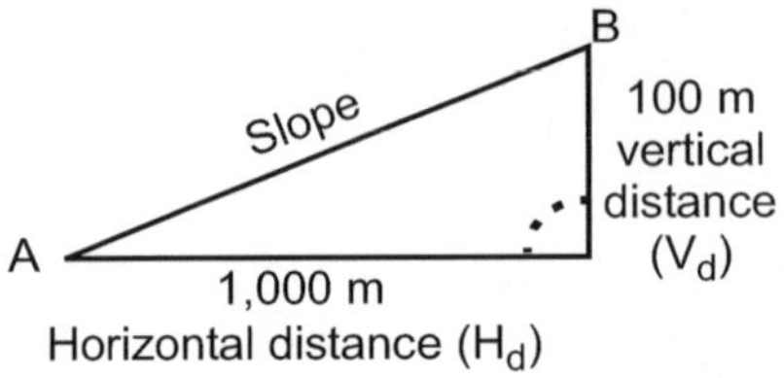


$$\text{Percent of slope} = \frac{V_d}{H_d} \times 100$$

$$= \frac{100\ m}{1,000\ m} \times 100$$

$$= +10\ \text{percent}$$

**Figure I-5. Calculate the Percent of Slope**

two hasty and most practical methods are the map and pace methods.

## Map Method

I-23. Use a large-scale map (such as 1:50,000) to estimate the percent of slope quickly. After identifying the slope on the map, find the difference in elevations between the top and bottom of the slope by reading the elevation contours or spot elevation. Then, measure and convert the horizontal distance (usually road distance) to the same unit of measurement as the elevation difference. Substitute the vertical and horizontal distances in the percent-of-slope formula and compute the percent of slope *(Figure I-6)*. This is a hasty method and is the less accurate of the two methods.

## Pace Method

I-24. The pace method is another quick way to estimate the percent of slope. Before using this method, accurately determine the height and pace of each member of the reconnaissance team. As a general rule of thumb, the eye level of the average soldier is 1.75 meters above the ground *(Figure I-7, page I-20)*. The pace of the average soldier is 0.75 meters. Perform the following steps to determine the percentage of slopes:

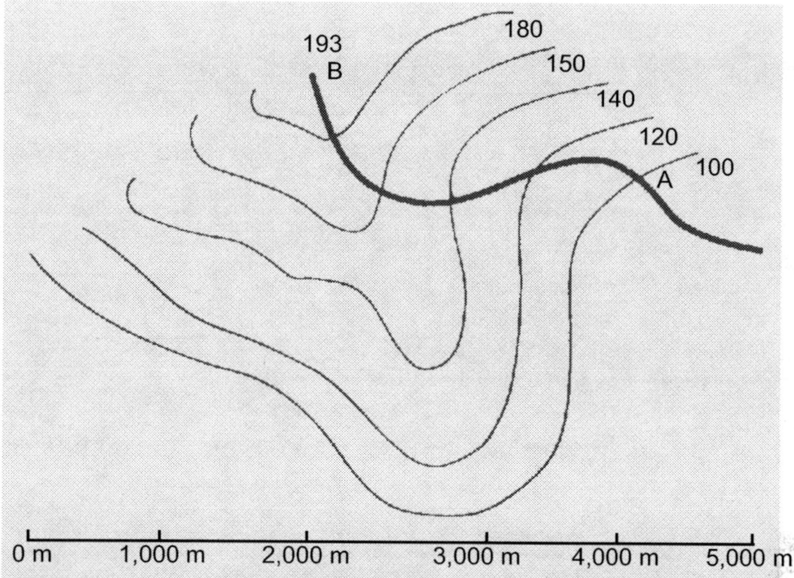


Find the difference in elevation ($V_d$)

| | |
|---|---|
| Spot elevation B | = 193 m |
| Spot elevation A | = 100 m |
| Difference in elevation ($V_d$) | = 93 m |

Find the horizontal distance ($H_d$)

Road distance A to B (use a piece of paper and a graphic scale as shown in FM 21-26) = 3,720 m

Use the percent-of-slope formula.

$$\frac{V_d}{H_d} \times 100$$

$$= \frac{93}{3,720} \times 100$$

$$+ 2.5 \text{ percent}$$

**Figure I-6. Determine the Slope with a Map**

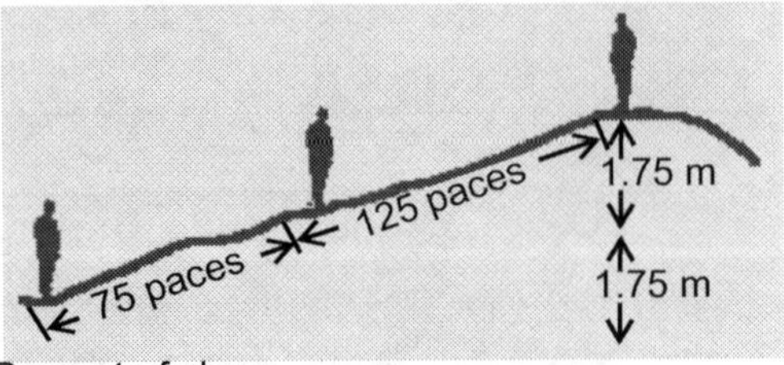


| | | | |
|---|---|---|---|
| Find: | Percent of slope | = | |
| Given: | Eye-level height | = | 1.75 m |
| | Pace | = | 0.75 m |
| Answer: | Vertical distance | = | 2 x 1.75 m |
| | | = | 3.5 m |
| | Horizontal distance | = | 75 paces +125 paces x 0.75 |
| | | = | 150 m |

$$Percent\ of\ slope = \frac{V_d}{H_d} \times 100 = \frac{3.5}{150} \times 100$$

**Figure I-7. Determine the Slope Using the Pace Method**

***Step 1.*** Stand at the bottom of the slope with your head and eyes level.

***Step 2.*** Sight an easily identifiable spot on the slope, horizontal to eye-level.

***Step 3.*** Walk forward and stand on the spot, recording the number of paces. Repeat this procedure until you reach the top of the slope. Estimate fractions of eye level.

***Step 4.*** Compute the vertical distance by multiplying the number of sightings by the eye-level height of 1.75 meters. Compute the horizontal distance by totaling the number of paces and converting them to meters by multiplying by the known pace-to-meter conversion factor of 0.75 meters.

***Step 5.*** Calculate the percent of slope by substituting the values into the percent-of-slope formula *(Figure I-7)*. Because this method considers horizontal ground distance and incline distance as equal, the reasonable accuracy is only for slopes less than 30 percent. This method requires repeated practice to achieve reasonable accuracy.

## FORDS

I-25. A ford is a location in a water barrier where the current, bottom, and approaches allow personnel and vehicles to cross and remain in contact with the bottom during the crossing. While engineers are fully trained and equipped in this area, MP are also capable of performing this task. In the course of a route reconnaissance, MP identify potential fording sites as a contingency for bridges that may become unusable. When identifying a ford on the reconnaissance overlay, include the following information in the ford symbol:

- The ford location, as indicated by an arrow from the ford symbol to the site on the overlay.
- A serial number assigned to each ford for reference. Follow the unit SOP in assigning serial numbers, ensuring that the same number is not duplicated on any one overlay.
- The type of ford, as determined by bottom composition, width, and water depth. Use V for vehicle, P for pedestrian, or VP for both. Approaches are not considered in determining the ford type.
- The stream's normal current velocity is expressed in meters per second. Seasonal limiting factors follow the stream velocity notation and are shown by the following letters:
  - X - No seasonal limitations except for potential sudden flooding of limited duration (flash flooding).
  - Y - Serious, regular, or recurrent flooding or snow blockage.

**NOTE: If the Y symbol is used, the route type in the route classification formula automatically becomes type Z.**

- The length of the ford (distance from nearshore to far shore), in meters.
- The nature of the bottom surface, using the most appropriate letter symbol—
  - M – mud.
  - C – clay.
  - S – sand.
  - G – gravel.
  - R – rock.
  - P – artificial paving.
- The normal stream depth at its deepest point, expressed in meters.

I-26. Separate all elements of the ford symbol by slashes. Substitute a question mark for any item of information that is unknown. Record all the information on the route reconnaissance overlay.

## Stream Width

I-27. There are many methods of determining (estimating) the stream's width. The most common and practical is the compass method. Begin by using a compass to take an azimuth from a point on the nearshore and close to the water's edge to a point on the opposite shore and close to the water's edge *(Figure I-8)*. On the nearshore, establish another point that is on a line and at a right angle to the azimuth selected. The azimuth to the same point on the far shore is plus or minus 45 degrees (800 mils) from the previous azimuth. Measure the distance between the two points on the nearshore. This distance is equal to the distance across the stream.

## Current Velocity

I-28. Current velocities vary in different parts of a stream. Velocity is usually slower near the shore and

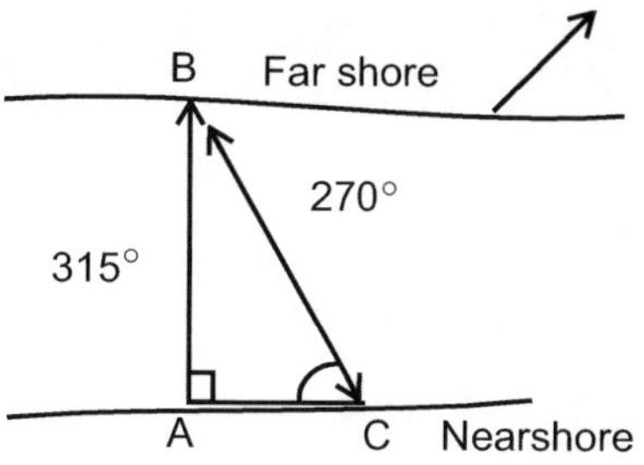


A - Point on nearshore
B - Point on far shore
AB - Distance to be measured
Azimuth of line AB is 315° (moving left, add 45°; moving right, subtract 45°).
Azimuth of line CB is 270°.
Difference between azimuth AB and azimuth CB is 45°.
Distance along AC equals distance along AB.

**Figure I-8. Determine Stream Width**

faster in the main channel of the stream. To estimate stream velocity—

- Measure a distance along the nearshore.
- Throw a light, floating object that will not be affected by the wind into the stream.
- Record the time that is takes for the object to travel measured distance.
- Repeat this procedure at least three times to establish an accurate average.
- Use the average time of the test in the following formula *(Figure I-9, page I-24)* to determine the stream's velocity.

*Velocity (in meters per second) = measured distance (in meters) / average time (in seconds)*

## TEMPORARY ROUTE SIGNING

I-29. A military route sign system, like the US highway sign system, can enable road users to reach their destinations by following route signs and road markings displayed along the roadside. MP patrols monitor signs on a routine basis, checking specific signs before critical

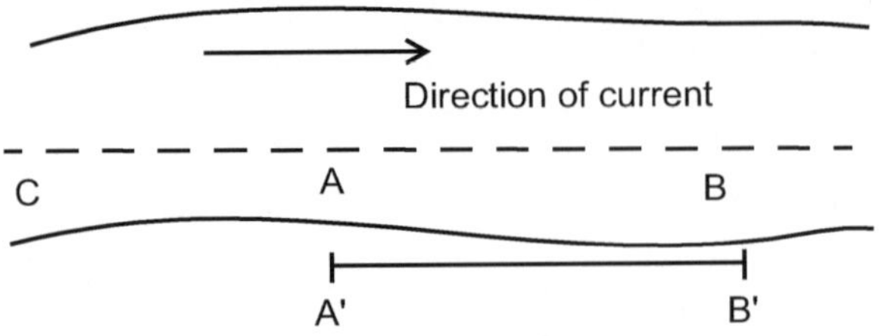


$$Velocity = \frac{A'B'\ (meters\ or\ feet)}{Time\ to\ float\ from\ A\ to\ B\ (seconds)}$$

Measure distance A'B'.
Throw a floating object into the stream at C.
Determine the time required for the object to float to distance from A to B.

**NOTE: This should be done three times, with the final velocity being the average of the three.**

**Figure I-9. Determine Stream Velocity**

moves. Engineers erect permanent signs, but these signs can be damaged, destroyed, or moved by weather, saboteurs, and battle.

I-30. When MP on patrol encounter immediate and temporary MSR obstructions, like blown bridges or NBC contamination, they construct and erect signs quickly to guide vehicles around the obstruction. Prepare and post temporary signs to—

- Identify routes.
- Reroute traffic around problem areas.

- Help convoys and units move quickly and easily to their destinations even on an unfamiliar route.
- Show the drivers the locations of staging areas, tactical AAs, detours, key units, and facilities.
- Give directions, distances, and general information.
- Help lost military personnel find their way to the closest MP element.

## TEMPORARY SIGNING

I-31. Obtain signing materials, signs, paint, and wire through the Army supply system. In an emergency, use boards, shingles, or cardboard. Use a portable sign-making kit to prepare signs when the signs will not conflict with standardization agreements (STANAGs).

I-32. Preplanned route signing for convoys and units traveling long stretches of MSRs and link routes must be planned well in advance. It is a time-consuming and manpower- and material-intensive operation. Signs used for this type of route are built by engineers or by MP battalion or company supply personnel. If tasked to take part in such an activity, ensure that after the signs have been constructed and coded for a particular route or operation, they are stored in a secure place to avoid compromise.

## TEMPORARY ROUTE SIGNING

I-33. Place signs where they will support the traffic control plan and the traffic circulation plan. Specific sign locations are shown on the traffic control plan overlay prepared at the PM operations section. Changes in sign locations are reflected on the traffic control plan overlay as signs are added and deleted.

I-34. **Plan**. When MP encounter immediate and temporary MSR obstructions, like blown bridges or NBC contamination, they use the squad route sign kit or construct and erect hasty signs quickly to guide vehicles around the obstructions. Planning considerations include the following:

- Identifying routes.
- Rerouting traffic around problem areas.
- Helping convoys and units move quickly and easily to their destinations, even on unfamiliar routes.
- Showing drivers the locations of staging areas, tactical AAs, detours, key units, and facilities.
- Giving directions, distances, and general information.

I-35. **Execute**. Place signs where they will support the traffic control plan and the traffic circulation plan. Specific sign locations are shown on the traffic control plan overlay. Changes in sign locations are reflected on the traffic control plan overlay as signs are added or deleted.

I-36. Often, one three-man team can place signs along a route. To post a sign, a team member dismounts and walks 50 to 100 meters up the road. The other team members provide security and check and confirm the sign's placement to ensure that the drivers will be able to see the sign.

I-37. MP squads use the route signing kit to place signs along a route within the squad's AO. One team posts the signs while the other teams provide security. The squad leader checks and confirms the sign placement by traveling the entire route. He develops an overlay with the location of each sign in his squad's AO. Squad overlays are consolidated and contribute to a platoon traffic control plan.

I-38. More teams, up to a squad, may be needed for signing tasks in urban areas, in areas where a thorough reconnaissance has not been conducted, and in areas where the threat is unknown or is thought to be great. One team erects signs while following teams provide overwatch security. Use the following guidelines when placing signs:

- Place hazard signs about 150 meters before the hazard.
- Place regulatory signs where a regulation takes effect.
- Place all signs on the side of the road facing the traffic flow, about 1 meter off the traveled roadway. Conceal them from air view. If no cover is available, slant the sign stake forward.
- Place signs 1 to 2 meters above road level. Place all the signs at the same height if possible. Sign height is governed by roadside foliage, by whether the route is in an urban or a rural area, and by day or night use. In urban areas, place signs so that they are not hidden by vehicles or pedestrians, do not hinder pedestrians, and can be seen at night with street lighting or vehicle headlights. In open country, a good sign height is between thigh and knee height. This usually makes signs visible by day and night. Be sure that signs are not obscured by foliage.
- Use the least number of signs needed to be effective. Ensure that every sign is necessary and specific.
- Use more signs in urban areas than in rural areas.
- Use more signs on night routes than on routes primarily used during the day.
- Use signs to inform drivers to follow the common route when one road is used for two signed routes. Use signs to inform drivers when the routes diverge.

I-39. Conceal all the signs so that they are seen only from the direction from which they are approached.

There is no exact rule stating the distance from which a sign should be visible. However, the distance should be no greater than security allows and not less than is reasonable for those receiving directions.

I-40. Carefully conceal illuminated signs. Ensure that the light source is just strong enough to light the sign, but not strong enough to be seen from the air. This entails masking and covering the light sources. Consider placing chemical light sticks on top of the signs.

I-41. Place temporary route signs where they will provide warning and reaction time for the drivers. Do not block existing civilian signs. Place warning panels at convenient distances from where a route regulation takes effect. This distance can be shown on the panel (for example, BLACKOUT 500 METERS). In areas where blackout drive is the rule and drivers are using night-vision equipment, employ IR emitters on signs along routes to help guide the units to their destination. Just like any other light source, ensure that the IR emitters cannot be seen from the air.

I-42. Place guide signs at road junctions to prevent confusion. Put signs on both sides of the road if needed. Place confirmation signs *(Table I-3)* 150 meters beyond the critical road junctions to let drivers know that they are on the correct route.

**Table I-3. Signs Identifying Military Routes and Locations**

| Type | Examples | Description | Purpose | Placement |
|---|---|---|---|---|
| Direction indicator |  | White, 30-centimeter diameter disk with black directional arrow (ID number or name is mounted below the disk)<br>or<br>White disk on black rectangular board with the identification number above | Indicates the direction and identifies the route or the name | At intersections where roads merge and where routes separate |

**Table I-3. Signs Identifying Military Routes and Locations (Continued)**

| Type | Examples | Description | Purpose | Placement |
|---|---|---|---|---|
| Guide signs (use together at important road junctions) |  | Rectangular with the symbols in white on black background; includes directional arrow and route number, name, and/or symbol | Indicates the locations, distances, directions, and routes | Where needed |
| Warning signs (use together at important road junctions) |  | Same as above | Shows the correct direction to take at route junctions | A sufficient distance (50 to 100 meters) before a junction to allow drivers to make the turn safely (On roads where speed is restricted, place the signs 25 meters before the junction.) |

**Table I-3. Signs Identifying Military Routes and Locations (Continued)**

| Type | Examples | Description | Purpose | Placement |
|---|---|---|---|---|
| Confidence signs |  | Same as above | Reassures the drivers that they are still on the correct route (Use them in urban areas to assure drivers that they are following the correct route. Use them on long stretches of road where it is unnecessary to use warning and confirmation signs for a considerable distance.) | Where needed |
| Confirmation signs |  | Same as above | Lets drivers know that they are on the correct route after changing direction | Just after turns, but visible while making the turn, if possible |

**Table I-3. Signs Identifying Military Routes and Locations (Continued)**

| Type | Examples | Description | Purpose | Placement |
|---|---|---|---|---|
| Countdown signs |   | Same as above | Warns of significant locations, such as SPs, RPs, starts and ends of routes, link routes, MSR junctions, and blackout areas—anything requiring a major change to the movement | Series of three signs at 100-meter intervals before the designated location |
| Regulatory signs |   | Military equivalent of civilian signs like STOP and YIELD | Regulates and controls the traffic on a route | Where needed (posted by engineers and considered permanent) |

**Table I-3. Signs Identifying Military Routes and Locations (Continued)**

| Type | Examples | Description | Purpose | Placement |
|---|---|---|---|---|
| Hazard signs |  | Yellow, diamond-shaped background with information printed in black | Indicates traffic hazards, such as dangerous corners, steep hills, or crossroads (rarely used in COMMZ, as civilian signs usually suffice. Use military signs in combat zones.) | Where needed (posted by engineers and considered permanent) |
| Military casualty evacuation route signs | 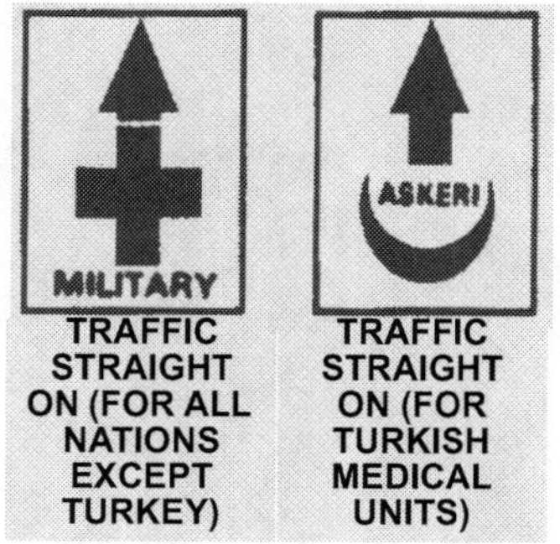 | Rectangular white background with a red directional arrow, cross, or crescent; the word MILITARY; the unit or subunit designation; and other information like national markings<br>or<br>A directional disk with four segments cut out to form a cross or a directional disk with a crescent cut out (include the same information as above.) | Indicates evacuation route for military casualties | Where needed |

**Table I-3. Signs Identifying Military Routes and Locations (Continued)**

| Type | Examples | Description | Purpose | Placement |
|---|---|---|---|---|
| Civilian casualty evacuation route signs | 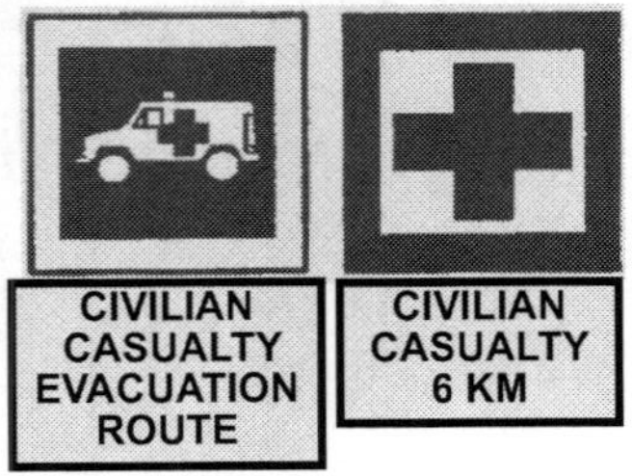  | Blue Geneva Conventions information sign (includes ambulance in white with a red cross or crescent and the words CIVILIAN CASUALTY EVACUATION ROUTE beneath the sign in the HN language) | Indicates civilian casualty evacuation routes | Along routes for civilian traffic (designated by the HN) |
| Blackout warning signs |   | Based on the Geneva Conventions hazard warning sign (Place the legend and the distance on a rectangular plaque beneath the warning sign.) | Indicates the beginning of a blackout area | Same as warning signs |

**Table I-3. Signs Identifying Military Routes and Locations (Continued)**

| Type | Examples | Description | Purpose | Placement |
|---|---|---|---|---|
| Blackout enforcement signs |   | Geneva Conventions prohibitory sign; plaque beneath says VEHICLE LIGHTS FORBIDDEN | Indicates that a blackout is in effect | Every 100 meters along the blackout route |
| Blackout relaxation signs |   | Same as a blackout warning sign | Indicates the end of a blackout area | At the end of the blackout route |

I-43. Place detour signs next to general traffic signs to identify the detour. Place the detour sign to the side (left or right) of the general sign that corresponds to the new direction to be taken.

I-44. Use signs to mark the entrance to a HQ or an installation along with a halt sign or other regulatory signs. Signs marking turnoffs and roads or tracks should include a directional disk or a directional arrow. Place signs to—

- Indicate where vehicles leave a signed route to get to the HQ or the installation.
- Mark the road or track leading to the HQ or the installation.
- Mark the entrance to the HQ or the installation.

## SIGNING FOR KEY POINTS ON ROUTES

I-45. Place countdown signs *(Table I-3, page I-29)* at the beginning and end of a route. Clearly mark SPs and RPs with the appropriate countdown signs. Feeder routes to the SP may require signs to help the convoys find it.

I-46. Mark route detours with countdown signs showing the distance to where the detour begins. The signs clearly indicate the route to be detoured.

I-47. Place countdown signs so that they give clear warning of the end of the detour. Mark the end of the detour with a sign reading DETOUR END. Erect a warning sign at the end of the detour to show how to return to the original route.

I-48. Position detour signs ahead of time where terrain will require a defile or the like. Position the signs off the road, facedown on the ground. To implement the detour, erect the signs.

**NOTE: The point where a link route meets the circuit is an ideal place for MP control. It is the point where the circuit and the link route begin and end.**

I-49. Place signs at junctions of axial and lateral MSRs from all four directions. Place countdown signs 300, 200, and 100 meters before the junction.

I-50. To keep the number of signs to a minimum, battalion-size and smaller groups use a directional sign. The space for the military symbol may contain the UIN or the unit map-marking symbol. Print the UIN or the symbol so that it can be read when the arrow is vertical. This allows the sign to be used as a disk direction.

I-51. Therefore, only one type of sign is needed for all purposes. Put unit signs in place immediately before a unit movement. Remove them as soon as possible after the move.

## MAIN SUPPLY ROUTE SIGNS

I-52. MSR signs identify MSRs by number, a pictorial symbol, or a name. Names and pictures—

- Are easier to identify and to remember than numbers.
- Prevent confusion with link-route signs marking the routes of units having three-figure identification numbers.
- Prevent security compromises by removing the chance use of an identification number already used on a map overlay.

I-53. There are two types of MSRs, axial and lateral, in a theater of operations. On the theater of operation's traffic circulation plan—

- Axial MSRs run to and from the forward edge of the battle area (FEBA) and are identified by odd numbers, like 87 or 215. Axial MSRs are shown as solid lines on the traffic circulation plan overlay. Axial routes are represented by a pictorial symbol on the route sign and are marked on the overlay as pictorial; for example,

MSR CLUB (PIC). On axial MSRs, *up* is toward the FEBA. It is shown on overlays and signs as a plain arrow. *Down* is away from the FEBA. It is shown as an arrow with a bar on the tail end.

- Lateral MSRs run parallel to the FEBA and are identified by even numbers, like 86 or 214. Lateral MSRs are shown as broken lines on the traffic circulation plan overlay. A named route sign represents lateral routes. The names are short, three- or four-letter words like *fox*, *ant*, or *hen*. *Up* and *down* on lateral MSRs show only general directions of travel. The general direction shown by *up* or *down* varies with the theater of operation and with the location of the FEBA. *Up* is usually to the north or east; d*own* is usually to the south or west. A plain arrow on an overlay or sign indicates *up*; a barred arrow shows *down*. To avoid confusion on lateral MSRs, use the letters N, E, S, W, NE, SE, NW, and SW on route signs to show the general direction of movement.

**This paragraph implements STANAG 2010.**

I-54. Ensure that all route signs are large enough to be read easily in poor light. Size is not specified, but—

- Signs for international use cannot be less than 40 centimeters by 33 centimeters.
- Bridge classification signs must conform to STANAG 2010.

## LINK ROUTE SIGNS

I-55. There is no set requirements for the design of link route signs *(Figure I-10)*. (Sometimes showing direction with a white arrow on a black background is enough.) A link route connects a unit or an activity to an MSR. A departing convoy follows the link route signs to the MSR. The convoy follows the MSR until guided off the MSR by signs warning of the need to exit and again follow link route signs until arriving at its destination. That part of a traveled route coinciding with an MSR will not have link route signs.

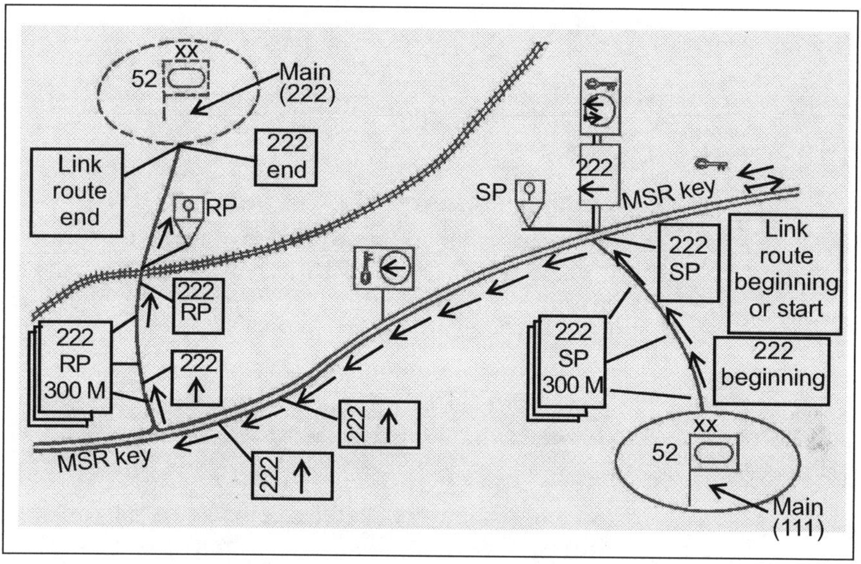


Figure I-10. Link Route Signing

## HEADQUARTERS AND INSTALLATION SIGNS

I-56. All signs must be used in HQ and parts of HQ, down to battalion or equivalent level. Examples are—

- Medical installations.
- Water points.
- Ammunition depots.
- TCPs.
- Decontamination stations.
- Fuel installations.

**This paragraph implements STANAG 1059.**

I-57. HQ and installation signs show the military symbol for the HQ or the installation. For HQ at brigade level and above, include the national distinguishing letters given in STANAG 1059. The sizes of the sign and the symbol are not standardized. The symbol and the background must be clearly contrasting colors. The actual colors used are left to the discretion of each nation.

## DIRECTIONAL SIGNS

I-58. A directional sign *(Table I-3, page I-29)* must be—

- A white disk displaying a black directional arrow accompanied by a route identification number or name that can be mounted beneath the disk.
- A white disk mounted or superimposed on a black rectangular board with the number or name on the board above the disk.

I-59. Directional disks cannot be less than 30 centimeters in diameter. Drill eight holes in the disks at equal intervals around the circumference so that they can be erected with the arrow pointing in the correct direction.

I-60. Directional signs show directions for axial and lateral routes on hastily prepared routes or very temporary routes like detours. The initial sign shows the use of disk directional signs and orders the drivers to follow the disk direction.

I-61. An MSR may require hundreds of signs. Make the signs in quantity without the black arrow. Stencil or stick the arrows on later. The superimposed signs are easiest to store and use. It is hard to secure the two elements of the circular two-piece signs, and it is hard to position that sign for display.

I-62. Use yellow instead of white for signs used during prolonged snowfall conditions or for signs permanently erected where there can be prolonged snowfalls.

# PORTABLE SIGN-MAKING KIT

I-63. The new portable sign-making kit (national stock number [NSN] 9905-01-478-3723) is lightweight and simple to use and contains tools and materials that, when used in conjunction with vehicle BII and locally available materials, produce temporary signs in support of MP operations. Since the kit is not fully self-contained, the user must plan ahead. The new, lighter kit gives MP units the ability to pack according to the METT-TC. Even though a significant difference exists between signs made with the kit and specifications in STANAGs (STANAG route signs generally have a black background with white letters, and the kit's sign background material is white), it does not differ from the kit it replaces. The soft-sided sign kit comes with white plastic bags that can be used to make signs up to 21 inches wide by 23 inches high. *Figure I-11,* shows a sample sign; and *Table I-4, page I-42,* shows a component listing for sign making.

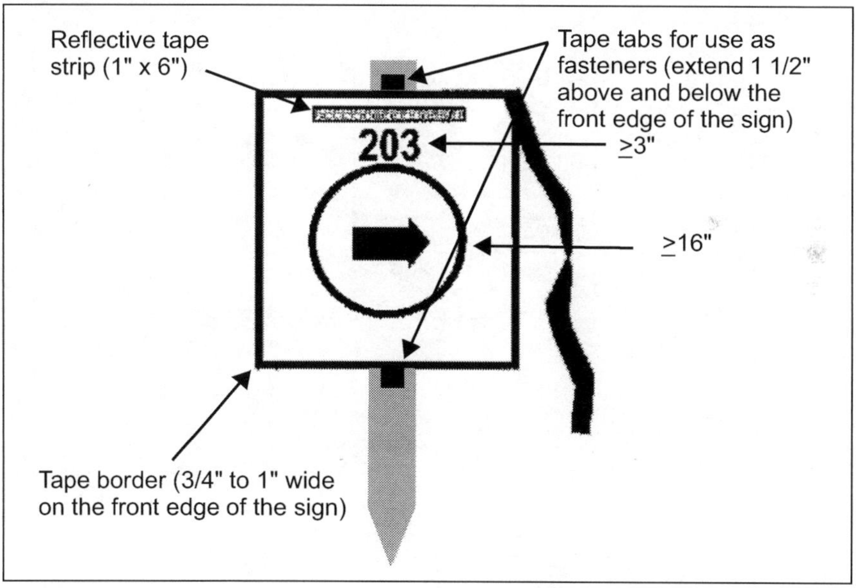


**Figure I-11. Sample Sign**

**Table I-4. Component Listing**

| Tools | NSN | |
|---|---|---|
| Tool bag | 5140-01-481-4460 | 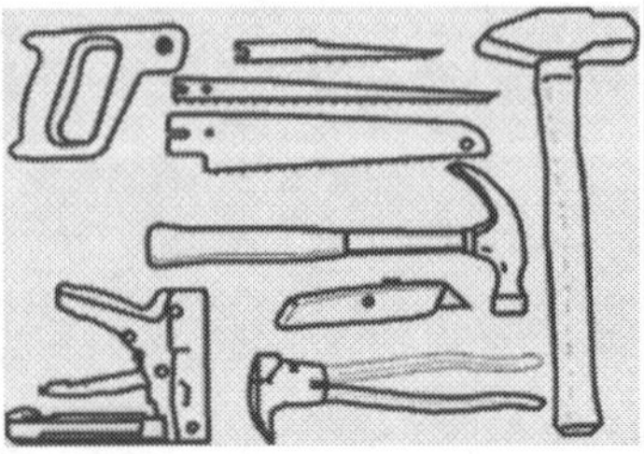 |
| Carpenter's hammer | 5120-01-112-8350 | |
| Utility knife<br>Replacement blades | 5110-00-892-5071<br>5110-00-293-2865 | |
| Linemans tool | 5120-00-293-3485 | |
| Blacksmith's/engineer's hammer (3 lb) | 5120-00-900-6103 | |
| Nested saw set | 5110-00-293-0090 | |
| Staple gun | 5120-00-889-1796 | |
| **Fasteners** | **NSN** |  |
| One pound of 1.75" steel screw-shank nails (concrete) | 5315-00-543-3023 | |
| Five pounds of 3.5" (10-12d) common nails | 5315-00-753-3885 | |
| One-half pound of 1.5" steel barbed-shank nails (roofing) | 5315-01-381-9609 | |
| Tape, pressure sensitive (2" x 160', 100 miles an hour) | 7510-00-074-5124 | |
| Staples, 1/2", box of 5,000 | 5315-00-889-2605 | |
| Cable ties, pkg of 100 | 5975-00-156-3253 | |
| Wire, nonelectrical, 1-lb roll | 9505-00-293-4208 | |

**Table I-4. Component Listing (Continued)**

| Marking Materials | NSN | |
|---|---|---|
| Black tube-type poster markers (2 each) | 7520-01-336-2856 | 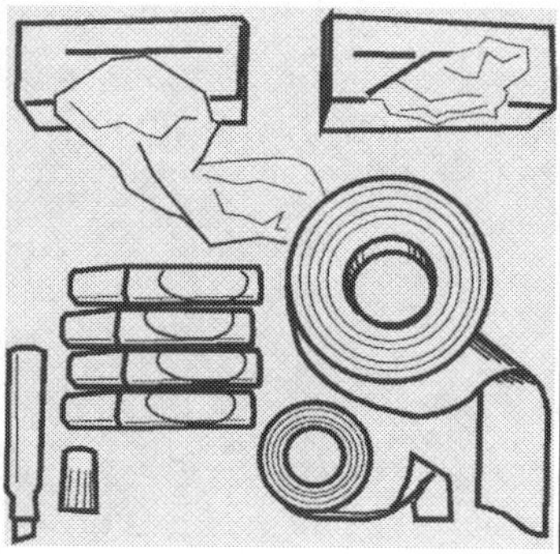 |
| Red tube-type poster marker | 7520-01-336-2857 | |
| Blue tube-type poster marker | 7520-01-336-2858 | |
| Bags, white plastic, trash, 8 gal, roll/box of 20 to 30 (2 each) | 8105-01-481-4225 | |
| Reflective tape (white 1" x 150') | 9390-00-949-8044 | |
| Ribbon, flagging, surveyors (arctic-glo pink) | 9905-00-194-9703 | |

## Appendix J

# Nuclear, Biological, Chemical Defense

The technology to produce and deliver chemical and biological agents and weapons of mass destruction is widely available. This technology provides military and nonmilitary leaders in unstable regions with a means to deter more technically advanced nations. Weapons of mass destruction can radically change the flow of battle, shift the balance of combat power, and place a superior force at risk. Areas targeted with these weapons may remain hazardous for days, even weeks after an attack. Additionally, nuclear reactor and chemical production facility accidents or sabotage can release radiological or chemical contaminants that can threaten friendly forces and military operations.

## OVERVIEW

J-1. Enemy forces may threaten sustainment areas to seize and maintain the initiative, while degrading or eliminating a unit's flexibility and capability to support operations. Sustainment area activities are lucrative targets for enemy NBC attacks. Attacking sustainment nodes weakens the main battle area's force effectiveness without adversely affecting an enemy's immediate maneuver. To achieve these aims, threat activities in sustainment areas target key critical support and logistic facilities. These facilities include—

- Special-weapons storage sites and delivery systems.

- $C^2$ facilities.
- Air defense artillery sites.
- Air bases.
- Seaports.
- MSRs.

## HAZARD DETECTION AND REPORTING

J-2. The best units to perform NBC reconnaissance are those units with the M93 NBC Reconnaissance System (FOX). Planners consider the disadvantages of conducting NBC reconnaissance with HMMWV-equipped units. The disadvantages of using a HMMWV include the following:

- The crew conducts operations in MOPP 4 while in contamination.
- The vehicle stops to conduct survey and detection.
- The vehicle has difficulty keeping up with the maneuver force over rugged terrain.
- The readings obtained using an M256A1 kit, detector paper, and a chemical agent monitor (CAM) are not as reliable as readings from the M93 (FOX).

J-3. Planners use the IPB and the METT-TC to determine the type of NBC reconnaissance technique and the type of equipment to be used for the NBC reconnaissance.

J-4. Early detection of NBC hazards is required for timely warning of units and personnel in the affected areas. NBC reconnaissance is an active contamination avoidance measure that provides commanders with information on NBC hazards in an AO.

J-5. NBC reconnaissance contributes to the overall intelligence collection effort and consists of the following five critical tasks: detecting, identifying, marking, reporting, and sampling. The four types of NBC reconnaissance include search, survey, sample, and surveillance.

J-6. MP may be tasked to conduct NBC reconnaissance while conducting other reconnaissance operations. They are equipped to detect and identify radiological and chemical hazards, mark the contaminated area, and report the hazard. They are not equipped to collect or handle samples.

J-7. NBC surveillance activities are planned to support the intelligence collection effort. MP conduct NBC surveillance by monitoring MSRs, critical facilities, and key terrain. They employ NBC detector equipment and measures that include the following:

- M8A1 automatic chemical agent alarm system.
- AN/VDR-2 RADIAC set.
- CAM.
- Visual observation for unexplained artillery or spray.
- M256A1, M8, and M9 paper.

J-8. If MP observe the indications of an NBC attack, they may be ordered to conduct an NBC survey to define the boundaries of the contamination, locate and mark clean bypass routes, and occupy OPs to observe the designated area. MP conduct an NBC survey, on order, to obtain detailed information concerning a contaminated area, including the location of the general boundaries of the contaminated area and the intensity of the contamination (nuclear) or the type of agent (chemical).

J-9. MP locate the boundaries and place warning markers around the contaminated area at all the entry points. They report information using an NBC 4 report and inform the road users of the contaminated area, and direct the traffic to alternate routes to avoid the contaminated area.

J-10. At all echelons, MP provide continuous information to higher HQ using NBC 1 and NBC 4 reports *(Figure J-1)*. These report formats provide a rapid means of disseminating information. Use the NBC 1 report to record the initial use and subsequent data concerning enemy NBC attacks. The initial NBC 1 report precedence is flash and all others are immediate. *Table J-1, pages J-6 through J-10,* gives an explanation of the various lines in an NBC report.

J-11. Use the NBC 4 report for radiation dose rate measurements. Usually, the unit submits two reports—one on initial contact and another for peak dose rate. Measure radiation dose rates in the open, 1 meter above the ground. If radiation dose rates are taken inside a vehicle or a shelter, at least one outside reading is necessary to determine the correlation factor. Other items of the report are optional. Refer to *FM 3-3* for more information about NBC reports.

J-12. As MP move throughout an AO, they monitor for the presence of an NBC hazard, mark contaminated areas, send NBC reports through operational channels, and direct traffic around or through hazards to ensure that the troops and the supplies get where they are needed on the battlefield.

## CONTAMINATION MARKING

J-13. Mark contamination so that unsuspecting personnel will not be exposed to it. When detecting or suspecting NBC hazards, mark all likely entry points into the area and report the contamination to higher HQ. The only exception to this policy occurs when marking the area would help the enemy. If the commander makes this exception, report the hazard to protect friendly units.

| NBC 1 Report (Observer's Report) | | |
|---|---|---|
| **NBC 1 (Nuclear) Report** | **NBC 1 (Biological) Report** | **NBC 1 (Chemical) Report** |
| B. NB062634 | B. LB206300 | B. LB200300 |
| C. 90-degree grid | D. 200410Z | D. 201405Z |
| D. 201405Z | E. 200414Z | E. 201412Z |
| G. Aircraft | F. LB206300, actual | F. LB206300, estimated |
| H. Surface | G. Aerial spray | G. Bomblets |
| J. 60 | H. Unknown | H. Nerve |
| L.15 degrees | | V. Airburst |

**NOTE: Line items B, D, H, and either C or F should always be reported. Use other line items if the information is known.**

| NBC 4 Report (Reconnaissance, Monitoring, and Survey Results) | |
|---|---|
| **NBC 4 (Nuclear) Report** | **NBC 4 (Chemical) Report** |
| Q. LB123987 | H. Nerve V |
| R. 35 | Q. LP200300, liquid |
| S. 201535Z | S. 170610Z |

**NOTES:**

**1. Line items H, Q, R, and S may be repeated as often as necessary.**

**2. In line R, descriptive words, such as initial, peak, increasing, decreasing, special, series, verification, or sumary may be added.**

**Figure J-1. Example of NBC 1 and NBC 4 Reports**

## Table J-1. Explanation of Line Items in NBC Report

| Line | Nuclear | Chemical and Biological | Remarks |
|---|---|---|---|
| A | Strike serial number | Strike serial number | Assigned by the NBC center |
| B | Position of the observer | Position of the observer | Grid coordinates (universal transverse mercator [UTM] or place). |
| C | Direction of the attack from the observer, to include the unit of measure | Direction of the attack from the observer | Nuclear: Use degrees magnetic north (DGM), degrees true north (DGT), degrees grid north (DGG), mils magnetic north (MLM), mils true north (MLT), or mils grid north (MLG) Chemical: Measure clockwise from the grid or magnetic north (in degrees or mils) |
| D | DTG of detonation | DTG for the start of the attack | Nuclear: Use Zulu time. Chemical: Designate time zone used. |
| E | NA | DTG of the end of the attack | Designate time zone used. |
| F | Location of the area attacked | Location of the area attacked | Use grid coordinates (or place). State whether the location is actual or estimated. |
| G | Suspected or observed event and the means of delivery or kind of attack | Kind of attack | State whether the attack was by artillery, mortars, rockets, missiles, bombs, or spray. |

### Table J-1. Explanation of Line Items in NBC Report (Continued)

| Line | Nuclear | Chemical and Biological | Remarks |
|---|---|---|---|
| H | Type of burst | Type of agent and type of burst<br>P = persistent<br>NP = nonpersistent | Nuclear: Specify air, surface, or subsurface. Chemical: State whether by air, ground, or spray attack. |
| I | NA | Number of munitions or aircraft | If known. |
| J | Flash-to-bang time | NA | Use seconds. |
| K | Crater present or absent and the diameter | Description of terrain and vegetation | Nuclear: Sent in meters. Chemical: Sent in NBC 6. |
| L | Cloud width at H+5 minutes | NA | State whether measured in degrees or mils. |
| M | Stabilized cloud top or cloud bottom angle or cloud top or bottom height at H+10 minutes | NA | Nuclear: State whether the angle is cloud top or cloud bottom and whether it is measured in degrees or mils. Chemical: State whether the height is cloud top or cloud bottom and whether it is measured in meters or feet. |
| N | Estimated yield | NA | Sent as kilotons. |
| O | Reference DTG for estimated contour line not H+1 | NA | Use when the contours are not plotted at H+1. |
| P | For radar purposes only | NA | None |

**Table J-1. Explanation of Line Items in NBC Report (Continued)**

| Line | Nuclear | Chemical and Biological | Remarks |
|---|---|---|---|
| PA | NA | Predicted hazard area (coordinates) | If the wind speed is 10 kilometers per hour or less, this item is 010 (the radius of the hazard area in kilometers). |
| PAR | Coordinates of external contours of the radioactive cloud | NA | Six-digit coordinates. Letter R identifies RADAR set. |
| PB | NA | Duration of the hazard in the attack and hazard area | State in days, hours, minutes, and so forth |
| PBR | Downwind direction of the radioactive cloud and the unit of measure | NA | DGM, MLM, DGT, MLT, DGG, or MLG. The letter R identifies a RADAR set. |
| Q | Location of the reading | Location of the sampling and the type of sample | Nuclear: UTM or the place.<br>Chemical: UTM or the place. State whether the test was air or liquid. |
| R | Dose rate or actual value of decay exponent | NA | State the dose rate in centigray per hour (cGyph). |
| S | DTG of the reading | DTG contamination detected | State the time of the initial identification test sample or when the reading was taken. |

**Table J-1. Explanation of Line Items in NBC Report (Continued)**

| Line | Nuclear | Chemical and Biological | Remarks |
|---|---|---|---|
| T | H+1 DTG | DTG of the latest contamination survey of the area | NBC 5 and NBC 6 reports only. |
| U | 1,000-cGyph contour line | NA | Plot in red. |
| V | 300-cGyph contour line | NA | Plot in green. |
| W | 100-cGyph contour line | NA | Plot in blue. |
| X | 20-cGyph contour line (30 cGyph contour line is used by other NATO forces) | Area of actual contamination | Nuclear: Plot in black.<br>Chemical: Plot in yellow. |
| Y | Direction of the left and right radical lines | Downwind direction of the hazard and the wind speed | Nuclear: Direction measured clockwise from grid north (GN) to the left and then right radial lines (degrees or mils, state which), four digits each.<br>Chemical: Direction is four digits (degrees or mils) and the wind speed is three digits (kilometers per hour). |

**Table J-1. Explanation of Line Items in NBC Report (Continued)**

| Line | Nuclear | Chemical and Biological | Remarks |
|---|---|---|---|
| Z | Effective wind speed<br>Downwind distance of Zone 1<br>Cloud radius<br>Unit of measure for each category | NA | Three digits—effective wind speed (kilometers per hour)<br>Three digits—downwind distance of Zone 1 (kilometers or nautical miles)<br>Two digits—cloud radius (kilometers or nautical miles) (the wind speed is less than 8 kilometers per hour, this line contains only a three-digit radius of Zone 1 [kilometers]) |
| ZA | NA | Significant weather phenomena | Air stability: Two digits<br>Temperature in centigrade: Two digits<br>Humidity: One digit<br>Significant weather phenomena: One digit<br>Cloud cover: One digit |
| ZB | Used to transmit correlation factors or transmission factors | Remarks for additional chemical and biological attacks only. | Include any additional information. |

## PROCEDURES FOR MARKING

J-14. Face markers away from the contamination. For example, if markers are placed 50 meters from the outside edge of a contaminated area to mark a radiological hot spot, they face away from the point of the highest contamination reading. Place markers at roads, trails, and other likely points of entry. When time and mission permit, use additional markers. The distance between the signs varies. In open terrain, place them farther apart than in hilly or wooded areas. Soldiers should be able to stand in front of a marker and see the markers to the left and right of it.

J-15. Units discovering a marked, contaminated area do not have to conduct elaborate, time-consuming surveys. The new unit checks the extent of the contamination and alters its plans if necessary. If the size of the hazard has expanded or decreased, relocate the signs. If the hazard is gone, remove the signs and report changes to higher HQ.

### TYPES OF MARKERS

J-16. US forces use standard NATO markers to make it easier for allies to recognize the hazards. These markers are in the standard NBC marking set. Colors and inscriptions on a marker indicate the type of hazard. Other contamination information is written on the front of the sign.

## RADIOLOGICAL CONTAMINATION DETECTION AND MONITORING

J-17. Conduct radiological monitoring using an AN/VDR2, and start on the order of the commander or as set by the SOP. Mark the area with radiological contamination markers. Report radiation dose rates and the times and locations of the readings to higher HQ using the NBC 4 nuclear report. All units start continuous monitoring when–

- Moving from one area to another on the battlefield.

- Getting a fallout warning.
- Knowing that a nuclear burst was reported, seen, or heard.
- Detecting radiation of 1 centigray per hour by periodic monitoring.
- Being ordered by the unit commander.

J-18. Continuous monitoring stops on order from the higher HQ or when the dose rate falls below 1 centigray per hour, except for units on the move.

## TECHNIQUES FOR MONITORING

J-19. Use both direct and indirect techniques when conducting radiological monitoring. Refer to *FM 3-3* for more information on radiological monitoring.

### Direct Monitoring

J-20. The direct technique is the simplest and the most precise. The soldier must—

- Stand at the desired location.
- Hold the RADIAC meter waist high and turn it slowly 360 degrees. Refer to *TM 11-6665-251-10*.
- Record the highest reading on *DA Form 1971-R*.
- Take the reading in the open, at least 10 meters from buildings or large structures, if possible.

### Indirect Monitoring

J-21. Use the indirect technique inside shelters or vehicles. The soldier must—

- Stand at the center of the shelter.
- Hold the RADIAC meter 1 meter above the floor and slowly rotate it 360 degrees.
- Record the highest reading.
- Take all readings from one selected position when monitoring from inside a vehicle or shelter.

## SURVEY TEAM

J-22. MP may be tasked to perform as part of a survey team. When essential contamination information cannot be obtained from monitoring reports, a radiological survey may be required. Radiological surveys are directed efforts to learn the extent and intensity of radiological contamination. A survey requires a control team and one or more survey teams. The HQ directing the survey usually provides the control team. In a team assigned to a survey—

- The team leader monitors the RADIAC meter and provides $C^2$.
- The driver operates the vehicle.
- The gunner provides security.

J-23. The control team instructs the survey teams on what to do. The MP team leader ensures that each team member receives the following information from the control team:

- Purpose. The survey determines the presence and the level of fallout or induced radiation near ground zero, in a specific area, along a route, at a specific location, or any combination of these.
- Start and finish times. The times during which a survey will be conducted.
- Survey execution. The route to be followed and the locations where readings are to be made or the distances needed between the readings are given.
- Radiation safety precautions. The control team tells the survey team the turn-back dose. The dose rate is monitored on the AN/VDR2. If the turn-back dose rate is reached as the survey team travels forward, the team immediately leaves the fallout area by the same route used to enter it, unless instructed otherwise. If the dose rate decreases as the survey team moves ahead, the team continues to perform its mission.

- Recording limitations. The survey team will not convert inside readings to outside readings. Record readings in the dose rate column only.
- Communications. The method and times to be used to communicate survey data to the chain of command.
- Special instructions. Any special instructions for marking the contaminated areas or special security precautions.

# CHEMICAL AGENT DETECTION

J-24. During routine operations, MP may detect chemical agents while making periodic checks for contamination. Or as previously discussed, teams may be tasked to conduct a chemical reconnaissance along an MSR or in a given area after an enemy attack. Some chemical agents are odorless, colorless, tasteless, and invisible. Teams must use chemical agent detection equipment and materials to be certain that all chemical agents are detected.

## PREPARATION

J-25. When assigned a mission or a task to locate and mark a suspected contaminated area, MP leaders must ensure that they prepare properly for the mission. Preparation for any NBC reconnaissance mission begins with the inspection of personnel and equipment. As a minimum, each squad must have the following equipment on hand:

- M8 paper.
- M9 paper.
- M256 and M256A1 detector kits.
- M8A1 alarm.
- CAM.
- Marking kit.
- M13 decontamination apparatus (DAP).

- M291 decontamination kit.
- MOPP gear.
- Mark 1 nerve agent auto injector kit (NAAK).
- VS-17 marking panels.

J-26. In addition to ensuring that the proper equipment is on hand, leaders must ensure that alarms and paper are properly mounted and functional and that all external equipment is stowed. The platoon leader includes a rehearsal of NBC reconnaissance techniques when preparing the mission. The platoon leader coordinates with the unit chemical officer for any special instructions, ensuring that thorough decontamination support is available at the conclusion of the mission.

## RECONNAISSANCE

J-27. When detecting for chemical contamination during NBC reconnaissance, the MP leader—

- Ensures that all members of the party wear MOPP 4 with M9 paper attached to their protective clothing and vehicle.
- Drives along the route until the chemical agent alarm sounds. By the time the alarm sounds, the team may be some distance from where the agent was first encountered. Keep this in mind when selecting the area to be searched.
- Gives special attention to shell craters, low-lying patches of woods, defiles, ravines, streambeds, and areas covered with high grass or underbrush.
- Has one MP remain in each vehicle to provide security and to monitor the radio.
- Has other patrol members dismount and check for contaminants.
- Ensures that one person uses M8 detector paper on the tip of a stick.

- Ensures that one person operates the M256 chemical agent detector kit. (Refer to *TM 3-6665-307-10* for detailed operating instructions for the M256, and refer to *FM 3-6* for more information.)

J-28. When MP find chemical contamination, they mark the area with NBC markers. They service the automatic chemical agent alarm to ready it for further operation. MP must be outside the contaminated area when the alarm is serviced. *TM 3-6665-312-12&P* provides instructions for operators and organizational personnel. MP remount the vehicles and continue the reconnaissance. While moving, they send an NBC 4 report by radio and follow this procedure each time a chemical agent is detected. The goal of the reconnaissance is to define the contaminated area only to the degree necessary to provide friendly forces with the information needed to avoid the area.

## BIOLOGICAL AGENT DETECTION

J-29. For warning of biological attacks or contamination outside a unit's immediate area, the unit relies on warnings relayed through its chain of command from division level or higher units. MP help provide warning of biological hazards by their reports of area activity passed on from their route, zone, and area reconnaissance.

J-30. Be alert to any sign that biological agents are being used. Promptly report any unusual occurrences of sickness in troops or civilians. Report any unusual actions of animals, birds, or large numbers of dead animals or birds not likely to have been caused by combat. Observe the types of weapons and munitions used during attacks. Compare them to the known characteristics of enemy biological attacks. Refer to *FM 3-3* for detailed information on biological agent detection, identification, and defense.

J-31. Biological attacks are most likely to occur at night or during extended twilight to avoid toxin exposure to direct sunlight. (Direct sunlight degrades biological agents.) Cloudy and foggy days are also favorable for the use of biological agents.

J-32. Watch for the following indications that a biological attack has occurred:

- A low-flying aircraft that appears to be producing a mist or a spray.
- The use of any type of spray device.
- The use of ammunition that does not seem to have any immediate effect, such as a bomblet.
- Unusual types of bomblets.
- Insect swarms that suddenly appear after an aircraft drops containers that do not seem to have any immediate effects.

## SELF-DEFENSE MEASURES

J-33. MP not only provide early warning for units in the sustainment area, but they also employ NBC defense measures to protect their own forces. NBC defense is a balance of the following three principles:

- Avoidance.
- Protection.
- Decontamination.

### AVOIDANCE

J-34. Avoiding NBC attacks and hazards is the key to NBC defense. Avoiding contamination involves bypassing contaminated areas and avoiding detection by the enemy. If forces avoid detection, they are less likely to be targeted for attack and will not need to apply protection and decontamination. *FM 3-3* provides detailed guidance on conducting contamination avoidance procedures.

J-35. NBC contamination avoidance is especially critical for light forces. Leaders conduct risk assessments to determine what protective measures they will direct. Refer to *FM 3-4* for detailed information on conducting a risk assessment.

## PROTECTION

J-36. A unit may become contaminated because of direct NBC attack or because the mission causes them to enter a contaminated area. In either case, NBC protection is an integral part of operations. Avoidance and protection are closely related. Techniques that work for avoidance also provide protection. There are four broad groups of activity that comprise protective measures and include the following:

- Hardening positions and protecting personnel.
- Assuming MOPP.
- Reacting to attack.
- Using collective protection.

J-37. Refer to *FM 3-4* for detailed guidance on conducting NBC protective procedures.

## DECONTAMINATION

J-38. The extent and timing of decontamination depend on the tactical situation, the mission, the degree, the type of contamination, and the resources available. Refer to *FM 3-5* for detailed guidance on conducting decontamination operations.

J-39. The primary purpose of decontamination is to stop erosion of combat power and to reduce the number of casualties. Combat power drops as soon as personnel don MOPP gear. The mask restricts the field of vision and reduces observation and target acquisition ability. Communication is more difficult. Mobility is reduced because personnel reduce their physical work rate to avoid heat stress. The longer a unit remains contaminated, the greater the chance of NBC casualties. The three levels of decontamination are immediate, operational, and thorough.

## Immediate Decontamination

J-40. Immediate decontamination is a basic soldier survival skill carried out by soldiers as soon as possible after they discover they are contaminated. Its basic purposes are to minimize casualties, save lives, and limit the further spread of contamination. Personnel treat any contact between chemical or biological agents and bare skin as an emergency. Some agents can kill if they remain on the skin for longer than a minute. The best technique for removing or neutralizing these agents is to use the M291 skin decontamination kit. Leaders must ensure that their soldiers are trained to execute this technique automatically, without waiting for orders.

J-41. Begin personal wipe down within 15 minutes of contamination. The wipe down removes or neutralizes contamination on the hood, mask, gloves, and personal weapon. For chemical and biological contamination, use mitts from the M295 individual equipment decontamination kit (IEDK). For radiological contamination, wipe off the contamination with a cloth or simply brush or shake it away.

J-42. Begin the operator's spray down of the equipment immediately after completing the personal wipe down. The spray down removes or neutralizes contamination on the surfaces that operators must touch frequently to perform their mission. For chemical and biological contamination, use on-board decontamination apparatuses, such as the M11, the M13, or the M295 IEDK to decontaminate surfaces where DS2 cannot be applied.

**NOTE: Wash decontamination solution number 2 (DS2) off surfaces no more than 30 minutes after application. If necessary, use 5-gallon water cans or other water sources to assist in removing DS2.**

J-43. For radiological contamination, brush or scrape away the contamination with whatever is at hand or flush it with water and wipe it away.

## Operational Decontamination

J-44. Operational decontamination allows a force to continue fighting and sustain its mission after being contaminated. It limits the hazard of transferring contamination by removing most of the gross contamination on equipment and nearly all the contamination on soldiers. This speeds the weathering process and allows clean areas (people, equipment, and terrain) to stay clean. Following operational decontamination, soldiers who have removed sources of vapor contamination from their clothing and equipment unmask temporarily in hazard-free areas to eat, drink, and rest.

J-45. Accomplish operational decontamination using the assets of the parent unit. Make use of the following decontamination techniques:

- Vehicle wash down.
- MOPP gear exchange.

J-46. Conduct these procedures at the same time. They are best performed at squad level. Vehicles and personnel that are not contaminated should not go through either technique. Conduct vehicle wash down as far forward as possible. It is normally performed using the power-driven decontamination equipment (PDDE) crew with assistance from the squad decontamination crew. It is most effective if started within 1 hour after contamination.

J-47. The two steps in vehicle wash down are as follows:

***Step 1.*** Button up the vehicle and secure the equipment.

***Step 2.*** Wash down the vehicle and the equipment with hot, soapy water for 2 to 3 minutes.

J-48. Because speed is important, do not check vehicles for contamination after vehicle wash down. Remove only gross contamination.

## Thorough Decontamination

J-49. Thorough decontamination reduces contamination to negligible risk levels. It restores combat power by removing nearly all contamination from the unit and individual equipment. This allows troops to operate equipment safely for extended periods at reduced MOPP levels. A contaminated unit conducts detailed troop decontamination (DTD) under the supervision of the chemical unit. Ordinarily, the chemical unit selects a site, sets it up, and performs the detailed equipment decontamination (DED) with assistance from the contaminated unit. A small risk from residual contamination remains, so make periodic contamination checks after this operation.

J-50. Conduct thorough decontamination as part of an extensive reconstitution effort in brigade, division, and corps support areas. Support sites at lower levels cannot provide the quantities of decontamination resources (water, decontaminants, and time) required for such an extensive process. In some cases, a contaminated unit could conduct a thorough decontamination operation with organic decontamination assets, but support from a chemical unit is usually required.

J-51. After thorough decontamination, the unit moves into an adjacent AA for reconstitution. Support elements from the brigade, division, or corps support areas replenish combat stocks, refit equipment, and replace personnel and equipment. The newly reconstituted unit leaves the AA fully operational and fit to return to battle.

J-52. Thorough decontamination does the best job of getting rid of contamination and its hazards, but it is often not possible. It requires large quantities of valuable resources that may not be immediately available. The next best solution is to decontaminate only what is necessary to sustain the force and continue to fight.

## MILITARY POLICE LEADERS' RESPONSIBILITIES

J-53. MP leaders ensure that all soldiers understand the concepts of NBC defense and the skills necessary to survive an attack. Individual NBC skills include–

- Recognizing hazards and taking cover.
- Using protective measures and MOPP gear.
- Knowing avoidance measures, first aid, and decontamination.

J-54. MP commanders also designate and train special-purpose teams. These special-purpose NBC teams include the following:

- Chemical agent detection teams.
- Radiological survey and monitoring teams.
- Decontamination teams.

J-55. To assist with NBC defense, MP commanders rely on the unit NBC officer and the NBC NCO. They—

- Provide technical assistance to the commander on NBC defense training and operations.
- Coordinate unit NBC defense activities.
- Give NBC defense instructions to achieve basic operating standards of proficiency for the unit and individuals.
- Plan and supervise NBC defense training aspects of operational training exercises and maneuver.
- Supervise the preparation of unit NBC defense SOPs.
- Supervise the operation and maintenance of NBC material.
- Determine by dosimetry or calculation (as appropriate) the total dose of radiation and the time spent in radiologically contaminated areas to avoid exceeding command exposure guidance.
- Prepare fallout prediction patterns.

- Plan NBC reconnaissance and advise commanders on the best routes to cross or bypass an NBC contaminated area.
- Plan and coordinate decontamination within the unit.
- Maintain records of the unit's radiation exposure.
- Estimate downwind hazards for chemical attacks.
- Report NBC data to the next higher HQs.
- Analyze the unit's vulnerability to an NBC attack.
- Evaluate individual and unit competence in NBC defense, and advise the commander on the unit's ability to survive and continue operations in an NBC environment.
- Advise the commander on all matters pertaining to unit NBC defense.

## MISSION-ORIENTED PROTECTION POSTURE LEVELS, ALARMS, AND SIGNALS

J-56. Soldiers on the integrated battlefield face a combination of NBC and conventional attacks. Individual and unit protection against chemical attack or contamination hinges on effective use of MOPP gear and on individual proficiency in basic NBC skills. All MP must be familiar with the standard MOPP levels shown in *Figure J-2, page J-24,* and *Table J-2, page J-25.*

J-57. When an NBC attack is recognized, every soldier receives the warning and assumes the appropriate MOPP level. Those in immediate danger need warnings they can see or hear. The alarm or signal must be simple and unmistakable for quick and correct reaction. Units not immediately affected need the information to prepare for the hazard or to change plans.

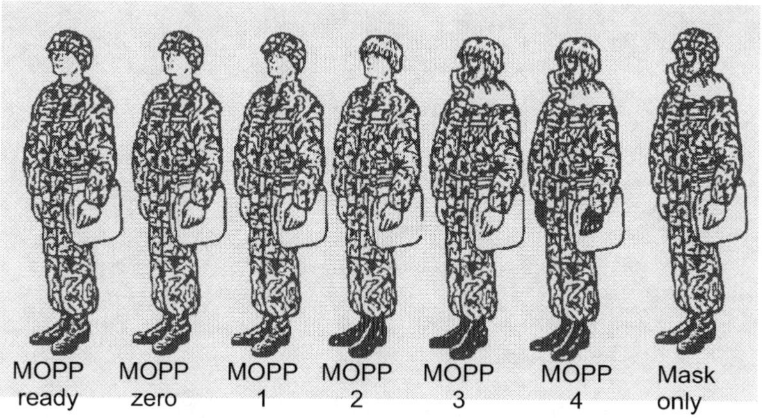


**Figure J-2. Illustrated MOPP Levels**

J-58. Mark the contaminated area if an NBC hazard has been located. The NBC Warning and Reporting System (NBCWRS) and the contamination markers contribute to the warning procedures for follow-on forces. In the immediate area of contamination, several methods (or a combination of methods) will allow quick reaction by all platoon members.

J-59. These methods include vocal alarms (shout of *GAS*), the M8A1 alarm, nonvocal alarms (horn blast or banging of metal-to-metal objects), and visual alarms (most commonly, hand-and-arm signals). The tactical situation may not allow for audible alarms; therefore, the platoon SOP should clearly detail the visual signals for contamination.

# BIOLOGICAL DEFENSE

J-60. The key protective measure against a biological attack is maintaining a high order of health, personal

## Table J-2. Standardized MOPP Levels

| MOPP Equipment | MOPP Levels | | | | | | Command |
|---|---|---|---|---|---|---|---|
| | MOPP Ready | MOPP Zero | MOPP 1 | MOPP 2 | MOPP 3 | MOPP 4 | Mask Only |
| Mask | Carried | Carried | Carried | Carried | Worn[1] | Worn | Worn |
| Overgarment | Ready[3] | Available[4] | Worn[1] | Worn[1] | Worn[1] | Worn | Available |
| Vinyl overboots | Ready[3] | Available[4] | Available[4] | Worn | Worn | Worn | Available |
| Gloves | Ready[3] | Available[4] | Available[4] | Available[4] | Available[4] | Worn | Available |
| Helmet protective cover | Ready[3] | Available[4] | Available[4] | Worn | Worn | Worn | Available |
| Chemical protective undergarment[2] | Ready[3] | Available[4] | Worn[2] | Worn[2] | Worn[2] | Worn[2] | Available |

[1]In hot weather, the coat or hood can be left open for ventilation.

[2]The chemical protective undergarment (CPU) is worn under the battle dress uniform (BDU) (applies to the special operation forces and armor vehicle crewman).

[3]Must be available to the soldier within 2 hours, and the second set must be available within 6 hours.

[4]Within arms reach of the soldier.

**NOTE: Refer to *Figure J-2* for an illustration of the MOPP levels.**

hygiene, and sanitation discipline. Biological attacks are hard to detect. If an attack occurs, chances of survival are better if soldiers are healthy, physically fit, keep all immunizations up to date, and maintain good personal hygiene. Keep your body clean to prevent ingestion of biological agents. Keep small cuts or scratches covered and germ-free by using soap, water, and first aid measures. Since insects carry biological agents, keep clothes buttoned and your skin covered to prevent insect bites.

## NUCLEAR ATTACK DEFENSE

J-61. MP can defend themselves against nuclear attack before, during, and after the event.

### DEFENSE BEFORE A NUCLEAR ATTACK

J-62. The best defense against a nuclear attack is to dig in. Prepare unit defensive positions, which vary from individual foxholes to improved positions, whenever the tactical situation permits. Keep individual weapons, equipment, clothing, and other issue items in vehicles. Secure equipment because the blast wave will convert unsecured items into lethal missiles. Disperse and protect supplies, explosives, and flammable materials. Turn off all electronic equipment and remove antennas from vehicles.

J-63. Reverse slopes of hills and mountains give some nuclear protection. The initial radiation, heat, and light from the fireball of a nuclear blast tend to be absorbed by hills and mountains. Using of gullies, ravines, ditches, natural depressions, fallen trees, and caves can reduce nuclear casualties.

### DEFENSE DURING A NUCLEAR ATTACK

J-64. Defense during a nuclear attack is either dismounted or mounted. When taking dismounted defensive actions, never run for cover. Immediately drop flat on the ground (face down) or to the bottom of a foxhole, with your head toward the blast. Cover exposed skin as much as possible and close your eyes. Remain down until the blast wave has passed and debris has stopped falling. Stay calm, check for injury, check weapons and equipment for damage, and prepare to continue the mission.

J-65. MP in vehicles should exit and move away from the vehicle and take dismounted defensive actions.

## DEFENSE AFTER A NUCLEAR ATTACK

J-66. Once the attack ends, forward an NBC 1 nuclear report, organize survivors, secure and organize equipment, assist casualties, improve protection against possible fallout, and begin continuous monitoring. If the radiation dose rate reaches a hazardous level after fallout is complete, be prepared to move, on order, to a less hazardous area.

### Fallout Warning

J-67. The first person to detect the arrival of fallout is usually the radiological monitor operating a RADIAC meter. As soon as he notes a dose rate of 1 centigray per hour or higher, he warns unit personnel. All personnel hearing the warning relay it to others. If the mission allows, move into a shelter with overhead cover and stay there until given an *all clear* signal or until directed to move. If the mission does not allow the unit to take cover, decontamination becomes more important and perhaps more difficult.

### Supervision of Radiological Monitoring

J-68. Perform radiological monitoring routinely with a AN/VDR-2 RADIAC meter to determine the presence and intensity of a radiation hazard. MP leaders must ensure that their teams are properly trained on this equipment. The two types of monitoring is periodic and continuous.

J-69. **Periodic Monitoring**. Periodic monitoring (once every hour) assures the platoon that the area is not contaminated or, if applicable, provides a warning when contamination is detected after the platoon arrives. Periodic monitoring is initiated under the following conditions:

- After the first use of nuclear weapons in the theater of war.
- When the platoon is out of contact with higher HQ.
- When ordered by higher HQ.
- When the platoon stops continuous monitoring.

J-70. **Continuous Monitoring**. Continuous monitoring is the surveillance for radiation in the platoon's area or position. Initiate continuous monitoring when any of the following situations occur:

- When a nuclear detonation is observed or reported in the area of operations.
- When an NBC 3 report is received and the platoon is in the predicted area of contamination.
- When ordered by higher HQ.
- When a dose rate of 1 centigray per hour is recorded in periodic monitoring.

## Supervision of Tactical Dosimetry Operations

J-71. Each MP team is issued a dosimeter. Selected soldiers wear them. Check all dosimeters to be used for the operation. Turn in dosimeters for recharging if they do not read zero. If a charger is not available, note the original reading. Ensure that the dosimeter readings are reported accurately. Collect readings at least once daily. Average these readings, round to the nearest 10, and report the average and the radiation exposure status (RES) to higher HQ.

J-72. Use individual actions for nuclear defense when operating in or crossing radiologically contaminated areas. Close vehicles tightly and cover cargoes with tarps or tenting. Mission permitting, keep the speed down to prevent dust. Vehicles should maintain adequate following distances to stay out of the dust raised by preceding vehicles.

J-73. After the unit exits a contaminated area, check personnel, equipment, and cargo for contamination and decontaminate them, if necessary. Monitor dose rates closely to ensure compliance with operational exposure guidance (OEG). Update the RES if appropriate.

## CHEMICAL ATTACK DEFENSE

J-74. MP must be prepared to defend themselves before, during, and after a chemical attack.

### DEFENSE BEFORE A CHEMICAL ATTACK

J-75. The best defense before a chemical attack is preparation. Use protective procedures. Ensure that all personnel have protective masks available, and ensure that each mask fits and functions properly. All personnel should wear the proper protective clothing according to the MOPP level designated by the commander. Protect all equipment and supplies from liquid chemical contamination by keeping them organized and covered.

J-76. Emplace an M8A1 automatic chemical agent alarm. The M8A1 is the primary means of detecting an upwind chemical attack. The system provides two essential elements of survival: detection of a toxic agent cloud and early warning to the troops in the monitored position. The platoon leader decides where to place the chemical alarm. Place the detector units no more than 150 meters upwind from the platoon's perimeter or position. Space the available detector units about 300 meters apart, and ensure that each detector unit is connected to the alarm unit by telephone cable (WD-1).

Position the alarm units near radiotelephone assets; this makes it easy to alert the unit of an attack. Blowing sand or dust, rain, sleet, snow, temperatures below 40 degrees Fahrenheit, and tropical conditions can affect the operation of the alarm.

J-77. If an NBC hazard cannot be avoided, soldiers must be prepared to protect themselves and equipment from the effects of exposure. Base the type and degree of protection required on the unit's mission and the hazard. Note that the line between contamination avoidance and protection is not distinct; many actions contribute equally to both.

### DEFENSE DURING A CHEMICAL ATTACK

J-78. Give the alarm. All unmasked soldiers put on their protective masks and other MOPP gear. All personnel move inside their vehicles and close all doors and hatches, if applicable. This aids in the protection from gross liquid contamination. The platoon leader directs the use of M256A1 detector kits and the CAM to determine the type of agent, and he submits an NBC 1 report. The platoon then continues the mission.

### DEFENSE AFTER A CHEMICAL ATTACK

J-79. Forward an NBC 1 chemical report, treat casualties, perform immediate decontamination as required, and mark the contaminated area.

## SYMPTOMS AND TREATMENT OF CASUALTIES

J-80. Potential adversaries may have access to a wide variety of biological and chemical agents on the modern battlefield. These agents can be dispensed alone or with other carriers or agents. Casualties resulting from exposure to biological or chemical agents require medical treatment as quickly as possible.

J-81. The first step in the treatment process is usually appropriate self-aid and buddy-aid measures. These vary depending on the agent. Soldiers mask to prevent inhaling or ingesting additional agents and remove agents from exposed skin by washing with soap and water or using the M291 kit. They use buddy-aid procedures to observe each other for early symptoms of toxic exposure and provide medical assistance.

J-82. The platoon leader selects separate casualty collection points for contaminated and noncontaminated casualties to prevent cross contamination. Decontaminate all contaminated casualties as thoroughly as the situation allows before being evacuated. The platoon includes the number of contaminated patients in its casualty evacuation request. This allows the evacuation team to send the proper number of vehicles for pickup.

J-83. Chemical agents fall into four major categories: nerve, blister, blood, and choking. The primary routes of attack on the body are through the respiratory system and the skin. These agents are especially dangerous because they can kill or incapacitate quickly. The first, and most important, step in dealing with them effectively is to recognize symptoms so that proper treatment can be administered. *Table J-3, page J-32,* shows protection and detection measures, symptoms, and treatment and decontamination procedures for chemical agents.

## UNMASKING PROCEDURES

J-84. Soldiers unmask as soon as possible except when a biological or chemical attack is expected. Use the procedures outlined in the following paragraphs to determine if unmasking is safe.

J-85. If an M256 or M256A1 detector kit is available, use it to supplement unmasking procedures. The kit

**Table J-3. NBC Protection and Detection Chart**

| Agent Type | Protection | Detection | Symptoms | Effects | First Aid | Deconta-mination |
|---|---|---|---|---|---|---|
| Nerve | Protective mask and suit | M8A1, M256A1, CAM, M8 paper, or M9 paper | Difficult breathing, drooling, nausea, vomiting, convulsions, or blurred vision | Incapacitates | Use Mark 1 NAAK and convulsant antidote for nerve agents (CANA). | Use an M291 kit and flush eyes with water. |
| Blister | Protective mask and suit | M256A1, M8 paper, M9 paper, or CAM | Burning eyes, stinging skin, and irritated nose (no symptoms with mustard or nitrogen mustard) | Blisters the skin and damages the respiratory tract | Use the same procedures as for second- and third-degree burns. Refer to FM 21-11. | Use an M291 kit and flush eyes with water. |
| Blood | Protective mask | M256A1 | Convulsions and coma | Incapacitates . | None | None |
| Choking | Protective mask | Odor only (resembles newly-mown hay or green corn). | Coughing, choking, nausea, headache, and tightness in the chest | Floods and damages the lungs | Avoid move-ment and keep warm. | None |

does not detect all agents; therefore, use the proper unmasking procedures, which take about 15 minutes. If all the tests with the kit (including a check for liquid contamination using M8 detector paper) have been performed and the results are negative, use the following procedures:

- The senior person selects one or two soldiers to start unmasking procedures. If possible, they move to a shady place. Bright, direct sunlight can cause pupils to constrict, giving a false symptom.
- The selected soldiers unmask for 5 minutes and then clear and reseal their masks.
- The leaders observe the unmasked soldiers for 10 minutes. If no symptoms appear, request permission from higher HQ to signal *all clear*.

J-86. The leaders watch the soldiers for possible delayed symptoms. Always have first aid treatment immediately available in case it is needed.

J-87. If an M256 or M256A1 kit is not available, unmasking procedures take about 35 minutes. When a reasonable amount of time has passed after the attack, find a shady area and use M8 paper to check the area for possible liquid contamination. Conduct unmasking using the following procedures:

- The senior person selects one or two soldiers. They take a deep breath and break their mask seals, keeping their eyes wide open.
- The soldiers clear and reseal their masks after 15 seconds. The leaders observe them for 10 minutes.
  - If no symptoms appear, the same soldiers break the seals, take two or three breaths, and clear and reseal their masks. Observe them for 10 minutes.

- If no symptoms appear after 10 minutes, the same soldiers unmask for 5 minutes and then remask.
- If no symptoms appear after 10 minutes, the leaders request permission from higher HQ to signal *all clear*. He continues to observe soldiers in case delayed symptoms develop.

J-88. Give the all-clear signal by word of mouth through the chain of command. Higher HQ initiates the all-clear signal after testing for contamination proves negative. If required, use standard sound signals, such as a continuous, sustained blast on a siren, a vehicle horn, or a similar device. When *all clear* is announced on the radio, it must be authenticated before compliance. The commander designates the specific all-clear signal and includes it in the SOP.

# Appendix K

# Civil-Disturbance Measures

This appendix discusses the crowd control measures used when there is a civil disturbance.

## OVERVIEW

K-1. The mission of the military forces in a civil disturbance is to apply the minimum force necessary to help local or HN authorities restore law and order. Most often, military personnel disperse unauthorized assemblages and patrol disturbed areas to prevent unlawful acts. Other missions assigned to MP or other military forces include—

- Maintaining the mechanics of essential distribution, transportation, and communications systems.
- Making a show of force.
- Setting up roadblocks.
- Cordoning off areas.
- Dispersing crowds.
- Releasing riot control agents (RCAs).
- Serving as security forces or reserves.
- Initiating needed relief measures, such as distributing food or clothing or establishing emergency shelter.
- Employing nonlethal munitions and equipment.

K-2. Military control force commanders must know what options are available to them. They select the option that is best for the specific physical and psychological environment. They must be able to reduce the intensity of the confrontation and restore order.

K-3. The commitment of military forces to civil-disturbance control operations does not automatically give these forces police power. There are legal and commonsense reasons to restrict the police power of military forces. All military leaders and planners must be familiar with laws, regulations, and policies that govern military involvement in civil disturbances.

K-4. In all contacts with the civilian population and the participants of the disturbance, military forces must display fair and impartial treatment and must adhere to the principle of minimum force. Whenever possible, civil police apprehend, process, and detain civil law violators. Military forces perform these functions only when necessity dictates and to the minimum extent required. The return these functions to civil authorities as soon as possible.

K-5. As the disturbance subsides, the commander takes steps to restore control to civil authorities. The control force gradually reduces the number and scope of its operations. It takes care not to give the impression that all controls have been removed. Withdrawal is not immediate. The disturbance may flare up again if the participants get the impression that authorities have abandoned the operations. The control force gradually withdraws in a phased return of control to civil authorities.

## CIVIL DISTURBANCES ON DEPARTMENT OF DEFENSE INSTALLATIONS OUTSIDE THE CONTINENTAL UNITED STATES

K-6. On DOD installations overseas and at US embassies and consulates, HN forces generally control disturbances targeted at US facilities because of the possible international political ramifications of foreign civil disturbances. Status of Forces Agreements define the legal considerations that guide and constrain actions by US military commanders. Commanders must have an effective liaison with HN authorities. Through close coordination with HN authorities, US commanders can determine the level of visibility and the involvement, if any, of US forces.

## CROWD BEHAVIOR

K-7. The presence or absence of social factors like leadership, moral attitudes, and social uniformity may influence crowd behavior. Leadership has a profound effect on the intensity and direction of crowd behavior. When blocked from expressing its emotions in one direction, a crowd's frustration and hostility can be redirected elsewhere. The first person to give clear orders in an authoritative manner is likely to be followed. Agitators can exploit a crowd's mood and convert a group of frustrated, resentful people into a vengeful mob. Skillful agitators using television, radio, and other communications media can reach large portions of the population and incite them to unlawful acts without having direct personal contact. In a civil-disturbance environment, any crowd can be a threat to law and order because it is open to manipulation.

K-8. Additionally, crowd behavior may be affected by emotional contagion or panic.

- Emotional contagion provides the crowd psychological unity. The unity is usually temporary, but it may be long enough to push a crowd to mob action. When emotional contagion prevails, normal law and authority are suppressed, increasing the potential for violence.
- Panic can occur during a civil disturbance when—
  - Crowd members perceive their safety is at risk and attempt to flee the area.

- Crowd members cannot disperse quickly after exposure to RCAs.
- Escape routes are limited.
- Escape routes are blocked or congested.

K-9. Control force members are also susceptible to crowd behavior. They may become emotionally stimulated during a tense confrontation. Commanders must counteract this. Control force members must exercise restraint individually and collectively. Rigorous training, firm and effective supervision, and complete awareness and understanding of ROE and ROI are necessary to offset the effect of crowd contagion on the control force.

## CROWD TACTICS

K-10. In civil disturbances, crowds employ any number of tactics to resist control or to achieve their goals. Tactics may be unplanned or planned and nonviolent or violent. The more purposeful the disturbance, the more likely the possibility of well-planned tactics.

K-11. Nonviolent tactics may range from name-calling to building barricades. Demonstrators may converse with control force members to distract them or to gain their sympathy. They may try to convince control force members to leave their posts and join the demonstrators. Demonstrators may use verbal abuse. Expect obscene remarks, taunts, ridicule, and jeers. Crowd members want to anger and demoralize the opposition. They want authorities to take actions that later may be exploited as acts of brutality.

K-12. Sometimes women, children, and elderly people are placed in the front ranks. This plays on a control force's sympathy to try to discourage countermeasures. When countermeasures are taken, agitators take photographs to stir public displeasure and embarrass the control force. Demonstrators may form human blockades to impede traffic by sitting down in roads or at the entrances to buildings. This can disrupt normal activity, forcing control personnel to physically remove the demonstrators. Demonstrators may lock arms, making it hard for the control force to separate and remove them. It also makes the control force seem to be using excessive force.

K-13. Groups of demonstrators may trespass on private or government property. They want to force mass arrests, overwhelm detainment facilities, and clog the legal system. Demonstrators may resist by going limp and forcing control force members to carry them. They may chain or handcuff themselves to objects or to each other. This prolongs the demonstration. Agitators may spread rumors to incite the crowd and try to force the control force to use stronger measures to control or disperse the crowd. The agitators want to make the control force appear to be using excessive force. Terrorist groups may try to agitate crowds as a diversion for terrorist acts. They also try to provoke an overreaction by the control force.

K-14. Violent crowd tactics, which may be extremely destructive, can include physical attacks on people and property, fires, and bombings. Crowd use of violent tactics is limited only by the attitudes and ingenuity of crowd members, the training of their leaders, and the materials available to them. Crowd or mob members may commit violence with crude, homemade weapons, or they may employ sophisticated small arms and explosives. If unplanned violence occurs, a crowd will use rocks, bricks, bottles, or whatever else is at hand. If violence is planned, a crowd can easily conceal makeshift weapons or tools for vandalism. They may carry—

- Balloons filled with paint to use as bombs.
- Bolt cutters to cut through fences.
- Picket signs to be used as clubs.

- Pipes wrapped in newspapers to throw as deadly missiles.
- Firecrackers dipped in glue and covered with BBs or small nails to use as deadly grenades.
- Plywood shields and motorcycle helmets to protect against riot batons.
- Safety goggles to protect against tear gas.

K-15. A crowd may erect barricades to impede troop movement or to prevent a control force from entering certain areas or buildings. They may use vehicles, trees, furniture, fences, or any other material. In an effort to breach barriers, rioters may throw grapples into wire barricades and drag them. They may use grapples, chains, wire, or rope to pull down gates or fences. Long poles or spears may be used to keep control forces back while removing barricades or to prevent the use of bayonets. They also may crash vehicles into gates or fences to breach them.

K-16. Rioters can be expected to vent their emotions on individuals, troop formations, and control-force equipment. Rioters may throw rotten fruits and vegetables, rocks, bricks, bottles, improvised bombs, or other objects from overpasses, windows, and roofs. In the past, troops, firefighters, and utility workers on duty during a civil disorder have been beaten, injured, or killed. Vehicles have been overturned, set on fire, or otherwise damaged.

K-17. Rioters may direct dangerous objects like vehicles, carts, barrels, and liquids at troops located on or at the bottom of a slope. On level ground, they may drive wheeled vehicles at the troops, jumping out before the vehicles reach the target. This tactic is also used to breach roadblocks and barricades.

K-18. Rioters may set fire to buildings and vehicles to block the advance of troops. Fires are also set to create confusion or diversion, destroy property, and mask looting and sniping. Rioters may flood an area with gasoline or oil and ignite it. On the other hand, they may pour gasoline or oil down a slope or drop it from buildings and ignite it.

K-19. Weapons fire against troops may take the form of selective sniping or massed fire. The fire may come from within the ranks of the rioters or from buildings or other adjacent cover. The weapons used can vary from homemade one-shot weapons to high-powered rifles. Snipers may try to panic control force members into firing a volley into the crowd. Innocent casualties make a control force appear both undisciplined and oppressive.

K-20. Explosives may be used to breach a dike, levee, or dam. Bombs can be exploded ahead of troops or vehicles so rubble blocks a street. They can be used to block an underpass by demolishing the overhead bridge. In extremely violent confrontations, bombs placed in buildings may be timed to explode when troops or vehicles are near. Demolition charges can be buried in streets and exploded as troops or vehicles pass over them. Explosive-laden vehicles can be rolled or driven at troops. Rioters may attach explosives to animals and force them toward troops. They then detonate the explosives by remote control. Even harmless looking objects like cigarette lighters and toys can be loaded with explosives and used as weapons.

## COMPANY LEVEL OPERATIONS

K-21. Normally, civil-disturbance operations are conducted at company level. The company operations section coordinates for special equipment that includes the following:

- Kevlar and face shield.
- Body armor.
- Shields, work gloves, and batons.
- Protective masks.

- Elbow pads.
- Shin guards.
- Weapons and bayonets.
- CS/OC spray.
- Smoke grenades.
- Breaching ramps.
- Scaling ladders.
- Portable radios.
- Portable lighting.
- Bullhorns.
- Video home system (VHS) cameras to videotape individuals in the crowd for identification.
- Transportation assets to move people to detention cells or detention areas.
- Hand and leg irons and flex cuffs.
- NVDs.

K-22. The MP operations center also coordinates for support, to include medical and emergency operations center (EOC) personnel, SRTs, hostage negotiators, MWD teams, PSYOP, civil affairs, local or HN law enforcement personnel, and apprehension or detention facilities for civilian personnel.

K-23. Rehearsals are imperative to the overall success of suppressing civil disturbances even before they occur. Rehearsals should cover the use and employment of nonlethal munitions, practice situations in which ROE or the use of force is applied, the use of riot dispensers, hand and arm signals or voice commands, and formations.

K-24. MP use appropriate crowd control measures that include formations such as—

- Using the line formation to push or drive a crowd straight back.
- Using the echelon formation to turn or move crowds away from buildings, fences, or walls.

- Using the wedge formation penetrate and split a crowd.
- Using the diamond formation to enter a crowd.

K-25. When a small unit, normally a squad, moves a dignitary through a crowd or an apprehension team into a crowd, they generally use the diamond formation. The other formations are trained at squad level and above but are normally performed by a platoon or company. The line, echelon left or right, and wedge formations are the basis for platoon and company formations. A squad must be skilled in the basic formations before practicing in platoon-size or larger formations.

K-26. Refer to *FM 19-15* for more information about civil-disturbance techniques and procedures.

## NONLETHAL MUNITIONS

K-27. Nonlethal munitions are explicitly designed and primarily employed to incapacitate personnel or material, while minimizing fatalities, permanent injury to personnel, and undesired damage to property and the environment. Unlike conventional lethal munitions that destroy their targets through blast, penetration, and fragmentation, nonlethal munitions employ means other than gross physical destruction to prevent the target from functioning.

K-28. The doctrine and concepts of operation for nonlethal weapons are designed to reinforce deterrence and expand the range of options available to commanders. They enhance the capability of US forces to accomplish the following objectives:

- Discourage, delay, or prevent hostile actions.
- Limit escalation.
- Take military action in situations where the use of lethal force is not the preferred option.
- Protect our forces better.

- Disable equipment, facilities, and personnel temporarily.

K-29. Nonlethal munitions are not required to have a zero probability of producing fatalities or permanent injuries, and complete avoidance of these effects is not guaranteed or expected. When properly employed, nonlethal weapons should significantly reduce these effects when compared with using lethal weapons to physically destroy the same target.

K-30. Nonlethal capabilities provide a wider range of options that augment but do not replace traditional means of deadly force. The option to resort to deadly force must always remain available when the commander believes it is appropriate to the mission.

K-31. MP normally use the lowest level of force necessary to control most incidents that involve noncombatants. RCAs, such as CS and MWD, and crowd control devices, such as the riot baton, are examples of nonlethal means currently used by MP forces. These options are currently listed in *AR 190-29*.

K-32. Training for the use of nonlethal munitions requires leaders and soldiers to understand the limited use of these systems in environments with restrictive ROE. Training will be continuous at all levels to ensure that nonlethal munitions are properly employed.

K-33. The countermeasures for thwarting virtually all nonlethal options are usually apparent, quickly learned, and readily available. Because they do not kill, nonlethal options teach an adversary what to avoid in the future. Commanders must be prepared to stay one step ahead of motivated belligerents. Many nonlethal options have both maximum effective and minimum safety ranges. Individuals struck short of the minimum safety range often suffer severe injuries or death, while the effects of most nonlethal devices are greatly mitigated at longer ranges. In order to be effective, an adversary must be engaged within the effective zone (beyond the minimum safety range and short of the maximum effective range).

K-34. Nonlethal munitions and equipment currently available are shown in *Appendix L*. Refer to *Figure 3-2, page 3-11,* for the range of the munitions.

## NONLETHAL MUNITIONS EMPLOYMENT CONSIDERATIONS

K-35. Nonlethal munitions employment must be well documented in ROE. Leaders must constantly ensure that soldiers understand when and how to effectively employ them. Incorrect application of these munitions can have significant operational and political ramifications. Employment considerations include the following:

- **Individual.** When possible, do not change individual weapons. Designate individuals as nonlethal shooters. Nonlethal shooters carry lethal munitions only for personal protection. Carry lethal rounds in a place to avoid confusing nonlethal rounds with lethal rounds.
- **Squad.** Squad leaders carry stun grenades and maintain their weapon loaded with lethal ammunition. Ideally, the squad does not change their task organization to accommodate the addition of nonlethal equipment. They should designate nonlethal shooters instead.
- **Patrols.** Commanders do not plan a nonlethal patrol, but they plan a combat and security patrol with a nonlethal attachment when the mission dictates. Carrying a shotgun limits the flexibility an individual has because of the time it takes to transition from nonlethal to lethal. Use shotguns only from a fixed position where adequate coverage is available.
- **Static position.** Individuals on static positions have their weapons loaded with lethal

ammunition, and additionally they carry nonlethal munitions. Both shotguns and M203s work well from static positions and are used together when possible.

- **Contact teams.** Teams of personnel whose primary responsibilities are to make personal contact with the adversary are established before deployment. Train team members in unarmed self-defense, open-hand control, handcuffing, and flex cuffing. Equip contact teams with personal protection gear, 9-millimeter pistols, and flex cuffs. Due to the physical nature of contact teams, do not arm contact team members with long rifles. If security for the contact team is an issue, attach a security element. MWD teams may also augment contact teams to help locate the adversary. Refer to *FM 90-40* for more information on nonlethal techniques and procedures.

# Appendix L

# Weapons and Equipment

MP must be rapidly deployable, versatile, and as lethal as the force that they support. MP are organized and equipped to provide functional battlefield capabilities that range from the deliberate attack (area security) to civil-disturbance control operations (L&O). Although not all inclusive, the listed data is intended to aid in planning MP operations. In any tactical situation, it is important that MP be able to properly identify threat weapons, vehicles, and aircraft. When threat is observed, MP report their location, activity, and direction of travel to higher HQ by the fastest means available.

## FRIENDLY VEHICLES

L-1. *Table L-1, page L-2,* gives MP leaders a quick look at the capabilities of the various vehicles in the Army inventory.

**Table L-1. Vehicle Capabilities**

| Vehicle Capabilities | M998 | M1025/1026 | M1114 | M1117 | Light Medium Tactical Vehicle (LMTV) |
|---|---|---|---|---|---|
| Weight (gross vehicle weight [GVW]) (in pounds) | 4,950 | 5,250 | 12,100 | 29,500 | 16,500 |
| Height (in inches) | 69 | 73/73 | 73 | 102 | 112 |
| Length (in inches) | 180 | 180/185 | 190.5 | 237 | 251 |
| Width | 85 | 85/85 | 85 | 101 | 96 |
| Fuel capacity (in gallons) | 25 | 25 | 25 | 50 | 58 |
| Maximum speed (in miles per hour) | 55 | 55 | 55 | 63 | 58 |
| Range (in miles) | 350 | 300 | 275 | 440 | 400+ |

**Table L-1. Vehicle Capabilities (Continued)**

| Vehicle Capabilities | M998 | M1025/1026 | M1114 | M1117 | LMTV |
|---|---|---|---|---|---|
| **Fording Capabilities** | | | | | |
| With kit (in inches) | 60 | 60 | 60 | NA | 60 |
| Without kit (in inches) | 30 | 30 | 30 | 60 | 60 |
| Pay load (maximum pounds) | 2,500 | 2,500 | NA | 3,360 | 5,000 |

# FRIENDLY WEAPONS

L-2. *Table L-2* gives MP leaders a quick look at the capabilities of the weapons used by MP units.

**Table L-2. Weapon Capabilities**

| Weapon Capabilities | M2 | M4 | M9 | M16A2 | MK-19 | M136 (AT4) | M203 | M249 |
|---|---|---|---|---|---|---|---|---|
| Weight (in pounds) | 84 | 7.5 | 2.6 | 8.7 | 76 | 14.8 | 11 | 15.5 |
| Length (in inches) | 66 | 29.75 closed, 33 open | 8.5 | 39 | 43 | 40 | 39 | 41.1 |
| **Range** | | | | | | | | |
| Maximum (in meters) | 6,765 | 3,600 | 1,800 | 3,600 | 2,212 | 2,100 | 400 | 3,600 |
| Arming (in meters) | NA | NA | NA | NA | 18 to 30 | 10 | 14 | NA |
| Minimum safe (in meters) | NA | NA | NA | NA | 28 | NA | 31 | NA |

### Table L-2. Weapon Capabilities (Continued)

| Weapon Capabilities | M2 | M4 | M9 | M16A2 | MK-19 | M136 (AT4) | M203 | M249 |
|---|---|---|---|---|---|---|---|---|
| **Effective Range** | | | | | | | | |
| Area target (in meters) | 1,830 | 600 | NA | 800 | 2,212 | NA | 350 | 800 |
| Point target (in meters) | 1,200 | 500 | 50 | 580 | 1,500 | NA | 160 | 600 |
| Moving target (in meters) | NA | NA | NA | 200 | NA | NA | NA | NA |
| **Rate of Fire (in Rounds Per Minute)** | | | | | | | | |
| Cyclic | 500 | 700 to 970 | NA | 700 to 800 | 375 | NA | NA | 800 |
| Rapid | 40* | NA | NA | NA | 60 | NA | 35 | 200* |
| Sustained | 40* | 12/15 | 60 | 16 | 40 | NA | 35 | 85 |
| *With barrel change. | | | | | | | | |

# FRIENDLY NONLETHAL EQUIPMENT AND MUNITIONS

L-3. The following descriptions and illustrations describe nonlethal equipment and munitions that are currently available.

## NONLETHAL EQUIPMENT

L-4. The following nonlethal equipment provides bodily protection for soldiers involved in a nonlethal operation and allows soldiers to capture the threat with a minimum of force.

### Nonballistic and Ballistic Riot Face Shields

L-5. The riot face shield *(Figure L-1)* provides soldiers with improved facial protection from thrown objects. The face shield is lightweight, adaptable to the current helmet, transparent, and scratch-resistant. It is adjustable to up and down positions. The mechanism for attaching and removing the shield from the helmet is robust and simple and requires no tools in the field.

**Figure L-1. Riot Face Shield**

L-6. The ballistic riot face shield is similar in design, but it is heavier and provides facial protection up to a 9-millimeter full metal jacket (FMJ) or 124 grains at 1,400 feet per second. This face shield is primarily used by SRTs in force-entry scenarios and can also be used for MOUT operations.

## Nonballistic and Ballastic Riot Body Shields

L-7. The nonballistic riot body shield *(Figure L-2)* is 24 inches by 48 inches by .157 inches. It provides soldiers with improved protection from frontal, side, and overhead assaults. This shield is lightweight, transparent, and scratch-resistant. The ballistic riot body shield *(Figure L-3, page L-8)* is 24 inches by 36 inches with a 4- by 16-inch window for viewing the threat. It is similar in design to the nonballistic shield but is heavier. It provides ballistic protection up to a 9-millimeter FMJ or 124 grains at 1,400 feet per second. This shield is primarily used by SRTs in forced-entry scenarios and can also be used for selected MOUT operations.

**Figure L-2. Nonballistic Riot Body Shield**

Figure L-3. Ballistic Riot Body Shield

## Nonballistic and Ballistic Shin Guards

L-8. Nonballistic shin guards *(Figure L-4)* provide soldiers with improved protection from thrown objects. They are lightweight and black in color.

Figure L-4. Nonballistic Shin Guards

L-9. Ballistic shin guards are similar in design to nonballastic guards, but they are heavier and provide protection up to a 9-millimeter FMJ or 124 grains at 1,400 feet per second. They are primarily used by SRTs in forced-entry scenarios and can also be used in selected MOUT operations.

## Baton

L-10. The standard 36-inch wooden riot baton *(Figure L-5)* is currently in use for riot control. It is used for self-defense and to keep rioters out of arm's reach of the soldiers conducting crowd control tactics.

**Figure L-5. Standard 36-inch Wooden Riot Baton**

## Portable Bullhorn

L-11. The portable bullhorn *(Figure L-6, page L-10)* is a critical communication device when conducting crowd control tactics. The bullhorn can facilitate communication with the crowd in conjunction with linguist or PSYOP support. It also assists in communicating commands to troops engaged in crowd control by projecting over the crowd's noise.

## Individual Voice-Amplification System

L-12. The individual voice-amplification system *(Figure L-7, page L-10)* is a critical communication device for conducting crowd control tactics when using RCAs and wearing a protective mask. This device facilitates oral communications and increases the user's ability to be heard on radios and other devices.

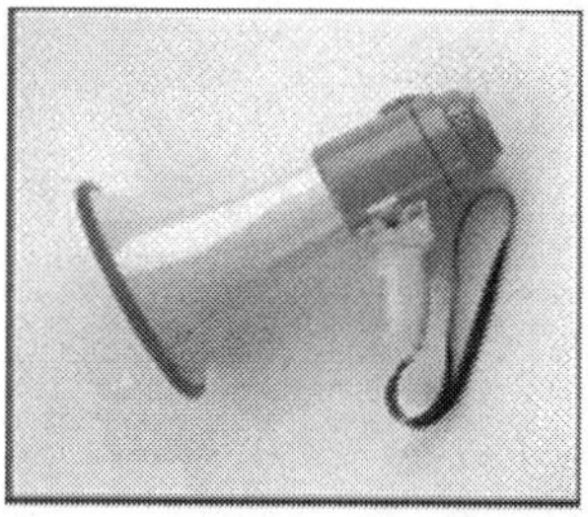

Figure L-6. Portable Bullhorn

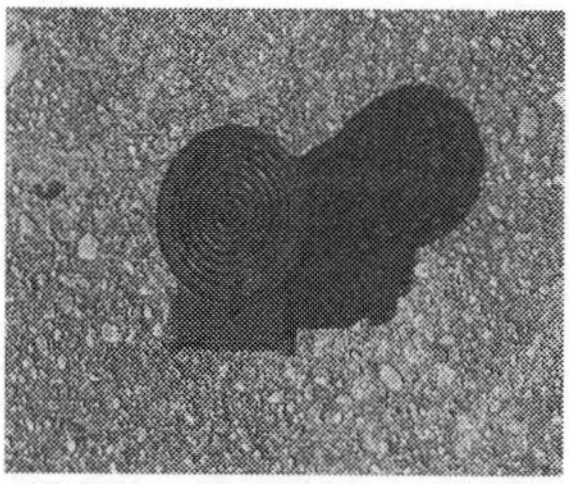

Figure L-7. Individual Voice-Amplification System

### Individual Oleoresin Capsicum and M36 Individual Chlorobenzyl Malononitrile Dispersers

L-13. These individual RCA dispersers *(Figure L-8)* are used primarily for self-defense and to keep rioters out of arm's reach of soldiers conducting crowd control tactics or engaged in missions were noncombatant threat exists.

### Midsize Riot Control Dispersers of Oleoresin Capsicum and Chlorobenzylidene Malononitrile

L-14. This RCA disperser *(Figure L-9)* is primarily used by formations conducting crowd control tactics, law

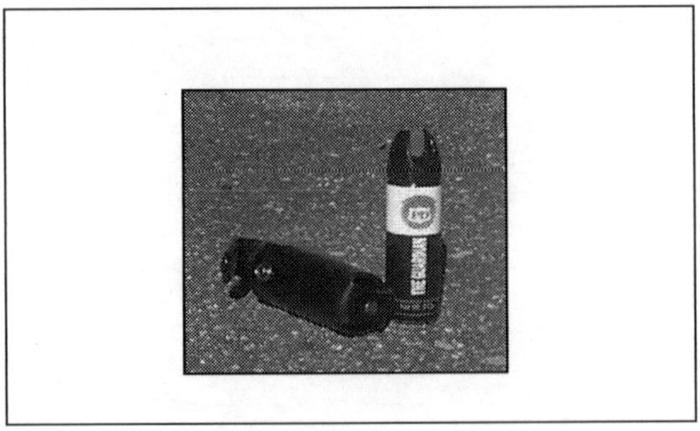

**Figure L-8. Individual OC and CS Dispersers**

enforcement, and I/R operations. It is lightweight, can be operated by one person, and is easily refilled and pressurized with available maintenance equipment or 2½- or 5-ton truck compressors. It is intended to provide a small unit with self-defense capabilities from large crowds out to 10 meters (more range is possible based on the wind) or offensively to clear crowds from critical areas (toward preplanned escape routes).

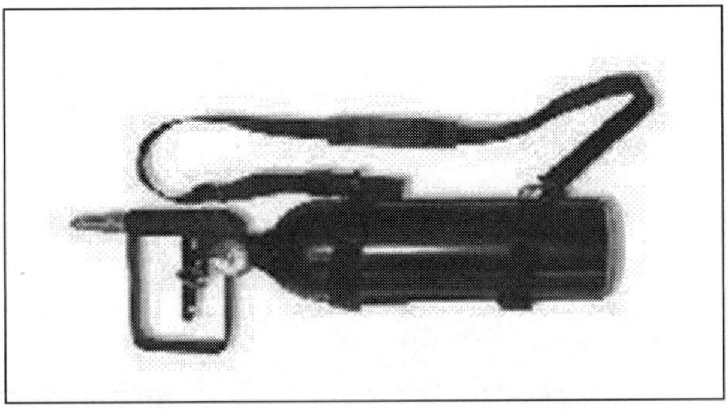

**Figure L-9. Midsize Riot Control Dispersers of OC and CS**

## M33A1 Riot Control Disperser Filled With Chlorobenzylidene Malononitrile or Dibenz (B, f)-1, 4-Oxazepine (CR)

L-15. This RCA disperser *(Figure L-10)* is primarily used by formations conducting crowd control tactics. It is twice as heavy as a midsized disperser, carries a larger payload, and dispenses powdered CS or liquid CR. One individual can operate it, and it can be refilled and pressurized with available M4 compressors. It is intended to provide a small unit with self-defense capabilities from large crowds out to 15 meters (100 meters is possible based on wind speed and direction). Use the M33A1 offensively to clear crowds from critical areas.

**Figure L-10. M33A1 Riot Control Disperser Filled With CS or CR**

L-16. Some training is required to operate the dispenser. The unit NBC NCO may conduct this training. The device requires up to 2,000 pounds per square inch of pressure to function properly, and it is maintained with an M254 maintenance kit.

## High-Intensity Xenon Searchlight

L-17. Use this individual, handheld searchlight *(Figure L-11)* for illumination in crowd control operations during darkness. Use it for general illumination of the operational area; to pinpoint agitators or threat, to reduce the ability of rioters to see troop formations and actions, and to enhance tactical deception techniques for units conducting crowd control operations.

**Figure L-11. High-Intensity Xenon Searchlight**

## Lightweight Disposable Restraints

L-18. Lightweight disposable restraints *(Figure L-12, page L-14)* are also known as flex cuffs. Individual soldiers can carry large quantities of these restraints to immobilize individuals being detained. When freeing a detainee, cut the restrains off him with the safe cutting device supplied with the restraints or with utility shears. A reusable, red restraint training device is available.

## Shotgun Munitions Carrier

L-19. The shotgun munitions carrier *(Figure L-13, page L-14)* is a 12-gauge ammunition carrier that straps to the stock of the M12 shotgun. It allows the firer to carry nonlethal ammunition that is readily available.

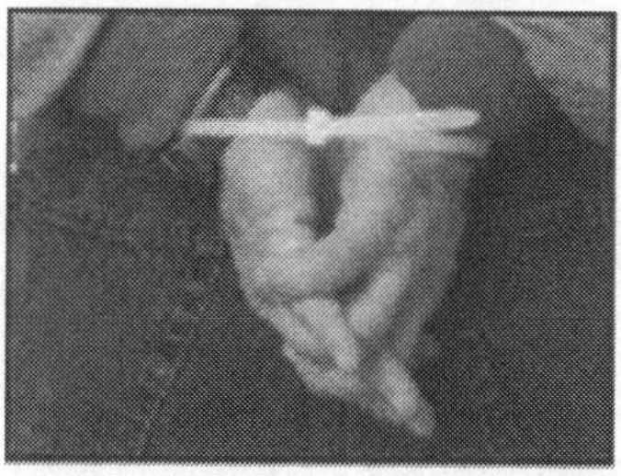

Figure L-12. Lightweight Disposable Restraints

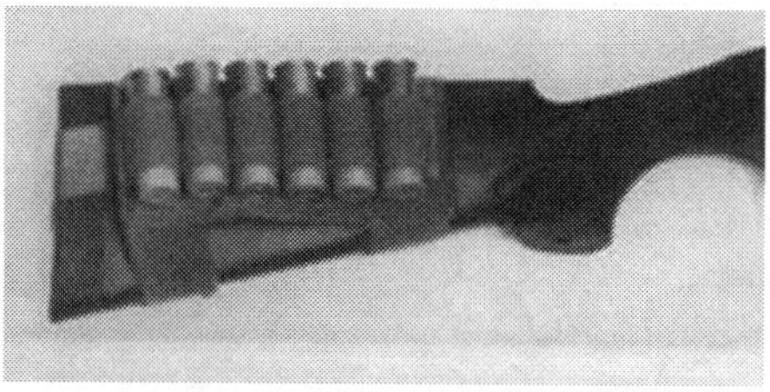

Figure L-13. Shotgun Munitions Carrier

## Portable Vehicle-Arresting Barrier (PVAB)

L-20. The PVAB *(Figure L-14)* is designed to assist with short-term physical security of critical facilities and for use at checkpoints. Use the PVAB to augment a vehicle checkpoint. It adds the ability to stop any light vehicle (up to 7,500 pounds) that attempts to flee without killing the occupants. This device may be set up by 2 or 3 soldiers in less than an hour.

L-21. When emplaced, the PVAB resembles a standard speed bump. When armed, the device operates within 1.5 seconds and deploys a high-tensile net that catches the vehicle and slows it to a stop with internal braking

Figure L-14. Vehicle Entrapped in a PVAB

mechanisms. The net and other features prevent occupants from fleeing the vehicle. The PVAB is best suited for mobile, short-term vehicle checkpoints in areas under US control where there is a threat of terrorist-type activity and where the threat uses the cover of noncombatants to infiltrate US AOs.

## NONLETHAL MUNITIONS

L-22. The following munitions provide soldiers with a nonlethal way to break contact, enforce a buffer zone, or stun an individual.

### 12-Gauge Nonlethal Point Target Cartridge Round (M1012)

L-23. The point target cartridge round *(Figure L-15, page L-16)* stuns individuals by delivering a strong blow to the body without penetrating it. This munition allows soldiers to enforce a buffer zone (standoff distance) with a violent crowd, break contact, or stun an individual target for possible detention by snatch teams. Fire the round at the center mass of an adult subject at ranges of 10 to 30 meters. Beyond 30 meters, the projectile loses accuracy and may no longer have the velocity required to stun an individual. This round has applications in law enforcement, I/R facilities, and US military detention facilities.

**DANGER**
**Shots fired at subjects closer than 10 meters or shots to the head or groin may cause serious injury or even death.**

**Figure L-15. 12-Gauge Nonlethal Point Target Cartridge Round**

## 12-Gauge Aerial Diversionary-Device Round

L-24. The aerial diversionary-device round *(Figure L-16)* provides multishot nonlethal capability to distract individuals or crowds. In crowd control, it delivers a flash bang projectile over the heads of a violent or potentially violent crowd, and is used to distract the crowd (in combination with other distraction devices and troop maneuvers). It allows other troop formations to maneuver to positions that are more advantageous

L-25. The round is designed to be fired at ranges of 75 to 100 meters and is placed about 5 meters above the crowd.

**DANGER**
**Shots fired directly at subjects or in enclosed areas may cause serious injury.**

**Figure L-16. 12-Gauge Aerial Diversionary-Device Round**

## 12-Gauge Nonlethal Area Target Cartridge Round (M1013)

L-26. The area target cartridge round *(Figure L-17, page L-18)* provides the capability to stun or deter two or three threats by delivering a strong blow to the body without penetrating it. This round has a wide range of capabilities for tactical, law enforcement, I/R, and US military detention operations.

L-27. The round is designed to be fired at the center mass of an adult threat at ranges of 10 to 30 meters.

Figure L-17. 12-Gauge Nonlethal Area Target Cartridge Round

Shots fired closer than 10 meters may cause serious injuries. The projectile loses accuracy when shot beyond 30 meters and may no longer be effective.

### 40-Millimeter Sponge Round (Point) (M1006)

L-28. The 40-millimeter sponge round *(Figure L-18)* delivers a strong, stunning blow to a threat's body without penetrating it. This round has a wide range of capabilities for tactical, law enforcement, I/R, and US military detention operations.

Figure L-18. M1006 40-Millimeter Sponge Round

L-29. The round is designed to be fired at the center mass of an adult threat at ranges of 10 to 50 meters.

**WARNING**
**Shots fired closer than 10 meters may cause injuries.**

### 40-Millimeter Crowd Dispersal Round (Area) (XM1029)

L-30. The 40-millimeter crowd dispersal round *(Figure L-19)* delivers a strong, stunning blow to a threat's body without penetrating it. This round has a wide range of capabilities for tactical, law enforcement, I/R, and US military detention operations. In crowd control, it provides a nonlethal capability that can be used to break contact and enforce a buffer zone (standoff distance) with a violent crowd.

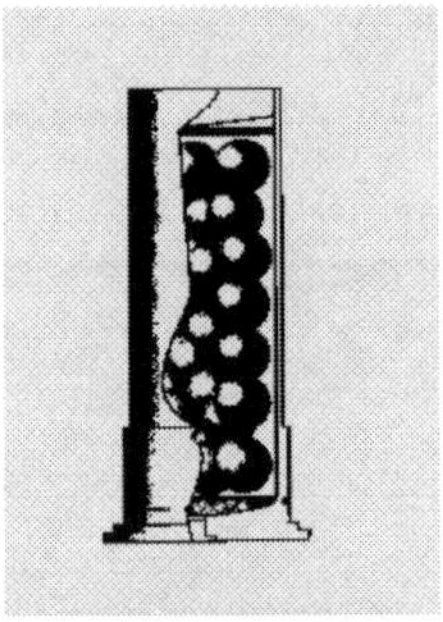

**Figure L-19. 40-Millimeter Crowd Dispersal Round (Area)**

L-31. The round is designed to be fired at the center mass of an adult threat at ranges of 10 to 30 meters.The projectile loses accuracy when shot beyond 30 meters and may not stun or deter the threat.

**WARNING**
**Shots fired closer than 10 meters may cause injuries.**

## 40-Millimeter Carrying Pouch

L-32. The 40-millimeter carrying pouch is slung over the shoulder of a soldier. It provides the ability to carry nonlethal ammunition separate from lethal rounds.

## M84 Stun Grenade (Diversionary Device, Hand-Thrown)

L-33. The M84 *(Figure L-20)* is a hand-thrown, flash bang, stun device used primarily by SRTs in forced-entry scenarios. It is used for selected MOUT or crowd control operations.

**Figure L-20. M84 Stun Grenade**

L-34. The M84 is designed to be thrown into a room (through an open door, a standard glass window, or other opening). It delivers a loud bang and a brilliant flash that temporarily disorients and detracts the occupants. Because of its reusable metal body, do not throw it into a crowd, as it may be returned to friendly troops in the form of a projectile. Instead, throw it into a controlled area in conjunction with other deception and distraction techniques.

## NONLETHAL, TURRET-MOUNTED, 66-MILLIMETER GRENADE LAUNCHER (M315) AND LIGHT VEHICLE OBSCURATION SMOKE SYSTEM, 66-MILLIMETER, NONLETHAL GRENADE (L96/97 XM98/99)

L-35. The LVOSS/M315 *(Figure L-21)* is a 66-millimeter smoke grenade-launching platform designed to give HMMWVs ASV M1117 the ability to obscure their position in the same manner as armored vehicles. These launchers can launch any of the smoke or nonlethal 66-millimeter munition.

**Figure L-21. LVOSS and M315**

L-36. The M315 turret-mounted, 66-millimeter, multipurpose, adjustable grenade launcher installation kit was developed to provide a vehicle-mounted nonlethal platform. The system is capable of delivering nonlethal payloads to support a variety of mission requirements and can be mounted on various vehicles (M1025, M1026, M966, M114, or ASV M1117) equipped with a machine gun or a TOW II missile system mount. The system electronically fires 66-millimeter cartridges from four adjustable firing tubes.

L-37. The LVOSS, 66-millimeter, nonlethal grenade (L96/97 XM98/99) is an area target munition that can be fired from the standard LVOSS/M315 launcher *(Figure L-22)*. It can be mounted on selected armament carrier HMMWVs and can be fired from any 66-millimeter, smoke-launching system found on most armored vehicles.

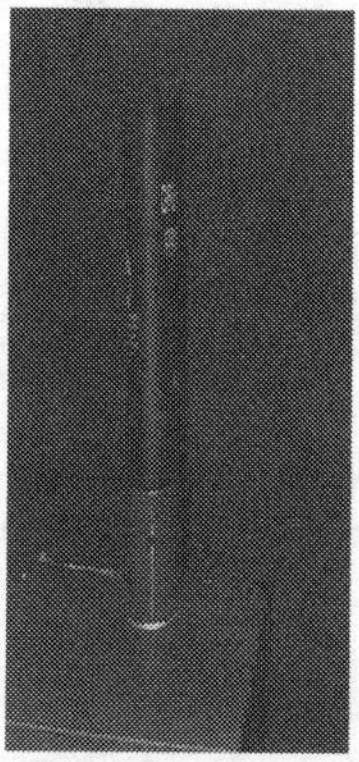

**Figure L-22. LVOSS/M315 66-Millimeter Nonlethal Grenade**

L-38. These rounds are designed to be fired from 80 to 100 meters. They deliver a flash bang diversionary warning (M98), a payload of rubber nonpenetrating projectiles (M99) (affecting a 10- to 20-meter circular area), or CS/cinnamic acid (CA) practice below the point of burst (L96A1/L97A1). These munitions lack the ability to be precision-delivered and are meant to affect a large number of people at long standoff ranges. This provides convoys and crowd control formations a long range support weapon to affect crowds beyond the range of shoulder-fired nonlethal weapons. Use the rounds to provide supporting nonlethal fires to crowd control formations.

## M5 Modular Crowd Control Munitions

L-39. The MCCM *(Figure L-23)* munition is similar in operation to a claymore mine, but it delivers nonlethal effects to the threat by delivering a strong, nonpenetrating blow to the body with multiple submunitions (600 rubber balls). This round has a wide range of capabilities for tactical, law enforcement, I/R, and US military detention operations. In crowd control, it provides a nonlethal counterpersonnel capability that can be used to break contact, enforce a buffer zone (standoff distance), or demonstrate a show of force.

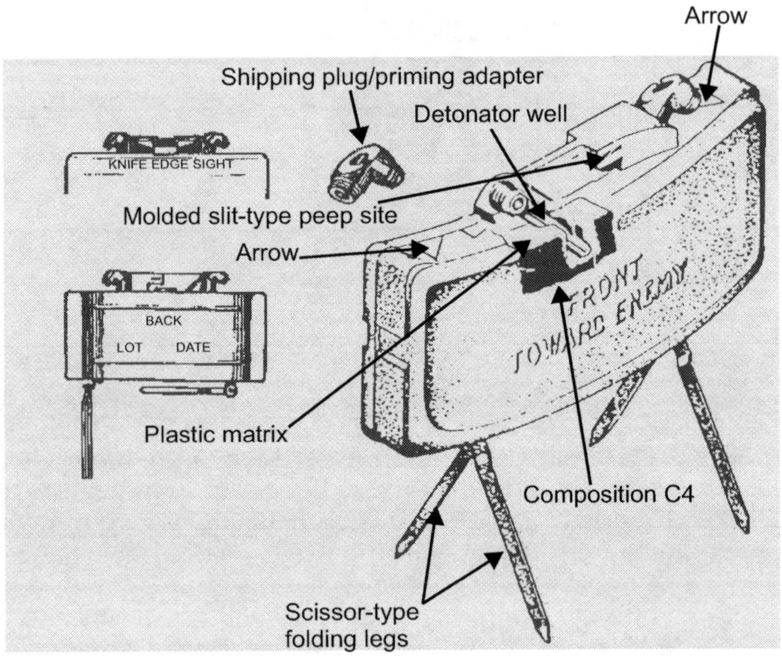


**Figure L-23. M5 MCCM**

L-40. This round is designed to be fired at the center mass of adult threats at ranges of 5 to 15 meters. It has a shot arc covering between 60 to 80 degrees (laterally).

**WARNING**
**Shots fired at subjects closer than 5 meters may cause injury.**

# FRIENDLY COMMUNICATION, SINGLE-CHANNEL, GROUND-TO-AIR RADIO SYSTEM (SINCGARS)

L-41. SINCGARS is the primary communication system for MP. It is a series of 2-way FM radio sets that use an 18-element keypad for tuning 2,320 channels. This includes 8 preset channels in the single-channel mode and 6 preset channels in the jam-resistant, frequency-hopping mode. *Table L-3* and *Table L-4* provide the planning ranges for SINCGARS.

**Table L-3. Voice Transmission Maximum Planning Ranges**

| Type of Radio | RF Switch Position | Planning Ranges |
|---|---|---|
| Manpack or vehicular | LO<br>M<br>HI | 200 to 400 meters<br>400 meters to 5 kilometers<br>5 to 10 kilometers |
| Vehicular only | PA | 10 to 40 kilometers |

L-42. To increase the transmission range of SINCGARS, connect the OE-254 antenna group. It is an omnidirectional, biconical antenna designed for broadband operation, without field adjustment, from 30 to 88 megahertz, up to 350 watts. The following is the tabulated data transmission range of the OE-254:

- Between two OE-254 antenna groups–
  - Average terrain: 36 miles (57.9 kilometers).

    - Difficult terrain: 30 miles (48.3 kilometers).
- Between an OE-254 antenna group and a vehicular whip antenna–
    - Average terrain: 30 miles (48.3 kilometers).
    - Difficult terrain: 25 miles (40.3 kilometers).

**Table L-4. Data Transmission Maximum Planning Ranges**

| Type of Radio | Baud Rate Used | RF Switch Position | Planning Ranges* (In Kilometers) |
|---|---|---|---|
| Manpack/ vehicular | 600 to 4,800 baud per second (BPS) | HI | 3 to 5 |
| (short range) | 16,000 BPS (16 kilobauds per second [KBPS]) | HI | 1 to 3 |
| Vehicular | 600 to 2,400 BPS | PA | 5 to 25 |
| (long range) | 4,800 BPS | PA | 5 to 22 |
| | 16,000 BPS (16 KBPS) | PA | 3 to 10 |
| ***Ranges are based on the line of sight and are average for normal conditions. Ranges depend on the location, the sighting, the weather, and the surrounding noise level, among other factors. The use of the OE-254 antenna increases the range for both voice and data transmissions. Enemy jamming and mutual interference conditions degrade these ranges. In data transmission, the use of a lower baud rate increases the range.** | | | |

## THREAT WEAPONS AND EQUIPMENT

L-43. In any tactical situation, it is important that MP be able to properly identify threat weapons, vehicles, and aircraft. MP report threat location, activity, and direction of travel to higher HQ by the fastest means available.

L-44. *Tables L-5* through *L-12, pages L-27* through *L-52,* outline the most common features of threat weapons, equipment, and vehicles that MP may encounter in a hostile environment. Special operations forces and airborne assault, reconnaissance, and insurgent units often use these weapons.

**Table L-5. Small Arms**

| Small Arms Characteristics | 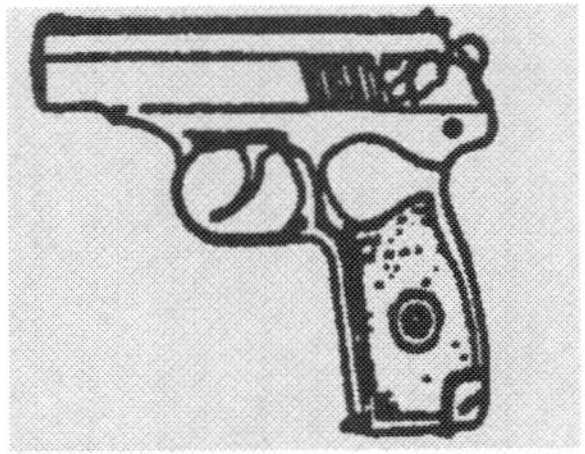 9-Millimeter PM Pistol |  5.45-Millimeter AK-74 Assault Rifle |
|---|---|---|
| Range, effective and maximum (in meters) | 50 | 500/1,000 |
| Rate of fire, practical and cyclic (rounds per minute) | 30 | 100/600 |
| Ammunition type | 9- by 18-millimeter ball | 5.45- by 39-millimeter ball,<br>ball tracer,<br>incendiary T |
| Fire mode | Semiautomatic | Selective semiautomatic or fully automatic |

## Table L-5. Small Arms (Continued)

| Small Arms Characteristics |  5.45-Millimeter Squad Machine Gun (RPK)-74 Light MG | 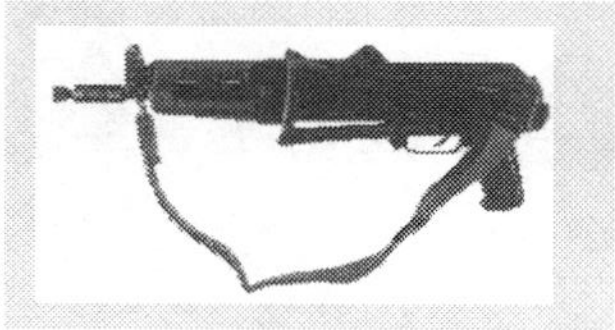 5.45-Millimeter AKSU-74 Submachine Gun (SMG) |  7.62-Millimeter Soviet Sniper Rifle (SVD) |
|---|---|---|---|
| Range, effective and maximum (in meters) | 800/2,500 | 300/1,500 | 1,300 with scope, (800 without)/ 3,800 |
| Rate of fire, practical and cyclic (rounds per minute) | 150/600 | 150/600 | 30 |
| Ammunition type | 5.45- by 39-millimeter ball, ball tracer, incendiary T | 5.45- by 39-millimeter ball, ball tracer, incendiary T | 7.62- by 54R-millimeter light or heavy ball, steel core, tracer, AT, incendiary "rose" sniper bullet |
| Fire mode | Selective semiautomatic or fully automatic | Selective semiautomatic or fully automatic | Semiautomatic |

## Table L-5. Small Arms (Continued)

| Machine Gun Characteristics |  RPK-74 | 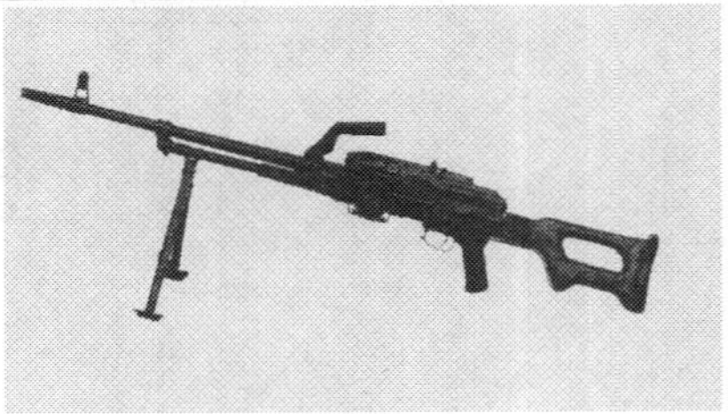 PKM |
|---|---|---|
| Range, effective and maximum (in meters) | 800/2,500 | 1,000/3,800 |
| Rate of fire, cyclic and practical (rounds per minute) | 600/150 | 650/250 |
| Ammunition type | 5.45- by 39-millimeter, rimless | 7.622- by 54-millimeter ball, ball tracer, API-T, incendiary |
| Fire mode | Selective | NA |
| Armor penetration (in millimeters) | NA | 8 millimeters at 500 |

**Table L-6. Grenade Launchers**

| Grenade Launcher Characteristics | 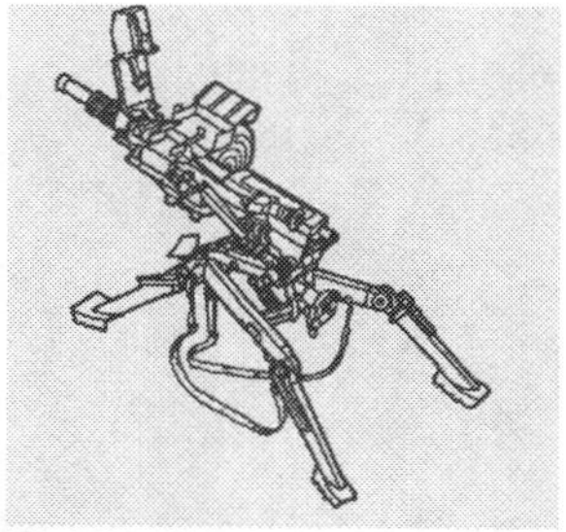 Armored Gun System (AGS)-17 | 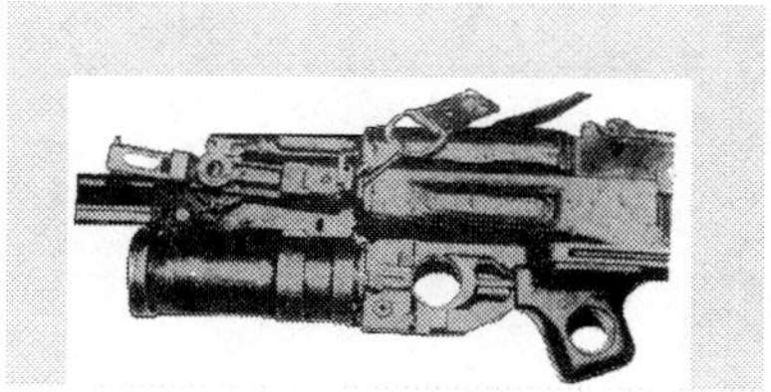 BG-15/GP-25 |
|---|---|---|
| Range, effective and maximum (in meters) | 1,200 indirect fire, 700 direct fire/1,730 | 4 to 400/400 |
| Rate of fire, practical and cyclic (rounds per minute) | 60 to 100/100 to 450 | 5 |
| Elevation (in degrees) | 7 to 87 | NA |
| Traverse (in degrees) | 30 | NA |
| Platform | Tripod, vehicle, or helicopter | Under barrel grenade launcher |
| Fire mode | Selective, semiautomatic, or fully automatic | Single shot |
| Feed | 29-round belt in drum magazine | Muzzle loaded |
| Ammunition type | 30-millimeter, HE fragment | 40-millimeter, HE fragment, bounding HE fragment |
| Caliber (in millimeters) | 30 | 40 |
| Crew | 3 | 1 |
| Remarks | 15-meter lethal area of burst | Bounding HE fragment strikes the ground and "bounds" up 1.5 to 2 meters before exploding |

## Table L-7. Rocket-Propelled Grenades

| AT Grenade Launcher Characteristics |  Rocket-Propelled Grenade (RPG)-7V | | | | |  RPG-16D |
|---|---|---|---|---|---|---|
| **Projectile** | PG-7 | PG-7M | PG-VR | PG-VL | OG-7/7V | PG-16 |
| Range, moving and stationary target (in meters) | 300/500 | NA | NA | NA | NA | 500/800 |
| Armor penetration (in millimeters) | 330 | 330 | >750 | >PG-7M <PG-VR | NA | 375 |
| Ammunition type | High-explosive, antitank (HEAT) | HEAT | Tandem HEAT | HEAT | HE fragment | HEAT |
| Warhead caliber (in millimeters) | 85 | 72 | 105 | 93 | NA | 58.3 |
| Rate of fire (rounds per minute) | 6 | 6 | 6 | 6 | 6 | 6 |
| Crew | NA | NA | NA | NA | NA | 2 |
| Tube caliber (in millimeters) | 40 | 40 | 40 | 40 | 40 | 58.3 |

**Table L-7. Rocket-Propelled Grenades (Continued)**

| **AT Grenade Launcher Characteristics** | 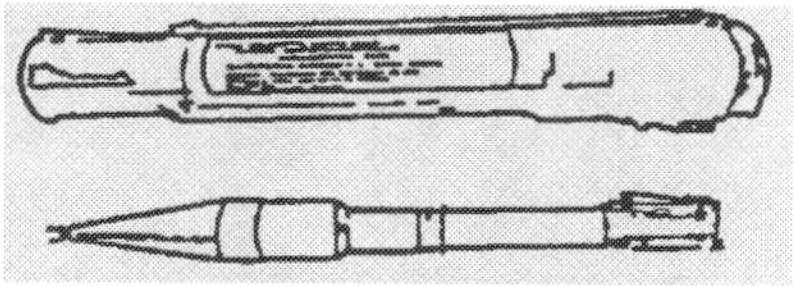 **RPG-18 and 22** | | 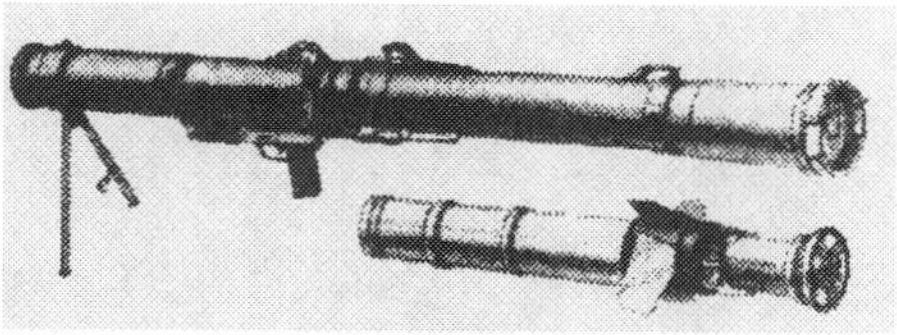 **RPO-A** |
|---|---|---|---|
| | RPG-18 | RPG-22 | RPO-A |
| **Projectile** | | | |
| Range, moving and stationary target (in meters) | 200 | 250 | 200 direct fire<br>1,000 indirect fire |
| Armor penetration (in millimeters) | 360 | 390 | NA |
| Ammunition type | HEAT | HEAT | FAE |
| Rate of fire (rounds per minute) | NA | NA | 2 |
| Crew | 1 | NA | NA |
| Tube caliber (in millimeters) | 64 | 73 | 93 |

**Table L-8. Recoilless Rifles**

| AT Gun and Rifle Characteristics | 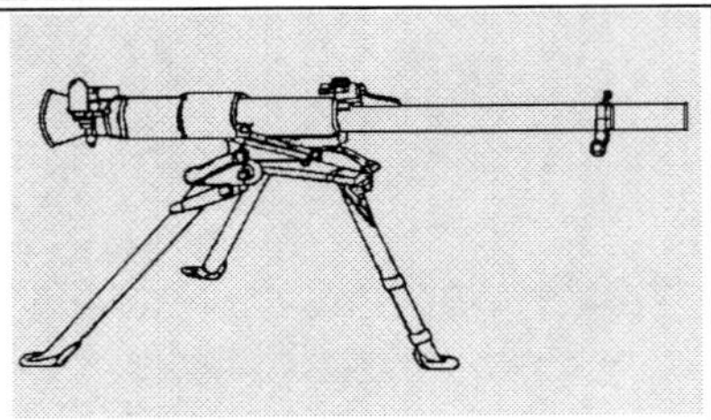 Self-Propelled Gun (SPG-9) Recoilless Rifle | 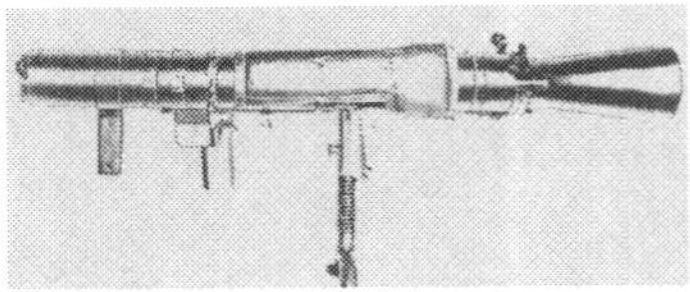 84-Millimeter Carl Gustaf Recoilless Rifle |
|---|---|---|
| Range, HEAT (in meters) | 1,000 | 500 |
| HEAT, rocket assist (RA) | NA | 250 |
| HE | NA | 500 |
| Indirect fire | NA | 1,000 HE, 1,300-meter smoke, 2,300-meter illumination |
| Armor penetration (in millimeters) | 400 | 500 HEAT, 900 HEAT, RA |
| Rate of fire, maximum and sustained (rounds per minute) | 6 | 6 |
| Fire control | Direct-view optics II (DVO) | 3x DVO |
| Ammunition type | HEAT, HE, rocket-assisted projectile (RAP), armor piercing (AP) AT hollow charge | HEAT, HEAT RA, illumination, smoke, high explosive, dual purpose (HEDP), HE |
| Crew | 3 or 4 | |

## Table L-8. Recoilless Rifles (Continued)

| **Infantry Fighting Vehicle (IFV) Characteristics** |  **Bronevaya Maschina Piekhota, Armored Vehicle, Infantry (BMP)-1P IFV** |  **BMP-2 IFV** |  **BMP-3 IFV** |
|---|---|---|---|
| **Main Armament (Caliber and Model)** | 73-millimeter 2A38 gun | 30-millimeter 2A42 gun | 100-millimeter 2A70 |
| Ammunition type | HEAT FS, HE fragment | High explosive incendiary (HEI), high explosive tracer (HET), APT | HE fragment, HE |
| Range, effective (in meters) | 800 | 1,500 light armor, 3,000 air, 2,500 ground | 4,000 HE fragment |
| Rate of fire (rounds per minute) sustained/maximum | 10 | 300/500 | 10/15 |

**Table L-8. Recoilless Rifles (Continued)**

| IFV Characteristics | BMP-1P IFV (Continued) | BMP-2 IFV (Continued) | BMP-3 IFV (Continued) |
|---|---|---|---|
| **Secondary Armament** | AT-4a/5a ATGM | AT-4a/5a ATGM | AT-10 ATGM |
| Range, effective (in meters) | 2,000/4,000 | 2,000/4,000 | 5,000 |
| Rate of fire, cyclic and practical (rounds per minute) | | | |
| Penetration (millimeters at meter range) | 600/650 | 600/650 | 660 |
| **Auxiliary Armament** | 7.62-millimeter PKT MG | | 30-millimeter 2A72 gun |
| Range, effective (in meters) | | | |
| Ammunition type | | | APT, HE, HET |
| **Auxiliary Armament** | | | |
| Model | | 7.62-millimeter PKT MG | 3- by 7.62-millimeter PKT MG |
| Commander | IR | IR | IR |
| Speed, road and off road (kilometers per hour) | 65/45 | 65/50 | 70/50 |

**Table L-8. Recoilless Rifles (Continued)**

| IFV Characteristics | BMP-1P IFV (Continued) | BMP-2 IFV (Continued) | BMP-3 IFV (Continued) |
|---|---|---|---|
| **Vehicle Characteristics (Continued)** | | | |
| Range, road and off road (in kilometers) | 600/570 | 600 | 600 |
| Ground clearance (in millimeters) | 390 | 420 | 190 to 510 |
| Armor, hull/turret (in millimeters) | 19/23 | 19/23 | |
| Dimensions (length by width by height, in meters) | 6.735 by 2.94 by 2.068 | 6.86 by 3.13 by 2.45 | 7.2 by 3.2 by 2.6 |
| Crew and passengers | 3/8 | 3/7 | 3/7 |

**Table L-8. Recoilless Rifles (Continued)**

| IFV Characteristics |  BMD-2 | 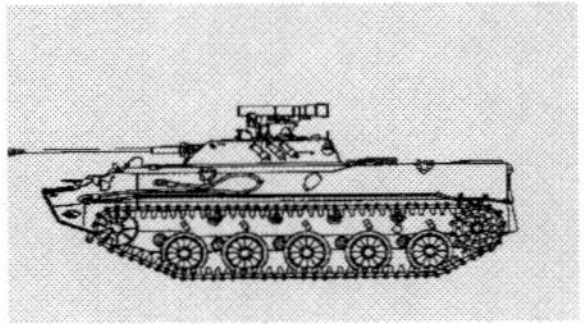 BMD-3 | 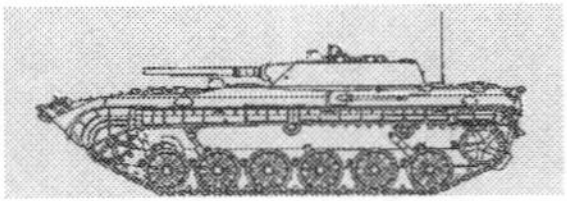 BRM |
|---|---|---|---|
| **Main Armament (Caliber and Model)** | 30-millimeter 2A42 gun | 30-millimeter 2A42 gun | 73-millimeter 2A38 gun |
| Ammunition type | APT, fragment T, HEI | APT, fragment T, HEI | HEAT FS, FE fragment |
| Range, effective (in meters) | 1,500 light armor, 2,000 air, 4,000 soft skin | 2,000 AP, 4,000 HE | 800 |
| Rate of fire (rounds per minute) sustained and maximum | 240/600 | 240/600 | 10 |
| Secondary Armament | AT-4/5 ATGM | AT-4a/5a ATGM | 7.62-millimeter PKT MG |
| Range, effective (in meters) | 2,000/4,000 | 2,000/4,000 | 1,000 |
| Rate of fire, cyclic and practical (rounds per minute) | | | 650/250 |
| Penetration (millimeters at meter range) | 600 at 650 | 600 at 650 | 8 at 500 meters |

**Table L-8. Recoilles Rifles (Continued)**

| IFV Characteristics | BMD-2 (Continued) | BMD-3 (Continued) | BRM (Continued) |
|---|---|---|---|
| **Auxiliary Armament** | | 30-millimeter, AG-17 grenade launcher | |
| Model | 2- by 7.62-millimeter PKT MG | 7.62-millimeter PKT MG | |
| **Vehicle Characteristics** | | | |
| Night sights | | Active IR | Second generation |
| Speed, road and off road (kilometers per hour) | 60/35/10 | 70/45/10 | 70/10 |
| Range, road and off road (in kilometers) | 500/350 | 500 | 500 |
| Armor, hull and turret (in millimeters) | | 16 | 10 |
| Dimensions (length by width by height, in meters) | 5.4 by 2.63 by 1.615 | 6.1 by 3.134 by 2.25 | 6.75 by 2.97 by 1.98 |
| Crew and passengers | 2/5 | 2/5 | 3/6 |

Table L-9. Armored Personnel Carriers (APCs)

| APC Characteristics |  **BTR-60PB** |  **BTR-70** |  **BTR-80** |
|---|---|---|---|
| **Main Armament (Caliber and Model)** | 14.5-millimeter Soviet 14.5-millimeter, heavy, MG (KPVT) | 14.5-millimeter MG (KPVT) | 14.5-millimeter MG (KPVT) |
| Range, effective (in meters) | 2,000 ground, 1,400 air | 2,000 ground, 1,400 air | 2,000 ground, 1,400 air |
| Rate of fire (rounds per minute) sustained and maximum | 80/600 | 80/600 | 80/600 |
| **Secondary Armament** | 7.62-millimeter PKT MG | 7.62-millimeter PKT MG | 7.62-millimeter PKT MG |
| Range, effective (in meters) | 1,000 | 1,000 | 1,000 |
| Rate of fire, cyclic and practical (rounds per minute) | 650/250 | 650/250 | 650/250 |
| Penetration (millimeters at meter range) | 8 at 500 meters | 8 at 500 meters | 8 at 500 meters |

**Table L-9. APCs (Continued)**

| APC Characteristics | BTR-60PB (Continued) | BTR-70 (Continued) | BTR-80 (Continued) |
|---|---|---|---|
| **Vehicle Characteristics** | | | |
| Commander | IR | IR | IR |
| Driver | IR | IR | IR |
| Range, road and off road (in kilometers) | 500 | 450 | 800 |
| Armor, hull and turret (in millimeters) | | 10/7 | |
| Dimensions (length by width by height, in meters) | 7.5 by 2.8 by 2.23 | 7.85 by 2.8 by 2.45 | 7.62 by 2.9 by 2.35 |
| Crew and passengers | 3/7 | 2/8 | 3/7 |

## Table L-9. APCs (Continued)

| APC Characteristics | 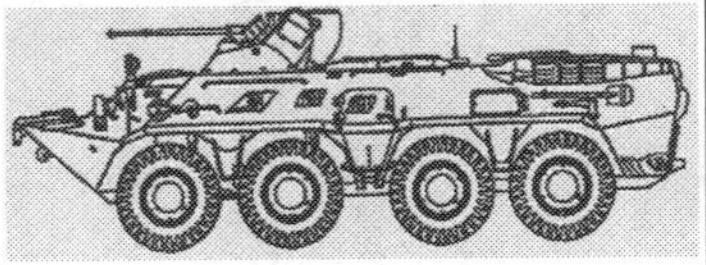 BTR-80A |  BTR-D | 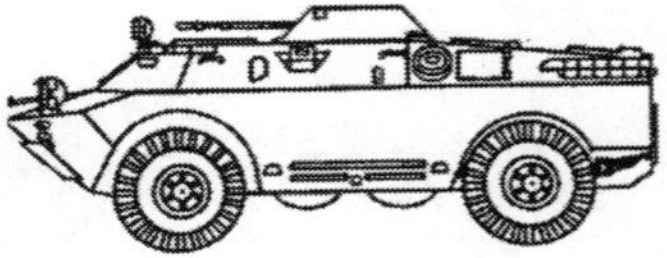 BRDM-2 Reconnaissance |
|---|---|---|---|
| **Main Armament (Caliber and Model)** | 30-millimeter 2A72 gun | 2 by 7.62 PKT MG | 14.5-millimeter KPVT MG |
| Range, effective (in meters) | 2,000 APT, 4,000 HEI, 800 night | 1,000 | 2,000 ground, 1,400 air |
| Rate of fire (rounds per minute) sustained and maximum | | 250/650 | 80/600 |
| **Secondary Armament** | 7.62-millimeter PKT MG | | 7.62-millimeter PKT MG |
| Range, effective (in meters) | 1,000 | | 1,000 |
| Rate of fire, cyclic and practical (rounds per minute) | 650/250 | | 650/250 |
| Penetration (millimeters at meter range) | 8 at 500 meters | | 8 at 500 meters |

**Table L-9. APCs (Continued)**

| APC Characteristics | BTR-80A (Continued) | BTR-D (Continued) | BRDM-2 Reconnaissance (Continued) |
|---|---|---|---|
| **Vehicle Characteristics** | | | |
| Night sights | | | |
| Range, road and off road (in kilometers) | 800 | 500/350 | 750 |
| Armor, hull and turret (in millimeters) | | | |
| Dimensions (length by width by height, in meters) | | 5.883 by 2.63 by 1.67 | 5.75 by 2.262 by 2.31 |
| Crew and passengers | 2/8 | 3/10 | 4 |
| Remarks | | Airborne assault. Can have AGS-17 grenade launcher. BTR-RD has 5 AT-4 antitank guided missiles (ATGMs), dismountable to 2 tripods. | |

**Table L-9. APCs (Continued)**

| APC Characteristics | 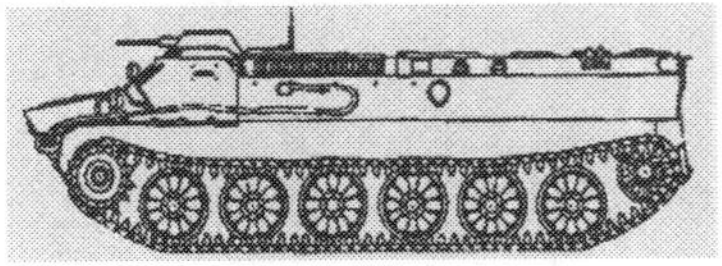 MTLB |
|---|---|
| **Main Armament (Caliber and Model)** | 7.62 PKT MG |
| Range, effective (in meters) | 1,000 |
| Rate of fire (rounds per minute) sustained and maximum | 250/650 |
| **Vehicle Characteristics** | |
| Night sights | |
| Range, road and off road (in kilometers) | |
| Armor, hull and turret (in millimeters) | 7/7 |
| Dimensions (length by width by height, in meters) | 6.45 by 2.85 by 1.87 |
| Crew and passengers | 2/11 |
| Remarks | |

**Table L-10. ATGM**

| ATGM Characteristics | AT-3c SAGGER 9K11 Malyutka | AT-4a/b SPIGOT 9K111/9K111M Fagot/Faktoriya | AT-5 a/b SPANDREL 9K133/9K133m Konkurs/ Konkurs-M | AT-6 a/b/c SPIRAL 9K114 Shturm |
|---|---|---|---|---|
| Range (in meters) | 500 to 3,000 | 70 to 2,000/75 to 3,500 | 70 to 4,000/75 to 4,000 | 400 to 5,000/400 to 5,000/400 to 6,000 |
| Flight time to maximum range (in seconds) | 23 to 26 | 11/19.5 | 19 | 13.3/16/18.6 |
| Guidance and command link | Wire MCLOS, 9M14-2: wire SACLOS | Wire SACLOS | Wire SACLOS | Radio freguency (RF) SACLOS |
| Warheads | HEAT, 9M14-2: tandem HEAT | HEAT | HEAT/tandem HEAT | HEAT or FAE |
| Remarks | The AT-3 continues to be improved and proliferated worldwide. The M14-2 missile is effective against reactive armor. | Thermal sight detection range is 3,600 meters, with identification at 2,000 meters. | Thermal sight detection range is 3,600 meters, with identification at 2,000 meters. | FAE ranges 400 to 5,000 meters. |

**Table L-10. ATGM (Continued)**

| ATGM Characteristics | AT-7 SAXHORN 9K115 Metis | AT-8 SONGSTER 9K112 Kobra | AT-9 9K120 Ataka | AT-10 STABBER 9K116-1/9K116 Bastion/Kastet |
|---|---|---|---|---|
| Range (in meters) | 40 to 1,000 | 100 to 4,000 | 400 to 6,000 | 100 to 5,000 |
| Flight time to maximum range (in seconds) | 6 | 10 | 14.5 | 15 |
| Penetration (rolled hardened armor (RHA)/behind extended range artillery (ERA), in millimeters) | 500 | 800 | 1,000/800 | 660 |
| Guidance and command link | Wire SACLOS | RF (30 gigahertz) SACLOS | | Laser beam rider SACLOS |
| Warheads | HEAT | HEAT | Tandem HEAT, blast, FAE, antihelicopter | HEAT |
| Remarks | Thermal sight detection range is 3,200 meters, with identification at 1,600 meters. Emplace and displace times are 12/20 seconds. | Fired through the main gun tube. | | Fired through the main gun tube. Off BMP-3 range is 100 to 4,000 meters, but it can fire on the move. |

**Table L-10. ATGM (Continued)**

| ATGM Characteristics | AT-11a/b SNIPER 9K119/9K119M Svir/Invar | AT-12 STABBER 9K116-2 Sheksna | AT-13 9K115-2 Metis-M | AT-14 Kornet |
|---|---|---|---|---|
| Range (in meters) | 100 to 5,000 | 100 to 5,000 | 80 to 1,500 | 100-5,500 |
| Flight time to maximum range (in seconds) | 15 | 12 to 4,000 | 8.4 | 22 |
| Penetration (RHA/ behind ERA in millimeters) | 770 AT-11a, 1,050 AT-11b | 800 | 1,000/800 | 1,200/980 |
| Guidance and command link | Laser beam rider SACLOS | Laser beam rider SACLOS | Wire SACLOS | Laser beam rider SACLOS |
| Warheads | HEAT/tandem HEAT | HEAT | Tandem HEAT, FAE | Tandem HEAT, FAE |
| Remarks | Fired through the main gun tube. T-72 variants have a 4,000-meter maximum range. | Fired from the halt through the main gun tube. | FAE has greater power than 152-millimeter artillery round. | Thermal sight range is 4,000-meter detection and 3,500-meter identification. Maximum altitude of engaged helicopters is 3,000 meters. |

**Table L-10. ATGM (Continued)**

| ATGM Characteristics | AT-16 9K121 VIKhR | Milan (France [FR]) 1/2/2T/3 | Tow (US) 1/ITOW/ 2a/2b | HOT (FR/Germany [GE]) |
|---|---|---|---|---|
| Range (in meters) | 1,000 to 10,000 | 20 to 2,000 | 3,750/3,750/65- 3,750/200 to 3,750 | 75 to 4,000 |
| Flight time to maximum range (in seconds) | 23 | 12.5 | 2a: 20, 2b: 21 | 16.5 |
| Penetration (RHA/ behind ERA in millimeters) | 1,050/900 | 600/1,060/1,200/ 1200 | 500, ITOW:800, 2:920, 2a 1,000 | 850, 2:950 |
| Guidance and command link | Laser beam rider SACLOS (lock on before launch) | 1 and 2: SACLOS wire SACLOS IR | SACLOS wire | SACLOS wire |
| Warheads | Tandem HEAT | | | |
| Remarks | Maximum airborne target speed is 800 kilometers per hour. | Milan 3 medium range ATGM has IR command link resistant to jamming and tandem warhead. | TOW 2b is fly over, shoot down, top attack, with self-forming penetrant 2nd generation forward looking infrared (FLIR). | HOT 3 long-range ATGM has a tandem warhead, bispectral day and night sights, and may be mounted on posts, vehicles, or helicopters. |

**Table L-11. Rotary Wing Aircraft**

| Rotary Wing Aircraft Characteristics | 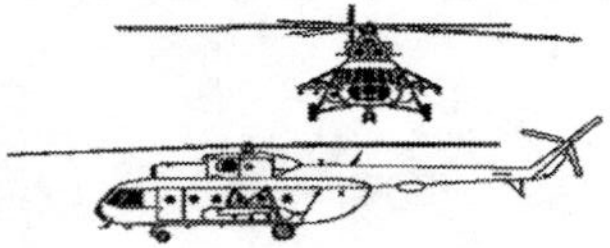 Mi-8 HIP | 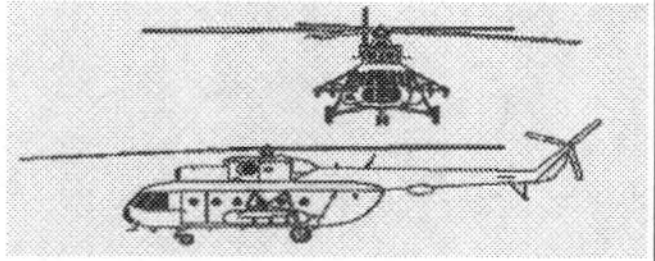 Mi-17 HIP H |  Mi-24 HIND D/E/F |
|---|---|---|---|
| Mission | Utility, transport | Utility, transport | Attack, close support |
| Gun | 12.7 millimeter | 12.7 millimeters | 30-millimeter HIND F<br>12.7-millimeter HIND D/E |
| ATGM | AT-2c | AT-2c, AT-3c | 4-16<br>AT-2c HIND D<br>AT-6c HIND E/F |
| Rockets | 57 millimeters | 57 millimeters | 57 millimeters or<br>80 millimeters |
| Bombs | 250 kilograms,<br>500 kilograms | 250 kilograms,<br>500 kilograms | 250 kilograms,<br>500 kilograms |

Table L-11. Rotary Wing Aircraft (Continued)

| Rotary Wing Aircraft Characteristics | Mi-8 HIP (Continued) | Mi-17 HIP H (Continued) | Mi-24 HIND D/E/F (Continued) |
|---|---|---|---|
| Range, normal payload (in kilometers) | 460 | 495 | 480 |
| Crew | 3 | 3 | 2 |
| Passengers | 24 | 24 | 8 |
| Remarks | Troops can fire personal weapons through windows. It has four external hard points. | Improved version of Mi-8MT, have upgraded engines and six external hard points. | |
| | 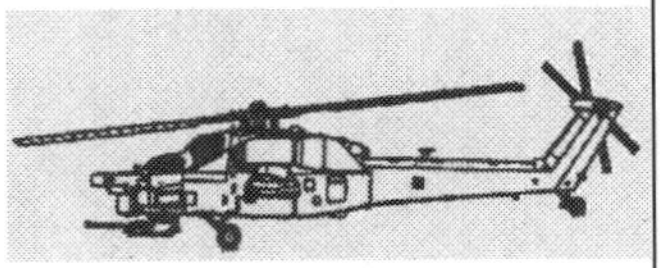 Mi-28N HAVOC | 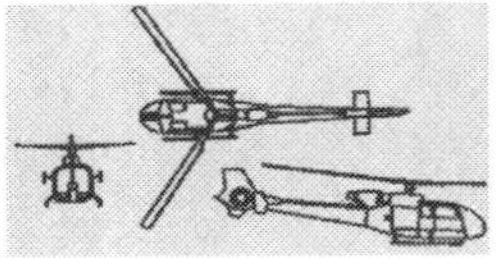 Gazelle (FR) | 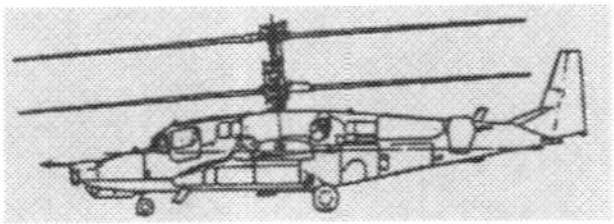 Ka-50 HOKUM "Black Shark" or "Werewolf" |
| Mission | Attack, close support | Light attack utility | Attack, close support |
| Gun | 23 millimeters<br>30 millimeters | 2 by 7.62 millimeters | 2A42 30-millimeter cannon |

**Table L-11. Rotary Wing Aircraft (Continued)**

| Rotary Wing Aircraft Characteristics | Mi-28N HAVOC | Gazelle (FR) | Ka-50 HOKUM |
|---|---|---|---|
| ATGM | 16 AT-6/9/16 | AT-3, HOT | AT-16 |
| Rockets | 57 millimeters | 5 millimeters<br>68 millimeters<br>2.75 inches | 80 millimeters |
| Bombs | NA | NA | 250 kilograms, 500 kilograms |
| Range, normal payload (in kilometers) | 470<br>1,100 with drop tanks | 710 | 455 |
| Crew | 2 or 3 | 1 or 2 | 1 |
| Passengers | NA | 2 | NA |
| Remarks | This system is not operationally fielded in any armed force. | NA | The armored cockpit withstands 23-millimeter fire and the windscreen withstands 12.7-millimeter fire. |

**Table L-11. Rotary Wing Aircraft (Continued)**

| Rotary Wing Aircraft Characteristics | MD 500 (US) | UH-1H (US) | BO-105 (GE) | Lynx (UK) |
|---|---|---|---|---|
| Mission | General purpose, light attack | Utility | General purpose, light attack | General purpose, attack |
| Gun | 7.62 millimeters | 7.62 millimeters | 20 millimeters | 2 by 20 millimeters |
| ATGM | TOW | NA | HOT<br>TOW | HOT<br>Hellfire<br>TOW |
| Rockets | 70 millimeters | NA | NA | Sura, 80 millimeters |
| Range, normal payload (in kilometers) | 540 | 465 | 318 | 540 |
| Crew | 2 | 2 | 2 | 2 |
| Passengers | 2 | 11 | 3 | 10 |

## Table L-12. Combat Aircraft

| Combat Aircraft Characteristics | 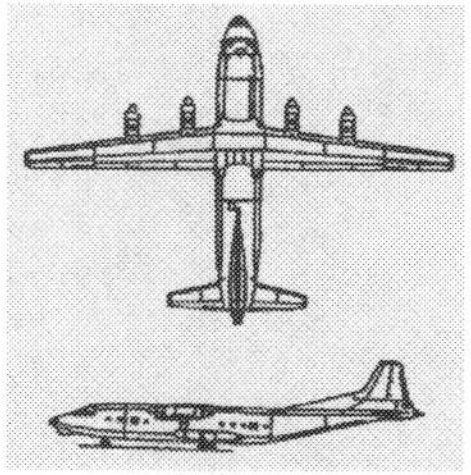 An-12 CUB | 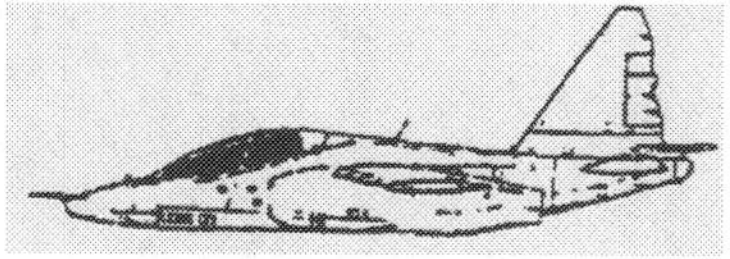 SU-25 FROGFOOT |
|---|---|---|
| Mission | Medium transport aircraft | CAS |
| Gun(s) | NR-23 23 millimeters in the tail turret | GSh-30-2, 30 millimeters (250 rounds), AO-17a, 30-millimeter pods |
| ATGMs | | 16x AT-16 or AT-9 |
| Rockets | | 57-millimeter S-5 pod, 80-millimeter S-8 pod (up to 8 pods) |
| Air-to-surface missiles | | AS-7, AS-10, AS-11, AS-14, AS-17 |
| Bombs | | 100 kilograms, 350 kilograms, 500 kilo-grams (up to 4,000 kilograms) |
| Combat radius (in kilometers) | 1,500 to 1,800 | 495 |
| Payload, paratroops | 90 | |
| Remarks | It can operate from dirt strips. | SU-25M: AT, SU-25UB: trainer, ceiling 10,000 meters, 4,344 kilograms maximum payload, 6.5 gram limit. |

**Table L-12. Combat Aircraft (Continued)**

| Combat Aircraft Characteristics | 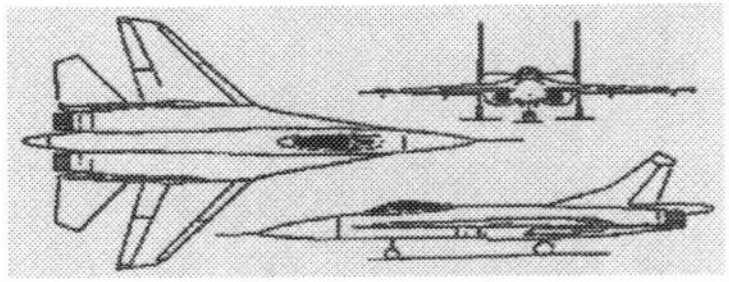 SU-27 Flanker B |
|---|---|
| Mission | Fighter/intercepter, fighter bomber variant |
| Gun(s) | GSh-301, 30-millimeter cannon |
| ATGMs | |
| Rockets | 80-millimeter S-8 or 120-millimeter S-25 |
| Air-to-air missiles | R-27, R-73, AA-10, AA-11 |
| Bombs | 100 kilograms, 250 kilograms, 500 kilograms |
| Combat radius (in kilometers) | 1,125/1,950 with tanks |

# Glossary

| | |
|---|---|
| **∞** | infinity symbol |
| **"** | inch(es) |
| **'** | foot; feet |
| **<** | more than |
| **>** | less than |
| **°** | degree |
| **?** | unknown information |
| **AA** | assembly area |
| **AAR** | after-action review |
| **AB** | A map's reference point. |
| **ABD** | air base defense |
| **ACE** | ammunition, casualty, and equipment |
| **ADC** | area damage control |
| **AFJI** | Air Force joint instruction |
| **AFM** | Air Force manual |
| **AFSF** | Air Force security force |
| **AGS** | armored gun system |
| **AI** | area of interest |
| **alt** | alternate |
| **ammo** | ammunition |
| **ANCD** | automated net control device |

| | |
|---|---|
| **AO** | area of operation |
| **AOE** | Army of Excellence |
| **AOR** | area of responsibility |
| **AP** | armor piercing |
| **APC** | armored personnel carrier |
| **APL** | antipersonnel land mines |
| **APT** | armor piercing tracer |
| **AR** | Army regulation; automatic rifle (SAW) |
| **AS** | area security |
| **ASAP** | all-source analysis product |
| **ASCC** | Army service component command |
| **ASI** | addition skill identifier |
| **ASP** | ammunition supply point |
| **ASV** | armor security vehicle |
| **AT** | antitank |
| **AT/FP** | antiterrorism/force protection |
| **ATGM** | antitank guided missiles |
| **ATTN** | attention |
| **AWOL** | absent without leave |
| **BCC** | battlefield circulation control |
| **BCOC** | base cluster operations center |
| **BDOC** | base defense operations center |
| **BDU** | battle dress uniform |

| | |
|---|---|
| **BHL** | battle handover line |
| **BII** | basic-issue item |
| **bldg** | building |
| **BMP** | Bronevaya Maschina Piekhota, armored vehicle, infantry |
| **BPS** | baud per second |
| **BSA** | brigade support area |
| **BT** | bomb trench |
| **c** | distance of the centerline of the road |
| **C** | clay; distance from the centerline |
| **$C^2$** | command and control |
| **C4** | composition C4 |
| **CA** | cinnamic acid; civil affairs |
| **CAC** | crossing area commander |
| **CAM** | chemical agent monitor |
| **CANA** | convulsant antidote for nerve agents |
| **CAS** | close air support |
| **CASEVAC** | casualty evacuation |
| **CCIR** | commander's critical information requirements |
| **cdr** | commander |
| **CERTEX** | certification exercise |
| **cGyph** | centigray per hour |
| **CI** | civilian internee |

| | |
|---|---|
| **CID** | criminal investigation division |
| ℄ | centerline |
| **cm** | centimeter |
| **co** | company |
| **COA** | course of action |
| **COMMEX** | communication exercise |
| **COMMZ** | communications zone |
| **CONEX** | container express |
| **CONUS** | continental United States |
| **CP** | command post |
| **CPOG** | chemical protective overgarment |
| **CPU** | chemical protective undergarment |
| **CR** | Dibenz (B, f)-1, 4-oxazepine |
| **CS** | combat support; chlorobenzul-malononitrile |
| **CSC** | crossing site commander |
| **CSS** | combat-service support |
| **CTOC** | corps tactical operations center |
| **CTUS** | customs territory of the United States |
| **DA** | Department of the Army |
| **DAP** | decontamination apparatus |
| **DC** | dislocated civilian |
| **DCG** | degrees grid north |

| | |
|---|---|
| **DD** | Department of Defense |
| **Dec** | December |
| **DED** | detailed equipment decontamination |
| **DFAS-IN** | Defense Finance and Accounting Service-Indianapolis |
| **DFC** | defense force commander |
| **DGG** | degrees grid north |
| **DGM** | degrees magnetic north |
| **DGT** | degrees true north |
| **DM** | designated marksman |
| **DNVT** | digital nonsecure voice terminal |
| **DOD** | Department of Defense |
| **DS** | direct support; decontamination solution |
| **DS2** | decontamination solution number 2 |
| **DSA** | division support area |
| **DTD** | detailed troop decontamination |
| **DTG** | date-time group |
| **DTO** | division transportation office |
| **DVO** | direct-view optics |
| **DZ** | drop zones |
| **E** | east |
| **EA** | engagement area |
| **EAC** | echelon above corps |

| | |
|---|---|
| **EEFI** | essential elements of friendly information |
| **EEP** | engineer equipment park |
| **EIC** | end item code |
| **ENDEX** | end of exercise |
| **EO** | executive order |
| **EOC** | emergency operations center |
| **EPW** | enemy prisoner of war |
| **ERA** | extended range artillery |
| **ERP** | engineer regulating points |
| **FA** | field artillery |
| **FAE** | fuel air explosive |
| **FCF** | field confinement facility |
| **FDC** | fire direction center |
| **FEBA** | forward edge of the battle area |
| **FFIR** | friendly force information requirements |
| **FIST** | fire support team |
| **FLIR** | forward looking infrared |
| **FM** | field manual |
| **FMFM** | fleet Marine force manual |
| **FMJ** | full metal jacket |
| **FN** | foreign nation |
| **FO** | forward observer |

| | |
|---|---|
| **FOX** | M93 NBC Reconnaissance System |
| **FP** | force protection |
| **FPF** | final protective fire |
| **FPL** | final protective line |
| **FR** | France |
| **FRAG** | fragment |
| **FRAGO** | fragmentary order |
| **FSO** | fire support officer |
| **G** | gravel; M203 |
| **G2** | Assistant Chief of Staff, G2 (Intelligence) |
| **G5** | Assistant Chief of Staff, G5 (Civil Affairs) |
| **gal** | gallon |
| **GB** | guard bunker |
| **GE** | Germany |
| **GM** | grid to magnetic |
| **GMG** | grenade machine gun |
| **GN** | grid north |
| **GP** | general purpose |
| **GPS** | Global Positioning System |
| **GS** | general support |
| **GTA** | graphic training aid |
| **GVW** | gross vehicle weight |

| | |
|---|---|
| **HAVOC** | Soviet mi-28 aircraft |
| **$H_d$** | horizontal distance |
| **HE** | high explosive |
| **HEAT** | high explosive antitank |
| **HEDP** | high explosive, dual purpose |
| **HEI** | high-explosive incendiary |
| **HET** | high-explosive tracer |
| **HHC** | headquarters and headquarters company |
| **HI** | high |
| **HIND** | Soviet helicopter |
| **HIP** | Soviet medium lift helicopter |
| **HMMWV** | high mobility multipurpose wheeled vehicle |
| **HN** | host nation |
| **HOKUM** | Soviet Ka-? aircraft |
| **HOT** | air-to-ground antitank missile, also call Euromi |
| **HTS** | highway traffic section |
| **hwy** | highway |
| **HQ** | headquarters |
| **IBCT** | initial brigade combat team |
| **ICITAP** | International Criminal Investigative Training Assistance Program |

| | |
|---|---|
| **ICM** | improved conventional munitions |
| **ID** | identification |
| **IEDK** | individual equipment decontamination kit |
| **IFV** | infantry fighting vehicle |
| **IPB** | intelligence preparation of the battlefield |
| **IPC** | interpersonal communication |
| **I/R** | internment and resettlement |
| **IR** | infrared |
| **IRIC** | Internment Resettlement Information Center |
| **JRA** | joint rear area |
| **Jul** | July |
| **KBPS** | kilobaud per second |
| **km** | kilometers |
| **KPVT** | Soviet 14.5-millimeter heavy machine gun |
| **L** | location |
| **L&O** | law and order |
| **LAW** | light antiarmor weapon |
| **lb** | pound(s) |
| **LC** | line of contact |
| **LCE** | load-carrying equipment |
| **LD** | line of departure |

| | |
|---|---|
| **ldr** | leader |
| **LMTV** | light medium tactical vehicle |
| **LO** | low |
| **LOC** | lines of communication |
| **LOGSEC** | logistics security |
| **LOTS** | logistics over the shore |
| **LP** | listening post |
| **LRA** | local reproduction authorized |
| **LVOSS** | Light Vehicle Obscuration Smoke System |
| **LZ** | landing zones |
| **m** | meter; perpendicular distance from the center of the tape to the centerline of the road |
| **M** | medium; mud |
| **MANSCEN** | Maneuver Support Center |
| **MBA** | main battle area |
| **MCCM** | modular crowd control munitions |
| **MCLOS** | manual command line of sight |
| **MCO** | Marine Corps orders |
| **MCRP** | Marine Corps warfighting publication |
| **MCT** | movements control team |
| **MCWP** | Marine Corps warfighting publication |
| **MDMP** | military decision-making process |

| | |
|---|---|
| **MEDEVAC** | medical evacuation |
| **MELIOS** | mini eyesafe laser infrared observation set |
| **MEVA** | mission-essential or vulnerable area |
| **METL** | mission-essential task list |
| **METT-TC** | mission, enemy, troops, terrain, time available, and civilian considerations |
| **MG** | machine gun |
| **MHZ** | mega hertz |
| **MI** | military intelligence |
| **MIJI** | meaconing; intrusion; jamming; interference |
| **mil** | military |
| **MILES** | multiple-integrated laser engagement system |
| **MILVAN** | military-owned demountable container |
| **min** | minute |
| **MLC** | military load classification |
| **MLG** | mils grid north |
| **MLM** | mils magnetic north |
| **MLT** | mils true north |
| **mm** | millimeter |
| **MMS** | maneuver and mobility support |
| **MOPP** | mission-oriented protection posture |

| | |
|---|---|
| **MOUT** | military operations on urbanized terrain |
| **MP** | military police |
| **mph** | miles per hour |
| **MPI** | military police investigation |
| **MRE** | meal, ready-to-eat |
| **MRO** | medical regulating officer |
| **MSR** | main supply route |
| **MTMC** | military traffic management command |
| **MTF** | medical treatment facility |
| **MTP** | mission training plan |
| **MWD** | military working dogs |
| **N** | north |
| **NA** | not applicable |
| **NAAK** | nerve agent auto injector kit |
| **NAI** | named area of interest |
| **NATO** | North Atlantic Treaty Organization |
| **NBC** | nuclear, biological, chemical |
| **NBCWRS** | Nuclear, Biological, Chemical Warning and Reporting System |
| **NCO** | noncommissioned officer |
| **NCOIC** | noncommissioned officer in charge |
| **NE** | northeast |
| **NEO** | noncombatant evacuation operations |

| | |
|---|---|
| **NGO** | nongovernmental organization |
| **NLP** | nonlethal procedure |
| **NLW** | nonlethal weapons |
| **no** | number |
| **NP** | nonpersistent |
| **NSN** | national stock number |
| **NTC** | National Training Center |
| **NVD** | night vision device |
| **NVG** | night vision goggle |
| **NW** | northwest |
| **NWP** | Naval warfare publication |
| **OB** | obstruction |
| **OC** | observer/controller; oleoresin capsicum |
| **OCOKA** | observation and fields of fire, cover and concealment, obstacles, key terrain, and avenues of approach |
| **OCONUS** | outside the continental United States |
| **Oct** | October |
| **OD** | other detainees |
| **OEG** | operational exposure guidance |
| **OF** | observed fire |
| **OIC** | officer in charge |
| **OP** | observation post |
| **OPCON** | operational control |

| | |
|---|---|
| **OPFOR** | opposing forces |
| **OPLAN** | operation plan |
| **OPNAVINST** | Chief of Naval Operations Instruction |
| **OPORD** | operation order |
| **OPSEC** | operations security |
| **ORP** | objective rally point |
| **OVM** | on-vehicle material |
| **P** | pedestrian; artificial paving; persistent |
| **PA** | power amplifier |
| **PAO** | Public Affairs Office |
| **PC** | point of curvature |
| **PCI** | precombat inspection |
| **PCIR** | police and criminal information requirements |
| **PDDE** | power-driven decontamination equipment |
| **PDF** | principal direction of fire; portable document format |
| **PERSCOM** | personnel command |
| **PEWS** | Platoon Early-Warning System |
| **PIAP** | police information assessment process |
| **PIO** | police intelligence operations |
| **PIR** | priority intelligence requirements |

| | |
|---|---|
| **PKT** | The standard vehicle with a mounted light machine gun and is the basic Soviet coaxial AFV weapon. |
| **PL** | phase line |
| **PLGR** | precision lightweight global positioning system receiver |
| **plt** | platoon |
| **PM** | provost marshal |
| **PMCS** | preventive-maintenance checks and services |
| **PMO** | Provost Marshals Office |
| **POE** | port of embarkation |
| **POL** | petroleum, oils, and lubricants |
| **PRC** | populace and resource control |
| **PSG** | platoon sergeant |
| **PSYOP** | psychological operations |
| **PT** | point of tangency |
| **PVAB** | portable vehicle-arresting barrier |
| **PVO** | private volunteer organization |
| **pyro** | pyrotechnics |
| **R** | radio; rock; radius of curve; radar set |
| **R&S** | reconnaissance and surveillance |
| **RA** | rocket assist |
| **RADAR** | radio detection and ranging |

| | |
|---|---|
| **RADIAC** | radiation detection, indication, and computation |
| **RAOC** | rear-area operations center |
| **RAP** | rocket-assisted projectile |
| **RCA** | riot control agents |
| **RES** | radiation exposure status |
| **RF** | radio frequency |
| **RHA** | rolled hardened armor |
| **RL** | release line; railroad bridge |
| **ROE** | rules of engagement |
| **ROI** | rules of interaction |
| **RORO** | roll-on/roll-off |
| **RP** | release point; retained person |
| **RPG** | rocket-propelled grenade |
| **RPK** | squad machine gun |
| **RTO** | radio telephone operator |
| **S** | size; sand; south |
| **S2** | Intelligence Officer (US Army) |
| **S3** | Operations and Training Officer (US Army) |
| **S4** | Supply Officer (US Army) |
| **S5** | Civil Affairs Officer (US Army) |
| **S&I** | standardization and interoperability |
| **SACLOS** | semiautomatic command line of sight |

| | |
|---|---|
| **SAGGER** | An AT-3 ATGM. |
| **SALUTE** | size, activity, location, unit, time, and equipment |
| **SAM** | surface-to-air missile |
| **SAW** | squad automatic weapon |
| **SAXHORN** | A nickname for the Soviet AT-7-ATGM. |
| **SE** | southeast |
| **SF** | security forces |
| **SINCGARS** | Single-Channel, Ground-to-Air Radio System |
| **SIR** | serious incident report |
| **SITREP** | situation report |
| **SJA** | Staff Judge Advocate |
| **SM** | soldier's manual |
| **SMCI** | senior military customs inspector |
| **SMG** | submachine gun |
| **SNAP** | size, nature, activity, protection |
| **SONGSTER** | The nickname for the Soviet AT-8 ATGM. |
| **SPANDREL** | An AT-5 ATGM. |
| **SPIGOT** | An AT-4 ATGM. |
| **SPIRAL** | A Soviet AT-6 ATGM. |
| **SOC** | security operations center |
| **SOFA** | Status of Forces Agreement |

| | |
|---|---|
| **SOI** | signal operating instructions |
| **SOP** | standard operating procedure |
| **SP** | start point |
| **SPG** | self-propelled gun |
| **SPOTREP** | spot report |
| **sqd** | squad |
| **SRT** | special-reaction team |
| **SSCO** | small-scale contingencies operations |
| **SSN** | social security number |
| **STABBER** | A nickname for the Soviet AT-10 ATGM. |
| **STANAG** | standardization agreement |
| **STX** | situational training exercise |
| **SVD** | Soviet sniper rifle |
| **SW** | southwest |
| **T** | time; snow blockage |
| **T&E** | traversing and elevating |
| **TA** | training area |
| **TACSOP** | tactical standing operating procedure |
| **TAI** | traffic accident investigation |
| **TC** | training circular |
| **TCF** | tactical combat force |
| **TCMD** | transportation control and movement document |

| | |
|---|---|
| **TCP** | traffic control post |
| **TEM** | training execution model |
| **TL** | team leader |
| **TLP** | troop-leading procedures |
| **TM** | technical manual |
| **tm** | team |
| **TO** | table of organization |
| **TOC** | tactical operations center |
| **TOE** | table(s) of organization and equipment |
| **TOW** | tube-launched, optically tracked, wire-guided missile |
| **TP** | training practice |
| **TRADOC** | United States Army Training and Doctrine Command |
| **TRL** | traffic regulating line |
| **TRP** | target reference point |
| **TRS** | transportation railway service |
| **TSC** | theater support command |
| **TTP** | tactics, techniques, and procedures |
| **U** | unit |
| **UCMJ** | Uniform Code of Military Justice |
| **UIN** | unit identification number |
| **UK** | United Kingdom |
| **UO** | urban operation |

| | |
|---|---|
| **US** | United States |
| **USACIDC** | United States Army Criminal Investigation Division Command |
| **USAF** | United States Air Force |
| **USAPA** | United States Army Publishing Agency |
| **USATSC** | United States Army Training Support Center |
| **USCG** | United States Coast Guard |
| **USDA** | United States Department of Agriculture |
| **USDOJ** | United States Department of Justice |
| **UTM** | universal transverse mercator |
| **UXO** | unexploded ordnance |
| **V** | vehicular |
| **$V_d$** | vertical distance |
| **VHS** | video home system |
| **VP** | both vehicle and pedestrian |
| **W** | flooding; west |
| **WO** | warning order |
| **Y** | limited, all-weather route, significant, serious, regular, or recurrent flooding or snow blockage |
| **X** | all-weather route |
| **Z** | Zulu; fair-weather route |

# Bibliography

AR 190-8. *Enemy Prisoners of War, Retained Personnel, Civilian Internees and other Detainees {OPNAVINST 3461.6; AFJI 31-304; MCO 3461-1}*. 1 October 1997.

AR 190-11. *Physical Security of Arms, Ammunition and Explosives.* 30 September 1993.

AR 190-12. *Military Police Working Dogs.* 30 September 1993.

AR 190-14. *Carrying of Firearms and Use of Force for Law Enforcement and Security Duties.* 12 March 1983.

AR 190-29. *Misdemeanors and Uniform Violation Notices Referred to US Magistrates or District Courts.* 1 March 1984.

AR 190-30. *Military Police Investigations*. 1 June 1978.

AR 190-40. *Serious Incident Report.* 30 November 1993.

AR 190-47. *The Army Corrections System.* 15 August 1996.

AR 190-51. *Security of Unclassified Army Property (Sensitive and Nonsensitive).* 30 September 1993.

AR 525-13. *Antiterrorism Force Protection (AT/FP): Security of Personnel, Information, and Critical Resources.* 10 September 1998.

AR 700-93. *Processing and Shipping DOD Sponsored Retrograde Materiel Destined for Shipment to the United States, Its Territories, Trusts, and Possessions.* 15 May 1980.

DA Form 1156. *Casualty Feeder Report.* 1 June 1966.

DA Form 1248. *Road Reconnaissance Report.* 1 July 1960.

DA Form 1249. *Bridge Reconnaissance Report.* 1 July 1960.

DA Form 1251. *Ford Reconnaissance Report.* 1 January 1955.

DA Form 1355-1-R. *Hasty Protective Row Minefield Record (LRA)* 00 October 1997.

DA Form 1594. *Daily Staff Journal or Duty Officer's Log.* 1 November 1962.

DA Form 1711-R. *Engineer Reconnaissance Report (LRA).* 00 May 1985.

DA Form 1911. *Altitude and Azimuth (SIN-COS)* 00 February 1957.

DA Form 1971-R. *Radiological Data Sheet-Monitoring or Point Technique (LRA).* 00 September 1994.

DA Form 2028. *Recommended Changes to Publications and Blank Forms.* 1 February 1974.

DA Form 2404. *Equipment Inspection and Maintenance Worksheet.* 1 April 1979.

DA Form 2823. *Sworn Statement.* 1 December 1998.

DA Form 3881. *Rights Warning Procedures / Waiver Certificate.* 00 November 1989.

DA Form 3975. *Military Police Report.* 00 December 1998.

DA Form 4137. *Evidence/Property Custody Document.* 01 July 1976.

DA Form 5367-R. *Personnel Status Report (LRA).* 00 December 1984.

DA Form 5517-R. *Standard Range Cards (LRA).* 00 February 1986.

DA Pamphlet 190-12. *Military Working Dog Program.* 30 September 1993.

DA Pamphlet 190-51. *Risk Analysis for Army Property.* 30 September 1993.

DD Form 1265. *Request for Convoy Clearance.* 1 January 1959.

DD Form 1408. *Armed Forces Traffic Ticket (Book, Consisting of 25 Three-Part Sets).* 00 December 1987.

DD Form 1911. *Material Courier Receipt.* 00 May 1982.

DD Form 2707. *Confinement Order (Available on DOD Website).* 00 November 1999.

DD Form 2708. *Receipt of Inmate or Detained Person* 00 November 1999.

DD Form 2745. *Enemy Prisoner of War (EPW) Capture Tag.* 00 May 1996.

DFAS-IN 37-1. *Finance and Accounting Policy Implementation.* January 2000.

*Geneva Convention Relative to the Treatment of Prisoners of War.* August 1949.

FM 100-14. *Risk Management.* 23 April 1998.

FM 101-5. *Staff Organization and Operations.* 31 May 1997.

FM 101-5-1. *Operational Terms and Graphics*

*{MCRP 5-2A}*. 30 September 1997.

FM 101-5-2. *US Army Report and Message Formats.* 29 June 1999.

FM 17-95. *Cavalry Operations.* 24 December 1996.

FM 17-98. *Scout Platoon.* 10 April 1999.

FM 19-10. *The Military Police Law and Order Operations.* 30 September 1987.

FM 19-15. *Civil Disturbances.* 25 November 1985.

FM 19-20. *Law Enforcement Investigations.* 25 November 1985.

FM 19-25. *Military Police Traffic Operations.* 30 September 1977.

FM 20-32. *Mine/Countermine Operations.* 29 May 1998.

FM 2-11. *First Aid for Soldiers.* 27 October 1988.

FM 21-26. *Map Reading and Land Navigation.* 7 May 1993.

FM 21-60. *Visual Signs.* 30 September 1987.

FM 21-75. *Combat Skills of the Soldier.* 3 August 1984.

FM 23-25. *Light Antiarmor Weapons.* 17 August 1994.

FM 25-100. *Training the Force.* 15 November 1988.

FM 25-101. *Battle Focused Training.* 30 September 1990.

FM 27-10. *The Law of Land Warfare.* 18 July 1956.

FM 3-19.1. *Military Police Operations.* 22 March 2001.

FM 3-19.30. *Physical Security.* 8 January 2001.

FM 3-19.40. *Military Police Interment / Resettlement Operations.* 1 August 2001.

FM 3-3. *Chemical and Biological Contamination Avoidance {FMFM 11-17}.* 16 November 1992.

FM 3-4. *NBC Protection {FMFM 11-9}.* 29 May 1992.

FM 3-5. *NBC Decontamination {MCWP 3-37.3}.* 28 July 2000.

FM 3-6. *Field Behavior of NBC Agents (Including Smoke and Incendiaries) {AFM 105-7; FMFM 7-11-11}.* 3 November 1986.

FM 3-34.2. *Combined-Arms Breaching Operations.* 31 August 00.

FM 34-2. *Collection Management and Synchronization Planning.* 8 March 1994.

FM 34-130. *Intelligence Preparation of the Battlefield.* 8 July 1994.

FM 41-10. *Civil Affairs Operations.* 14 February 2000.

FM 44-8. *Combined Arms for the Air Defense.* 1 June 1999.

FM 5-170. *Engineer Reconnaissance.* 5 May 1998.

FM 55-10. *Movement Control.* 9 February 1999.

FM 55-20. *Rail Transport in a Theater of Operations.* 1 June 2000.

FM 6-20. *Fire Support in the Airland Battle.* 17 May 1988.

FM 6-30. *Tactics, Techniques, and Procedures for Observed Fire.* 16 July 1991.

FM 7-8. *Infantry Rifle Platoon and Squad.* 22 April 1992.

FM 7-10. *The Infantry Rifle Company.* 14 December 1990.

FM 90-10-1. *An Infantryman's Guide to Combat in Built-Up Areas.* 12 May 1993.

FM 90-40. *Multiservice Procedures for the Tactical Employment of Nonlethal Weapons (NLP) (S&I, CDR, USATSC, ATTN: ATIC-TMSD-T (Bldg {MCRP 3-15.8; NWP 3-07-31; USCG Pub 3-07.31}*. 6 October 1998.

*Geneva Convention Relative to the Treatment of Prisoners of War.* August 1949.

GTA 19-7-1. *Enemy Prisoner of War (EPW) Basic Commands.* 1989.

*Section 660 of the Foreign Assistance Act.*

STANAG 1059. *Distinguishing Letters for Geographical Entities for Use in NATO.* 17 April 2000.

STANAG 2010. *Military Load Classification Markings.* 18 November 1980.

TC 25-10. *A Leader's Guide to Lane Training.* 26 August 1996.

TM 11-5860-202-10. *Operator's Manual for Mini Eyesafe Laser Infrared Observation Set (MELIOS) AN/PVS-6 (NSN 5860-01-35-8551) (EIC: N/A).* 1 September 1995.

TM 11-6665-251-10. *Operator's Manual for Radiac Set AN/VDR-2 (NSN 6665-01-222-1425).* 1 March 1988.

TM 3-6665-312-12&P. *Operator's and Organizational Maintenance Manual Including Repair Parts and Special Tools List for M8A1 Automatic Chemical Agent Alarm (NSN 6665-01-105-5623) and*

*Auxiliary Equipment M10 Power Supply (6665-00-859-2225); M10A1 Power Supply (6665-00-093-2739); M228 High Profile Mounting Kit (6665-00-859-2212) and M182 Low Profile Mounting Kit (6665-00-110-9492) {TO 11H2-17-1}*. 17 March 1985.

TM 5-6665-298-10. *Operator's Manual for AN/PSS-12 Mine Detecting Set {TM 5-6665-298-10}*. 28 February 1995.

# Index

Field Manual
No. 3-19.17

Headquarters
Department of the Army
Washington, DC, 6 July 2005

# Military Working Dogs

## Contents



**Distribution Restriction:** Approved for public release; distribution is unlimited.

# Figures



# Tables

# Preface

The last Army dog field manual (FM) was published in 1977. It reflected military working dog (MWD) doctrine developed during the Vietnam era. Although useful at that time, much of the information has since become obsolete. Today, MWD teams are employed in dynamic ways never before imagined.

Today's MWD team is a highly deployable capability that commanders have used around the world from Afghanistan to Africa and from the Balkans to Iraq. These specialized teams aid commanders in stability and support operations as well as in warfighting. Being modular and mobile makes these teams very agile. As situations dictate, MWD teams are quick to arrive and able to conduct various operations. Their versatility allows for effective transformation at all echelons among readiness for deployment and operations on the ground, through redeployment and back to readiness.

The highly aggressive dog tactics of the 1960s and 1970s are long gone. Today's MWD program effectively employs expertly trained and motivated handlers coupled with highly intelligent breeds of dogs. These teams are continuously rotating between their assigned duties and deployments worldwide to perform joint operations, multiechelon tasks, and interagency missions.

This FM addresses the current capabilities of the Military Police Working Dog Program as well as the potential for future applications. As technology and world situations change, the MWD team will continue the transformation process and give commanders the full-spectrum capabilities needed to be combat multipliers on the battlefield as well as persuasive force protection and antiterrorism assets.

*Appendix A* complies with current Army directives, which state that the metric system will be incorporated into all new publications.

This publication applies to the Active Army, the Army National Guard (ARNG)/the Army National Guard of the United States (ARNGUS), and the United States Army Reserve (USAR).

The proponent of this publication is United States Army Training and Doctrine Command (TRADOC). Send comments and recommendations on *Department of the Army (DA) Form 2028 (Recommended Changes to Publications and Blank Forms)* directly to Commandant, United States Army Military Police School, ATTN: ATSJ-DD, 401 MANSCEN Loop, Suite 2060, Fort Leonard Wood, Missouri 65473-8926.

Unless otherwise stated, masculine pronouns do not refer exclusively to men.

***Note***: For the purposes of this manual, a dog team refers to one handler and one MWD unless specified otherwise.

# Chapter 1

# The Military Working Dog Program

Dogs have been used for the protection of life and property since ancient times. From these beginnings, dog training and employment has been continuously refined to produce a highly sophisticated and versatile extension of the soldier's own senses. Even the most complex machines remain unable to duplicate the operational effectiveness of a properly trained MWD team.

## THE ROLE OF THE MILITARY WORKING DOG

1-1. MWDs provide a valuable asset to military police, infantry, special forces, the Department of Defense (DOD), and other government agencies. The MWD's senses of sight, smell, and hearing enhance his detection capabilities and provide commanders with a physical and psychological deterrent to criminal activity. Properly trained MWDs can prevent an intruder or suspect from escaping. When necessary, the MWD provides an added dimension of physical force as an alternative to the use of deadly force. Public knowledge of MWD team capabilities provides military police and various security forces with a formidable deterrent wherever the MWD team is employed.

1-2. MWDs are key resources for use in the military police combat support role. MWDs are trained for scouting, patrolling, and performing building and area searches. Commanders can use *Table 1-1* to determine what MWD type best suits their needs. Some MWDs have also been trained to track although this is not a required area of expertise. All of these skills contribute to the successful completion of the five military police functions across the full spectrum of military operations.

**Table 1-1. MWD Mission-to-Type Matrix**

| *Mission* | *MWD Type* | | |
|---|---|---|---|
| | **PEDD** | **PNDD** | **PD** |
| **A. Area Security Operations** | | | |
| (1) Area and reconnaissance | X | X | X |
| (2) Screening and surveillance | X | X | X |
| (3) Base/air-base defense | X | X | X |
| (4) Cordon and search | X | | |
| (5) Checkpoints | X | X | X |
| (6) Roadblocks | X | X | X |
| (7) Response force | X | X | X |
| (8) Critical site, asset, HRP security | X[1] | X | X |
| **B. Maneuver and Mobility Support Operations** | | | |
| (1) Maneuver support | X | X | X |
| (2) Mobility support | X | X | X |
| **C. I/R Operations** | | | |
| (1) EPW and CI | X[2] | X[2] | X[2] |
| (2) Evacuation | X | X | X |
| **D. Law and Order Operations** | | | |
| (1) Force protection | X | X | X |
| (2) Military police investigations | | X | |

**Table 1-1. MWD Mission-to-Type Matrix (Continued)**

| *Mission* | *MWD Type* | | |
|---|---|---|---|
| | **PEDD** | **PNDD** | **PD** |
| (3) Customs support | X | X | |
| (4) Redeployment operations | X | X | |
| (5) Suspicious/unattended packages | X[3] | | |
| (6) Health and welfare inspections | X | X | |
| (7) Crowd control | X[4] | X[4] | X[4] |
| (8) Alarm responses | X | X | X |
| (9) Bomb threats | X | | |
| (10) Public MWD demonstrations | X | X | X |
| **E. PIO** | X | X | X |

1 PEDDs will not be used to provide security in ASPs, as the ammunition and explosives contained at the ASP will detract from the MWD's detection ability and may distract the MWD from his patrol function.

2 MWDs will not be used in a correction/detention facility to ensure custody of prisoners.

3 PEDDs can be used to search the area around a suspicious/unattended package for secondary devices. At no time will a PEDD or handler be used to search the package itself.

4 Direct confrontation with demonstrators is not recommended, but is authorized with the commander's approval when lesser means of force have been unsuccessful.

1-3. MWDs are a unique item; they are the only living item in the Army supply system. Like other highly specialized equipment, MWDs complement and enhance the capabilities of the military police. MWD teams enable the military police to perform its mission more effectively and with significant savings of manpower, time, and money.

# HISTORICAL OVERVIEW OF THE MILITARY WORKING DOG

1-4. Dogs have served in active service at the sides of their handlers for decades. They have been heroes, showing bravery under fire, saving lives (often losing their own), and bringing comfort to the injured and infirmed. The first recorded American use of military dogs was during the Seminole War of 1835 and again in 1842. In Florida and Louisiana, the Army used Cuban bred bloodhounds for tracking. During the US Civil War, dogs were used as messengers, guards, and unit mascots.

1-5. The Army Quartermaster Corps began the US Armed Forces' first war dog training during World War II. By 1945, they had trained almost 10,000 war dogs for the Army, Navy, Marine Corps, and Coast Guard. Fifteen war dog platoons served overseas in World War II. Seven platoons saw service in Europe and eight in the Pacific.

1-6. MWDs were trained at Fort Carson, Colorado, organized into scout dog platoons, and used in the Korean conflict for sentry duty and support of combat patrols. In 1957, MWD training moved to Lackland Air Force base (LAFB), Texas, with the Air Force managing the program.

1-7. Throughout the Vietnam Conflict, the Military Police Corps used dogs with considerable success. Most of these were sentry dogs used to safeguard critical installations such as ports and airfields. A new dimension in canine utilization was realized when marijuana detector dog teams were trained and deployed to assist military police in suppressing illicit drug traffic. Sentry and marijuana detector dog teams were then deployed worldwide in support of military police. An important outgrowth of the conflict was the development of canine research and development efforts. These ongoing efforts were able to initiate the first steps toward developing a more intelligent and stronger military dog, training dogs to detect specific drugs and explosives, developing multiple-purpose dogs, and employing tactical dogs by electronic remote control.

1-8. In the 1990s and early 2000s, MWDs were deployed around the globe in military operations such as Just Cause, Desert Shield and Desert Storm, Uphold Democracy, and Enduring Freedom and Iraqi Freedom. These teams were effectively utilized to enhance the security of critical facilities and areas, as well as bolster force protection and antiterrorism missions, allowing commanders to use military police soldiers and other assets more effectively elsewhere. MWD operations in Haiti showed the first large-scale utilization of MWDs since Vietnam. In the Balkans, military police deployed with MWDs to provide force protection and antiterrorism support and area security throughout the US base camps. MWDs trained as explosive detector dogs conducted vehicle sweeps and building searches to ensure the safety of US soldiers. US Army MWD teams operated beside Iraqi police officials to help in law and order missions and restore the war torn country into a peaceful region.

1-9. MWDs are force multipliers. Installation commanders should include MWDs when planning for force protection and antiterrorism countermeasures. They expand the individual soldier's effectiveness in the face of those that would come against him. They provide an immense physical and psychological effect when used as a show of force in day-to-day operations. The various uses of MWDs have been effectively employed in many aspects of military police missions. MWDs are utilized effectively at gates, camps and bases, and checkpoints and for random searches for narcotics and explosive devices. MWDs are also utilized for other missions in support of combat, combat support, and combat service support units.

1-10. MWD teams are assigned to perform high-risk personnel (HRP) security missions. MWD teams perform narcotic detection missions on the Mexico and US border to assist US Customs. They have been deployed on the United States and Canada border to perform law and order operations. The canine can easily detect what his handler cannot see, hear, or smell. The MWD's controllable aggressive behavior, coupled with the physical and psychological effects he creates, makes him and his handler an impressive and unmatchable team.

# ORGANIZATION AND STRUCTURE OF MILITARY WORKING DOG UNITS

1-11. The US Army MWD program is an organization with structure and responsibilities that reach from the DOD level down to the handler level. For information concerning organization and responsibilities above the local level, see *Army Regulation (AR) 190-12*.

## KENNEL MASTER TEAM

1-12. The kennel master (KM) team is comprised of two noncommissioned officers (NCOs) (one sergeant first class [SFC], who is the KM and one staff sergeant [SSG], who is the plans NCO). The KM team is modular and provides technical supervision of MWD teams assigned to the military police brigade. This team may be assigned to the corps military police brigade, echelon above corps (EAC) military police brigade, or the internment and resettlement (I/R) brigade. In addition to technical supervision, this team plans and coordinates MWD operations and support requirements. The KM team is also responsible for MWD team proficiency training. This team supervises the storage and accountability of the narcotic and explosive training aids as well as all assigned equipment.

## PATROL EXPLOSIVE DETECTOR DOG TEAM

1-13. The patrol explosive detector dog (PEDD) team is comprised of three NCO MWD handlers (one SSG, and two sergeants [SGTs]) and three PEDDs. This team detects explosives and contraband in support of combat, combat support, and combat service support units and the five military police functions. It provides MWD patrol support (including I/R operations) when not employed with explosive detection. This team is normally assigned to the corps military police brigade, EAC military police brigade, military police I/R brigade, or the I/R command and control (C2) detachment. This entire modular team can be deployed together as a unit with all assigned equipment or, based on the specific mission, a handler and an assigned MWD may be deployed apart from the team.

## PATROL NARCOTIC DETECTOR DOG TEAM

1-14. The patrol narcotic detector dog (PNDD) team is comprised of three NCO MWD handlers (one SGT and two specialists [SPC]) and three PNDDs. This team detects narcotics and other drugs in support of combat, combat support, and combat service support units and the five military police functions. It provides MWD patrol support (including I/R operations) when not employed with narcotics detection. This team is normally assigned to the corps military police brigade, EAC military police brigade, military police I/R brigade, or the I/R C2 detachment. This entire modular team can be deployed together as a unit with all assigned equipment or, based on the specific mission, a handler and an assigned MWD can be deployed apart from the team.

## INDIVIDUAL DUTIES AND RESPONSIBILITIES

1-15. There are five levels of responsibility in the MWD program. The terminal level is the program manager and the beginning level, the handler. This is not to be confused with the operational chain of command or standards of grade. These individuals have specific duties in managing and performing the MWD program.

### PROGRAM MANAGER

1-16. The program manager executes major Army command (MACOM) level responsibilities for the MWD program. He is the equivalent of an operations NCO and is an appointed certification authority whose duties include—

- Conducting MWD team patrol and detector certifications.
- Requisitioning personnel and MWDs.
- Providing policy and guidelines for the MACOM.
- Forecasting annual operational budgets.
- Coordinating MWD support for the office of the Secretary of Defense, US Secret Service (USSS), US Customs Service (USCS), and Department of State missions.
- Serving as the MACOM provost marshal's (PM's) advisor for the employment of MWDs in contingency operations, force protection and antiterrorism support, and homeland defense.

### KENNEL MASTER

1-17. The KM is the senior NCO in charge of the installation's MWD program, consisting of 6 to 30 handlers and respective MWDs. KM duties are the equivalent of a platoon sergeant's duties when 9 or more handlers and MWDs are assigned. When 8 or fewer handlers and MWDs are assigned, the duties are equal to those of a squad leader. In addition to traditional leadership responsibilities, KM duties include—

- Monitoring all training to ensure that MWD teams are ready for annual certifications and worldwide deployments in support of various operational missions.
- Advising the commander on the employment of MWD teams.
- Ensuring that team proficiency is maintained.
- Preparing for MACOM and veterinary inspections.
- Forecasting annual operational budgets.
- Assisting the MACOM program manager in requisitioning personnel, MWDs, and equipment.
- Managing the daily maintenance and upkeep of kennel facilities.
- Monitoring MWDs' health and welfare.
- Managing procurements and accounts for explosive and narcotic training aids.
- Advising installation and unit force protection and antiterrorism officers on the use of MWDs.
- Coordinating MWD support for the USSS, the USCS, DOD protective services units, combat support operations, the Army Recruiting Command, and other civilian agencies.

### Military Working Dog Plans Noncommissioned Officer

1-18. The MWD plans NCO is part of the KM team and performs the duties of the maneuver mobility support operations (MMSO) and training NCOs. The plans NCO is selected for his ability and overall knowledge of training, mentoring, and deploying MWD assets. This NCO is responsible for—

- Developing training plans.
- Coordinating and planning the unit mission-essential task list (METL).
- Supervising the execution and tracking of the deployment plan.
- Tracking the movement of MWD teams while they are deployed.
- Ensuring that deployed soldier's issues are resolved.
- Assisting in the supervision of the installation MWD program.
- Acting as the KM when the KM is absent.
- Preparing short- and long-term training outlines and risk assessments.
- Validating and supervising proficiency training.
- Overwatching preparation for the deployment of MWD teams.

### Senior Military Working Dog Handler

1-19. The senior MWD handler is the senior NCO of the PEDD or PNDD teams. His duties are the equivalent of the squad leader of the PEDD team or the team leader of the PNDD team. In addition to traditional leadership responsibilities, he is responsible for ensuring that—

- Annual certification training is conducted.
- MWD teams are ready for deployments.
- Team proficiency is maintained.
- Handlers are proficient with their assigned MWDs.
- Coordination for MWD team support for the USSS, the USCS, combat support operations, the Army Recruiting Command, and other civilian agencies is completed.

### Military Working Dog Handler

1-20. The handler provides the daily care and grooming for his assigned MWD. He ensures that proficiency skills are maintained in his assigned MWD. The handler and MWD perform either PNDD operations or PEDD operations. The handler and MWD are all trained for patrol but the variant for narcotics detector or explosive detector depends on the MWD team's training. The handler also—

- Fills out training records daily.
- Grooms the assigned MWD daily.
- Trains the assigned MWD daily.
- Maintains MWD kennels daily.
- Performs additional kennels duties.

***Note***: For more information about additional skill identifier (ASI) Z6 standards of grade, refer to *DA Pamphlet (Pam) 611-21*.

## FUNDAMENTALS FOR THE USE OF MILITARY WORKING DOGS

1-21. An MWD can be trained to respond consistently to certain sensory stimuli (sights, sounds, odors, or scents) and alert the handler. Under most circumstances, a properly trained MWD can smell, hear, and visually detect movement better than a person.

1-22. Commanders and supervisors should consider the capabilities of the MWD team when identifying and evaluating possible MWD employment. Not all MWD teams are able or equipped to perform all missions equally (see *Table 1-1,* page 1-1).

## USE OF PATROL DOGS

1-23. All MWD teams are trained to perform patrol missions. One benefit of employment is the deterrence of criminal activity. The obvious presence and well-published activities of MWD teams can successfully deter trespassers, vandals, violent persons, and other would-be criminals. When the desired task is to deter unauthorized or suspicious individuals, an MWD team should be assigned to a location during day or night.

1-24. The patrol dog's (PD's) contribution to the law and order effort is most effective when the team is utilized as a walking patrol. The law enforcement duties that a PD team can perform as a walking patrol include checking or clearing buildings and patrolling parking lots, family housing areas, and troop billet areas. Mobility significantly increases the potential area of coverage. The best and most effective method is the ride awhile, walk awhile method. This concept helps cover large patrol areas, and the periodic exercise helps keep the team alert. All PD teams can be utilized for—

- Antiterrorism operations.
- Force protection and antiterrorism efforts.
- Special-reaction team (SRT) operations.
- Assistance in military police investigations.
- Walking and mobile patrols.
- Alarm responses.
- Building checks and searches.
- The identification and apprehension of individuals.
- The protection of HRPs.
- The protection of funds.
- Open-area searches.
- Civil disturbances.
- Perimeter security.
- External intrusion detection.
- Response forces.
- Access control points (ACPs) support.
- High-risk targets (HRT) support.
- Mission-essential vulnerable areas (MEVA) support.
- I/R operations.

1-25. Many of the tasks above can be performed in combat support or force protection and antiterrorism operations. In addition, PD teams can and should be utilized as a tool for area defense, providing warning and response. PD teams are effective on combat patrols and as listening posts and reconnaissance team members. The employment of PD teams should not be overlooked during I/R or enemy prisoner of war (EPW) operations. PDs are trained to apprehend suspects at or near a crime scene, stop those who may attempt to escape, and protect their handlers from harm. Some PDs may be able to locate suspects who have left the scene of a crime.

1-26. The PD team should be utilized to provide a psychological deterrent and a show of additional force protection and antiterrorism measures. PD teams should be placed on roving patrols at ACPs, vulnerable areas, and the location of HRP. These missions should increase as force protection and antiterrorism levels increase.

## USE OF PATROL NARCOTIC DETECTOR DOGS

1-27. Military police commanders, planners, and supervisors have a very effective means of detecting the possession and/or use of controlled substances with the employment of a PNDD team. PNDDs are trained to recognize the scent of certain illegal substances through a program of practice and reward.

1-28. In addition to PD missions, PNDD dogs can be utilized for the following:

- Criminal Investigation Division (CID) and military police investigator drug investigations.
- USCS inspections.
- Health and welfare inspections.
- Barracks searches.
- Workplace searches.
- Random gate inspections.
- Area searches.
- Vehicle searches.
- Postal inspections.
- I/R operations.
- Aircraft and luggage searches.
- Predeployment and redeployment searches.
- Other US government agency support.

1-29. When the PNDD responds on any of the substances he is trained to detect, his handler takes the appropriate action. PNDD teams perform a valuable service by helping to rid the military community of illegal drugs and the problems associated with drug and controlled substance abuse.

## USE OF PATROL EXPLOSIVE DETECTOR DOGS

1-30. PEDD teams are very useful in many searches or investigations involving explosives. The search for explosives is critically important to ensure that the community is safe and protected from would-be terrorists. PEDD teams are particularly useful if there is a need to locate one or more of these explosive items, which may have been hidden in an area.

1-31. The deterrent value and the detection capabilities of PEDD teams are a very effective countermeasure to terrorism. Public knowledge that PEDD teams are in use is a deterrent to persons who may try to use explosives illegally. The knowledge that explosives can be detected by PEDD teams at entry points or in places where explosives have been hidden can prevent a person from attempting to commit terrorist acts.

1-32. In addition to PD missions, PEDD dogs can be utilized for—

- Bomb threat incidents.
- Suspicious/unattended package incidents.
- ACP searches.
- Checkpoint searches.
- Random gate searches.
- CID and military police investigations section.
- Health and welfare inspections.
- Building and area searches.
- Postal inspections.
- Aircraft and luggage searches.
- Predeployment and redeployment searches.
- Protective service missions.
- Other US government agency support.

1-33. A well–trained PEDD team can conduct a significantly more effective search of an area or facility in a much shorter time than a number of people can. Using PEDD teams helps reduce the potential risk to persons who would otherwise have to do the search without the benefit of the dog's superior sense of smell. More detailed information concerning the employment of PD, PNDD, and PEDD teams is discussed in *Chapters 5, 6,* and *7.*

# Chapter 2

# Administrative Requirements

This chapter contains information that is useful in planning an MWD program or adding MWD teams to an existing program. The guidelines in this section are not meant to be absolutes for utilization, training, or maintenance. They give general guidance and information that should be considered when planning a program. Leaders will find the following information helpful for writing authorization documentation statements and for planning the local MWD program. It is important to remember that all MWDs are trained as PDs in addition to the specific type of detector dog (narcotic or explosive).

## DETERMINATION OF A NEED FOR MILITARY WORKING DOGS

2-1. The installation provost marshal (PM), military police unit commander, or task force commander must take the initiative to establish the local MWD program if it does not already exist. The specific needs must be determined and the costs of the program must be justified. The decision process involves a thorough risk and crime analysis, an accurate evaluation of the requirements of the entire military police mission, and the use of the working dog yardsticks located in this chapter.

2-2. Some of the factors to consider when determining the need for MWD teams include—

- The mission of the team.
- The size of the area of responsibility of the team.
- The size of the installation or the area population.
- The number of personnel to be served by the team.
- Incident rates for all crimes.
- The present capability and the commitment of a portion of the manpower resources as handlers.
- The types of terrain on the installation or in the probable areas of deployment.
- The types of combat support missions for which MWD teams can be used.
- The number of critical facilities or areas.
- The capabilities of each type of team.
- The number of HRP on the installation or in the area of responsibility.

## VERIFICATION OF AUTHORIZATIONS

2-3. Initial MWD authorizations should be probationary using the following criteria:

- Adjustments must be made based on actual use data.
- A baseline ratio of 24 hours of MWD utilization to 8 hours training per week (4 hours patrol and 4 hours detection) must be maintained.
  - Utilization includes all missions (force protection/law enforcement, security, and combat support), whether patrolling, detecting drugs or explosives, or performing other functions (in combination), when a handler and dog are being employed together as a team.
  - Training includes all training activities to maintain, improve, regain, or develop dog skills such as obedience, controlled aggression, patrolling, scouting, detection of persons, and detection of drugs or explosives.
- An automatic review must be made by the respective MACOM after two years.

2-4. There is a trade-off when adjusting between the training and utilization necessary to reinforce basic skills or add new skills. For example, four hours per week may be adequate to maintain detector dog proficiency or to train in new skills, but additional training may be justified at times and training hours may be increased accordingly. This increase in training time should be accompanied by an equivalent reduction in utilization time. However, training hours should not be increased merely as a substitute for lack of utilization.

# MISSION REQUIREMENTS

2-5. The mission requirements for the following support missions should be considered. These requirements should be used only as guidelines.

## INSTALLATION FORCE PROTECTION AND LAW ENFORCEMENT SUPPORT

2-6. In order to provide continuous 24-hour, year-round MWD protection of the installation, nine MWDs are needed. Six of these dogs will be PEDD and three will be PNDD. This requirement is based on each MWD team being used a minimum of 24 hours per week. This does not include hours needed for training, grooming, or health care requirements. This does allow for temporary duty (TDY), handler leave, and illness.

2-7. The goal is to have one PEDD per installation available at all times for bomb threats or explosive type missions (not counted as part of the working force protection requirement). Installations with a three- or four-star headquarters require one additional PEDD authorization. This is due to the increased workload, facility criticality, facility vulnerability, location, past history of terrorist target selections, and present threat assessments.

## HIGH-RISK PERSONNEL SECURITY MISSIONS

2-8. DOD tasks the Army to provide PEDD support for presidential, vice presidential, Secretary of State, and Secretary of Defense visits throughout the world. Currently, there is not a published formula to predetermine mission frequencies or requirements for these missions. On average, 40 teams inside the MACOM support the various missions above. These figures do not include travel or mission preparation time.

## CUSTOMS SUPPORT

2-9. While no formula exists to quantify requirements for military customs support, this remains an ongoing mission that must be viewed as a requirement for additional MWD authorizations. MWDs are used to screen household goods shipments, privately owned vehicles (POVs), container express (CONEX) containers, military equipment, US mail, and port operations. Workload varies by installation population, location, unit missions and deployments, and the time of year.

## CONTINGENCY OPERATIONS SUPPORT

2-10. Currently, approximately 50 MWD teams are supporting various contingency operations. This requirement will continue until mission completion. This figure represents 25 percent of Army MWD authorizations. Support for contingency operations has continued to rise since 1990 and will continue to do so. This increase must be factored into planning for MWD authorizations. A conservative average of 100 MWDs is needed for contingency operations on a yearly basis. This number is to ensure that MWD teams do not perform back-to-back deployments and that they receive proper proficiency training according to *AR 190-12* (4 hours per week in each specialty and 95 percent accuracy for PEDDs or 90 percent accuracy for PNDDs).

## KENNEL CONSTRUCTION

2-11. When new MWD programs are started at locations where there are no facilities in place, kennel construction must be complete before the dogs arrive. The MWDs should be ordered through the military standard transportation and issue procedures (MILSTRIP) process approximately one year before the expected completion date of the kennel facility. In operations where MWD teams will be deployed for long periods of time and will be rotated, a permanent facility must be constructed within one year. Deployable field kennels may be used in the interim. The joint service MWD committee states that the DOD's design guide for MWD facilities must be referenced in planning the construction or renovation of a kennel facility.

2-12. Once the need and the justification for an MWD program are established, the PM, military police commander, or task force commander initiates a request to the installation facilities engineer to develop the kennel facility design. Before the kennel design has been finalized, the activity or installation veterinary corps officer (VCO), the physical security inspector, safety personnel, and the KM must be consulted to ensure that the minimum requirements are included in the final design. After kennel designs are completed and construction costs are estimated, the kennel project is submitted to the installation commander for approval and funding. The request for authorization of MWD teams must include the statement that the commitment has been made to fund and build kennels within one year prior to the arrival of the MWDs at the installation. Refer to *Appendix B* and *DA Pam 190-12* for details on kennel construction.

## PROCUREMENT

2-13. All MWDs must be procured from the procurement section at the 341st Training Squadron, Lackland AFB, Texas. Local procurement of an MWD is prohibited. Requisition of an MWD team (one dog and one handler) requires two separate transactions. Handlers are obtained as a result of a personnel action requesting military police soldiers with the ASI Z6 in the required pay grades. MWDs are obtained as a result of a supply action. Commanders will assign an appropriately qualified handler to every MWD on hand. Assignment will be consistent with the policy of "one dog and one handler" as outlined in *AR 190-12*.

2-14. When the replacement of an MWD and not a handler is needed, requisitions are done according to *AR 700-81*. Initial requisitions for dogs should be submitted when positions for handlers are authorized. The MWDs in the Army supply system are—

- Dog, patrol: narcotic detector; national stock number (NSN) 8820-00-435-7542, line item number (LIN) D33800.
- Dog, patrol: explosives detector; NSN 8820-00-188-3880, LIN G33732.

2-15. The method of how to replace an MWD that has died, been adopted, or has undergone euthanasia is to submit a requisition in the MILSTRIP format that includes a statement that the unit or installation already has a trained handler on station. The MILSTRIP should also be submitted when the VCO has stated in writing that an MWD only has a 12-month life or work expectancy and/or is placed in deployment category IV (permanently nondeployable).

2-16. KMs should send requisitions for new or replacement dogs, in the MILSTRIP format, to their local property book office and forward a copy to their respective MACOM. The headquarters of the Air Force Security Police Agency will review all requisitions for dogs prior to processing by the $341^{st}$ Training Squadron, Procurement Section.

## TRAINING REQUIREMENTS

2-17. Training is needed at each installation to ensure that teams remain proficient. This realistic and challenging training should be scenario driven and relate to all of the possible operations in which MWD teams will operate according to the unit METL. Properly trained MWD teams are better prepared to accomplish all facets of MWD missions.

## INITIAL TRAINING

2-18. MWD teams are initially trained at the 341st Training Squadron, Lackland AFB, Texas. This training is done at a basic level for both the handler and the dog and is not intended to prepare either the handler or the dog for immediate duty in an MWD team. The training is only intended to give a basic overview of the MWD program, the health and care of MWDs, and proper training techniques. It is the responsibility of the gaining unit to train a new handler or dog on all missions conducted by Army MWD teams. Proficiency training is continuously required to ensure that teams are reliable for use in patrol, narcotic detection, and explosive detection.

## PROFICIENCY TRAINING

2-19. The effectiveness of MWDs trained in patrol, narcotic detection, or explosive detection depends on the continual reinforcement of the abilities of the dog team through proficiency training. This training is mandatory and if it is not conducted properly, an appointed certification authority can decertify the MWD team. Failure to meet the minimum training standard for at least the 35 days prior to a certification action will result in MWD teams not being evaluated for certification.

2-20. The minimum training standard is at least four hours of training each week in patrol and at least four hours of training each week in narcotic or explosive detection. The minimum standard of proficiency to maintain certification as a detector team is 90 percent or higher (with no more than a 10 percent false response rate) for narcotic detector dogs, and 95 percent or higher (with no more than a 10 percent false response rate) for explosive detector dogs.

2-21. Failure to maintain an average that meets or exceeds the minimum standard for three consecutive training events will result in the automatic decertification of the MWD team. Decertified teams may not be used in any way, and must undergo intensive retraining. These teams may be recertified only after retraining and a consistent demonstration to an appointed certification authority. Proficiency training should be conducted both on and off leash.

## PROFICIENCY TRAINING OF PATROL DOG TEAMS

2-22. The effectiveness of PD teams depends on the continual reinforcement of mandatory proficiency training. PD teams must receive a minimum of 4 hours of patrol proficiency training each week. The training environment should approximate the military police working environment as closely as possible. Training should occur at varying times throughout the day and night and on varying days of the week, including weekends.

2-23. The emphasis in training is to develop the skills of both the handler and the dog so that they complement each other and the team becomes a working unit. As a proficiency evaluation criterion, the correct performance of the required task verifies that the handler understands how to control his PD and is able to do so.

2-24. Train the PD to protect with or without command based on any perceived threat to the handler. Because of the potential of serious bodily injury in this specialty, the certification of the handler and the dog as a team is very stringent. Ensure that the proficiency training of PDs includes an emphasis on critical tasks. There are no waivers for failure to accomplish a critical task by a PD team.

## WEAPON FIRE TRAINING

2-25. Training on weapon firing ranges is essential for the PD to become proficient and not be deterred from attacking agitators during gunfire. The dog must not attack the handler during gunfire. The firing of all weapons assigned to the handler should be done with the MWD present whenever possible. PDs can be desensitized with the firing of many different types of weapons. This can often be accomplished by arranging for the handlers to take the dogs to weapon ranges of different units.

2-26. Determine the dog's reaction to the sound of gunfire. It may be necessary for the handler to use counterconditioning techniques until the desired proficiency is achieved. Counterconditioning techniques include starting at distances of 300 meters and slowly bringing the gunfire closer to the dog (or as safety allows, bringing the dog closer to the weapon). The goal is for the dog not to bark or show any signs of aggression when the handler fires all assigned weapons. This can be a slow process over the course of several days.

## OBEDIENCE COURSE TRAINING

2-27. The obedience course exposes the PD to various obstacles (see *Figure 2-1*) that simulate walls, open windows, tunnels, ramps, or steps. The dog's exposure to these obstacles reduces the amount of time required to adapt dogs to different environments. The dog learns to negotiate each of the obstacles. When confronted with a similar obstacle in the working environment, the dog is not deterred from completing his mission. The obedience course also develops the handler's ability to control the dog's behavior both on and off leash. The obedience course is not a substitute for exercise. A dog should never be required to negotiate the obedience course until he has been warmed up by proper exercise.

**Figure 2-1. Obedience Course Obstacles**

## PROFICIENCY TRAINING OF DETECTOR DOG TEAMS

2-28. Continuous training of detector dogs is required so they maintain proficiency in the ability to seek and find drugs or explosives, depending on their specialty. Ensure that the detector training is always documented properly. Documentation and training are critical to both maintaining team certification and to establishing probable cause for a search based on a detector dog's number of correct responses.

2-29. There are several differences in the training requirements for PNDD and PEDD. Proficiency training scenarios for both types of detector dogs must be varied to avoid conditioning the dogs to a repetitious training pattern. The following factors will be varied for each proficiency training exercise:

- The general training exercise area.
- The number of training aids planted.
- The specific locations of the training aids in the area being used for training.
- The amount of explosives or drugs used in the training aid plant.
- The type of explosives or drugs used.
- The time of day or night of the training.
- The type of training aid container.
- The type of distraction material planted with the training aid.
- The length of time the training aid is left in place before the detector dog team search.
- The person used to handle and/or plant the training aid.
- The height above or below the floor level of the training aid plants.
- The size of the room or area in which the training aids are planted.

## DECOYS

2-30. The decoy is the person who role-plays the primary adversary for training and evaluating the MWD team. The decoy may be a suspect, a subject, an attacker, an agitator, a drunk, an escapee, an enemy, or any of a number of other persons an MWD team may expect to encounter while performing military police duties. A decoy may also be neutral or an ally, such as another military policeman, a supervisor, a lost juvenile, or an innocent person passing through an identification check. Use both males and females for decoys, agitators, and suspect role-players. Vary the clothing the role-players are dressed in to expose the dogs to civilian, military, and the ethnic dress of different countries, if possible.

2-31. Decoys must be taught how to handle the arm protector sleeve or wrap and other aggression tools such as the bite suit (*Figure 2-2*), the hidden wrap, and leg wraps for self–protection when the dog attacks. The sleeve or wrap used should be a hidden sleeve rather than the full, heavy padded sleeve. The hidden sleeve can be concealed under a sweater or field jacket to achieve more realistic conditions for training the MWD.

**Figure 2-2. Bite Suit**

2-32. KMs and plans NCOs teach decoys how to understand and use the different drives the MWD shows. Once understood, these drives or needs of the MWD can be exploited to cause the desired result in training. Decoys must learn the type of pattern to take to and from the MWD when training on aggression tasks or the bite. For safety and optimum training effect, decoys will learn how to target the MWD into the bite gear and what actions to take during a partial and full bite.

# VALIDATION

2-33. The KM or plans NCO will conduct a validation prior to requesting certification. Validations will then be conducted quarterly to verify the patrol and detection accuracy rates recorded on *DA Form 2807-R (Military Working Dog Training and Utilization Record)* and/or *DA Form 3992–(R) (Narcotics or Explosives Detector Dog Training and Utilization Record)*. Additional validations will be given to any patrol or detector MWD whose demonstrated proficiency appears to differ significantly from the rating recorded on his *DA Form 2807-R* and/or *DA Form 3992–R*. The validation will verify proficiency or identify the specific patrol/detector weaknesses that need corrective training.

2-34. Validations include at least 30 trials for a narcotic detector dog and at least 40 trials for an explosive detector dog. These trials will be divided equally with each substance the dog is trained to detect. Training aids will be distributed in at least five different areas over a five–day period. These areas should include (if available) vehicles, barracks, open areas, warehouses, theaters, planes, and luggage. A written record of the evaluation and validation will be maintained on record for a minimum of two years.

# CERTIFICATION OF MILITARY WORKING DOG TEAMS

2-35. *AR 190-12* defines the standards and specific requirements for all certification of MWD teams. Certification of an MWD and handler is valid for one year and is immediately nullified when—

- The dog is assigned to another handler.
- The handler and MWD have been separated or have not conducted training for thirty-five or more consecutive days.
- The team fails to meet the minimum detector standard of 95 percent accuracy (explosives) and 90 percent accuracy (narcotics) with no more than a 10 percent false response rate for three consecutive months.

## SCOUTING OR PATROLLING

2-36. The PD must be able to detect a person by scent, sound, and sight (on leash only). This task is rated critical. The following tasks must be accomplished to demonstrate proficiency in scouting or patrolling:

- Daytime standards.
  - Detect and respond to the scent of a person who is hidden within 75 yards, and follow the scent to the person's location.
  - Detect and respond to a sound made by a person 100 yards away, and follow this sound to the source.
  - Visually detect and respond to a person in view 100 yards away, and pursue the person on command.
  - Demonstrate proficiency in pursuit, attack, and release.
- Nighttime standards:
  - Detect and respond to the scent of a person who is hidden within 75 yards, and follow the scent to the person's location.
  - Detect and respond to a sound made by a person 100 feet away, and follow this sound to the source.
  - Demonstrate proficiency in pursuit, attack, and release.

***Note.*** There is no nighttime standard for sight due to the dog's natural ability to revert to other senses in the dark.

## VEHICLE PATROL

2-37. The PD must be able to ride quietly and calmly inside a vehicle with the handler until provoked or commanded. This is a semicritical task.

## TRACKING

2-38. Tracking is the only noncritical task on which PDs are trained. Tracking is noncritical because not all PDs have tracking potential. The local KM should know which dogs have been trained to track. PDs that demonstrate a potential for tracking are identified by the KM or plans NCO at each unit, and the dogs are then further developed in this task. Although tracking is a valuable skill for the PD team, improper performance of the task does not degrade the overall effectiveness of the PD team to perform the military police mission.

2-39. PDs selected for tracking training should be trained and employed for tracking on a tracking harness. The harness serves as a useful cue to the PD that it should focus his senses on tracking. At the start of the track, the handler will give the command "Track" and begin casting for the scent. Initially, scent pads are identified and then an item with the tracklayer's scent on it will be used. The dog is required to track and alert on any items left by the tracklayer.

2-40. The minimum level of proficiency a PD trained for tracking must attain is to be able to follow a human scent at least one hour old, for 1 mile over varied terrain, and on a course with several turns. Given suitable tracking conditions, a skilled tracker dog team can follow the natural wanderings of individuals or groups of persons for at least 3.1 miles over rugged and varied terrain and on a scent track that is at least 12 hours old. While tracking, the dog is also capable of alerting his handler to the presence of tripwires and ambushes.

2-41. Evaluation of tracking tasks must include various ages and lengths of the track. The areas used to conduct tracking training and evaluations will be rotated frequently. Tracks will be at least one-third of a mile long but no longer than one mile. The age should vary from ten minutes to one hour. Tracks will include turns, but evaluators should ensure that the person laying the tracks is not trying to deceive the dog with undue tracklaying. Articles will be placed on the track for the dog to find.

# INSPECTIONS

2-42. Inspections help ensure that the MWD section is in compliance with Army standards. *Appendix C* contains an example inspection memorandum to assist the inspection officials on what areas to inspect at an MWD section.

## DEPARTMENT OF THE ARMY INSPECTIONS

2-43. Personnel from headquarters, Department of the Army (HQDA), conduct annual inspections of the MACOM MWD program. These inspections are used to assist the MACOM PM and program manager in moving the MACOM MWD program forward. This inspection will be on all aspects of the MACOM MWD program (see *Appendix C*).

## MAJOR ARMY COMMAND INSPECTIONS

2-44. MACOM program managers conduct annual inspections of handlers, MWDs, training, team use, team proficiency, equipment, and kennel facilities. This ensures compliance with *AR 190-12*.

## PROVOST MARSHAL AND COMMANDER INSPECTIONS

2-45. PMs and/or commanders conduct monthly inspections of handlers, MWDs, training, team use, team proficiency, equipment, and kennel facilities. These inspections are conducted at least monthly on one or more areas of the kennels and quarterly on all aspects of the installation MWD section to ensure compliance with *AR 190-12*.

2-46. A written record of the monthly and quarterly inspection is maintained on file with the appropriate corrective action (both in memorandum format), stating the corrective action taken for each deficiency. This written record is maintained for one year past the inspection date.

2-47. The responsible VCO will conduct an inspection of the kennel facility at least quarterly to ensure health and welfare for both the handlers and dogs assigned to the MWD section. A copy of this inspection is maintained on file for a minimum of two years. The responsible PM and/or commander takes all actions necessary to correct any deficiencies noted by the VCO. Veterinary inspections are to be conducted monthly, if possible.

# TRAINING RECORDS

2-48. There are two forms that must be used to annotate MWD training, *DA Forms 2807-R* and *3992-R*. These forms are official military documents and must be maintained daily with factual information. Many times these documents are used as court documents, so the importance of the completion and accuracy of these documents cannot be stressed enough. Example forms and the procedures for filling these forms out are in *Appendix D*.

## PROBABLE CAUSE FOLDERS

2-49. Probable cause folders are one of the most important methods of proving an MWD team's reliability. Legal requirements to conduct a search of private property based on the response of an MWD do not allow the probable cause folder a margin for error. All the items contained in the probable cause folder are copies. The following original records must be kept on file at the kennel facility:

- ***Department of Defense (DD) Form 1834 (Military Working Dog Service Record).*** *DD Form 1834* must be maintained on the MWD and kept as the initial form in the probable cause folder. This form must be accurate and signed on a change of a handler. When the MWD is trained on new substances, ensure that they are listed on the *DD Form 1834*. The 341st Training Squadron at Lackland AFB, Texas, must approve new substances.
- ***Lackland AFB Form 375 (Patrol Dog Certification).*** *Lackland AFB Form 375* follows the *DD Form 1834* in the probable cause folder and lists patrol certification information for the MWD.
- ***Lackland AFB Form 375a (Detector Dog Certification).*** *Lackland AFB Form 375a* follows the *Lackland AFB 375* (if dual certified), and lists postgraduation certification information for the MWD.
- ***Air Force (AF) Form 1256 (Certificate of Training).*** A copy of the handler's *AF Form 1256* follows *Lackland AFB Form 375a*. The handler provides a copy of the *AF Form 1256* for the probably cause folder.
- **Certification Letter**. This letter attests to the fact that the appointed certification authority has witnessed a demonstration of the capabilities of the dog team. This letter, at a minimum, should list the handler and dog, the whelp, the type of MWD (PEDD or PNDD), the social security number (SSN) of the handler, and the date of certification. This letter should be maintained in the probable cause folder after the handler's graduation certificate until the next certification of the team or a change of handler.
- **Quarterly Validation and Evaluation by the KM**. This document is used to show that the KM has conducted a quarterly evaluation on the MWD team. If the KM is not available, the plans NCO can also conduct this evaluation. This document follows the certification letter in the probable cause folder.
- **Training and Utilization Records**. The last items that should be maintained in the probable cause folder are the last three months of training and utilization records for the MWD team. These records include *DA Forms 2807-R* and *3992-R*. Once an additional month has passed, the latest month will be transferred to the MWD team folder in the filing system.

## TRANSPORTATION REQUIREMENTS

2-50. Every unit, installation, and activity with MWD teams should assign enough support vehicles so that all missions and kennel support activities can be properly conducted and maintained. The minimum standard for transportation is one support vehicle per two MWD teams (two dogs and two handlers). Vehicle requirements should be determined by mission requirements and the overall number of MWD teams. Some modifications are necessary so that the MWD can be seen and can provide a visual deterrence during law and order or force protection and antiterrorism missions. These modifications will maximize the mission effectiveness of the dog while providing the dog with a comfortable, supportive, and protective area inside the vehicle. This type of modification is normally a kennel crate or cage secured in the rear of the vehicle; the crate or cage should have a nonslippery surface.

2-51. MWD handlers perform complex missions. Four-by-four type passenger vehicles should be used to ensure that MWD teams can get to any mission during inclement weather or on rough terrain.

2-52. Pickup trucks are essential for the transportation of explosive training aids. *AR 190-12* lists the requirements for explosives transported in government vehicles. However, pickup trucks should not be used for transporting MWDs for training or during road patrol type duties. This vehicle offers limited protection for the MWD from inclement weather and innocent bystanders.

2-53. Warning signs stating "CAUTION—MILITARY WORKING DOGS" should be placed on all sides and the rear of any and all vehicles used to transport MWDs. The signs should also be printed in the host nation language. The only exception to this is the use of unmarked vehicles for President of the United States and Vice-President of the United States missions.

## NARCOTICS AND EXPLOSIVE TRAINING AIDS

2-54. Detector dog training and evaluations require the frequent use of narcotic and explosive training aids. Due to the potential safety hazards and risks involved with these aids, it is absolutely mandatory for all MWD personnel and commanders to know and follow the regulatory guidance. MWD leaders and handlers must participate in periodic review and training on the requirements noted in *AR 190-12*. These requirements are clearly stated in *AR 190-12*.

## Chapter 3

# Managing Military Working Dog Operations

Managing an MWD program requires attention from the program manager, the KM, and the plans NCO. This chapter will help these individuals better understand their specific requirements and concerns regarding the scheduling of various missions.

## SCHEDULING AND EMPLOYMENT

3-1. The KM or plans NCO assists and advises the commander or operations officer in scheduling and employing MWD teams. The KM and plans NCO are the most qualified to determine the duty schedules, work cycles, and hours of operation to support each mission. All MWD teams should be utilized at least 24 hours per week to perform nontraining related duties.

3-2. Scheduling must allow for a minimum of four hours of patrol proficiency training per week and four hours of detector training per week. Details concerning proficiency training requirements are discussed in *Chapter 2*. Daily scheduling should include enough time for feeding, grooming, and exercising the MWDs as well as maintaining equipment and kennel facilities.

## PREDEPLOYMENT OPERATIONS

3-3. MWD leaders and planners need to ensure that handlers are prepared for all possible missions before the moment of execution. Careful planning and concurrent training is an ongoing process that involves all soldiers in every kennel section. No MWD team should ever be sent on a mission without the proper preparation.

### PROGRAM MANAGERS PREDEPLOYMENT RESPONSIBILITIES

3-4. MACOM program managers play a key role in MWD team readiness. The careful management of information concerning each kennel enables leaders and planners to keep the unit in a constant state of readiness for any possible deployment or no-notice mission.

3-5. Program managers keep track of the—

- Total number of personnel at each kennel in the MACOM.
- Certification status of all teams at each kennel in the MACOM.
- Deployability status of all MWDs and handlers (see *Chapter 8*).
- Total number and location of deployed MWD teams.
- Status of kennel facilities at deployment locations.
- Status of VCO support at deployment locations.

### KENNEL MASTER PREDEPLOYMENT RESPONSIBILITIES

3-6. The KM's responsibility is to ensure the daily accomplishment of all missions. The readiness of all assigned MWD teams is a process that requires continuous effort from all leaders. KMs must maintain statistics and make the necessary corrections to have all personnel and equipment available for future deployments and missions.

3-7. The KM is responsible for—

- Managing the number of personnel assigned to each kennel.
- Verifying the status of certified teams and the dates of certification.
- Maintaining the deployability status of MWDs and handlers.
- Ensuring that soldiers have soldier readiness packets prepared according to *AR 600-8-101*.
- Conducting monthly and quarterly counseling.
- Maintaining accurate and detailed training records that will give the forward deployed KM an understanding of strengths and weaknesses.
- Ensuring that MWD handler equipment for short-notice deployments is prepared at all times.
- Conducting annual 9-millimeter handgun qualification. More than one qualification should be conducted (if possible) to better train MWD handlers on the use of the 9-millimeter handgun and to ensure that handlers are always qualified.
- Ensuring that handlers meet the standards outlined in *AR 600-9* and *FM 21-20* by conducting height and weight testing and physical fitness tests.
- Ensuring that the mandatory common military training and common task tests are conducted according to *AR 350-1*.
- Training soldiers in tasks associated with deployments (for example, kennel setup, utilization, and missions).
- Ensuring that MWD handlers have received any needed driver's licenses or transportation motor pool (TMP) licenses before a deployment order arrives.
- Ensuring that soldiers have received the needed individual readiness training in preparation for deployment in a combat zone.
- Conducting mine awareness training.

## Plans Noncommissioned Officer Predeployment Responsibilities

3-8. Deployment planning must include coordination with the supporting military VCO. The plans NCO must consider the expected duration of the deployment and the types of facilities anticipated at the destination.

3-9. The plans NCO is responsible for ensuring that the handlers and MWDs in the MWD team or kennel section are technically proficient. The plans NCO is best able to evaluate and recommend to the KM which MWD team is most qualified and prepared for deployments and missions. Through continuous and realistic predeployment training, the plans NCO ensures that the MWD teams are proficient in—

- Crowd control training.
- Building searches.
- Tactical movements (through obstacles and from point to point).
- Gunfire training (handler and decoy).
- Defense training.
- Controlled aggression.
- Explosive detection (area searches, bunker searches, and tactical vehicle searches).
- Narcotics detection (area searches, bunker searches, and tactical vehicle searches).

## Additional Predeployment Considerations

3-10. Army mobility requirements necessitate that each MWD has a shipping crate. Shipping crates are used to transport MWDs, simplify handling and loading, and give MWDs adequate space and ventilation during transport. Shipping crates may also be used as temporary kennels at intermediate sites or at the deployment destination.

3-11. The average water requirement for a single MWD is 10 gallons a day. Enough dog food must be taken to last 90 days or until resupply can be established. MWDs must be fed the standard high-performance dog food, contracted and supplied by the General Services Administration (GSA), unless otherwise directed by the supporting VCO. MWDs are fed twice daily unless the supporting VCO directs another meal frequency.

3-12. First aid kits must accompany the MWD team on every mission. Information on the deployment destination, climate, and terrain help determine the specific contents of the first aid kits. For example, if MWDs are employed on rough terrain, a pad toughener is included in the first aid kit.

3-13. All mission-related equipment must travel and arrive with MWD teams so that they can be fully operational as quickly as possible. Any equipment that is not sent with the MWD teams should be sent by other means to arrive at the location within 60 days of the MWD team's arrival. An example equipment list is in *Appendix E.*

# DEPLOYMENT CONSIDERATIONS

3-14. There are many important factors to consider when leaders and MWD planners are preparing for deployments. To ensure that every deployment results in mission success, every aspect from notification to redeployment must be considered. These aspects include, but are not limited to, field kennels, training, and operations.

## FIELD KENNELS AND SUPPORT FACILITIES

3-15. A suitable kennel site must be carefully selected. Before deployment, some possible locations may be selected on a map from a leader's reconnaissance. As soon as possible after arrival, the sites should be inspected to choose the most suitable location. When selecting a site, consult a VCO (whenever possible) about possible health hazards.

3-16. Locating the kennel in a congested area may be necessary to protect it from enemy attack. A temporary screen or physical barrier may be needed to block the dog's sight so he will be able to rest.

3-17. When shipping crates are used as temporary kennels, the interior, exterior, and ground beneath the crate must be cleaned daily to prevent the accumulation of moisture and waste or insect infestation. It must be raised four to six inches off the ground to allow for adequate drainage and to reduce parasite breeding places. In hot climates, place crates under trees, a tarpaulin, or plywood to provide shade and ventilation. Spread gravel (if available) around and under the crates to allow for drainage and the easy removal of solid waste. A large tent or prefabricated building can serve as a temporary storage location for equipment, rations, and other supplies.

3-18. A second tent or building erected at the kennel site to house handlers and the KM provides the kennel with sufficient manpower for security of the kennel area. Although conditions around a temporary kennel may be primitive, a high standard of sanitation is essential to prevent diseases and parasite infections. Permanent kennel facilities must be erected within one year of the arrival of MWD teams to the location of operational use. Modular expandable MWD transport containers (NSN 961120, 961130, or an equivalent) (*Figure 3-1*, page 3-4*)* with integrated air conditioning, heating, and equipment storage are recommended for use as temporary kennels in field environments.

## DEPLOYMENT TRAINING

3-19. As soon as possible after arriving at the deployment destination and while making sure that site security is established, deployment training must begin. Deployment training allows the MWD teams to quickly develop familiarity with their new location. This is also known as right-seat ride time. The KM or plans NCO needs to evaluate or reevaluate the degree of skill of each of the MWD teams so that the KM and NCO are fully prepared to support the deployed force. Initial evaluation and training should consist primarily of field problems such as attack, search, re-attack, and scouting and basic obedience. Training should also include the handler firing his assigned weapon while maintaining control of the MWD. Some training should be done at dusk, and then in darkness, so that the handler may see (for safety) and the MWD may adjust to the noise and flash of weapons.

Figure 3-1. Modular, Expandable MWD Transport Container

## MANEUVER AND MOBILITY SUPPORT OPERATIONS

3-20. MMSO consist of measures necessary to enhance combat movement and the ability to conduct movement of friendly resources in all environments. These measures ensure that commanders receive personnel, equipment, and supplies as needed. MMSO is conducted across the full spectrum of military operations. The primary focus of military police during MMSO is to ensure the swift and uninterrupted movement of combat power and logistical support. KM teams will coordinate MWD operations with the plans and operations officer or task force commander.

## AREA SECURITY OPERATIONS

3-21. When setting up listening posts or conducting combat patrols, consider the availability and suitability of PDs for the particular mission. The KM or plans NCO must be briefed early regarding the mission. This will ensure that the most suitable PD team can be selected and that the handler can prepare for the mission. The handler should check the MWD for any limiting ailments and conduct training rehearsals to become familiar with the mission. The KM or plans NCO and selected handlers must be involved in planning conferences and briefings involving the utilization of an MWD team.

3-22. In defensive operations, combat patrols are used to provide early warning, to confirm intelligence information, and to detect or deter enemy action. PD teams greatly enhance the security of reconnaissance and combat patrols.

3-23. On combat patrols, a PD works at maximum efficiency for only two or three hours. During inclement or extreme hot or cold weather, this time will vary. The KM, plans NCO, and VCO should advise the commander on the MWDs' particular capabilities. The team is most effective in uninhabited areas. If an MWD frequently alerts on friendly forces and is continuously taken off his alert, he soon loses interest and reliability. Only the best teams should be selected.

### Defiles and Route Reconnaissance/Surveillance Operations

3-24. MWD teams can be used to enhance the security at defile and route reconnaissance/surveillance operations. Consider the following to ensure mission success:

- The mission at hand (the possible threat scenario).
- The size of the area to be covered.

- The type and number of MWD teams.
- The duration of the mission.
- Sustainment supplies.
- Communications.

3-25. Operational planning must ensure that the proper numbers and types of teams are requested to perform the mission. Planning considerations must allow for proper rotations and rest-to-work ratios for the MWD teams.

### Checkpoints

3-26. MWD teams are often called on to enhance the effectiveness of checkpoints. After receiving the mission to perform checkpoint duty, planners and handlers need to consider the following:

- The force protection condition (FPCON) (see *Appendix F* for details).
- The type and number of teams required versus those available.
- Sustainment equipment.
- Communications.
- Available support elements.

### Area Defense

3-27. Areas that cannot be covered by static defensive posts because of vegetation, terrain, or some other peculiar feature can be secured using PD teams. Prior to positioning any MWD team in a defensive posture, ensure that intelligence preparation of the battlefield (IPB) details any possible placement of friendly or enemy mines. When relieving others from established defensive positions, use range cards to familiarize the team with the area.

3-28. A PD team can detect personnel using concealed avenues of approach and provide an early warning to defenders. Each MWD team's area of responsibility should be large enough so that the team can move to take advantage of the prevailing wind directions. A fixed position should be established to provide covering fire for each MWD team in case the team must withdraw closer to the defensive perimeter after the MWD has alerted on an attempted hostile probe or penetration.

## Warning and Response Procedures

3-29. Several procedures should be developed for warning friendly forces of a PD's response because circumstances may prevent the use of a radio. Also, different types of responses tailored to fit a variety of potential situations should be developed. For example, it may be desirable to have the MWD team follow the alert and locate the cause. At other times, combat intelligence may require the team to maintain the alert position until assistance arrives or to withdraw to a more advantageous position. Whatever action is taken, the handler should not release an MWD unless it is necessary to defend—

- Himself.
- Other personnel.
- Protected resources.

## Internment and Resettlement Operations

3-30. I/R operations consist of those measures necessary to guard, protect, and account for people that are captured, detained, confined, or evacuated by US forces. In any military operation involving US forces, accountability and the safe and humane treatment of detainees is essential. US policy demands that all persons who are captured, interned, evacuated, or held by US forces are treated humanely. This policy applies from the moment detainees become the responsibility of US forces and continues until the time they are released or repatriated. (See *ARs 190-8* and *190-47* and *FMs 3-19.40* and *27-10*.)

3-31. MWDs offer a psychological and actual deterrent against physical threats presented by personnel housed in an I/R facility. They cannot be used as security measures against US military prisoners. MWDs reinforce security measures against penetration and attack by small enemy forces that may be operating in the area. MWDs also provide a positive, effective alternative to using firearms when preventing disturbances or escapes by personnel.

3-32. MWD handlers will not use their MWDs to guard prisoners inside prisons or detainee holding facilities and confinement facilities. In addition, MWD handlers will not use their MWDs to degrade, torture, injure, or mistreat EPWs, detained personnel, civilian internees (CIs), or other detainees in US custody.

***Note:*** MWDs are not used during interrogation operations.

### Demonstrations

3-33. Conduct periodic demonstrations in full view of personnel housed in an I/R facility to increase the psychological deterrent of an MWD. Emphasize how easily and quickly an MWD can overtake a fleeing individual. Highlight the MWD's ability to attack and overcome physical resistance. Demonstrate an MWD's scouting ability. To ensure a successful demonstration, use only the most qualified MWD teams.

### Perimeter Security

3-34. Use an MWD team as an addition to perimeter security by making periodic, unscheduled patrols around the perimeter fence during periods of darkness. During inclement weather, a temporary blackout or an electrical failure increases the number and frequency of patrols.

**WARNING**

**Never set patterns or establish routines when conducting security patrols. Enemy observing security measures can exploit set times by attacking when the patrol ends or before it begins.**

### Narcotic and/or Explosives Detection

3-35. Walk an MWD team through living areas to search for contraband. PEDD teams can be utilized to detect explosive devices and components in the I/R facility and PNDD teams can be utilized to detect narcotics.

### Work Details

3-36. Position an MWD team between the work detail and the area offering the greatest avenue of escape. MWD teams provide a valuable addition to work detail guards, particularly those employed in areas offering the greatest potential for escape. However, MWD teams are deterrents and must not be used as guards.

## Police Intelligence Operations

3-37. *AR 525-13* directs all military law enforcement to develop foreign and domestic criminal intelligence. MWD teams are an integral part of the police intelligence operations (PIO) process. PIO are conducted in conjunction with all functions in which the MWD teams perform. For example, a PEDD team working at a checkpoint to detect explosives will also be gathering intelligence. This holds true for all MWD functions.

3-38. Leaders and planners of MWD operations must develop procedures to ensure that the MWD teams gather and submit all intelligence in a timely manner. Standardized reports are an effective means that allow for the accurate and detailed submission of intelligence. Contact the supporting intelligence section (S2), provost marshal office (PMO), or CID office to ensure that all PIO coordination is in place prior to performing MWD functions.

## LAW AND ORDER OPERATIONS

3-39. Law and order operations are those that MWD teams perform to enforce laws, orders, and regulations and to maintain discipline and safety. Law and order operations can be performed in both combat and peacetime environments

## RANDOM ANTITERRORISM MEASURES PROGRAM

3-40. The random antiterrorism measures program (RAMP) is designed to be used at all FPCON levels to increase physical security awareness throughout an installation. RAMP involves the random use of measures, identified in *AR 525-13* and *TC 19-210*, for FPCON levels to make installation operations less predictable and therefore more secure.

3-41. The use of MWD teams during RAMP provides for more thorough detection of narcotics and/or explosives and for a psychological deterrent. An MWD presence can also increase confidence in installation security in general.

3-42. KMs coordinate with the PMO or S3 to ensure that RAMP checks are successful. Coordination will include—

- The reconnaissance of patrol areas (map reconnaissance as a minimum).
- Rehearsals and inspections before operations.
- Transportation.
- Equipment needed on site.
- A threat briefing.
- Rules of engagement.
- Use-of-force criteria.
- After-action review (AAR) requirements (see *Appendix D*).

## TEMPORARY DUTY MISSIONS (UNITED STATES CUSTOMS SERVICE/HIGH-RISK PERSONNEL SECURITY MISSIONS)

3-43. USCS safeguards the internal security of the United States through border searches. Because of this, they have broad powers in conducting searches. By utilizing an MWD team, all vehicles, ships, luggage, and baggage can be checked at border clearance crossings or ports. MWD teams are called upon to support local or federal authorities in detection operations. The Joint Enforcement Military Customs Operation places MWDs at border clearance ports to conduct searches of vehicles and cargo entering the United States. The MWD team takes no law enforcement action. The MWD team only searches and reports. *AR 90-12* gives a more detailed explanation of civil support by MWD teams.

3-44. The Air Force has been designated to serve as the primary service point of contact for all very important person (VIP) missions received from the Office of the Secretary of Defense, USSS, and/or Department of State. The VIP coordination officer who (under authority from the Secretary of Defense) tasks and coordinates all missions performed by DOD PEDD team personnel, regardless of service affiliation, issues all mission tasks.

3-45. MWD leaders and planners ensure that the certification of the team selected to perform the mission is well documented and current. Other considerations in planning for USCS support include—

- The travel arrangements of the MWD team.
- Government travel credit cards.
- Lodging for the handler.

- The kennel and boarding.
- VCO support.

## MILITARY WORKING DOG PUBLIC DEMONSTRATIONS

3-46. Public demonstrations are conducted to show the abilities of MWD teams. These demonstrations are coordinated through the KM to ensure that the appropriate team is available for the target audience. The goal of public demonstrations is to deter criminal activity and to promote public acceptance of the MWD.

3-47. At no time should demonstrations make an MWD a form of entertainment. The KM or plans NCO should ensure that the details of sensitive police training techniques are not explained to the public. Avoid lengthy or frequent public demonstrations that allow MWDs to tire easily or become bored.

## HEALTH AND WELFARE DEMONSTRATIONS

3-48. Initially, the MWD team should provide the responsible commander with a basic demonstration of the MWD's ability. The KM or plans NCO places a training aid in an undisclosed location, and the handler enters the search area with the MWD. The handler presents and conducts a systematic search of the area. When the MWD alerts, the appropriate reward is given. This demonstrates the MWD team's proficiency.

3-49. After the demonstration has been conducted, the unit should ensure that the area being searched is free of soldiers and that all rooms are unlocked. MWD teams are a tool the commander uses to find any illicit narcotic or explosive devices. It is the responsibility of the unit to properly search the room after the MWD has cleared the room. If the MWD team responds, military police personnel are called to continue the search. The MWD handler conducting the search should not search for the illicit narcotics or explosives. Unit commanders should also ensure that no personnel enter the building or area until the search is complete. This is usually accomplished by placing NCOs at various entranceways to stop personnel from entering the area.

## COMPETITIVE EVENTS

3-50. MWD teams should be encouraged and allowed to participate in competitive events, seminars, or conferences conducted by outside agencies or police canine organizations. Exposure to competitive events can challenge MWD teams to develop skills or learn advanced employment and training techniques. These events may require new tasks or methods that the handler can use to identify and apply to military missions. In this way, the team's proficiency and effectiveness are enhanced.

## RECOVERY OPERATIONS

3-51. After every deployment or extended off-site mission, MWD leaders need to ensure that vital actions are conducted upon the return of the MWD team. These actions include—

- Conducting an AAR after every MWD deployment or extended mission to assist in capturing lessons learned. See *Appendix G* for detailed information on AARs.
- Conducting veterinary inspections.
- Returning any explosive or narcotic training aids.
- Updating training requirements and records.
- Recertifying teams.

# Chapter 4
# Legal Considerations

The use of MWDs has the same legal considerations as the use of a military police soldier. Each situation and location may be subject to various Status of Forces Agreements (SOFAs), general orders, and other legal restrictions. The following chapter addresses some fundamental areas. It is important to consult with the Staff Judge Advocate whenever legal questions pertaining to an MWD arise.

## USE OF FORCE

4-1. Procedures for the use of force are found in *AR 190–14*. This regulation states that personnel engaged in law enforcement or security duties will use the minimum amount of force necessary to control the situation. This regulation further details the levels of force and encourages commanders to substitute nonlethal measures for firearms when they are considered adequate for safely performing force protection, antiterrorism, and law and order functions. The regulation states that in evaluating the degree of force required for specific situations, the following options should be considered in the order listed:

- Verbal persuasion.
- Unarmed defense techniques.
- Chemical aerosol irritant projectors (subject to host nation or local restrictions).
- The military police baton.
- MWDs.
- The presentation of a deadly force capability.
- Deadly force.

4-2. Although high in the order of measures, it is clear that the release of the MWD as a use of force is a nonlethal option to be used before deadly force is considered. It must be understood that the release of an MWD to pursue, bite, and hold a person is likely to result in an injury to the suspect. This injury can be very minor such as scratches or bruises; however, serious bodily harm is possible. For this reason, it is required by *AR 190-14* that all lesser means of force must be used first whenever possible.

4-3. The use of force differs between MWD handlers and regular military police because of the uniqueness of the MWD. Positive control of the MWD must be maintained at all times throughout all of the use-of-force measures. Often, just the presence or arrival of a handler with his MWD can control a situation. If the desired level of control is not achieved, verbal persuasion is the first degree of force to be used. If verbal persuasion is not enough force to maintain or gain control, the next measure may be considered. If the situation allows, military police may use unarmed defense techniques, spray irritants, and the baton before MWD handlers use the MWD as a measure of force.

4-4. A handler is not likely to use unarmed defense techniques when holding an MWD on leash. The handler must remain in positive control of the MWD; therefore he will not be able to get close enough to a suspect to employ unarmed defense techniques. The use of chemical aerosol irritant projectors or the baton would also hinder the handler's ability to keep positive control of the MWD. Therefore, the options are not reasonable measures of force available to the handler.

4-5. Military police and task force commanders, along with PMs, establish clear policies regarding the release of MWDs according to *ARs 190-12* and *190-14*. If the minimum amount of force necessary to control the situation requires the use of MWD force off leash, the handler must ensure that the following steps are followed:

- Ensure that the MWD has identified the same target as the handler before releasing him to bite.
- Give the warning order, "Halt, halt, halt, or I will release my dog." If assigned in a foreign country, give the order in the primary language of the host nation.
- Warn all bystanders to cease all movement.
- Call the MWD off the pursuit and regain control of him if the suspect indicates surrender or stops moving.

4-6. Use extreme caution when removing an MWD from a suspect. Regain and maintain leash control of the MWD until he has become calm enough to obey the commands "Heel" and "Stay." Do not release an MWD if the suspect is out of sight, unless an off-leash building search is authorized. Do not release MWDs in areas where children are present.

4-7. MWD teams are not used for crowd control or direct confrontation with demonstrators unless the responsible commander determines it to be absolutely necessary. When it is necessary, dogs are kept on a short leash to minimize the danger to innocent persons. Dogs are not released into a crowd. Civil disturbance contingency plans include specific criteria for the use of MWD teams that are consistent with *FM 3-19.15, AR 190-12,* and *AR 190–14.*

# SEARCHES

4-8. Rules 311 and 313 through 316, *Manual for Courts-Martial (MCM),* prescribe HQDA policies concerning the conduct of searches and inspections and the disposition of property seized in conjunction with those activities. Property refers to property of the United States or of nonappropriated fund activities of an armed force of the United States wherever it is located.

4-9. Commanders and law enforcement personnel must be aware of the distinction between searches, seizures, inspections, and inventories and the provisions for the execution of these actions as outlined in the *MCM.* Knowledge of this information is necessary to ensure the legal use of a PNDD or PEDD.

4-10. Probable cause is a reasonable belief that a crime has been committed and that the person, property, or evidence sought in connection with the crime is located in the place or on the person to be searched. Probable cause must be established prior to searching any persons, property, or residence. Probable cause can be established in the following ways:

- Personal observation.
- Information obtained from a reliable informant or from the witness who observed the crime.
- Use of a detector dog team.
- Any combination of these three.

4-11. The detector dog team is considered a "reliable informant," according to Rule 315 (f)(2), *MCM.* The reliability of a team is established through the initial and annual certifications of the detector dog team by the certification authority and through proper documentation of the MWD's training records.

4-12. Certification establishes the foundation for an officer to grant a search authorization or warrant. Detector dogs are first certified after finishing their basic training. To meet the legal requirements permitting their operational use, each detector dog team must undergo an annual certification and a validation.

4-13. Each detector dog team undergoes a certification when first assigned to duty. A designated certification authority conducts the certification. The detector dog team must perform a demonstration for the person who authorizes any search. This may be any commander from company level up to the post commander. The guideline is that the person who witnesses the demonstration can only authorize a search of units he has command authority over. This person must also review the probable cause folder of the dog and must validate the team as reliable, in writing. This validation is good as long as the commander is assigned to that command.

4-14. The certification authority should witness a demonstration of the dog's capabilities. All substances the dog has been taught to respond on should be used. Additionally, the dog's capability to detect past/present odor should be demonstrated. At the completion of the demonstration, the results are documented in letter form, signed by the commander, and entered into the dog's probable cause folder. The MWD handler is responsible for preparing and maintaining the probable cause folder and the contents (see *paragraph 2-49*, page 2-9*)*. The probable cause folder accompanies the handler each time the dog is used on an actual search. Each document in the folder should be explained to the commander requesting a detector dog search. The commander with command authority should witness a simple demonstration of the dog's capabilities prior to a search. The above procedure should be followed whenever possible. This procedure establishes the dog's reliability on a continuing basis and is essential in establishing probable cause.

4-15. PEDD and PNDD teams must maintain high standards of accuracy to provide the desired level of confidence in their searches. PNDD teams must maintain a 90 percent accuracy rate for the number of trials conducted, while PEDD teams must maintain a 95 percent accuracy rate for the number of trials conducted. The installation commander authorizes the support of local and federal law enforcement agencies. *DOD Instruction 3025.12* governs the use of all PEDD teams. It governs the use of PNDD teams and ensures compliance with *Section 1385, Title 18, United States Code (USC) (The Posse Comitatus Act) (18 USC 1385 [The Posse Comitatus Act])*. Complete certification, recertification, and decertification details for narcotic and explosive detector MWD teams can be found in *AR 190-12*.

# Chapter 5
# Patrol Dogs and Characteristics

PDs can be used in traditional law and order functions, but commanders and planners should not overlook the tactical enhancements that a PD team offers to combat support operations. All aspects of military police functions can be enhanced by the employment of MWD teams. This chapter discusses many practical applications of the PD team.

## EMPLOYMENT TECHNIQUES DURING COMBAT SUPPORT OPERATIONS

5-1. PD teams can perform many combat support operations. These teams enable the tactical commander to free up soldiers and employ their resources in other areas. This force multiplier is especially valuable in area security and force protection and antiterrorism operations.

### PERIMETER SECURITY

5-2. PD teams are useful for perimeter (*Figure 5-1,* page 5-2) and distant support posts because they are usually located away from normal activity and large numbers of people. These posts also usually enclose large areas, and it would take a large number of single sentries to secure the area effectively. The large size of these areas allows the PD team to change positions to take advantage of the prevailing wind direction. These posts may be secured only during periods of advanced security and high threat. Occasional random posting of these areas is recommended, especially during periods of limited visibility (rain, snow, dust, smoke, and fog) or night conditions. Barriers and obstacles (such as fences, buildings, gullies, and streams) must also be identified and considered in security post planning. Place the PD team so that these obstacles offer the least interference to security.

5-3. Perimeter security, depending on the area to be covered, should use a two-team concept. The teams overlap each other while walking around the perimeter. PD teams should not come into direct contact with each other. Considerations include—

- The commander's intent.
- Priority intelligence requirements.
- Rules of engagement.
- Use-of-force criteria.
- The wind direction and distracters in the area (see *Chapter 8*).
- Friendly and enemy forces around the area.
- Rest areas within the perimeter to allow the MWD to relieve himself.
- Watchtowers or other security forces in the area.

5-4. The PD team should initiate the perimeter security operation by walking in a zigzag pattern around the exterior perimeter. The team should be allowed to go outside of the security area to ensure that no unwanted intruders are present.

**Figure 5-1. Perimeter Security**

## Intrusion Detection System Augmentation

5-5. Continuing development and use of Intrusion Detection Systems (IDSs) for security make it necessary to reassess the role of PD teams in the security of priority resources where IDSs are used. Although IDSs are designed to greatly reduce the probability of a successful area penetration, they are not designed to replace PD teams. IDSs cannot see beyond the outer clear zones, counter intrusions, or scout the progress of an intruder force after penetration. They are also subject to malfunction and breakdown. PD teams cannot counter all potential problems, but they can be used selectively to counter some of the IDS limitations. With the advice of the KM, security planners should develop plans that set an effective balance between sensor systems and PD teams. Some suggested functions for integrating PD teams into IDS-augmented security systems are as follows:

- PD teams may be used inside or outside critical sites or locations of HRP type personnel areas to give immediate response to areas not clearly seen by closed circuit television (CCTV) or to support security response teams in their role of alarm assessment
- PD teams may be used to scout hidden intruders from the point of penetration to the hiding location inside protected areas.
- PD teams may be posted outside critical sites during periods of increased threat to expand the protected area beyond the range of the IDS. This use is particularly desirable when there is heavy vegetation beyond the perimeter clear zone or when protected areas are located on or near installation perimeters and may, therefore, be easily accessible to the public.
- PD teams may be used as an element of the security response team or as an element of other designated response forces.
- PD teams may be used to compensate for decreased IDS effectiveness when poor visibility or other environmental conditions adversely affect the IDS equipment or when the IDS is not functioning.

## Secondary Response Forces

5-6. Secondary response forces are typically used as blocking forces and for sweeps and counterattacks. PD teams can greatly improve the capabilities of the secondary response force. Some factors that must be considered are that—

- Blocking positions must be selected downwind from the PD team's position when PDs are used with blocking forces. In this type of mission, PDs should only be used to alert the forces of possible intruders.
- The secondary response force moves into the wind with the PD team in the lead position when the PDs are used for sweeps. It is imperative that security be placed on the PD teams to ensure that they are protected.
- A PD team can often pinpoint the exact location of individual intruders.

5-7. A PD team can help detect, alert, and catch hidden intruders while clearing a position that has been counterattacked successfully. PEDDs can be used to help clear the area of explosives. (See *Chapter 7*.)

## Listening Posts

5-8. On a listening post, the PD team should be positioned forward of the tactical area to reduce distractions to the PD, yet close enough to maintain contact with friendly forces. In selecting a location, consider the primary detection senses of the PD. Whenever possible, PD team listening posts should be located downwind from any potential enemy position or dismount avenue of approach. Other locations force the PD to rely only on his hearing and sight for detection, reducing his effective use of scent.

## Combat Patrols

5-9. In defensive operations, combat patrols are used to provide early warning, to confirm intelligence information, and to detect or deter enemy action. PD teams greatly enhance the security of reconnaissance and combat patrols.

5-10. On combat patrols, a PD works at maximum efficiency for only two or three hours. The team is most effective in uninhabited areas. If a PD frequently alerts on friendly forces and is continuously taken off his alert, he soon loses interest and reliability.

5-11. Patrol leaders and members must be briefed on actions to take when a dog handler is seriously wounded or killed. Dogs that have worked closely with patrol members and have developed a tolerance for one or more of them will usually allow one of the patrol members to return it to the kennel area. However, the dog may refuse to allow anyone near his handler, and other handlers may need to be called to help. The canine must be separated from the handler by—

- Coaxing the dog away with friendly words or food.
- Covering the dog with a poncho to immobilize it.
- Staking out the dog or leashing him.

5-12. A human will always take precedence over an animal. If no effort is effective, the dog may be destroyed. However, all efforts should be made to avoid this; an MWD is an expensive and valuable asset to replace.

5-13. If a situation arises where a canine is injured but the handler is not, the handler must be allowed to accompany the canine. If wounded or killed, an MWD should be evacuated using the same assets and should receive the same consideration as that given to a soldier under the same circumstances.

## Tactical Patrols

5-14. PD teams can be used in tactical patrols to detect enemy presence, to help the patrol avoid discovery, and to locate enemy outposts. When the PD alerts, the handler should signal the patrol to halt until the cause of the alert can be identified and the patrol can proceed safely. If a firefight develops while the PD team is at the point position, the PD team should respond to fire team's directions and act as a regular member of the patrol.

5-15. Generally, the best locations for PD teams are directly in front of the patrol or on the flanks to enhance the flexibility and security of the formation. Prevailing wind direction should be used to improve the patrol's chances for early warning of enemy forces. If the wind direction is from the rear of the patrol, the PD is forced to rely on sight and sound and may not be as effective. The handler must concentrate on the PD to read his alert and will not be able to use a weapon rapidly. Therefore, members of the patrol should be assigned to protect the PD team.

5-16. PD teams can be used to enhance the capabilities of an ambush patrol. A PD team should be placed in front of the patrol to minimize distractions, yet close enough to maintain contact with other patrol members. PDs on ambush patrols must remain silent and not respond aggressively to approaching enemy forces.

## Internment and Resettlement Operations

5-17. PD teams are used during I/R operations for camp perimeter security and to guard against escapes. The PD's keen senses of sight, smell, and hearing assist the handler and camp authorities in detecting unauthorized activity. A PD team's presence also acts as a deterrent to escapes. In the event of an escape, the PD is a valuable asset that can assist in the search and recovery of internees.

### PD Duties and an Internment and Resettlement Camp in Iraq

On 13 April 2003, SGT Carey A. Ford and his MWD Rex responded to a riot at Camp Bucca, Iraq. It was believed that an Iraqi internee had taken a pistol belt and some personal equipment from a guard. The prisoner had taken this equipment back to his tent and the 600 other internees refused to cooperate with the guard force. SGT Ford and MWD Rex went into the I/R camp unarmed. The MWD team moved through the disruptive and unruly crowd providing a secure environment for the quick-reaction force to extract the prisoner and retrieve the stolen pistol belt and personal equipment. SGT Ford related that EPWs would not come anywhere near his MWD Rex. "We work that fear to our advantage."

Four days later, SGT Ford and Rex were called on to scout for and apprehend an escaped EPW. The fugitive had a substantial lead on them, having made his escape two hours prior to SGT Ford's notification. SGT Ford entered a possible unexploded ordnance and/or antipersonnel minefield with Rex, who was also trained to detect explosives. The MWD team was determined to locate the dangerous fugitive and prevent access to any possible weapons caches since many had been located in the area. After hours of relentless searching, SGT Ford and Rex located and apprehended the escapee. Their actions provided the psychological deterrent necessary to prevent future escape attempts.

## Riot/Crowd Control

5-18. Extreme caution should be applied before committing PDs to a potential crowd control situation. Although the presence of a PD can be a powerful statement to an enemy or terrorist, it can elicit a reverse and potentially adverse effect on individuals in a riot/crowd control situation.

5-19. It is difficult to employ PDs properly during demonstrations or riots due to the high levels of confusion and excitement and the large number of potential antagonists. PDs should be trained to remain undistracted by loud noises and crowds. A crowd's reaction to a PD in a riot environment may be dependent on the location. In some areas of the world, the appearance of PDs at an angry crowd scene results in an escalation of violence. The crowd will often challenge soldiers to use the PDs as a measure of force. This is particularly true if a situation can be manufactured or provoked so as to be interpreted as an unreasonable use of force by the military. The presence of PDs can cause extreme fear (deterrence) in some cultures and regions. This fear can be advantageous and should be exploited when needed.

5-20. Committed PD teams may be used as an alternative to the use of deadly force to gain control of a riot situation. If the commander located on the ground directs the employment of PD teams for direct confrontation with demonstrators, PD teams should be held in reserve, out of the sight of the crowd. As the situation escalates, PD teams may be moved forward to within the sight of the crowd, but well away from the front lines.

5-21. When committed to the front lines in direct confrontation, PDs are allowed to bite only under the specific circumstances authorized by the responsible commander. PDs should never be released freely into a crowd. Other riot control force personnel should be positioned approximately three meters from any PD team when conducting any form of riot control. This is to prevent unintentional bite injuries to friendly, riot control force personnel that may happen to get between a PD and an instigator participating in the riot or demonstration.

5-22. PD teams must move a safe distance from the crowd to ensure the safety of the PD. Use of PD teams for direct confrontation with demonstrators is not recommended. According to the Army's use-of-force policy, the application of physical force against a person who cannot reasonably be suspected of committing an offense is not appropriate. Presence in a crowd does not necessarily constitute an offense or a situation where it would be reasonable to apply the degree of physical force of a PD. A PD would not be used unless the person is engaged in some other readily apparent criminal behavior such as theft, destruction of property, or assault.

## EMPLOYMENT TECHNIQUES DURING FORCE PROTECTION AND ANTITERRORISM OPERATIONS

5-23. PD teams contribute significantly to the military police force protection and antiterrorism effort. PDs are directly responsible for executing specific security-related measures and for providing support to other government agencies that contribute to force protection.

5-24. MWD support is accomplished under the leadership of the respective command PM. Force protection is accomplished through an active role in physical security, HRP security, law enforcement, and antiterrorism efforts. Force protection and antiterrorism operations consist of those actions that prevent or mitigate hostile actions against DOD personnel (to include family members), resources, facilities, and critical information. During force protection operations, offensive and defensive measures are coordinated and synchronized to enable the joint force to perform while degrading opportunities for the enemy.

5-25. In operations, PD teams are used to enhance the detection capabilities of the combat support force and to provide a psychological deterrent to hostile intrusions. If the PD teams are properly located in a hostile environment, the PD's alert provides an initial warning to the presence of a hostile force. In past combat operations, PD teams often provided warning of attacks early enough to allow a response force time to deploy and to prevent enemy forces from reaching their objectives. PD teams have also helped clear protected areas of hostile persons, explosives, and weapons after attacks.

> **Crowd Control in Kosovo**
>
> During Task Force Falcon (2A) in Kosovo, LTC Cloy (the battalion commander of 1-36 INF, 1 Bde, and 1AD) requested MWD support to assist the Alpha company clear a main supply route (MSR) that had been blocked by 10,000 Kosvar Albanians. LTC Cloy had already deployed both quick-reaction forces to assist Company A because the crowd would not stay off the MSR. LTC Cloy called the battalion operations center and directed that the MWD teams be sent to the scene.
>
> SFC Barnes, SSG Rogers, SSG Barters, and SGT McClintock responded with their dogs Queeny, Danny, Sandor, and Fedor and linked up with an infantry squad. The MWD teams and infantry squad moved dismounted to the town of Gjilani. The crowd was starting to become violent. LTC Cloy was in the middle of the crowd, standing in the center of the city at the critical MSR intersection. LTC Cloy attached the MWD teams to Company A. CPT Curtain, the company commander, told SFC Barnes to move the crowd away from his area and to establish a perimeter. The MWD teams moved into the diamond formation with the dogs on guard. Splitting the crowd in half, the MWD teams moved through the crowd and reached the city center MSR intersection in the center of the city. The handlers then used the dogs to move the crowd back out of the road. Company A and the four MWD teams established a perimeter line to keep the MSR open and traffic moving. They also apprehended troublemakers in the crowd.
>
> Later that evening when the dog handlers entered into the DFAC, LTC Cloy stood up, gained everyone's attention, and stated, "SFC Barnes, you and your handlers were like Moses. Your dogs parted the Red Sea!"

## Access Control Points

5-26. A PD team may be used to enhance force protection and antiterrorism efforts at ACPs. The PD team's function at an ACP is to deter unauthorized access, crime, and terrorism. The handler that is assigned to the ACP should use the PD so that the public can observe and hear the dog. Care must be taken to ensure that the handler constantly keeps the PD under control to prevent incident or injury to innocent personnel.

## Building and Area Searches

5-27. A PD team is especially effective in checking and searching buildings such as commissaries, post exchanges, finance offices, banks, and warehouses. The PD team can check buildings visually while patrolling and can stop and dismount so that the handler can physically check doors and windows. These checks should be made with the PD on leash. To take maximum advantage of the PD's scenting ability, the handler should approach buildings from downwind.

5-28. If a building is open or forced entry is evident, PD teams may be used to track hidden intruders from the point of penetration to their location. PDs may be released to apprehend an intruder suspected of committing a serious offense when the only alternatives are escape or the use of deadly force.

5-29. Prior to searching a building, the handler will determine whether to make the search with the PD on or off leash. Some of the factors that must be considered include the following:

- Friendly forces or innocent bystanders in the area.
- The time of day or night.
- Evidence of forced entry.
- The type and size of the building or area to be searched.
- The danger to any innocent persons in the area.
- The PD's ability to work off leash.
- The threat level and IPB.

5-30. Prior to releasing the PD inside a building or enclosed area, the handler will give a clear voice warning to any persons inside the building or in the area. They will be told to come out immediately and that failure to comply with these instructions will result in the release of a trained MWD. An interpreter may be needed to assist the PD team in giving the warning in the host nation language. All persons will be warned that the PD may attack without warning and that they could receive physical injuries. The handler maintains voice control of his PD throughout the search.

---

***Note***: A warning similar to the one for off-leash searches will precede on-leash searches. The searching team should be followed by at least one other military policeman to protect the search team.

---

## APPREHENSION OF SUBJECTS

5-31. The use of PDs to attack and hold an individual enemy or terrorist should not be overlooked. A PD can be sent into a building, confined space, or cave to find, attack, and hold an individual until the handler and friendly forces can arrive to place the subject into custody. Although disadvantages do exist (injury or death of the PD), the PD is better-suited than a human to immediately find and subdue a subject in an area where the interior structure is unknown to friendly forces.

## WALKING PATROLS

5-32. The PD team's contribution to MWD team law enforcement efforts is most effective when the team is on foot. Some of the law enforcement duties that a PD team can do as a walking patrol include checking or clearing buildings and patrolling parking lots, dependent housing areas, and troop billet areas.

5-33. Since it is easily seen, a PD used during daylight in congested areas provides a good psychological deterrent to certain crimes. PDs are tolerant of people, and the presence of a large number of people does not significantly reduce the PD's effectiveness except when conducting detection missions.

5-34. Using the PD team during darkness or periods of reduced visibility and in areas where there are fewer distractions can enhance the PD's detection ability. Military police may not detect a person fleeing a crime scene at night, but a PD will normally be able to detect and locate a fleeing suspect. When necessary and appropriate, the PD is also much more effective at pursuing and holding the suspect.

5-35. Periodic use of PD teams around on–post schools, especially when school is starting and dismissing, may help deter potential vandalism, child molestation, exhibition, and illegal drug activities. PD teams can also provide effective security for communications facilities, military equipment, and command posts (CPs).

## ALARM RESPONSES

5-36. In responding to an alarm condition at facilities such as clubs, finance offices, or banks, the PD team should be among the first military police on the scene. The PD team may be used to search and clear the building and immediate area and help with any apprehensions. If the PD must be used to find a suspect, other persons should attempt to avoid contaminating the area with scents or tracks that may confuse him.

## VEHICLE PARKING LOTS

5-37. PD teams are effective in detecting and apprehending thieves and vandals in vehicle parking lots. The presence of the PD team may deter potential acts of theft and vandalism. The PD team can be most effective by alternating between vehicle and foot patrols of the parking lots. During hours of darkness, when there is no activity in a parking lot, the PD team should approach the lot from the downwind side. If the PD alerts, the handler should locate and challenge the suspect for identification.

5-38. If a suspect tries to escape or avoid apprehension and the handler is reasonably certain the suspect has committed or is attempting to commit a serious offense, the handler should give a clear warning to the suspect to halt. When lesser means of force have failed to stop the fleeing suspect, the PD may be released to pursue, attack, and hold. If it is not safe to release the PD, the handler may attempt to follow the suspect if the PD has the ability to track.

# Chapter 6

# Narcotics Detector Dogs and Characteristics

MWDs are trained to perform dual roles. In addition to the uses and characteristics of the PD, the PNDD is trained to detect the odor and presence of specific drugs. This specialized ability to detect certain illegal drugs makes the PNDD a valuable tool to help commanders and other government agencies maintain law and order. The PNDD can be employed in combat and/or force protection and antiterrorism operations worldwide.

## EMPLOYMENT FOR COMBAT SUPPORT OPERATIONS

6-1. The PNDD provides commanders with unique capabilities in combat support environments. The narcotic detection capabilities of these dogs can be used to maintain the order and discipline of US soldiers as well as combatants and other persons involved in operations (EPWs, CIs, refugees, or other detained/interred persons [during time of war/conflict]). In addition to narcotic detection functions, PNDD teams should not be overlooked for their PD functions as discussed in *Chapter 5*.

### INTERNMENT AND RESETTLEMENT OPERATIONS

6-2. In addition to perimeter and other security duties, PNDD teams assigned to I/R camps can be used to ensure that no illegal drugs are at the I/R camps by conducting searches of all areas. Commanders and planners should consider having all inbound supplies or equipment searched by PNDD teams. Inbound and outbound mail and vehicles should also be considered as a means for the movement of illegal drugs and, as such, subject to a search by PNDD teams.

### POSTAL OPERATIONS

6-3. Postal inspections, including mail storage areas and drop boxes, can be conducted in conjunction with a military customs inspector (MCI) and according to *DOD Directive 5030.49*. The postmaster or MCI can request PNDD support from the area PM or military police brigade S3 with coordination through the KM team.

6-4. A PNDD team can be used to search mail that is received or shipped. Searches can be conducted as the mail comes off or before it goes on the truck prior to being shipped. The PNDD team can search up and down the conveyer belt as the mail is removed from or placed on the truck. If a conveyer belt is not being used, the handler can instruct postal personnel how to arrange the mail in a secure area so that the PNDD team can search it more effectively. When a PNDD team is used in the close quarters of a mailroom or delivery area, all nonessential personnel should remain clear of the area while the team is actively working.

6-5. When the PNDD responds to a parcel indicating the possible presence of illegal drugs, the CID or the military police investigations section will be notified for further action. At this point the PNDD team's actions are finished.

### AIRCRAFT AND LUGGAGE SEARCHES

6-6. Before beginning the search, ensure that all personnel have exited the aircraft and all compartments are opened. The PNDD team searches the aircraft before searching any luggage or cargo. When approaching the aircraft, conduct a fresh-air perimeter search from the downwind side. Once a perimeter search has been conducted, the team boards the vessel at either the front or rear. Once the area has been cleared to search, the handler ensures that only one door is open to limit draft and wind current. A systematic search of all areas within the aircraft is conducted from the top to the bottom. The handler advises other military police on the scene where the PNDD displayed a positive response, and the augmenting military police conduct a detailed search of the area surrounding the response.

6-7. When conducting inbound baggage inspections, the items should be aligned in a sterile area in rows with at least six feet between rows. The handler conducts an on-leash search, presenting each piece of baggage. When searching cargo, every effort should be made to search the cargo before it is palletized. The cargo should be aligned in rows, with at lest six feet between rows, in a sterile area (same as baggage). The handler conducts an on-leash search, presenting cargo as required. Once the PNDD team has searched the cargo, it may be palletized while remaining in the sterile area prior to embarkation.

6-8. Conducting debarkation operations is the same as inbound operations. The host unit (to which the cargo belongs or is being delivered) provides a representative (E6 or above) and a work detail to assist the PNDD team. When pallets are removed from the aircraft, they are placed in a sterile area. Once in the sterile area, the work detail unloads the pallets and arranges the cargo, as required by the PNDD handler.

## Force Protection and Antiterrorism Operations

6-9. PNDD teams are extremely useful in force protection and antiterrorism operations. Commanders and other government agencies use PNDD teams to protect the community from the harmful effects of illegal drugs. PNDD teams protect the force and surrounding communities by deterring criminal activity and enhancing law enforcement capabilities.

## Health and Welfare Inspections

6-10. PNDD teams conduct health and welfare inspections when requested by the commander and coordinated through the KM, PMO, and S3. The commander contacts the KM directly to schedule a health and welfare inspection. The KM, in conjunction with the installation PM, the CID special agent in charge, and the military police investigations supervisor, briefs all levels of the command on the availability and capabilities of the PNDD teams available to them on the installation.

6-11. When scheduling the inspection, the commander should have the following information available:

- The requested time of inspection.
- The areas to be inspected.
- The number of living quarters.
- The number and size of common areas.
- Other pertinent information.

*Appendix H* provides a guide to assist handlers with health and welfare inspections.

6-12. Once the coordinating instructions have been established, the KM coordinates a time and location to brief the commander and demonstrate the PNDD team's capabilities. It is important that the unit requesting the health and welfare inspection practices operational security (OPSEC) and informs only those personnel who have an immediate need to know. This usually consists of the commander and first sergeant or sergeant major. The KM advises the PM and the military police investigation supervisor when the inspection is scheduled and if any additional support will be needed.

6-13. A systematic inspection of the living quarters and common areas is conducted. If the PNDD responds to a locked wall locker, the soldier assigned to that wall locker is summoned to open it. Once the wall locker has been opened, unit representatives conduct a detailed search of the wall locker and its contents.

## Department of Defense Schools and Drug Abuse Resistance Education Program Inspections

6-14. KMs may receive requests from Drug Abuse Resistance Education (DARE) officers or DOD school principles to conduct a health and welfare inspection of a school. When conducting school inspections in conjunction with DARE, basic building inspection techniques apply. The primary concern when inspecting schools is the safety of the students and the secrecy of the search.

6-15. When possible, searches should be conducted in conjunction with a school activity, such as an assembly or fire drill. This will allow the handlers to conduct the search with a decreased risk of student involvement. Another option is to search the school after normal school hours or on a weekend or holiday. In any event, the use of the PNDD team on school grounds must be cleared by the school administration and the PM must be notified in advance.

### ASSISTANCE TO MILITARY POLICE INVESTIGATIONS SECTION AND/OR THE DRUG SUPPRESSION TEAM

6-16. The military police investigation section or the drug suppression team (DST) may contact the KM to schedule an inspection for a number of reasons. PNDD teams may be requested to conduct an inspection of a residence or building where illegal drugs have already been found. In addition, unit commanders may request a PNDD team to inspect the quarters of a soldier who has shown a positive response on a urinalysis. The PNDD is not used to verify or test a substance that has already been found. PNDD handlers contact the military police investigations section or the DST in any circumstance that a PNDD has responded positively according to local standard operating procedures (SOPs).

### MILITARY CUSTOMS MISSIONS

6-17. The US Army has been designated as the proponent service for the military customs program. According to current regulations, PNDD teams are used to inspect aircraft cargo areas and airframes when the military customs inspector program manager (MCIPM) directs these inspections. When conducting military customs missions, the handler should first inspect the vessel, then if deemed necessary by an MCI, continue with baggage.

6-18. When scheduling military customs missions, the MCIPM coordinates with the KM according to local SOPs. At a minimum, the MCIPM notifies the KM of—

- The requested time and date of the search.
- The number and types of vessels to be searched.
- The amount of cargo to be searched.
- The number of passengers and crew requiring the MCI's clearance.
- The determination on whether the search is a high-risk or selective search.

6-19. PNDD teams may inspect vessels such as military or civilian aircraft, boats or ships, vehicles, rail cars, or any other vehicle capable of carrying passengers and/or cargo according to the military customs program. These inspections are primarily conducted outside the continental United States. However, some missions in the continental US may be necessary for units returning from foreign theaters where the installation is their first stop within the customs territory of the United States. In addition, the military customs program may require preclearance for units or individuals deploying to areas outside the customs territory of the United States where an MCI has not yet been assigned or is not available. When conducting preclearance of mobilized or deploying soldiers' personal baggage while in formation, the MCI must ensure that the soldiers do not distract the PNDD team.

## VESSELS, BAGGAGE, AND CARGO INSPECTIONS

6-20. When establishing sterile areas, the MCI coordinates with the PNDD handler and ensures that the area is of sufficient size with minimal distractions to ensure the optimal working environment for the PNDD. When scheduling working parties for embarkation or debarkation, the PNDD handler coordinates with the MCI to ensure that sufficient personnel will be present to accomplish the mission in a timely manner, but not so many soldiers that they might present a distraction to the PNDD. If an examination has been deemed necessary by the MCI, the handler presents each piece of baggage or equipment to the PNDD.

## UNITED STATES CUSTOMS SERVICE MISSIONS

6-21. Missions in support of the USCS come from the MWD program manager via the KM. Handlers must have—

- Active government credit cards.
- Current passports.
- Appropriate civilian attire (or a civilian clothing allowance).
- Complete transportation requirements.

6-22. When temporarily assigned to the USCS, local USCS agents, policies, and procedures direct the PNDD team. Weapons are not authorized for handlers assigned to the USCS. The uniform of the day is determined by the USCS and is normally civilian attire.

6-23. Prior to deploying, every attempt should be made to conduct pretraining exercises with the USCS. These pretraining exercises will allow the MWD to acclimatize to larger quantities of contraband, which may be found while on the mission. The USCS will provide large training aids and determine the PNDD saturation point prior to deploying.

6-24. PNDD teams can expect to perform searches on aircraft, baggage, and tractor trailers as well as vehicle lines at US borders. When assigned to the USCS, the handler does not conduct physical searches of any kind. If the PNDD responds to an area, the handler advises a USCS agent and continues with the search, leaving the agent to conduct the actual physical search of the area.

6-25. Tractor trailers are searched in a USCS sterile area. Once the cargo has been offloaded and placed in standard lines, the PNDD team performs an exterior inspection of the vehicle before moving to all interior compartments and concluding with a systematic search of the cargo.

6-26. While assigned to a US border, the PNDD may be employed in vehicle lines. Vehicle lines are generally 10 to 20 vehicles in a row (placed bumper to bumper) that are inside a sterile area. The PNDD team travels up one side of the vehicle line and back down the other. The handler presents all productive areas. In the event that a USCS agent selects a vehicle for an individual vehicle inspection, that vehicle will be placed in a sterile area and the driver will open all the compartments of that vehicle. Once all occupants are clear of the vehicle, the PNDD will conduct a systematic inspection of the entire vehicle.

6-27. The PNDD will only be used in his narcotics detector role while deployed in support of the USCS or other federal agency. The PD capabilities of the PNDD will not be used for any reason. If the handler is instructed to use his PNDD in any method other than that of a narcotics detector dog, he must immediately notify his KM.

**PNDD Support to US Customs Counterdrug Missions**

In the summer of 2003, SGTs Payne and Mason from Fort Knox, KY, and SPC Johnson from Fort Sill, OK, were tasked to support a US Customs counterdrug mission along the southwest border between Mexico and the United States. After an overview and train up, the PNDD handlers and their assigned dogs went to work with the customs agents to detect illegal drugs at various points of entry into the country.

These teams became very proficient in detecting large quantities of illegal drugs. The PNDD teams detected narcotics in false floors, dashboards, gas tanks, seats, and hidden panels of many vehicles.

After three months of searching vehicles at checkpoints, the PNDD teams were responsible for detecting over 4,800 pounds of marijuana, over 95 pounds of cocaine, and other undisclosed contraband.

## RANDOM VEHICLE SEARCHES

6-28. The use of a PNDD at ACPs and installation gates can increase installation antiterrorism capabilities by providing a visual deterrent to terrorist groups and other criminal elements. The installation commander, task force commander, or base camp commander may task the PM to conduct random vehicle inspections according to local SOPs and the FPCON. When this occurs, the PM submits a schedule of random vehicle inspections to be conducted within a time frame that includes times, dates, and locations of inspections to be conducted. The schedule will be approved or disapproved and the inspection parameters will be dictated (will every vehicle be inspected or will vehicles be inspected at random according to the guidance provided) by the commander and returned to the PM. The PM then notifies the KM accordingly and provides him with a copy of the schedule with the commander's signature and inspection parameters. The PNDD's patrol capabilities may be used at ACPs to search for stowaways and personnel trying to gain access to the installation or a controlled area by hiding in delivery vehicles or other large cargo type vehicles. Extreme caution must be used when using the team in this manner. The commander must be aware that once the PNDD is given the command to search for personnel the MWDs aggression level will increase, thus increasing the chances of the PNDD biting personnel.

## ON-CALL DRUG RESPONSE

6-29. PNDD handlers will normally be placed in an on-call status. On call means the handler must remain within response time and distance according to the local SOP. When on call, PNDD handlers do not consume alcoholic beverages; however, they are in a duty position that allows them to wear appropriate civilian attire as well as leave the post according to the local SOPs.

6-30. When on call, PNDD handlers maintain positive control of a secondary communication device issued by the command. The secondary communication device may be a digital pager or cellular telephone as determined by the command. When recalled, the PNDD handler returns to the kennel in the most expedient legal manner to retrieve his PNDD. The PNDD will only be transported in an approved government owned vehicle (GOV). At no time will an MWD be transported in a POV. The handler may respond in appropriate civilian attire according to the local SOP.

6-31. KMs provide insight and guidance to the PM on technical matters of the SOP. The KM should provide recommendations as to acceptable distances which may be traveled while on call, required response times, the required secondary mode of communication, and any installation-specific items of interest, such as an alternate uniform based upon the customary civilian attire that is prominent in the area.

## Chapter 7

# Explosive Detector Dogs and Characteristics

In addition to the uses and characteristics of the PD, the PEDD is also trained to detect the odor and presence of explosives. This specialized ability to detect explosives makes the PEDD a valuable asset to help commanders and other government agencies maintain law and order, while enhancing safety procedures. PEDD teams can be employed in combat and/or force protection and antiterrorism operations worldwide. PEDD teams are valuable assets. The task force commander, the PM, and military police planners should consider the use of PEDD teams during combat support operations. This chapter details sound employment techniques and the capabilities of the PEDD teams that enable leaders to develop plans.

## INTERNMENT AND RESETTLEMENT OPERATIONS

7-1. The Army is the DOD executive agent for all EPW/CI operations. The military police is tasked with coordinating shelter, protection, accountability, and sustainment for I/R operations. The I/R function addresses military police roles with US military prisoners, EPWs, CIs, refugees, or other detained persons (during time of war/conflict). Although the combat support military police unit initially handles EPWs/CIs, modular military police (I/R) battalions with assigned military police guard companies and supporting MWD teams are equipped and trained to perform I/R operations for sustained amounts of time (see *FM 3-19.40*).

7-2. PEDD teams may be used for searching internees' belongings or confinement cells. PEDD teams should not be used to search a person. A PEDD is trained to attack with or without command; any sudden movements the internee may make could be perceived by the PEDD as an aggressive act. Safety must remain a priority for both the handler and military personnel.

7-3. The commander can also use PEDD assets to search the property of attached military personnel when needed. This is especially helpful when new units are rotating or replacing old units. These types of searches ensure that no explosives are taken out or brought in by US forces.

## POSTAL OPERATIONS

7-4. Postal inspections, including mail storage areas and drop boxes, can be conducted in conjunction with an MCI and according to *DOD Directive 5030.49*. The postmaster or MCI can request PEDD support from the area PM or military police brigade S3 with coordination through the KM team.

7-5. The PEDD team can be used to search mail that is received or shipped. Searches can be conducted as the mail comes off or before it goes on the truck prior to being shipped. The PEDD can search up and down the conveyer belt as the mail is removed from or placed on the truck. If a conveyer belt is not being used, the handler can instruct postal personnel how to arrange the mail in a secure area so that the dog can search the mail more effectively. When a PEDD team is used in the close quarters of a mailroom or delivery area, all nonessential personnel should remain clear of the area while the team is actively working.

7-6. When the dog responds to a parcel, indicating the possible presence of explosives, the area is evacuated immediately. Military police secure the area and set up a temporary CP a safe distance from the suspected explosives. Use the stand-off distance according to *Appendix I,* as applicable. Use the PEDD team to search the safe area for any possible secondary explosives. Explosive ordnance disposal (EOD) personnel are requested to dispose of the explosives or render them safe. EOD personnel may give explosives and improvised explosive device (IED) components to federal agencies for further investigation or turn over the scene to federal agencies after explosive hazards have been neutralized. The PEDD team remains at the temporary CP and resumes detection operations once EOD personnel remove the explosives.

# CHECKPOINTS AND ROADBLOCKS

7-7. PEDD teams can be used at checkpoints and roadblocks to supplement the control and movement of vehicles, personnel, and materiel (*Figure 7-1*). PEDD teams are added to security forces to detect explosives and to prevent illegal actions that may aid the enemy. Using these control measures deters terrorist activities, saboteurs, and other threats.

**Figure 7-1. Checkpoint Search**

7-8. When using PEDD teams at hasty or deliberate checkpoints and roadblocks, a search area must be designated for vehicles and a separate area must be designated for the vehicle occupants or any dismounted personnel. The PEDD team conducts a search of the vehicle only after security forces at the checkpoint ensure that all of the occupants have exited the vehicle and that a threat to friendly forces does not exist. Security forces at the checkpoint provide overwatch while the PEDD team searches for explosives.

***Note:*** At no time should a PEDD handler be used to search the interior of a vehicle. PEDD handlers are trained on searching techniques with a PEDD and should not be used for security force tasks.

7-9. In the event that the dog indicates the presence of explosives, the PEDD team withdraws immediately from the vehicle. The area is evacuated and EOD personnel and the chain of command are notified. The PEDD team stays behind a barrier at the safe-distance location in case the EOD team leader requests further detector assistance. (See *Appendix I* for stand-off distances.)

***Note:*** Be sure that radio operations are stopped until the PEDD team and security forces reach a safe distance from the suspected explosives.

# CORDON AND SEARCH OPERATIONS

7-10. Military police leaders and other task force commanders should consider using PEDD teams when conducting cordon and search operations. PEDDs are trained and able to greatly improve the combat force's success in uncovering hidden explosives and people.

7-11. Once the military police or other security forces have established the cordon, the PEDD teams should move in and operate with the search elements. During the search, the PEDD teams provide warning of hidden explosives or persons while the rest of the search element employs overwatch and search techniques.

7-12. The search element leader directs the actions of the team if people or explosives are detected. The PEDD handler should be prepared to advise the commander on the capabilities of the dog and actions required by the rest of the team to ensure the PEDD team's safety. It is important that all members of cordon and search teams rehearse together before conducting operations.

# FORCE PROTECTION AND ANTITERRORISM OPERATIONS

7-13. PEDD support for force protection and antiterrorism operations can be requested anywhere in the world and in different circumstances. In forward deployed areas or hostile environments, PEDD teams are often requested to ensure that any suspected explosives are detected before harm can come to US Army, joint, and allied forces. In peacetime environments around the world, PEDD teams perform the detection operations much the same as in combat support operations. The environment surrounding explosive detection operations does not interfere with the explosive detection capabilities of the PEDD team.

# SUSPICIOUS/UNATTENDED PACKAGE RESPONSES

7-14. Once a suspicious/unattended item that may contain explosives has been discovered, the military police or security forces on the scene request a PEDD team. Normally, military police evacuate the area and establish a CP a safe distance away from the threat using the stand-off distance listed in *Appendix I* of this manual. The PEDD team checks the perimeter and secures the area for secondary explosives. EOD personnel should respond to the scene and make contact with the PEDD team to be briefed on the location of the package. The PEDD team moves to within 50 meters of the package and begins an area search up to package. The PEDD team also scans the immediate area for secondary devices, when applicable.

> ***Note:*** At no time should a PEDD team be used to search the package itself. EOD personnel have received special training in handling suspicious/unattended packages and determining if a suspicious/unattended package contains explosives.

7-15. A PEDD handler must not present, touch, open, or physically disturb a package that is suspected of concealing explosives. Once the MWD team has discovered a suspicious item, EOD personnel assume responsibility for the area and take control of the situation in order to render the package safe or dispose of it as necessary.

> **PEDD Operations in Support of Operation Iraqi Freedom (OIF)**
>
> In April of 2003, SGT Raymond Rivera was performing explosives detection missions with his PEDD Kevin in support of OIF near Baghdad, Iraq. SGT Rivera was requested to sweep an area that a support element was in the process of clearing. There was so much debris that a forklift was being used to move large amounts of furniture and building supplies.
>
> The PEDD team would search a portion of the area and then the forklift would come in and haul off the debris searched. After the third load was hauled off, the dog responded to a mattress on the ground with an office table on it. Expecting an explosive devise, SGT Rivera notified the NCOIC of the detail and the area was cleared of all personnel (to a safe distance) and secured. EOD personnel were notified of the suspected explosives and responded to the scene.
>
> SGT Rivera was standing by at the safe area while EOD personnel uncovered two hand grenades that were buried in the sand just below the surface. SGT Rivera said, "Even though it was only two grenades, I was amazed that Kevin found them under all the debris. I'm glad we found them and not our soldiers."

# BOMB THREATS

7-16. A PEDD team should be requested for assistance in detecting explosives whenever bomb threats are made against persons or property for which the Army is responsible. The PEDD team's search for explosives begins only after the area or building has been evacuated of all personnel and security has been established. EOD personnel only respond when a suspect item is found; they do not conduct or participate in searches except in support of the USSS. EOD units place a team on standby once the unit is alerted of a bomb threat. PEDD teams coordinate this with the local EOD unit prior to any operation.

7-17. The PEDD team, EOD personnel, and site leadership collaborate to resolve the situation at a nearby CP (see *Appendix I* for stand-off distances). When PEDD teams are located on an installation, their use must be included in the bomb threat planning guidance.

7-18. A systematic search by the PEDD team should include the assistance of a spotter to ensure safety and accuracy. PEDD teams are to be used only to detect the possible presence of explosives.

> **WARNING:**
>
> **At no time will the handler touch, open, or otherwise disturb a possible explosive device. IEDs are easily disguised and can be triggered to explode by motion, sound, light, heat, or remote devices.**

7-19. Handlers and spotters must not change or otherwise disturb anything in the search. If the lights or any electric device are off or on, do not touch the switch until a thorough search is conducted.

7-20. Ensure that the handler and spotter are provided with available safety equipment including—

- A flack vest (in addition to body armor).
- Flashlights or helmet-mounted lights.
- Gloves.
- Safety goggles.

**WARNING:**

**Always keep track of how much time has elapsed when using a PEDD team at bomb threat incidents. Sufficient time must be maintained to ensure that the team exits the building or area well before the announced or suspected detonation time.**

7-21. If the PEDD responds to the presence of explosives, the PEDD handler and spotter note the location and leave the area. EOD personnel then begin measures to render the device safe or remove it. On the request of EOD personnel, the PEDD team may then continue the search until another suspected explosive is found or until the area is cleared by EOD personnel.

7-22. If the search is 30 minutes or longer in duration, explosive training aids should be planted periodically so that the dog can find them. This helps to keep the dog interested in the search. Training aids should only be planted in areas that have already been searched by the PEDD team.

## HIGH-RISK PERSONNEL SEARCHES

7-23. Military police provide the task force commander with valuable area security capabilities to protect C2 headquarters, equipment, and services essential for mission success. HRP security is an important task within the area security function. PEDD teams may be employed to enhance explosive detection and protection capabilities.

7-24. The KM or plans NCO determines which PEDD teams will perform the mission. The number of PEDD teams for any mission is determined by the threat, the number of PEDD teams available, any conflicting missions, and other environmental considerations noted in *Chapter 8*.

7-25. PEDD teams may be used as part of a combined arms operation or as an element of a larger security force. When used as part of a combined arms team, the PEDD teams need to be involved will all troop-leading procedures, including rehearsals. All personnel must have a complete understanding of the mission in order to create a safe and secure environment for the HRP. The PEDD teams provide explosive detection and personnel detection capabilities. PEDDs are also a deterrent to hostile forces or terrorists operating within the area of operations.

7-26. Prior to an HRP visit to an area, the PEDD team assists with force protection actions by searching for explosives and IEDs. PEDD also conduct searches of and provide security for outer perimeter areas. During HRP operations, the handler—

- Briefs the on-scene commander of the dog's capabilities.
- Communicates with EOD personnel.
- Obtains the HRP's itinerary and determines locations to be searched.
- Requests a spotter or helper to assist.
- Ensures that areas are secure after searches are done.
- Informs the commander of results and continues follow-on missions.

## ACCESS CONTROL POINTS

7-27. Using PEDD teams at ACPs can increase an installation's or base camp's force protection and antiterrorism capabilities by providing a visual deterrent to terrorist groups and other criminal elements. The PM notifies the KM of the requirement to perform vehicle searches at ACPs.

7-28. In the event that the PEDD detects the presence of suspected explosives, the military police forces at the ACP secure and evacuate the area (see *Appendix I* for stand-off distances). EOD personnel are then notified to render the item safe or remove it. PEDD handlers do not attempt to remove or otherwise disturb any suspected explosives.

7-29. PEDD team patrol capabilities may be used at ACPs to search for stowaways and personnel trying to gain access to the installation or a controlled area by hiding in delivery vehicles or other large cargo type vehicles. Extreme caution must be used when utilizing the team in this manner. The commander must be aware that once the PEDD is given the command to search for personnel the potential for injury exists.

# HEALTH AND WELFARE INSPECTIONS

7-30. PEDD teams conduct health and welfare inspections when requested by the commander and coordinated through the KM, PMO, and S3. The commander contacts the KM directly to schedule a health and welfare inspection. The KM, in conjunction with the installation PM, the CID special agent in charge, and the military police investigations supervisor, briefs all levels of the installation command on the availability and capabilities of the PEDD teams available to them.

7-31. When scheduling the inspection, the commander should have the following information available:

- The requested time of the inspection.
- The areas to be inspected.
- The number of living quarters.
- The number and size of common areas.
- Other pertinent information.

*Appendix H* provides a guide to assist handlers with health and welfare inspections.

7-32. Once the coordinating instructions have been established, the KM coordinates a time and location to brief the commander and demonstrate the PEDD's capabilities. It is important that the unit requesting the health and welfare inspection practice OPSEC and inform only those personnel who have an immediate need to know. This usually consists of the commander and 1SG or sergeant major. The KM advises the PM and the military police investigations supervisor when the inspection is scheduled and if any additional support is needed.

7-33. A systematic inspection of the living quarters and common areas is conducted. If the PEDD responds to a locked wall locker, the soldier assigned to that wall locker is summoned to open it. Once the wall locker has been opened, unit representatives conduct a detailed search of the wall locker and its contents.

# AIRCRAFT AND LUGGAGE SEARCHES

7-34. Before beginning the search, ensure that all personnel have exited the aircraft and that all compartments are opened. The PEDD team should search the aircraft before searching any luggage or cargo. The PEDD team conducts a fresh-air perimeter search from the downwind side when approaching the aircraft. Once a perimeter search has been conducted, the team boards the aircraft at either the front or rear. Once the area has been deemed clear, the handler ensures that only one door is open to limit draft and wind current. A systematic search of all areas within the aircraft is conducted from the top to the bottom.

7-35. If the dog alerts the handler to a possible presence of explosives, military police or security forces evacuate and secure the scene. The PEDD team moves a safe distance away from the suspected explosives but remains available to further assist EOD personnel, if requested. Once EOD personnel clear the area, the PEDD team may resume the search until another suspected explosive is found or until the area is cleared by EOD personnel.

7-36. When conducting inbound baggage inspections, the items should be aligned in a sterile area in rows with at least six feet between them. The handler should conduct an on-leash search, presenting each piece of baggage. When searching cargo, every effort should be made to search the cargo before it is palletized. The cargo should be aligned in a sterile area in rows with at least six feet between them (the same as baggage). The handler should conduct an on-leash search, presenting cargo as required. Once the PEDD team has searched the cargo, it may be palletized while remaining in the sterile area prior to embarkation.

7-37. Conducting debarkation operations is the same as inbound operations. The host unit (to which the cargo belongs or is being delivered) provides a representative (E6 or above) and work detail to assist the PEDD team. When pallets are removed from the aircraft, they are placed in a sterile area. Once in the sterile area, the work detail unloads the pallets and arranges the cargo as required by the PEDD handler.

# Chapter 8
# Veterinary Support

At most locations, veterinary support is provided by an assigned Army VCO. When there is no assigned VCO, an attending VCO from a nearby military installation is assigned responsibility for providing veterinary support. The Army provides veterinary support for all MWDs as authorized by *DOD Directive 6400.4* and *ARs 40-1, 40-3*, and *40-905*. Emergency, civilian veterinary support is authorized for MWDs when there is not a VCO available and/or care is beyond the capability of the veterinary treatment facility (VTF). The responsible VCO and Army Veterinary Service chain of command are responsible for establishing standard procedures for approval, provision, and payment for emergency, civilian veterinary support.

## VETERINARY RESPONSIBILITIES

8-1. The US Army Surgeon General provides professional veterinary support for the entire MWD program through the Army Veterinary Corps. This responsibility includes—

- Medical and surgical care at training facilities, bases, and installations.
- Inspections to ensure that the kennel facilities are sanitary.
- A professional review of plans for new construction and modification of kennels, support buildings, and sites.
- Prescribing an adequate feeding program.
- Instructing handlers and supervisors in all matters related to the health of MWDs.
- Conducting research to improve the MWD program.

8-2. The responsible VCO provides treatment for the diseased or injured MWD at the kennel site or at the installation VTF. The Army Veterinary Corps is responsible for equipping this facility and for providing medical and surgical supplies. The VCO is responsible for the MWD veterinary treatment record (VTR) and related information about examinations, immunizations, and treatment.

8-3. The VCO instructs handlers about dog health, care, feeding, and first aid. This instruction helps the handler to develop a better understanding of the MWD's health needs and improves the handler's ability to care for the MWD. The VCO prescribes an appropriate feeding program based on the MWD's health, the climate, and working conditions.

8-4. The VCO is involved in the review of all plans for new kennel construction or kennel modifications. This ensures that potential health and safety hazards can be corrected before construction begins.

### VETERINARY INSPECTIONS

8-5. At least quarterly, the VCO inspects the MWD kennel facility to make sure that proper sanitary standards are being maintained. It is recommended that an animal care specialist conduct a monthly check of the kennels to ensure that no major safety violations are present. The VCO ensures that insect and rodent control efforts are adequate and that the handlers are properly maintaining the general health of the MWD. The quarterly inspection also includes the following elements:

- An examination of the kennel facilities for safety hazards and distractions that may interfere with the rest and relaxation of the MWDs.

- A review of the adequacy of the kennel structure (particularly for environmental conditions) and feeding and watering schedules. The VCO should make recommendations to help prevent disease and injury.
- A review of kennel sanitation and cleaning procedures and MWD food storage procedures to minimize the risk of disease transmission to the MWD.

## MANDATORY TRAINING

8-6. The local VCO is required to provide training to the MWD handlers upon initial assignment and at least annually thereafter. More frequent training of handlers is recommended but not required. The Department of Defense Military Working Dog Veterinary Services (DODMWDVS) recommends that handler training be conducted at least quarterly. Topics include the following:

- Procedures on when and how to obtain veterinary support (an SOP is appropriate for this).
- General animal health.
- Grooming.
- Prevention of environmental injury (such as heatstroke or cold injuries).
- Emergency first aid (for bloat [gastric dilatation-volvulus], heat stroke, snake bites, and cold injuries).
- Administration or application of routine medications.
- Proper MWD feeding.
- Feed sanitation (proper storage, preparation, and cleanup).
- MWD chemical, biological, radiological, nuclear, and high-yield explosives protective measures.

8-7. It is best to schedule mandatory training so that seasonal problems are addressed prior to their occurrence and appropriate refresher training is provided prior to a deployment. This training is not meant to turn handlers into animal health technicians, but to give them the tools needed to provide emergency care until they can acquire veterinary support.

## WEIGHT CHECKS

8-8. Each month the MWDs body weight must be recorded and sent to the responsible VCO. The KM may communicate this to the VTF. Body weight is also measured and recorded monthly in order to monitor body condition and detect important trends in weight loss or gain that may indicate illness. The KM and/or handler documents data such as food intake, stool consistency, and vomiting in the monthly records.

## RECORDS/HEALTH CERTIFICATES

8-9. VTRs are maintained by the local VTF. It is very important that MWD handlers ensure that all health information is annotated in these records.

8-10. An examination is given just before shipping an MWD from one installation to another. The MWDs are examined to detect injury or disease and to support the safe shipment and continued good health and performance of the MWD. A health certificate is issued for the state or country into which the MWD is being shipped for TDY and must accompany the MWD. When the TDY period exceeds 10 days, a health certificate may be required for the MWD's return to his home station. A VCO signs this certificate no more than 7 days prior to reembarkation. If the TDY location is in a different Army VCO's area of operation, the gaining VCO is notified prior to the MWD's departure from the home station. An examination is given as soon as possible, but not more than 10 days after the arrival of an MWD at a new assignment, TDY, or deployed location and within 3 days of the MWD's return to his home station. It is important that the handler who deployed with the MWD is present for the examination so that the VCO can get accurate and complete information on any health issues while the MWD was deployed.

## Feeding and Food Control

8-11. MWDs require a diet that is significantly different from that of pet dogs. Their work demands much higher levels of energy and larger quantities of essential nutrients. The standard high-performance diet, contracted and supplied by the GSA for MWDs, meets these nutritional needs and is the only approved feed unless otherwise directed by the attending veterinarian.

8-12. Special diets may be procured and fed to individual MWDs when the VCO determines that other than the standard diet is required. Normally, these diets are available from the VTF, but the MWD unit is responsible for procurement costs.

8-13. Some MWDs have been trained using a food reward schedule that also requires special food. Food reward is used only if an MWD fails to respond to any other type of reward. Procure the special food through normal supply channels or authorized local purchase. The amount of food each MWD should be fed depends on his weight, the amount of activity, and the climate. Local VCOs determine the proper amounts to feed. The responsible VCO must be told when an MWD is trained using food.

8-14. The VCO also prescribes the time of day each MWD is to be fed. This depends on the MWD's duty schedule and the schedule of other kennel activities. MWDs should be allowed 2 hours to eat; leftover food is disposed of within another two hours and feeding pans are cleaned and put away. Never leave uneaten food in the kennel past the authorized 2-4 hour feeding period. If the MWD finishes his meal prior to the end of the feeding period the feed pan may be removed.

## Parasite Control

8-15. MWDs are at risk of becoming infested with internal parasites (intestinal worms and heartworms), and external parasites (fleas, ticks, and mange). Some of these parasites may also transmit infectious diseases to the MWD. In order to prevent infestation and disease transmission, all MWDs must be on a routine parasite prevention program supervised by the responsible veterinarian. At a minimum, this must include the administration of a monthly oral (pill or chewable) heartworm preventative and internal parasite preventive and the administration of a topical, veterinary-prescribed external parasite preventative against fleas and ticks.

8-16. When deployed to areas with high parasite risk, additional preventive measures such as prescription flea and tick collars and the oral antibiotic doxycycline will be prescribed by the supporting veterinarian. Doxycycline must be given with food. The handler must ensure that adequate supplies of all preventives are issued by the VTF prior to departing for TDY or deploying. Flea and tick collars will only be used when prescribed by the veterinarian and must be removed when the MWD is not under direct physical control (when off the choke chain and leash) of the handler to prevent accidental ingestion.

## Food Storage

8-17. Food should be stored in a sealed container that will not allow contamination or rodents to enter the food (for example, pail, utility 7 gallon, NSN 7240-01-411-0581). Bulk storage of food must be in an area separate from the storage of other equipment, cleaning supplies, or materials that may contaminate the food. Bags should be elevated 4-6 inches above the ground, and neatly stacked with a 4-inch space between the stacks and the walls. The sanitation standards for the storage of MWD rations are identical to those required for the storage of bulk rations for personnel in a troop issue subsistence activity, exchange, or commissary.

## Deployability Categories

8-18. The purpose of reporting MWD deployability status is to assist the MWD program managers, PM, commanders, and KMs with the selection of MWD teams for deployment and TDY and to rapidly identify MWDs with limited duty requirements. These guidelines take medical and physical fitness factors into consideration including—

- The anticipated MWD duty stress at deployed locations compared to the home location.
- The availability of veterinary support at the deployed location.

- The climate and environment of the deployed location.
- Any diagnosed or suspected medical conditions that may affect the MWD's performance.
- The MWD's physical conditioning and stamina as observed by the KM and handler.

8-19. It should be noted that some determinations are and will remain subjective; thus the need for prudent judgment by all individuals involved in these assessments remains. This is particularly true regarding the effect of age on an individual MWD. Old age is not a disease and MWDs are not downgraded just due to age; however, with increased age comes increased incidence of illness or disability. These problems may warrant a change in deployability status.

8-20. Deployability status is initially determined after the VCO examines the MWD at the receipt physical or semiannual physical exam. The KM determines training and proficiency status. The VCO and KM should discuss their findings and then categorize each MWD's status and forward the listing through the KM to the PM and MWD program manager. Updates to deployability status lists are performed at least monthly or as needed as changes occur in an individual MWD's status.

### Category 1—Unrestricted Deployment (Both Outside the Continental United States and in the Continental United States)

8-21. The MWD is medically fit for any contingency or exercise. There are no limiting or compromising factors. Medical factors may exist but will not limit performance. Special diets, controlled drugs, or special medications can be sent with the MWD (or be carried by the handler) in a quantity sufficient to last the full duration of the anticipated deployment.

### Category 2—Limited Deployment (Continental United States or Home Theater of Operation)

8-22. The MWD is medically fit for regions and missions with a minimal requirement for acclimation to heat or physical stress where complete veterinary support is available. The MWD is fit for short duration deployments. There are no significant limiting or compromising factors. Medical problems may exist which slightly limit performances but are controllable. Special diets or medications are required; however they are available in the deployed location or can be sent with the MWD. The reason for limited deployability must be reported.

### Category 3—Temporarily Nondeployable

8-23. A medical condition exists that impedes daily duty performance and is under diagnosis, observation, or treatment. The MWD may still be allowed to work at the home station but in a limited capacity.

### Category 4—Permanently Nondeployable

8-24. An unresolved medical or physical problem exists that frequently or regularly impedes daily duty performance and an early retirement date cannot be given. The reason for nondeployability must be reported. Medical or physical conditions warrant replacement within one year. The responsible veterinarian and KM should initiate an MWD disposition process at the time an MWD is declared category 4.

## Kennel Sanitation Procedures

8-25. Cleanliness is one of the most important factors for the good health of an MWD. The KM must enforce sanitary measures in and around the kennel area. A good standard of sanitation is the result of the cooperation of the handlers, KMs, supervisors, and the VCO.

8-26. The VCO and the PM or military police commander should set the standard of sanitation. This standard must be maintained by each of the handlers and the KM.

8-27. Kennels must be sanitary, in a good state of repair, and thoroughly cleaned every day. Kennels should be disinfected at least once every week using only those disinfecting products approved by the VCO. Kennels should also be disinfected whenever an animal is removed from a kennel so that the kennel will be ready to be occupied by another animal.

8-28. Stool must be removed from the runs as often as necessary as but no less than twice daily to prevent the MWD from becoming dirty. The method of disposing of stool depends on local conditions and the type of sewage system present. If stool must be carried from the area in cans, the cans must be cleaned and disinfected after each use.

8-29. The entire kennel area must be free of refuse and garbage that could attract rats and insects. Mosquito control measures must be used in ditches and swampy areas in the vicinity of the kennels. Disinfectants and disinfectant procedures should be used only with the approval of the VCO.

### MEDICATION

8-30. At times, a VCO prescribes special medication (to be given separately or mixed with food) for a sick or injured MWD. The handler must know how to administer these medications in both pill and liquid form.

8-31. When any foreign substance is placed directly into a dog's mouth, his first reflex is to spit it out. The handler must learn how to administer medication properly so that the MWD is forced to swallow. With the VCO's approval, pills or liquids can be mixed with a small amount of canned dog food or other soft food and fed to the MWD. To administer capsules or tablets—

- Place the fingers of your left hand over the MWD's muzzle and insert your left thumb under the lip and between his upper and lower teeth right behind the canine tooth, pressing your left thumb against the roof of his mouth to open it.
- Place the capsule or tablet into the MWD's throat at the extreme rear of the tongue to prevent him from spitting it out. Quickly remove your hand, close the MWD's mouth, and gently massage his throat until he swallows. The entire procedure must be quick and smooth to ease the MWD's apprehension and resentment.

8-32. Liquid medication is best administered with the help of another person. The handler—

- Holds the MWD's upper and lower jaws together with his left hand. The assistant pulls the MWD's lips away from the teeth at one corner of the mouth with his right hand. The MWD's nose is then pointed slightly upward, forming a natural funnel by the lip, and the assistant pours the liquid into this funnel.
- Elevates the MWD's head only slightly above the horizontal when giving liquid medicine. If the MWD's head is raised higher, he has difficulty in swallowing. Give the MWD adequate time for swallowing to prevent the liquid from getting into the trachea, nose, or lungs. Use extreme caution in giving oily liquids.
- Allows the MWD to lower his head and rest before proceeding if any signs of distress, such as coughing or struggling, appear. Do not give oral medications or any liquids if the MWD is unconscious or cannot swallow.

## FIRST AID

8-33. Normally, the handler's early recognition of symptoms of illness or injury allows sufficient time to get assistance from a VCO. However, situations may arise when medical help is not immediately available. The seriousness of the incident may require that the handler take emergency actions to protect the life or health of the MWD. The first aid instructions that follow are commonly used to save life, prevent further injury, and reduce suffering from pain. In all emergency situations, notify the VCO as soon as possible so that the MWD can receive professional medical attention.

8-34. First aid kits should always be available in the kennel area and at training sites. The handler should also carry a first aid kit as part of his or her equipment for the MWD on all operational missions. The VCO will determine the contents of first aid kits. When items are used from the first aid kit, they should be replaced with new items immediately.

## MUZZLES

8-35. When an MWD has been injured, the first consideration is to calm and immobilize the animal. Pain and distress, however, may cause the MWD to respond to the handler in an unpredictable manner. The MWD may not respond to verbal commands and may attempt to bite the handler and anyone helping the handler. Whether to apply a muzzle (*Figure 8-1)* or not depends on the nature of the emergency. If the MWD is unconscious, shows any difficulty in breathing, or has suspected head injuries, a muzzle should not be used. Otherwise, a muzzle should be used for safety.

**Figure 8-1. Muzzle Use**

8-36. There are several types of muzzles, but the leather basket muzzle is the best and most comfortable. It allows for free breathing and causes the least alarm and apprehension. The MWD can still inflict a wound while wearing this type of muzzle, so caution is required.

8-37. An improvised muzzle can be made using the MWD's leash and is called a leash muzzle. To apply the leash muzzle, tighten the choke chain on the MWD's neck by pulling the leash tightly with the right hand. Place the left hand, palm up, under the choke chain on the neck. Grasp the leash tightly as it passes through the palm of the left hand, wrap the leash once around the MWD's neck, and bring it up and across the left side of his head. Finally, wrap the leash twice around the muzzle and grab it tightly with the left hand.

**CAUTION**

The leash muzzle may be used when the leather muzzle is not available or when it would not provide adequate safety. Do not use the leash muzzle when the MWD is overheated, is having difficulty breathing, or is indicating that he may vomit. Do not leave it on for long periods in hot weather.

## FRACTURES

8-38. If a fracture occurs, restrain the MWD and encourage it to lie down and remain quiet. If the face and jaw are not injured, put the MWD in a loose muzzle. Notify the veterinarian immediately when a fracture occurs or is suspected. If there is any bleeding, treat the MWD according to *paragraph 8-40,* then proceed with splinting and moving the MWD to the veterinarian if he is not immediately available.

8-39. Prior to moving the MWD, the fracture area needs to be immobilized with a splint. This should be possible for most leg fractures. The leg should be splinted in the same position it is placed by the MWD when it is hurt. Do not try to straighten or reduce the fracture. Splinting an MWD's leg is similar to splinting a soldier's leg and may be done with the same materials: bandages, torn cloth, splints from the medic's bag, or other available material including sticks, boards, and rolled newspaper. After the leg is splinted, or if the fracture cannot be splinted, gently move the MWD onto a litter or improvised litter, such as a piece of plywood, or into a kennel and carry it to a vehicle for transport. It is best to slide the litter under the MWD rather than lifting the MWD onto the stretcher. The MWD may need to be strapped to the litter to allow safe movement. Make sure to support the injured leg or area during movement.

## BLEEDING WOUNDS

8-40. Bleeding must be quickly controlled, particularly wounds in the foot or leg that are prone to bleed freely. Applying pressure directly on the wound may control bleeding; use a sterile bandage or clean handkerchief or pinch the edges of the wound with your fingers. As soon as possible, apply a pressure bandage.

## BURNS

8-41. Most burns occur when an animal comes into contact with hot water, grease, tar, or other scalding liquids. MWDs may also get electrical burns by chewing electrical wires. If an MWD is trapped in a burning building, it may suffer from smoke inhalation in addition to surface burns. The handler can prevent all of these situations through positive control of the MWD. If a burn does occur, applying cold-water soaks or ice packs to the burn for approximately 20 minutes may treat minor burns. The cold helps to reduce the pain. The hair around the burn should be clipped away and the burn should be washed gently with a surgical soap. Blot it dry with sterile or clean soft gauze or cloth. Protect the burned area from rubbing by applying antibiotic ointment and a loose fitting gauze dressing.

## SHOCK

8-42. An animal may go into shock after injuries to internal organs, excessive bleeding, or trauma. Shock can be recognized by the following symptoms:

- A glassy look to the eyes.
- A rapid or weak pulse or rapid shallow breathing.
- A dropping body temperature. The lips and feet may feel cold.
- A paleness of the membranes of the mouth and eyes.
- Slow capillary refill time. To determine capillary refill time, press firmly against the MWD's gums until they turn white. Release the pressure and count the number of seconds until the gums return to their normal color. If it is more than two to three seconds, the MWD may be going into shock. Failure to return to the red-pink color at all indicates that the MWD may be in serious trouble and needs immediate assistance.

8-43. If a handler suspects that an MWD has internal injuries or is going into shock, request help from a VCO immediately. Keep the MWD warm and quiet, and lower his head to prevent possible brain damage. If it is necessary to move the MWD, use a litter. Administration of intravenous fluids by appropriately trained personnel will help to prevent permanent injury from shock.

## ARTIFICIAL RESPIRATION

8-44. There are conditions that may cause an MWD to stop breathing. When this happens, do not panic. Follow these procedures:

- Open the MWD's mouth and check for any obstructions. Extend the MWD's tongue and examine his throat.
- Clear the MWD's mouth of any obstructions then, close his mouth, and hold it gently closed.
- Inhale, and then cover the MWD's nose and mouth with your mouth.
- Exhale gently. Do not blow hard. Carefully force air into the MWD's lungs and watch for his chest to expand. Repeat every five to six seconds or at a rate of 10 to 12 breathes per minute.

### Snakebites

8-45. The bite of a poisonous snake can cause serious illness and death if not treated immediately. Keep the MWD quiet and calm, and request veterinary assistance or move the MWD as quickly as possible to a treatment facility. Panic or exertion causes snake venom to move more rapidly through the bloodstream. If possible, kill the snake so it can be shown to the VCO. The handler should be careful not to be bitten by the snake or come into contact with the snake venom. Bites may occur on the face or neck of the MWD. When this happens, remove the choke chain and loosen or remove the collar and muzzle. Swelling occurs rapidly after a snakebite, and this equipment may restrict breathing. Position the MWD's head extended from his body to allow him to get maximum airflow. An ice pack applied to the bite area helps to slow the flow of blood and helps to keep the venom from spreading.

**CAUTION**

A tourniquet should not be applied. Constricting the blood flow could increase tissue death.

### Foreign Objects in the Mouth

8-46. An MWD may get a stick or other foreign object stuck in his mouth or throat. The most common foreign object that becomes stuck is a Kong® or ball reward that is too small for the MWD. The MWD may cough, gag, have difficulty swallowing or breathing, paw at the mouth, and drool. If the MWD is having difficulty breathing, cautiously and gently open his mouth. Look for any abnormal object in the throat, under the tongue, between the teeth, in the gums, or in the roof of the mouth. If the MWD is having trouble breathing, removing the object is an emergency. The ball, Kong, or other object can be removed using pliers or clamps. If the object is not causing breathing problems, it is best to leave it in place and get veterinary assistance.

### Poisonous Substances

8-47. There are many toxic chemicals an MWD may come in contact with and/or ingest. These include insecticides, herbicides, rodenticides, and antifreeze. The symptoms of poisoning vary. Unless the handler is certain the MWD has eaten a poison, do not treat for poison. Occasionally, MWDs accidentally swallow narcotics or explosives training aids. Exposure to poisons and accidental ingestion of training aides should be prevented by positive control of the MWD by the handler. If the handler knows that the MWD has eaten a poison, narcotic, or explosive, take the following actions:

- Request veterinary assistance immediately.
- Determine the type and quantity of poison, chemical, narcotic, or explosive that has been swallowed. If any part of the substance or container is available, keep it for the VCO to examine.
- If the Army VCO is not available, immediately make contact with a civilian veterinary to determine if vomiting is appropriate. Causing an animal to vomit is not recommended in all types of poisoning as it may cause more harm to the animal. When vomiting is appropriate, the animal may be induced to vomit by giving him one-quarter to one-half of a cup of hydrogen peroxide.
- Keep the MWD quiet and warm until the VCO arrives.

8-48. For future reference, keep a copy of the material safety data sheet for any training aide that may be ingested available. MWD first aid should be according to the material safety data sheet for people.

### Overheating

8-49. Overheating results when an MWD is unable to eliminate body heat rapidly enough. This condition requires immediate action by the handler to save the MWD's life. During hot, humid weather, an MWD may easily become overheated during training, during operations, or while being transported. A body temperature of 105° Fahrenheit or more, combined with poor response to commands, weakness, unsteady movement, vomiting, difficult or labored breathing, convulsions, and collapse are all symptoms of overheating. It is helpful to take the MWD's temperature in order to monitor the severity of overheating and progress in cooling, but temperature alone does not determine the need for or method of treatment.

8-50. For mild to moderate heat injury, the MWD's temperature is generally lower than 104° Fahrenheit. The MWD may have mild symptoms, to include uncontrolled panting, a rapid but somewhat weak pulse, mild to moderate difficulty breathing (but not cyanotic [blue]), and exercise intolerance or loss of stamina. This is equivalent to heat exhaustion in people.

8-51. When the symptoms of overheating occur, loosen the leash, collar, and muzzle as much as possible without loosing control of the MWD. This allows for more efficient panting and cooling. Follow these guidelines at the onset of symptoms:

- Carry the MWD to the nearest shade and fan him if no wind is present.
- Try to quickly lower the MWD's body temperature by running and sponging cool water over the MWD's head, body, and legs. If a body of water is available, allow the MWD to stand in the water but make sure the MWD's head remains above water. Running water works better than standing water.
- Place the MWD in an air-conditioned kennel, vehicle, or shelter.
- Apply ice packs to the MWD's groin, armpits, and head to increase speed of cooling.
- Dampen, but don't soak, the MWD's feet, ears, and abdomen with rubbing alcohol. Allow it to evaporate and then reapply more.
- Do not let the MWD over drink. He should not consume more than one cup of water until he is calm and his temperature returns to normal.
- Monitor the MWD's temperature every 5-10 minutes, and stop cooling efforts when it reaches 100-103° Fahrenheit. Keep monitoring his temperature for an additional 60 minutes.

8-52. If the MWD's temperature remains normal (about 100-103° Fahrenheit), treatment is complete; however the MWD must be examined by a VCO and a heat injury must be reported in his VTR. If his temperature climbs again, restart cooling efforts.

8-53. If the MWD's temperature drops below 100° Fahrenheit, begin warming by drying and wrapping him with sheets and blankets. Remove the heat when his temperature goes above 101° Fahrenheit.

8-54. For severe heat injuries, the MWD's temperature is generally over 104° Fahrenheit combined with poor response to commands, weakness, unsteady movement, vomiting, difficult or labored breathing, uncontrolled panting, convulsions, collapse, a dull or depressed attitude and behavior, a rapid but somewhat weak pulse, exercise intolerance or loss of stamina, and cyanosis (blueness) or dark red mucous membranes. This is equivalent to heatstroke in people. If this occurs take the following actions:

- Initiate all cooling and monitoring efforts as noted in *paragraph 8-51*.
- Administer 1 liter of intravenous fluids to the MWD as quickly as possible if appropriately trained personnel are available.
- Continue cooling and monitoring until a veterinarian can evaluate the MWD.

## BLOAT (GASTRIC DILATION-VOLVULUS)

8-55. Bloat is an acute stomach enlargement and twisting that may be due to gas, food, or water. It may occur if the MWD is fed immediately before or after hard exercise or when he is returned to his kennel after work or exercise and allowed to drink too much water.

8-56. An enlargement may be seen just behind the ribs, primarily on the left side. The MWD may also be restless and show signs of pain in the abdominal region. He may attempt to vomit or have a bowel movement. His breathing may be difficult or labored due to pressure from the enlarged stomach.

8-57. Handlers should notify the VCO and stop all watering and feeding. Walking may enable the MWD to relieve himself through bowel movements or passing gas. Most cases require extensive treatment by the VCO.

8-58. To prevent bloat, MWDs should not be fed within a two-hour period before or after hard work or vigorous exercise. It is also safer to provide multiple small meals each day, feeding twice daily, rather than one large meal. In hot weather, give small amounts of water during training or working MWDs to prevent excessive thirst. For the first hour after working or training, only three inches of water in the bucket should be available. After this cooling-off period, more water may be given.

### First Aid Kit

8-59. The responsible VCO determines the contents of MWD first aid kits. The VCO should consider the skill of the handlers, access to veterinary support, and emergency situations that may occur. The DODMWDVS does not have a standard first aid kit supply list because needs vary. The local VCO and KM are responsible for developing and putting together an appropriate kit and training the handlers how to use it. The kit should contain supplies for cleaning and dressing minor wounds, cleaning and protecting an eye injury, bandaging or splinting minor breaks, and removing mild superficial foreign bodies (splinters or glass). It should also contain peroxide, antibacterial solutions, and a small supply of any prescription medications needed for individual MWDs.

## ENVIRONMENTAL EFFECTS ON THE MILITARY WORKING DOG

8-60. Just as the environment affects soldiers performing their duties, the MWD is affected by his surroundings. It is important for planners and leaders to become aware of the many different conditions that can affect the performance of MWDs. Not all MWDs are affected in the same manner. Acclimation times will vary between conditions and MWDs. Planning considerations should always include the VCO when the MWD will be used or trained in environments that he is not accustomed to.

### Wind

8-61. Always consider the direction and velocity of the wind when using an MWD for any detection operations. This consideration holds true even indoors. There may be a slight breeze that will move a scent away from the location of its source. Fans, air conditioning units, and objects in motion can create winds that will affect the performance of a detector dog.

8-62. In addition to these operational concerns, wind may create medical issues (such as blowing sand) that are negative. Wind also creates increased cooling, which is positive in hot weather but negative in cold weather.

### Inclement Weather

8-63. Precipitation can degrade the MWD's performance of detection operations; however, do not assume that an MWD will not be able to perform satisfactorily. The amount of moisture accumulated and the amount of time elapsed since the scent was left are variables. MWDs have been proven to be effective even in inclement weather. MWDs that become wet and are returned to their kennels or crates while wet are at increased risk of skin injury and infection.

### Heat

8-64. When operating in locations of extreme heat, the MWD should be allowed sufficient time to acclimate. Consult with the VCO at the location to determine how long an MWD should be allowed to acclimate. Care should be given to ensure that the MWD does not sustain injuries to the pads of his paws due to excessive walking on concrete or pavement in extreme heat conditions. Always ensure that the MWD has a fresh and constant source of drinking water.

### Cold

8-65. Extreme cold temperatures should not deter MWD operations. Special care should be taken to enable an MWD to perform in cold locations. The accumulation of moisture on the MWD's paw pads can freeze and injure him. Using paw pad protectors to aid in the prevention of injury due to cold weather may be helpful, but the MWD must be trained to wear them. The MWD should be allowed to rest and warm up on the same cycle as the handler.

## TERRAIN

8-66. The safety of the handler and MWD must be given the highest consideration when operating in hazardous terrain. Care should be taken when employment will take place in rocky or mountainous areas. Unfamiliar locations can pose threats in the form of caves, holes, loose footing, and other hazards. Whenever possible, conduct a reconnaissance of the area before employing an MWD.

## DISTRACTIONS

8-67. There are numerous distractions that can affect MWD operations. These are conditions that can cause the MWD to focus his attention away from the desired task at hand. Planning and consideration should be given to eliminate or reduce distractions that hinder operations. Distractions are natural and man-made. Some examples of distracters are:

- Other dogs (both MWD and indigenous).
- Other animals.
- Human presence and activities.
- Noise.
- Smells.

## Appendix A

# Metric Conversion Table

This appendix complies with current Army directives, which state that the metric system will be incorporated into all new publications. *Table A-1* is a conversion chart.

**Table A-1. Metric Conversion Chart**

| *US Units* | *Multiplied By* | *Equals Metric Units* |
|---|---|---|
| | ***Length*** | |
| Feet | 0.30480 | Meters |
| Inches | 2.54000 | Centimeters |
| Inches | 0.02540 | Meters |
| Inches | 25.40010 | Millimeters |
| Miles (nautical) | 1.85320 | Kilometers |
| Miles (statute) | 1.60930 | Kilometers |
| | ***Area*** | |
| Square inches | 6.45160 | Square centimeters |
| Square feet | 0.09290 | Square meters |
| | ***Volume*** | |
| Cubic feet | 0.02830 | Cubic meters |
| Fluid ounces | 29.57300 | Milliliters |
| | ***Temperature*** | |
| Degrees Fahrenheit | Subtract 32, multiply by 5/9 | Degrees Celsius |
| | ***Weight*** | |
| Ounces | 28.34900 | Grams |
| Pounds | 0.45359 | Kilograms |
| ***Metric Units*** | ***Multiplied By*** | ***Equals US Units*** |
| | ***Length*** | |
| Centimeters | 0.39370 | Inches |
| Kilometers | 0.53960 | Miles (nautical) |
| Kilometers | 0.62137 | Miles (statute) |
| Meters | 3.28080 | Feet |
| Meters | 39.37000 | Inches |
| Millimeters | 0.03937 | Inches |
| | ***Area*** | |
| Square centimeters | 0.15500 | Square inches |
| Square meters | 10.76400 | Square feet |
| | ***Volume*** | |
| Cubic meters | 35.31440 | Cubic feet |
| Milliliters | 0.03380 | Fluid ounces |

**Table A-1. Metric Conversion Chart (Continued)**

| ***Metric Units*** | ***Multiplied By*** | ***Equals US Units*** |
|---|---|---|
| | ***Temperature*** | |
| Degrees Celsius | Multiply by 9/5, add 32 | Degrees Fahrenheit |
| | ***Weight*** | |
| Grams | 0.03527 | Ounces |
| Kilograms | 2.20460 | Pounds |

# Appendix B
# Kennel Construction

This appendix is intended to be a starting point in the design process with the identification of requirements and spaces. It should serve as a supplement to the installation design guide and not as a replacement. This appendix provides commanders and planners with guidelines and specifications to consider for the construction and equipment of kennel facilities. The floor plan (*Figure B-1*) is an example of how an MWD facility works, how areas relate to each other, and what is required for each space. It is not intended to serve as a rigid model for duplication.

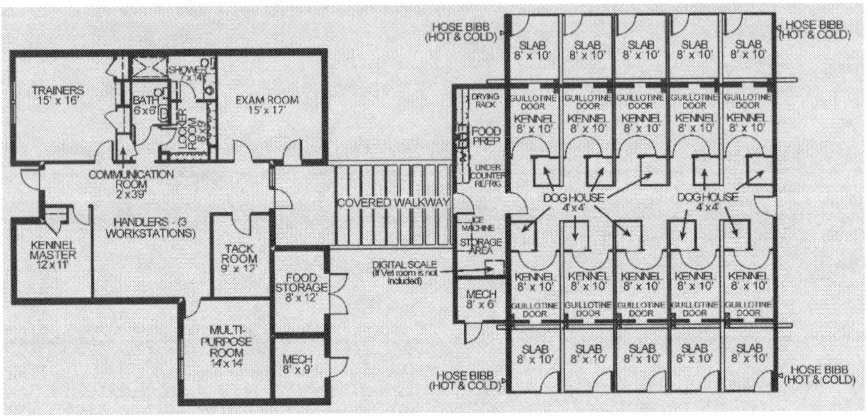


**Figure B-1. Example of an MWD Kennel Floor Plan**

## TEMPORARY KENNELS

B-1. There are two major types of kennels, temporary (*Figure B-2*, page B-2) and permanent. According to *AR 190-12*, temporary kennels are to be used during deployments for no longer than one year, to allow sufficient time for permanent kennels to be built. Temporary kennels may be constructed out of any available resources. The design is not standard and varies from site to site depending on the location, mission, and duration of operations. Using a concept similar to constructing fighting positions, the temporary kennels will be improved continuously as needed or until a permanent facility is completed.

B-2. Using tents and shipping crates may be a hasty method of setting up temporary kennels. Although this method is not optimal for sustained operations, it may be useful until other accommodations can be coordinated. Erect the tents at a location that allows the MWDs a safe place to rest away from other troop activities after operations. Whenever possible, use terrain or buildings to place a barrier between troop activity and the temporary kennel. Never place the temporary kennel site near motor pools, dining facilities, or high noise areas.

**Figure B-2. Temporary Kennel**

B-3. In some areas of operation, it may be possible to obtain the use of a building or other structures for temporary kennels. In these cases, leaders must ensure that the building is well ventilated, safe, and structurally sound. Ensure that there are no hazards (such as chemicals, bare wires, holes, or debris) that can cause injury to the MWDs or handlers.

B-4. Engineers are often used to assist in the improvement of temporary kennels. In addition, fences should be erected and dog runs created.

---

***Note*:** From the start of operations, sanitation measures must be employed in all temporary kennel configurations.

---

# PERMANENT KENNELS

B-5. The following information and figures give specific details for the construction of a permanent kennel facility and adjacent areas. These guidelines may be altered to allow for special needs or uses unique to each location.

## EXTERIOR AREAS

B-6. An MWD kennel complex (*Figure B-3*) consists of kennels, a support building, an obedience course, an exercise area, an MWD break area, exterior storage, parking areas, drives, and walks. Often, as a result of the location, utility services are limited at the site. Construction should provide the following utilities:

- Water (potable with backflow prevention).
- A sanitary sewer.
- Electric.
- Telephone.
- Network and communication service to the site.

Other systems should be incorporated according to local procedures.

B-7. The entire complex should be enclosed, at the minimum, with a heavy-duty, 8-foot-high chain-link fence with three strands of straight wire (no barbed) at the top to prevent an MWD from climbing or jumping out. A 10-foot (minimum) vehicle gate that can be padlocked is to be installed to allow for food deliveries to the kennel and other access requirements. A personnel entry gate that is visible from the KM's office is also required. The personnel gate should have a cipher lock and a notification buzzer that sounds in the KM's office, the obedience course, and the kennel. The buzzer in the kennel should have the capability to be turned on when personnel are in the area and turned off to allow the MWDs to rest when personnel are absent. All gates should be self-closing and self-latching.

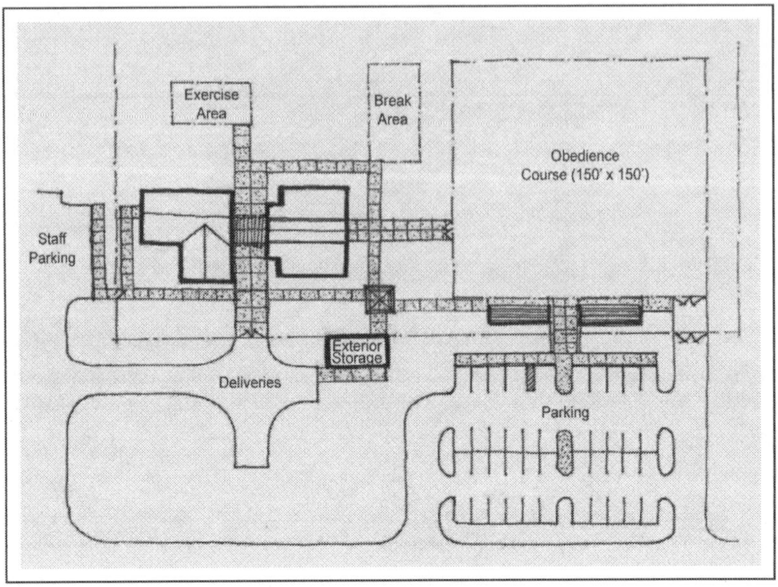


**Figure B-3. MWD Kennel Complex**

B-8. Security lighting for the area should be according to local policy. At a minimum, the lighting should be on from dusk to dawn and placed on all corners of the fencing area. Security lights should be placed in areas that the other lights do not cover.

B-9. Warning signs will be posted on the exterior fencing and buildings of the MWD kennel and exercise area. Signs should contain the following words: "DANGER—OFF LIMITS–MILITARY DOG AREA." Personnel approaching the kennel area should be able to see and read the warning signs under normal daylight conditions from a distance of 50 meters.

B-10. Locate a trash dumpster for the complex outside of the perimeter fence. Screen the dumpster and provide low-maintenance landscaping for the complex according to local policy.

## Exterior Storage, Parking Areas, Drives, and Walks

B-11. An enclosed, 400-square-foot exterior storage building is required to store maintenance equipment, portable kennels, and obstacles. Interior/exterior lighting, power, and a water faucet should be included.

B-12. Staff parking should be a paved parking lot large enough for eight POVs, three GOVs, two visitor vehicles, and one handicap parking space. Pave access drives to accommodate parking, exterior storage, food deliveries, emergency vehicles, and the transportation of MWDs.

B-13. Provide paved walks to all areas. When designing, give careful consideration to creating a one-way MWD traffic system. To prevent confrontations between the MWDs, avoid situations that allow MWDs to meet head on.

## Obedience Course (150 Feet by 150 Feet)

B-14. The obedience course plays an important role in maintaining the MWD's agility and stamina as well as in reinforcing obedience and proficiency training. The course should be grassed and free of hazards (trees, large rocks, holes, and burrs) that may be harmful to MWDs and handlers. The site should be graded for drainage but minimally sloped to provide a level field for training. The area should be enclosed with an 8-foot chain-link fence. Gates should be self-closing and self-latching with a minimum width of 5 feet to allow for lawn maintenance equipment. The area should be well lighted to eliminate shadows. Light poles should be located on the exterior of the fence. On and off switches should be manual. Weatherproof outlets should be provided at each light to provide power at the site. Speaker units connected to the chosen public address system should also be provided. Water should be at the course site with water faucets installed to provide water for MWDs and lawn watering. An irrigation system is desirable for long-term lawn maintenance.

B-15. Obstacles for the course (*Figure B-4*) include barrels, tunnels, steps, jumps, framed windows, and dog walks. Each obstacle is 15 to 20 feet from the previous obstacle; the course is run in a sequence. Obstacles should be constructed to the following specifications:

- Barrel 1 is 35 inches long with a 23-inch opening.
- Barrel 2 is 70 inches long with a 23-inch opening.
- Barrel 3 is 105 inches long with a 23-inch opening.
- The tunnel is 146 inches long with a 19-inch opening.
- The steps are constructed out of wood and are 94 inches high with 5 steps on each side. Each step is 43 inches wide by 16 inches high by 24 inches long. The top platform is 48 inches long by 43 inches wide.
- Jumps 1, 2, and 3 are constructed of wood and are 36 inches high at the maximum raised level. There are 6 removable boards that are 51 inches long by 6 inches high and 1 inch thick.
- The framed window is constructed of wood and is 8 feet tall. The maximum height for the MWD to negotiate is 36 inches. The window opening is 48 inches wide. Removable boards are 5 1/7 inches wide by 51 inches long and 1 inch thick.
- The A-frame is constructed of wood. Its two sides are 8 feet long by 4 feet wide by 1 inch thick with 1-inch by 2-inch boards every 6 inches to secure footing. It has a maximum raised height of 6 feet with a base spread of 10 feet.
- The dog walk is constructed of wood and is 26 inches high by 224 inches long and 11 inches wide with 1-inch by 2-inch boards every 4 inches to secure footing. Each end has a 45° incline ramp that is 11 inches wide by 49 inches long (included in the 224 overall length measurement).

***Note:*** See *DA Pam 190-12* for more complete construction guidelines including the bracing necessary for the various obstacles.

## Military Working Dog Break Area (10 Feet by 20 Feet)

B-16. An MWD break area should be located near the kennels. A break area allows the handler to release the MWD immediately after exiting the kennels or before entering the kennels so that the MWD can relieve himself. The area should be enclosed with an 8-foot chain-link fence. Gates should be self-closing and self-latching and a minimum of 5 feet wide to allow for lawn maintenance equipment. The ground can be a sandy area for cleaning ease. A water faucet is required to maintain the area.

## Exercise Area (20 Feet by 40 Feet)

B-17. The exercise area is a space where the MWD can be released without the handler being present. The exercise area should not be a shared area with the obedience course, as it would conflict with training objectives. The exercise area should be visually separated from the obedience course to prevent MWD from being distracted during training. The exercise area should be grassed, hazard free, and graded slightly for drainage purposes.

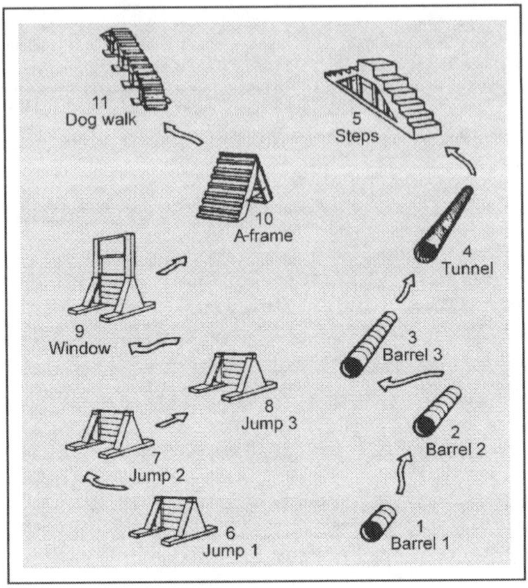


**Figure B-4. Obstacle Course Layout**

## INTERIOR AREAS

B-18. Interior areas have heating, ventilation, and exhaust systems; general-purpose lighting; and power outlets in all spaces. Provide phone, fax, and local area network (LAN) connections to support each area and hot and cold water to all latrines, sinks, and showers. Construct the facility to meet all fire protection and life safety feature requirements.

### Administration Areas

B-19. Since MWDs are likely to visit the administration area (*Figure B-5*, page B-6) at some point, finishes should be of durable material, require minimal maintenance, and be easily repaired or replaced in case of damage. (For example, flooring should be covered by carpet.)

#### *Kennel Master's Office (132 Square Feet)*

B-20. The KM's office serves as the nerve center for the complex. Ideally, the office should be located at the front of the administrative area with exterior windows that view the entry gate to the complex. It includes a lockable closet for supply storage. Furnishings include a desk, a chair, two visitor chairs, a four-drawer file cabinet, and a bookcase. Equipment includes a computer, a printer, a fax/copier, and a public address system or central intercom console.

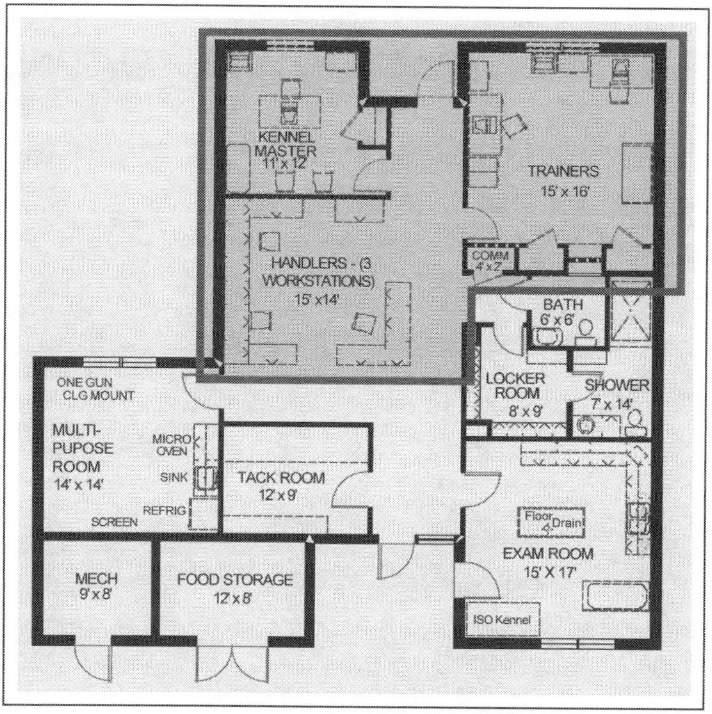


**Figure B-5. Kennel Administrative Areas**

***Trainer's and Plans Noncommissioned Officer's Office (240 Square Feet)***

B-21. The trainer office should provide workspace for two people. Lockable closets are required to store training aides for each trainer. Furnishings include a desk, a chair, and a four-drawer file cabinet for each trainer and a common bookcase. Equipment includes a computer, a printer, a copier, and a 4- by 8-foot dry erase board.

***Handlers' Area (210 Square Feet)***

B-22. The area for the handlers is an open area for general office functions. It includes—

- Three common workstations, depending on the size of the kennels, for handlers to update records and complete daily reports.
- Space for document storage.
- Furnishings (three workstations, file cabinets [one four-drawer vertical file per workstation or the equivalent], and a distribution counter.
- Equipment (computers and a printer).

As a general rule, one workstation should be available for every three handlers assigned to an MWD section.

### *Controlled-Substance Storage Room*

B-23. This room is inside the administrative area and allows sufficient room for the storage safe in which controlled-substance training aids are stored. The doors are secured using approved locking devices. Lightweight doors are replaced with solid metal or wood doors or covered with a 9- to 12-gauge security screen or 16-gauge sheet steel. They are fastened with smooth-headed bolts and nuts and peened in place. All windows providing access to a storage room that is not staffed 24 hours a day are protected by a 9- to 12-gauge security screen or 3/8-inch or larger diameter steel bars spaced no more than 6 inches apart. The frames holding the screen or bars must be fastened to the window frame with smooth-headed bolts.

## Special-Use Areas

B-24. Special-use areas (*Figure B-6,* page B-8) are the transition spaces between the administration area and the kennels. They are important in the day-to-day operations of the MWD facilities. They are not normally occupied on a daily basis. As such, these areas provide a secondary function of serving as a noise buffer between the kennels and the administrative areas. The special-use areas provide a workspace for the VCO and accommodate the unique storage requirements for the kennels.

### *Veterinary Treatment Room*

B-25. A veterinary treatment room is used to perform health care and first aid for MWDs since many locations do not have veterinary treatment areas in close proximity to the kennels. In order to maintain sanitary conditions, treatment rooms require a higher degree of cleanliness and durability than the rest of the support building. Basic finish requirements are—

- Seamless floors with an integral base.
- Washable walls.
- Painted gypsum board ceilings.
- Cabinets faced with plastic laminate.
- Solid-surface countertops.

B-26. Interior partitions should provide a sound transmission class rating of 50 to 55. Floor drains are centered in the room to aid in cleaning, and a dedicated floor drain is located under the isolation kennel. Wall and base cabinets with drawers are installed, including stainless steel dual washbasins with hot and cold water. One section of wall cabinets should be lockable so that prescription medicines for MWDs may be stored in the room. In addition to the normal electrical outlets required by code, a 220-volt outlet is required to support other portable equipment that may be used. Equipment for the room includes the following:

- **An MWD isolation kennel.** The entire structure should be made of uniform, nonglare type 304 stainless steel. The frame, cross members, and vertical spokes should be welded together at every intersection. Latches and hinges are made of 12-gauge stainless steel. The floor of the pen is a raised, removable grate. The floor grate is a heavy-gauge stainless steel, plastic-coated section installed 1 3/4 inches off the floor. A dedicated floor drain is required for this kennel.
- **Standard table tub**. The tub is made of type 304 stainless steel with a minimum depth of 15 inches. The tub slopes to the drain end. The design includes a removable, bathing rack made of plastic-coated stainless steel that fits into the bottom of the tub to give MWDs better footing during treatments. Faucets include a flexible, stainless steel hose that will extend the length of the tub and spray powerfully to reach all areas of the tub.
- **Exam room light.** A ceiling-mounted exam room light with an arm extension and swivel capacity of 360° is recommended.
- **Walk-on platform scale.** The scale is a type 304 stainless steel platform with a wall- or post-mounted light-emitting diode (LED) display. The scale is able to perform calibration and weigh animals to the nearest 0.1 of a pound or kilogram.
- **Stationary exam table.** The table top is type 304 stainless steel die-formed from one piece of 20-gauge steel. It should have raised edges to prevent fluid runoff and be permanently attached

to a heavy base. There should be a minimum 35-inch work height from the floor to the top of the table. The heavy base should have adjustable leveling screws.

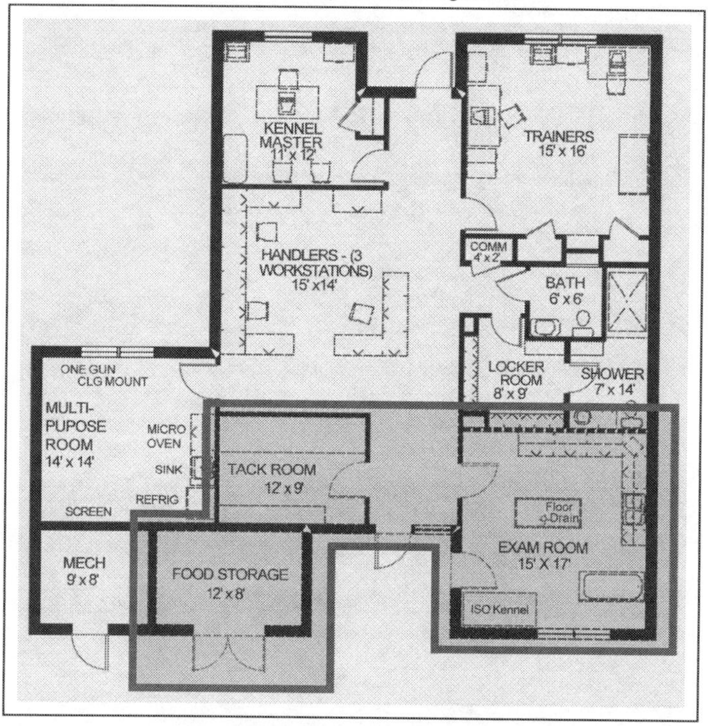


**Figure B-6. Kennel Special-Use Areas**

***Tack Room (108 Square Feet)***

B-27. The tack room is for the storage of extra MWD equipment, such as bite suits and portable kennels. The tack room is located near the entrance to the kennels but not the main building entrance. Requirements include 12 linear feet of shelves that are 18-inches wide and 36 inches deep and allow for 36 inches above each shelf; a 24-inch space with hooks to hang items on; and 6 linear feet of shelves that are 18 inches wide and 12 inches deep and allow for 18 inches above each shelf.

***Food Storage Room (96 Square Feet)***

B-28. The food storage room is normally for the bulk storage of a 30-day supply of food for the MWDs. The room is adjacent to the kennels with an exterior entry. The entry is a pair of 36-inch doors to allow for palletized food delivery and storage. Doors require weather stripping and thresholds to aid in insect and rodent control. Temperature and humidity control equivalent to the office area is required to control food spoilage. A minimum separation of 2 inches is required between the walls and the stored food to allow for air circulation. Shelving should be installed to meet local delivery operations.

## SUPPORT AREAS

B-29. The support areas (*Figure B-7*) provide space for the common-use areas of the building. Areas for the multipurpose room, latrines, communications, and mechanical necessities form the core of the support areas. Support areas should be centrally located and easily accessible for building occupants or maintenance personnel in the case of the mechanical room.

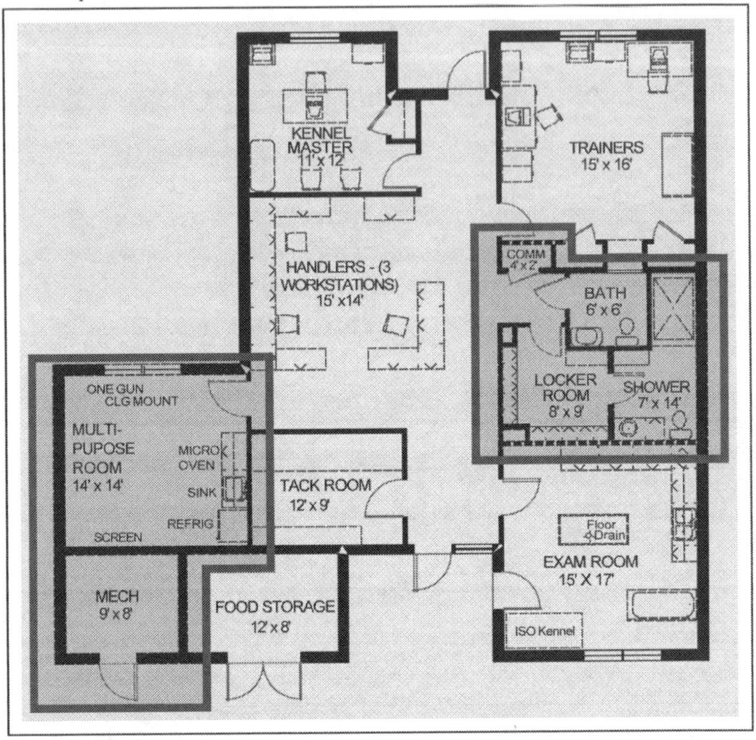


**Figure B-7. Kennel Support Areas**

### Latrine, Shower, and Lockers (206 Square Feet)

B-30. A unisex area is divided into three rooms for the latrine, shower, and lockers. One room is a handicapped accessible latrine with a water closet and lavatory for general use. The locker room includes 10 full-height metal lockers and a bench for MWD handlers' use. The shower room includes a shower, bench, water closet, and lavatory. Do not neglect to include other accessories such as mirrors, soap dispensers, paper towel dispensers, disposal units, grab bars, and hooks.

### Heating, Ventilation, Air Conditioning, Mechanical, Electrical, and Communications Systems

B-31. Heating, ventilation, and air conditioning (HVAC) is a process to add hot or cold air in a facility. It also introduces fresh air in to the facility to maintain indoor air quality and eliminate sick building syndrome. Design HVAC systems to optimize energy use and meet the design criteria for the area in which you are located. Place the hot water heater in the mechanical room and with a capacity that will meet the demands of the latrine and shower; multipurpose room; and veterinary treatment room. Locate electrical panels in this area as well. Although a recommended size is shown on the plans, the room should be big enough to support the equipment required for the location.

B-32. While the number of instruments and computers is not large, the communications closet is vital to the success of the organization. Provide a minimum of one prewired telephone and LAN outlet for each workstation, one in each common area, and one in the kennel food preparation area. Include telephone and LAN outlets for fax machines and printers. Conceal conduit in the walls.

### Multipurpose Room (196 Square Feet)

B-33. The multipurpose room serves as a combination conference room, classroom, and break room. Food preparation requirements are a light, a countertop, and base and upper wall cabinets for storage. Lighting should have a variable-level control for use as a conference room and/or classroom. Furnishings include either a conference table for fifteen or three five-person tables; chairs; a bulletin board; and a 4- by 8-foot dry erase board. Equipment includes a video projection system, a projector screen, a sink, a microwave oven, and a refrigerator.

## KENNEL AREA

B-34. The kennel area provides the daily living environment for the MWDs, giving each MWD a private place for eating and resting. Kennel areas should be built to accommodate large-breed dogs and designed as modular structures to allow for the future expansion of runs.

B-35. There are three types of kennel areas; indoor kennels, outdoor kennels, and indoor/outdoor kennels. The preferred standard is the combination indoor/outdoor kennel; however, there are factors that must be taken into consideration when selecting the kennel type. Cost is always a consideration, but climate also influences the selection. In cold weather climates, the benefit-to-cost ratio will drive the selection to an indoor kennel while the opposite is true for hot weather climates. The more temperate the climate the more suited the indoor/outdoor kennel is for use. Each type of kennel includes the same space requirements. A kennel includes kennel runs with doghouses, a food preparation area, a storage area, and a mechanical room.

B-36. The kennel should be separated from the administrative building by a minimum of 20 feet. This provides separation between different functions and aids in noise control. As a result of the constant interaction between the two facilities, a covered walkway is required to connect the facilities and provide protection from the elements. Care should be taken in the layout of the kennel in relation to the administrative building; a one-way system should be maintained in the movement of MWDs. The objective is to avoid situations where MWDs meet face to face and to prevent confrontations. Consider installing mirrors at corners, intersections, and blind spots to alert handlers of potential wrong-way traffic.

B-37. All entries and exits from the kennel area are required to be self-closing and self-latching. Doors exiting the kennel runs should open inward to aid in preventing MWD escape. Proper ventilation is important in the kennels to prevent the spread of diseases and to control odors. It is recommended that the ventilation standard be to cycle out the room air 10 to 15 times per hour. Kennel temperature should range from 45° to 85° Fahrenheit with humidity in the range of 40 percent to 70 percent. Temperature should be maintained within 10° of the exterior temperature. MWDs work more effectively and are more alert when the kennel temperature is close to the temperature of their working environment.

B-38. Potable water is required at the kennels. Hot and cold water lines are installed at ceiling level down the length of the 5-foot wide center corridor connecting to hose reels at each end of the corridor. A high-pressure washer is required to assist in sanitizing and cleaning kennel runs.

B-39. Since the kennel is basically a wet environment, all electrical receptacles should be provided with ground fault circuit interrupters and all-weather covers. Receptacles mounted at 48 inches above the finished floor are required at each end of the central corridor.

B-40. Each kennel run should have a waterproof light fixture that has an individual switch at the corridor entrance. Other areas require general-purpose lighting according to industry standards. All drain lines in the kennel should be a minimum of 6 inches in diameter and be designed to sustain flow velocities that will maintain a self-cleansing action. Due to the nature of the waste, cleanouts should be numerous and easily accessible. Floor drains should be included in the central corridor to aid in the cleaning of kennels.

## Indoor/Outdoor Kennel

B-41. A combination indoor/outdoor kennel (*Figure B-8*, page B-12) simply implies that there is an interior run and an exterior run linked with each MWD living area. A guillotine type door will connect the interior run and the exterior run. Other door types may be used, but the idea is to permit isolation of the MWD during cleaning operations. Ideally, operation of the door is from the central corridor and from the outside of the exterior run. At 6 feet above the finished floor, use translucent wall panels in the wall separating the runs to provide natural light inside the kennel.

B-42. The roof should extend over the exterior runs and include gutters and/or downspouts to prevent excessive water in the runs. Exterior framing should be enclosed with a soffit to prevent birds from roosting. Exterior water faucets (with freeze protection where required) with both hot and cold water are required at each corner of the building. These are necessary for cleaning exterior runs. Floor drains (*Figure B-9*, page B-12) should be located in the corner of each run adjacent to the common wall separating exterior and interior runs. Drains should tie-in to a common waste line. Drain covers should be flush with the floor. Kennel floors should slope (at least 1/4- inch per foot) toward the drains to allow for quick water drainage and drying.

## Indoor or Outdoor Kennel

B-43. The plan for the indoor or outdoor kennel (*Figure B-10*, page B-13) is the same for either style. The obvious difference is the wall that encloses the kennels. For indoor kennels, the exterior walls along the runs should have windows beginning at 6 feet above the finished floor to allow for natural light. Windows should be operable and hinged at the sill to tilt inward to prevent the escape of MWDs. For outdoor kennels, the exterior wall is simply an 8-foot chain-link fence. Water faucets with hot and cold water are required at each corner of the building. These are required to clean the gutter and drain system (*Figure B-11*, page B-13) for the kennel. A gutter and drainage system is provided at the back of the kennel run for either type of kennel for sluicing wastewaters from cleaning operations. Kennel run floors should slope to the gutter and the gutter should slope to the drain. Gutters should be sufficient to allow for easy cleaning. A minimum of 3 feet should be allowed between the end of the kennel and the wall or fence.

## Kennel Run

B-44. Runs should be arranged so that openings are staggered in order to avoid MWDs facing each other across the corridor. Full-height partitions are required between runs for all kennel types. The first 6 feet of wall above the floor should be sealed concrete. For interior runs, steel clad acoustical panels should be considered from the top of the concrete partition to the ceiling. Steel clad acoustical ceiling panels should also be considered above the kennels.

B-45. Heavy-gauge chain-link fence is used for exterior runs and is also an option for interior runs. Because of the noise generated in the enclosed area of an interior run, it is important that noise-reducing materials be introduced to prevent stress to the MWD and hearing loss to personnel.

B-46. As a minimum, end walls should be full-height, galvanized chain-link fencing. Take care to prevent sharp edges from bolts, hinges, and tie wires on the interior of the runs where MWDs may be cut or harmed. Self-latching entry gates should swing 180° and be able to cover the opening of the MWD house. This allows the MWD to be penned in the house during run-cleaning operations. A stainless steel water bucket is required for each kennel, including a holder that prevents the MWD from overturning the bucket.

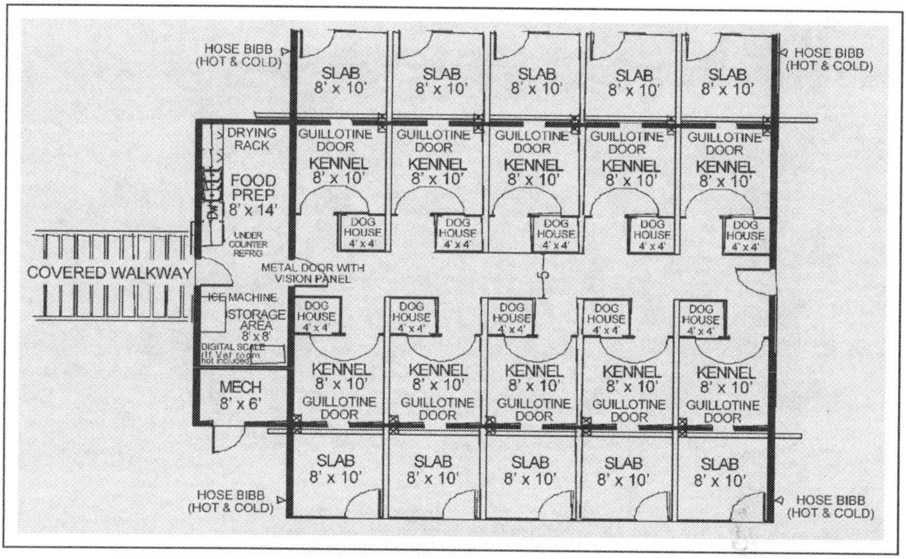


Figure B-8. Combination Indoor/Outdoor Kennel

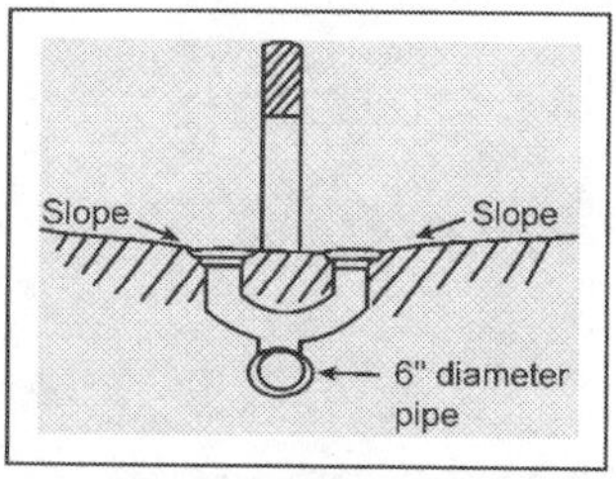


Figure B-9. Example of a Combination Indoor/Outdoor Kennel Floor Drain

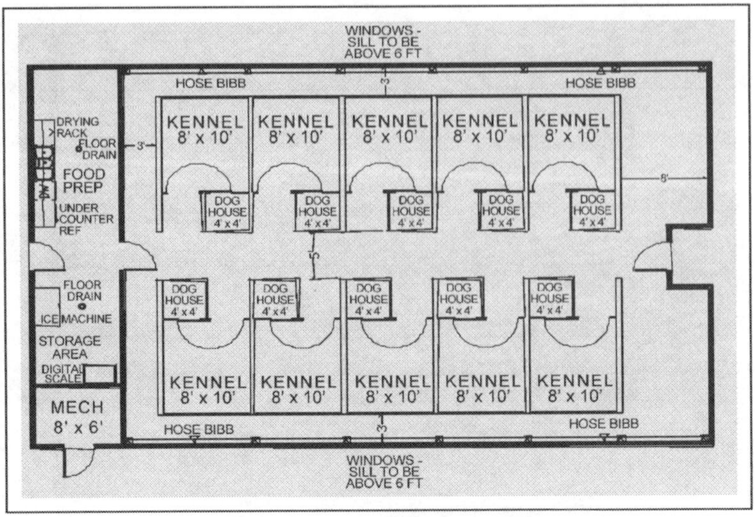


**Figure B-10. Indoor or Outdoor Kennel**

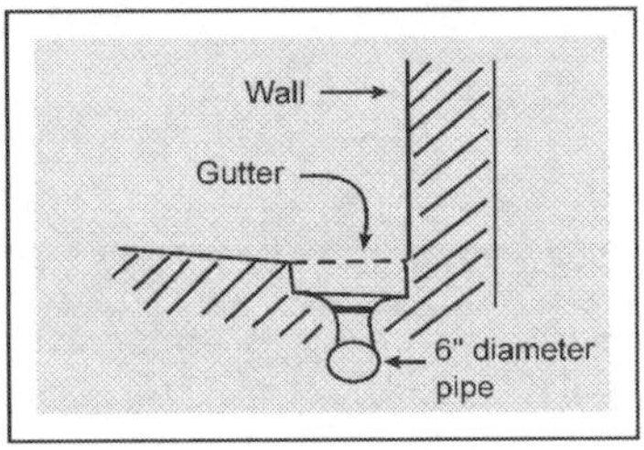


**Figure B-11. Example of an Indoor or Outdoor Kennel Floor Drain**

B-47. Flooring should be concrete and slope toward floor drains and/or gutters to provide rapid runoff of water and drying. Since runs are subject to frequent wash downs to clean and prevent disease, a smooth epoxy-type finish is needed. Temperature-controlled floors should be emplaced for locations that have inclement weather.

B-48. All materials in areas accessible to MWDs should be resistant to damage by scratching, biting, and chewing. Materials should be durable and easy to sanitize. Avoid using material such as angle iron that rusts over a period of time and generates sharp edges that pose a danger to MWDs.

B-49. MWD houses are 4 feet by 4 feet by 4 feet sealed, concrete boxes with a metal top. Tops should hinge toward the end wall in order to be raised to clean the house or treat an MWD. Once lifted, a holding device should be available to keep the top open. Wall opening dimensions are shown in *Figure B-12*, page B-14. The house should be a minimum of 8 inches and a maximum of 12 inches above the run floor. The house floor should have positive drainage toward the wall opening.

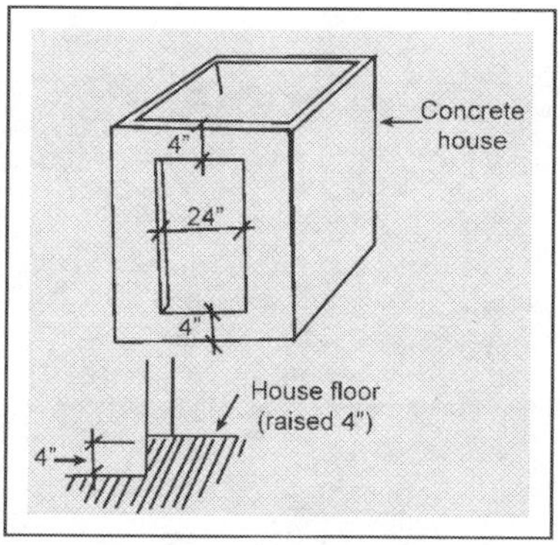


**Figure B-12. MWD House**

## Storage and Mechanical Area

B-50. Additional storage for the kennel is provided. A commercial quality ice machine is required to aid in cooling down MWDs. Include a floor drain for the area. If a veterinary treatment room is not included, a digital scale should be located in this area. Design mechanical systems to optimize energy use and meet the design criteria for the area where the kennel is located. Include the hot water heater in the mechanical room and with a capacity to meet the demands of the kennel. Allow for the location of electrical panels in this area as well.

## Food Preparation Room (112 Square Feet)

B-51. The food preparation area is where feedings are prepared and feeding pans/bowls are stored and cleaned. Food preparation for MWDs requires a large countertop with a deep, three-compartment stainless steel sink. Upper wall cabinets are required for general storage. An under the counter refrigerator, a dishwasher, and space to store food for daily use are necessary for operation. A 2-foot space at the end of the counter should accommodate a storing and drying rack for pans and bowls. Lighting and receptacles and a floor drain should be installed per the code. A telephone, bulletin board, feeding chart, and large garbage can should also be provided. Optional equipment includes a garbage disposal unit and a stackable washer and dryer.

## Appendix C

# Inspection Memorandum

This appendix provides a sample memorandum (see *Table C-1*, page C-2) for documenting the results of inspections. Commanders and program managers will find this useful in preparing for inspections. KMs, plans NCOs, and handlers will find this list helpful in conducting daily duties and adhering to requirements for inspections. Maintain completed memorandum records on file to assist in future inspections.

**Table C-1. Sample Inspection Memorandum**

**DEPARTMENT OF THE ARMY**
**UNIT**
**ADDRESS OF UNIT**

OFFICE SYMBOL 15 May 2004

MEMORANDUM FOR RECORD

SUBJECT: MACOM Technical Assistance Visit or Provost Marshal Inspection

**1. SECTION I. GENERAL INFORMATION.**

UNIT: 529th Military Police Company LOCATION: Kaiserslautern
DATE: 3-14-05 TIME: 0900

INSPECTOR (S): LTC Jeffery S. Captain ESCORT: SFC James Majors

**TYPE OF VISIT: __X__ ANNOUNCED______ UNANNOUNCED**

| TYPE | AUTHORIZED | ON HAND |
|---|---|---|
| **Patrol** | | |
| Patrol Explosive | 6 | 4 |
| Patrol Narcotic | 3 | 2 |
| Explosive Detector | 0 | 0 |
| Narcotic Detector | 0 | 0 |



| **2. SECTION II. PERSONNEL.** | **YES** | **N/A** | **NO** |
|---|---|---|---|
| a) Are all authorized MWDs on hand (unit TDA/MTOE document)? | ___ | ___ | ___ |
| b) Are all MWD handlers certified by the DOD Dog Center? | ___ | ___ | ___ |
| c) Are ASI trained handlers being used as specified in AR 614-200? | ___ | ___ | ___ |
| d) Is the assignment policy of one dog and one handler being followed? | ___ | ___ | ___ |
| e) Are all MWDs assigned to a handler? | ___ | ___ | ___ |
| f) Has an excess of dogs been reported to the MACOM? | ___ | ___ | ___ |
| g) Does the KM meet the minimum criteria to be appointed as a KM? | ___ | ___ | ___ |
| h) Are handlers being assigned duties that take away from their handling responsibilities (excluding PT, staff duty, field duty, and unit training)? | ___ | ___ | ___ |
| **3. RESPONSIBILITIES OF THE PROVOST MARSHAL AND COMMANDER.** | | | |
| a) Is there an updated SOP signed by the PM explaining policies, procedures, and responsibilities for the direction, management, and control of the kennels? | ___ | ___ | ___ |
| b) Has the PM initiated an MWD program based on the evaluation of the installation support, customs, HRP support, and combat support missions? | ___ | ___ | ___ |

**Table C-1. Sample Inspection Memorandum (Continued)**

| OFFICE SYMBOL<br>SUBJECT: MACOM Technical Assistance Visit or Provost Marshal Inspection | | **YES** | **N/A** | **NO** |
|---|---|---|---|---|
| c) | Has the PM issued written guidelines governing the release of an MWD to apprehend a suspect? | ____ | ____ | ____ |
| d) | Are there specific criteria in contingency plans governing the use of dog teams during civil disturbance situations? | ____ | ____ | ____ |
| e) | Does the PM provide guidance on procedures for explosive detector dog support to civil authorities? | ____ | ____ | ____ |
| f) | Is the proficiency of each MWD team evaluated periodically by the PM? | ____ | ____ | ____ |
| g) | Has the PM and/or unit commander checked one or more of these areas monthly: kennels, kennel support building, training and exercise areas, equipment, handlers, dogs, training, utilization and team proficiency? | ____ | ____ | ____ |
| h) | Has the PM and/or unit commander inspected all areas of the MWD kennels quarterly, and is a written record on file? | ____ | ____ | ____ |
| i) | Are MWD requisitions submitted according to *AR 190-12* and *AR 700-81*? | ____ | ____ | ____ |
| j) | Are MWDs placed on the unit property book by NSN, name, and tattoo number? | ____ | ____ | ____ |
| k) | Are MWDs assigned to the KM on a *DA Form 3161* (*Request for Issue or Turn-In)* hand receipt (according to *DA Pam 710-2-1*)? | ____ | ____ | ____ |
| **4.** | **VCO RESPONSIBILITIES.** | | | |
| a) | Are the dogs examined by the VCO every six months (according to *AR 40-905*)? | ____ | ____ | ____ |
| b) | Does the VCO inspect the kennel facility and area for proper sanitary standards on a quarterly basis? | ____ | ____ | ____ |
| c) | Is the issued dog food and diet checked periodically? | ____ | ____ | ____ |
| d) | Is there a program in effect for insect and rodent inspection? | ____ | ____ | ____ |
| e) | Does the VCO concur with the kennel standards/facilities? | ____ | ____ | ____ |
| | If not what course of action has been taken? ________________________________ | | | |
| f) | Does the VCO provide regular training to handlers on MWD health care, feeding, and first aid? | ____ | ____ | ____ |
| g) | Does the VCO provide a list to the KM concerning the deployable status of all MWD? | ____ | ____ | ____ |
| h) | Does the VCO issue health certificates for dogs pending shipment? | ____ | ____ | ____ |

**Table C-1. Sample Inspection Memorandum (Continued)**

| OFFICE SYMBOL<br>SUBJECT: MACOM Technical Assistance Visit or Provost Marshal Inspection | **YES** | **N/A** | **NO** |
|---|---|---|---|
| **5. WARNING SIGNS.** | | | |
| a) Are there warning signs with the following words posted at each instillation entrance, "**CAUTION—THIS AREA PATROLLED BY MILITARY WORKING DOGS**?" | ___ | ___ | ___ |
| b) Are kennels located away from disturbing influences and adequately posted with "**DANGER—OFF LIMITS—MILITARY DOG AREA**?" | ___ | ___ | ___ |
| c) Are the shipping crates stenciled with "**CAUTION—MILITARY WORKING DOG—DO NOT TAMPER WITH ANIMAL?"** | ___ | ___ | ___ |
| d) Are shipping crates stenciled with "**LIVE ANIMAL**?" | ___ | ___ | ___ |
| e) Are removable or permanent markings placed on the sides and rear of vehicles with the following phrase "**CAUTION—MILITARY WORKING DOGS**?" | ___ | ___ | ___ |
| **6. KENNEL AREA.** | | | |
| a) Has a risk analysis been conducted on the kennels? | ___ | ___ | ___ |
| b) Is the kennel manned both day and night? | ___ | ___ | ___ |
| c) Does the kennel have an IDS/CCTV system if not manned both day and night? | ___ | ___ | ___ |
| d) Are approved locks and padlocks used to secure government equipment? | ___ | ___ | ___ |
| e) Have primary and alternate key custodians been appointed, in writing, by the PM or commander? | ___ | ___ | ___ |
| f) Are personnel authorized to issue and receive keys listed on an access roster? | ___ | ___ | ___ |
| g) Are all keys controlled and accounted for on a *DA Form 5513-R* (*Key Control Register and Inventory)?* | ___ | ___ | ___ |
| h) Is the *DA Form 5513-R* secured in an approved, lockable container? | ___ | ___ | ___ |
| i) Is the key depository box permanently affixed to a wall? | ___ | ___ | ___ |
| j) Are keys inventoried semiannually, and is the last report on file? | ___ | ___ | ___ |
| k) Are all keys inscribed with a serial number? | ___ | ___ | ___ |
| l) Are kennels used to house or care for other animals? | ___ | ___ | ___ |

**Table C-1. Sample Inspection Memorandum (Continued)**

| OFFICE SYMBOL<br>SUBJECT: MACOM Technical Assistance Visit or Provost Marshal Inspection | YES | N/A | NO |
|---|---|---|---|
| m) Are adequate training areas adjacent to the kennel areas? | ____ | ____ | ____ |
| 1) Is the training area fence 8 feet high? | ____ | ____ | ____ |
| 2) Does the overhanging fence face inward? | ____ | ____ | ____ |
| 3) Is the fence staked to the ground? | ____ | ____ | ____ |
| n) Does the kennel area have an adequate water supply? | ____ | ____ | ____ |
| o) Are the kennels of a standard design? | ____ | ____ | ____ |
| **7. EQUIPMENT.** | | | |
| a) Is all authorized equipment on hand? | ____ | ____ | ____ |
| b) Are Air Force shipping crates returned within 10 days after receiving a new or replacement MWD? | ____ | ____ | ____ |
| c) Is there one shipping crate available for each dog authorized? | ____ | ____ | ____ |
| d) Is leather equipment properly maintained and cleaned? | ____ | ____ | ____ |
| e) Is all metal equipment free of rust? | ____ | ____ | ____ |
| f) Is the web equipment cleaned with proper cleaning material to prevent deterioration? | ____ | ____ | ____ |
| g) Are there adequate medical supplies on hand for first aid treatment and training? | ____ | ____ | ____ |
| **8. USE OF MWDs.** | | | |
| a) Are the dog teams being employed a minimum of 24 hours per week? | ____ | ____ | ____ |
| b) Are PDs employed in all areas of the installation during both day and night? | ____ | ____ | ____ |
| **9. NARCOTICS.** | | | |
| a) Is the MWD section registered with the Drug Enforcement Administration? | ____ | ____ | ____ |
| b) Is a *DEA 223 (Controlled Substance Registration Certificate)* on file and current? | ____ | ____ | ____ |
| c) Have all personnel been designated, in writing, by the PM to handle training aids? | ____ | ____ | ____ |
| d) Has primary and alternate narcotic training aid custodians been appointed, in writing, by the PM? | ____ | ____ | ____ |

**Table C-1. Sample Inspection Memorandum (Continued)**

| OFFICE SYMBOL<br>SUBJECT: MACOM Technical Assistance Visit or Provost Marshal Inspection | | YES | N/A | NO |
|---|---|---|---|---|
| e) | Have primary and alternate narcotic custodians been cleared by a favorable Crime Records Center name check? | ___ | ___ | ___ |
| f) | Is the Crime Records Center name check maintained on file? | ___ | ___ | ___ |
| g) | Have all personnel authorized to handle or use training aids been briefed at least annually? | ___ | ___ | ___ |
| h) | Is a written record of the briefing being maintained on file? | ___ | ___ | ___ |
| i) | Are the monthly/quarterly inventories being conducted by a disinterested E7 or above? | ___ | ___ | ___ |
| j) | Are the monthly/quarterly inventories being maintained with the controlled substances? | ___ | ___ | ___ |
| k) | Are the monthly/quarterly inventories being maintained for at least two years? | ___ | ___ | ___ |
| l) | Are joint inventories being conducted on the change of primary or alternate custodians and is the inventory maintained on file? | ___ | ___ | ___ |
| m) | Is a daily issue/return green logbook on hand? | ___ | ___ | ___ |
| n) | Is the property book properly maintained to account for all authorized explosive training aids? | ___ | ___ | ___ |
| o) | Is an *SF 702 (Security Container Check Sheet)* posted on the safe? | ___ | ___ | ___ |
| p) | Are *SFs 702* maintained for 30 days? | ___ | ___ | ___ |
| q) | Is a controlled-substance accountability folder on hand? | ___ | ___ | ___ |
| r) | Is the controlled-substance accountability folder properly maintained? | ___ | ___ | ___ |
| **10.** | **EXPLOSIVES.** | | | |
| a) | Have primary and alternate explosive custodians been appointed, in writing, by the PM? | ___ | ___ | ___ |
| b) | Have primary and alternate custodians been cleared by a favorable Crime Records Center name check? | ___ | ___ | ___ |
| c) | Are records of the Crime Records Center checks on file? | ___ | ___ | ___ |
| d) | Are all personnel authorized to receive and handle explosive training aids designated, in writing, by PM? | ___ | ___ | ___ |
| e) | Is annual explosive safety training being conducted by EOD personnel or other qualified personnel? | ___ | ___ | ___ |

**Table C-1. Sample Inspection Memorandum (Continued)**

| OFFICE SYMBOL<br>SUBJECT: MACOM Technical Assistance Visit or Provost Marshal Inspection | | YES | N/A | NO |
|---|---|---|---|---|
| f) | Is a written record of the safety training being maintained on file for a minimum of two years? | ____ | ____ | ____ |
| g) | Is there an SOP that outlines safety procedures in training areas, transportation of explosives, severe weather conditions, and accidents? | ____ | ____ | ____ |
| h) | Are the monthly/quarterly inventories being conducted by a disinterested E7 or above? | ____ | ____ | ____ |
| i) | Are the monthly/quarterly inventories maintained on file for at least two years? | ____ | ____ | ____ |
| j) | Are all required explosive training aids available to ensure that proficiency standards are met? | ____ | ____ | ____ |
| k) | Are chlorates being replaced quarterly? | ____ | ____ | ____ |
| l) | Is each type of explosive being stored in a separate vapor proof container? | ____ | ____ | ____ |
| m) | Is each vapor proof container properly marked? | ____ | ____ | ____ |
| n) | Has the vehicle that is used to transport explosive training aids been approved by safety personnel? | ____ | ____ | ____ |
| o) | Does the installation have an explosive detector training plan that has been approved by the installation commander which lists buildings and facilities authorized for training the PEDD teams? | ____ | ____ | ____ |
| p) | Is a daily issue/return green logbook on hand? | ____ | ____ | ____ |
| q) | Is the logbook properly maintained to account for all authorized explosive training aids? | ____ | ____ | ____ |
| r) | Is a *DA Form 7281-R (Command Oriented Arms, Ammunition, & Explosives Security Screening and Evaluation Record)* on hand for each handler that is authorized to handle or receive explosive training aids? | ____ | ____ | ____ |
| **11.** | **TRAINING.** | | | |
| a) | Does the KM (or handlers) give training to military police on MWD team functions and operational considerations? | ____ | ____ | ____ |
| b) | Are the MWDs receiving a minimum of 4 hours of patrol and 4 hours of narcotic/explosives proficiency training each week? | ____ | ____ | ____ |
| c) | Have all certified narcotic/explosive detector teams received quarterly validation testing by the KM? | ____ | ____ | ____ |
| d) | Is the minimum proficiency standard for assigned patrol/narcotic detector dogs 90 percent and patrol/explosive detector dogs 95 percent? | ____ | ____ | ____ |
| e) | Was the dog decertified when his proficiency standard was less than 90 percent and less than 95 percent for more than three consecutive months? | ____ | ____ | ____ |

**Table C-1. Sample Inspection Memorandum (Continued)**

OFFICE SYMBOL
SUBJECT: MACOM Technical Assistance Visit or Provost Marshal Inspection

| | YES | N/A | NO |
|---|---|---|---|
| **12. RECORDS.** | | | |
| a) Is the handler's information being recorded on the back of *DD Form 1834*? | ___ | ___ | ___ |
| b) Have all *DA Forms 2807-R* been completed properly? | ___ | ___ | ___ |
| c) Have all deficiencies for the MWDs been annotated and explained on DA Forms 2807-R? | ___ | ___ | ___ |
| d) Is a *DA 3992-R* being maintained to record the MWD's reliability? | ___ | ___ | ___ |
| e) Does each MWD team have a probable cause folder? | ___ | ___ | ___ |

**13. THE RESULTS OF THIS INSPECTION ARE—**

____ EXCELLENT

____ SATISFACTORY

____ UNSATISFACTORY



**14. REMARKS.** All areas noted with a deficiency will be annotated in the remarks section. A copy of this technical assistance visit will be provided to OPMG, the MACOM program manager, and the KM. All records of technical assistance visits will be maintained on file at the installation for a minimum of two years after the inspection date with the corrective action memorandum.

**15. POINT OF CONTACT.** The point of contact for this action is the undersigned at COMM: 555-555-1511 or e-mail address.

(Original Signed)
JEFFREY S. CAPTAIN
LTC, USA
MACOM PM

## Appendix D

# Training Records

This appendix provides samples and details of how to properly use MWD training records. Records must be maintained for the life of the MWD. At least one year of historical documents must be maintained on file at each kennel for every MWD. Records can be discarded two years after the death, adoption, or euthanasia of the MWD. Nearly all MWDs have a dual purpose (PD and PEDD or PNDD); therefore, every MWD should have a PD and a PEDD/PNDD form on file. Kennels that have internet access can obtain records via the Military Working Dog Management System (MWDMS).

## DEPARTMENT OF THE ARMY *FORM 2807-R*

D-1. *DA Form 2807-R (Military Working Dog Training and Utilization Record)* (*Figure D-1,* page D-3) is a standard form used by all MWD personnel. It is used to record the training and utilization of PDs. Use the following instructions when preparing a *DA Form 2807-R*:

### ADMINISTRATIVE DATA

D-2. Enter the following information in the appropriate blocks:

- The date.
- The MWD's name and type, tattoo number, and age.
- The handler's name and grade.
- The organization and the location.

### TRAINING

D-3. The training section of the form provides complete information on all of the training tasks that must be accomplished for a PD to maintain proficiency. It is not necessary to train on all of the controlled-aggression tasks during the same day; however, all of the controlled-aggression tasks must be trained every week.

D-4. Entries for the amount of time spent training on specific tasks during a particular day of the month are made in minutes. The monthly total for each specific task will be in hours and will be automatically entered in the far right-hand column titled "Total Hours." The handler's or trainer's evaluation of the MWD's daily rating is entered as S or U. Remarks are mandatory for any area the MWD fails to perform to standard. These remarks will be annotated on the reverse side of the form using the following format:

- Deficiency (DF). The MWD's shortcoming.
- Corrective action (CA). The handler's response to the MWD's deficiency.
- Corrective response (CR). The MWD's response to the CA.

D-5. Enter the appropriate information on—

- **Line 1.** Record the number of minutes per day that the MWD is trained on obedience on leash.
- **Line 2.** Record the number of minutes per day that the MWD is trained on obedience off leash.
- **Line 3.** Record the number of minutes per day that the MWD is trained on the obedience course.

- **Line 4.** Record the daily rating of the MWD on controlled aggression (either S or U). Write an explanation of any U rating on the reverse side of the form. In 4a to 4e, record the number of minutes per day that the MWD trained on each controlled-aggression task.
- **Line 5.** Record the number of minutes per day that the MWD trained on building searches on line 5.
- **Line 6.** Record the number of minutes per day that the MWD trained on gunfire with the handler firing the weapon. The number of rounds will be annotated on the reverse side of the form.
- **Line 7.** Record the number of minutes per day that the MWD is trained on gunfire with a third party firing a weapon. The number of rounds will be annotated on the reverse side of the form.
- **Line 8.** Record the number of minutes per day that the MWD is trained on scouting/patrolling. Record the distance (in feet) for scent, sight, and sound detection on lines 8a, b, or c.
- **Line 9.** Record the number of minutes per day that the MWD is trained on vehicle patrol.
- **Line 10.** Record the number of minutes per day that the MWD is trained on tracking.
- **Line 11.** Record the daily training rating (S or U). Explain deficiencies on the reveres side of the form. Any failure of a critical task requires a U daily training rating. See *DA Pam 190-12* for critical tasks.

## UTILIZATION

D-6. The utilization section of the form provides a daily record of the time spent performing military police duties in the three general categories of combat support operations, patrol–law enforcement, and patrol–security.

D-7. The combat support operations category may be used to record time spent performing military police missions (actual or training) in support of combat units. Such operations include field training exercises, CP exercises, mobilization exercises, and other activities related to the tactical and strategic missions of the Army, provided that these missions are accomplished with a handler and an MWD together as a certified team.

D-8. The patrol–law enforcement category may be used to record time spent performing law enforcement or force protection type missions. These missions include but are not limited to—

- Random vehicle searches.
- Law enforcement patrols.
- Other duties performed while working in a law enforcement type mission.

D-9. The patrol–security category may be used to record time spent performing security type missions such as—

- Dignitary security.
- Critical asset patrolling in a peacetime environment.
- USSS, Department of State, or other explosive type searches.

D-10. The daily rating of the MWD's performance on military police duties is recorded as S or U. All U performance annotations will have an explanation for the DF and have the CA taken recorded on the reverse side of the form and continuation sheets, if necessary. The total monthly utilization hours are recorded in the last column, titled "Total Hours."

D-11. When training is conducted during the actual use and employment of an MWD, the time used for training may also be reported in the training section. This training is not an automatic entry every time the team has utilization hours. Time annotated as training during utilization hours will be actual time training on patrol or detector tasks. It is unethical to count all or even half of an MWD team's utilization hours as both utilization and training time for every shift, and unrealistic numbers will not be accepted. When time is recorded as utilization and training, a notation is made on the reverse side of the form explaining the double time entry and the type of training conducted. This will make it possible to differentiate between training time, training time while on duty, and duty (utilization) time. Although additional training during utilization hours is acceptable, it will not be accepted as the primary amount of mandatory training hours each week.

## MILITARY WORKING DOG TRAINING AND UTILIZATION RECORD

For use of this form, see AR 190-12; proponent agency is ODCSOPS

| NAME/TYPE OF DOG | TATTOO/NUMBER | AGE | NAME OF HANDLER | GRADE | ORGANIZATION AND LOCATION | MONTH AND YEAR |
|---|---|---|---|---|---|---|
| NATZ/PEDD | E309 | 2 | BOOKSTEIN, DAVID B. | SGT | HHC 98TH ASG, BAMBERG, GE | SEP 2003 |

DAILY RATINGS: S - SATISFACTORY, U - UNSATISFACTORY (Explain deficiency and corrective action on reverse)

(Use reverse side of this form for any remarks or notes)

| TRAINING | 1 | 2 | 3 | 4 | 5 | 6 | 7 | 8 | 9 | 10 | 11 | 12 | 13 | 14 | 15 | 16 | 17 | 18 | 19 | 20 | 21 | 22 | 23 | 24 | 25 | 26 | 27 | 28 | 29 | 30 | 31 | TOTAL HOURS |
|---|---|---|---|---|---|---|---|---|---|---|---|---|---|---|---|---|---|---|---|---|---|---|---|---|---|---|---|---|---|---|---|---|
| | DAY OF MONTH | | | | | | | | | | | | | | | | | | | | | | | | | | | | | | | |
| 1. ON LEASH OBEDIENCE | 15 | 15 | 15 | 15 | 30 | | 30 | 40 | 10 | 15 | | | 20 | 15 | | 15 | 20 | 30 | 15 | | 15 | 15 | 15 | 15 | | | | | 15 | 15 | | 6.50 |
| 2. OFF LEASH OBEDIENCE | 20 | 20 | 15 | 15 | 40 | | 30 | 40 | 10 | 15 | | | 20 | 30 | | 15 | 30 | 30 | 15 | | 15 | 20 | 20 | 20 | | | | | 15 | 15 | | 7.50 |
| 3. OBEDIENCE COURSE | | | | | 20 | | 20 | | | 15 | | | | | | | | | | | | | | | | | | | | | | 0.92 |
| 4. CONTROLLED AGGRESSION (S OR U) | | S | S | | | | | S | S | | | | S | S | | | | | | | | | | | | | | | | | | |
| a. FALSE RUN | | 5 | 5 | | | | | 15 | 5 | | | | 5 | 5 | | | | | | | | | | | | | | | | | | 0.67 |
| b. ATTACK | | 10 | 10 | | | | | 20 | 10 | | | | 10 | 10 | | | | | | | | | | | | | | | | | | 1.17 |
| c. SEARCH AND ATTACK | | 5 | 5 | | | | | 15 | 10 | | | | 10 | 10 | | | | | | | | | | | | | | | | | | 0.92 |
| d. STAND OFF | | 10 | 15 | | | | | 20 | 5 | | | | 10 | 10 | | | | | | | | | | | | | | | | | | 1.17 |
| e. ESCORT | | 5 | 5 | | | | | 15 | 5 | | | | 5 | 5 | | | | | | | | | | | | | | | | | | 0.67 |
| 5. BUILDING SEARCH | | 15 | | | | | | 15 | | | | | | | | | | | | | | | | 20 | | | | | | | | 0.83 |
| 6. GUNFIRE - HANDLER | | | | | | | | 10 | | | | | | | | | | | | | | | | | | | | | | | | 0.17 |
| 7. GUNFIRE - DECOY/AGITATOR | | | | | 10 | | | | | | | | | | | | | | 30 | | | | | | | | | | | | | 0.67 |
| 8. SCOUTING/PATROLLING (TIME) | | | | | 15 | | | | | | | | | | | | | | | | | | | 15 | | | | | 15 | | | 0.75 |
| a. SCENT DETECTION (DISTANCE) | | | | | 40 | | | | | | | | | | | | | | | | | | | 30 | | | | | 30 | | | |
| b. SIGHT DETECTION (DISTANCE) | | | | | | | | | | | | | | | | | | | | | | | | | | | | | | | | |
| c. SOUND DETECTION (DISTANCE) | | | | | | | | | | | | | | | | | | | | | | | | | | | | | | | | |
| 9. VEHICLE PATROL | | | | | | | | | | | | | | | | | | | | | | | | | | | | | | | | 0.00 |
| 10. TRACKING | | | | | | | | | | | | | | | | | | | | | | | | | | | | | | | | 0.00 |
| 11. DAILY TRAINING RATING (S OR U) | S | S | S | S | S | | S | S | S | S | | | | | | S | S | S | S | | S | S | S | S | | | | | S | S | | 21.92 |

| UTILIZATION | 1 | 2 | 3 | 4 | 5 | 6 | 7 | 8 | 9 | 10 | 11 | 12 | 13 | 14 | 15 | 16 | 17 | 18 | 19 | 20 | 21 | 22 | 23 | 24 | 25 | 26 | 27 | 28 | 29 | 30 | 31 | TOTAL HOURS |
|---|---|---|---|---|---|---|---|---|---|---|---|---|---|---|---|---|---|---|---|---|---|---|---|---|---|---|---|---|---|---|---|---|
| | HOURS PER DAY | | | | | | | | | | | | | | | | | | | | | | | | | | | | | | | |
| 1. COMBAT SUPPORT OPERATIONS | | | | | | | | | | | | | | | | | | | | | | | | | | | | | | | | 0 |
| 2. PATROL - LAW ENFORCEMENT | 8 | 8 | 8 | 8 | 8 | | | 8 | 8 | 8 | | | 8 | 8 | 8 | 8 | | | | 8 | 8 | 8 | 8 | | | | | | 8 | 8 | | 144 |
| 3. PATROL - SECURITY | | | | | | | | | | | | | | | | | | | | | | | | | | | | | | | | 0 |
| 4. DAILY UTILIZATION RATING (S OR U) | S | S | S | S | S | | | S | S | S | | | S | S | S | S | | | | S | S | S | S | | | | | | S | S | | 144.0 |

| DAILY FEEDING | 1 | 2 | 3 | 4 | 5 | 6 | 7 | 8 | 9 | 10 | 11 | 12 | 13 | 14 | 15 | 16 | 17 | 18 | 19 | 20 | 21 | 22 | 23 | 24 | 25 | 26 | 27 | 28 | 29 | 30 | 31 | WT OF DOG DATE | LBS |
|---|---|---|---|---|---|---|---|---|---|---|---|---|---|---|---|---|---|---|---|---|---|---|---|---|---|---|---|---|---|---|---|---|---|
| | QUANTITY OF FOOD (BY WEIGHT) | | | | | | | | | | | | | | | | | | | | | | | | | | | | | | | | |
| 2 CU S/D | A | A | A | A | A | A | A | A | A | A | A | A | A | A | A | A | A | A | A | A | A | A | A | A | A | A | A | A | A | A | A | 1ST | 75 |
| 2 CU S/D | A | A | A | A | A | A | A | A | A | A | A | A | A | A | A | A | A | A | A | A | A | A | A | A | A | A | A | A | A | A | A | 15TH | 73.5 |

DA FORM 2807-R, Oct 84 EDITION DEC 72 IS OBSOLETE, AND REPLACES DA FORMS 2810-R AND 2815-R, DEC 72



Figure D-1. Sample DA Form 2807-R

DA Form 2807-R
Continuation Sheet

| DATE | REMARKS |
| --- | --- |
| 2-Sep-03 | **DF1:** MWD WAS DISTRACTED BY OTHER HANDLER; WOULD NOT FOCUS ON COMMANDS. |
| | **CA1:** CONDUCTED PROPER AVOIDANCE TRAINING TECHNIQUE. |
| | **CR1:** AFTER MWD DID NOT RESPOND TO INITIAL AVOIDANCE TRAINING TECHNIQUE, A PINCH COLLAR WAS USED, RESULTING IN REGAINED FOCUS. |
| | **DF2:** MWD BIT SLEEVE ON STAND OFF. |
| | **CA2:** CONDUCTED PROPER AVOIDANCE TRAINING TECHNIQUE. |
| | **CR2:** MWD PERFORMED STAND OFF CORRECTLY. |
| 5-Sep-03 | GUNFIRE TRAINING: 5 ROUNDS FIRED BY THIRD PARTY. |
| | **DF1:** MWD REPEATEDLY BROKE POSITION (SIT TO DOWN AND DOWN TO SIT). |
| | **CA1:** CONDUCTED PROPER AVOIDANCE TRAINING TECHNIQUE. |
| | **CR1:** MWD REMAINED IN POSITION. |
| | **DF2:** MWD WAS HESITANT TO OUT REWARD (BITE SLEEVE). |
| | **CA2:** CONDUCTED PROPER AVOIDANCE TRAINING TECHNIQUE. |
| | **CR:** MWD OUTED REWARD ON COMMAND. |
| 7-Sep-03 | MWD HAS BEEN BREAKING POSITION AFTER SHORT TIME DURING OFF/L OB WHEN HANDLER IS POSITIONED AWAY FROM HIM. TODAY MWD REMAINED IN THE SIT FOR 1 MIN 50 SEC, BROKE SIT, WAS VERBALLY CORRECTED AND REMAINED IN SIT FOR AN ADDITIONAL 25 SEC. |
| | **DF:** MWD BROKE FROM DOWN SEVERAL TIMES DURING PARADE REST. |
| | **CA:** CONDUCTED PROPER AVOIDANCE TRAINING TECHNIQUE. |
| 8-Sep-03 | GUNFIRE TRAINING: 10 ROUNDS FIRED BY HANDLER. |
| | **CR:** AFTER MWD DID NOT RESPOND TO AVOIDANCE TRAINING, ESCAPE TRAINING WAS CONDUCTED. EVENTUALLY MWD REMAINED IN DOWN. MWD IS SO EXCITABLE THAT HE HAS A TENDENCY TO CUE OFF HANDLER'S MOVEMENTS TO BREAK FROM POSITION. HANDLER IS CONCENTRATING ON TEACHING MWD NOT TO BE SO DISTRACTED BY OUTSIDE STIMULUS. |
| | **DF1:** MWD WAS GETTING DISTRACTED AND NOT HEELING IN THE PROPER POSITION. |
| | **CA1:** CONDUCTED PROPER AVOIDANCE TRAINING TECHNIQUE (HEEL-A-WAYS). |
| | **CR1:** MWD HEELS CORRECTLY. |
| | **DF2:** MWD WAS BITING SLEEVE DURING STAND OFF. |
| 11-Sep-03 | **CA2:** CONDUCTED PROPER AVOIDANCE TRAINING TECHNIQUE. |
| | **CR2:** MWD PERFORMED STAND OFF CORRECTLY. |
| | NO TRAINING DUE TO MWD GOING TO VET. MWD HAS BEEN DIAGNOSED WITH A POSSIBLE PULLED/STRAINED MUSCLE IN HIS RIGHT LEG, WHICH HE IS FAVORING AS HE WALKS. X-RAYS SHOW NO BREAKS OR FRACTURES. MWD CANNOT DO ANYTHING STRENUOUS FOR THE NEXT WEEK (NO RUNNING OR JUMPING UNTIL 18 SEP) INCLUDING BITE WORK OR OB COURSE. MWD CAN DO BASIC OB STARTING TUES, SEP 16. |
| 19-Sep-03 | GUNFIRE TRAINING: 15 ROUNDS FIRED BY DECOY. |



**Figure D-1. Sample DA Form 2807-R (Continued)**

D-12. Record the appropriate utilization information on—

- **Line 1.** Record the hours per day that the MWD is used for combat support operations.
- **Line 2.** Record the hours per day that the MWD is used for patrol–law enforcement.
- **Line 3.** Record the hours per day that the MWD is used for patrol–security.
- **Line 4.** Record the daily utilization rating (S or U). Explain any U on the reverse side of the form.

## TOTAL HOURS

D-13. Record the total hours of training and use in the last column on the right side of the form. The time should be recorded in hours and minutes. The MWDMS makes this entry automatically.

### Daily Feeding

D-14. Record the quantity of food that the MWD eats every day in this section. If the MWD is fed once daily, the type and amount of food will be noted in the first block to the left. Under quantity of food, the amount eaten for each feeding will be annotated as follows:

- "A" if all food was eaten.
- "3/4" if three-quarters was eaten.
- "1/2" if half was eaten.
- "1/4" if one-quarter was eaten.
- "0" if none of the food was eaten.

D-15. If the MWD is fed twice daily, the second block is used to record the second daily feeding (which could be a different type of food). If the MWD is fed once daily, the second block may be used to record a change in the diet of the MWD, directed by the veterinarian. Block 2 may also be used to indicate the type and amount of food used daily for MWDs on a food reward system. Monitoring MWD food consumption is a critical portion of the required daily health checks.

D-16. The last column, titled "Wt of Dog," is used to record a semimonthly weight for the MWD. The weight checks occur on or around the first and the fifteenth day of each month. There is sufficient room under the headings "Date" and "Lbs" to record both the date when the MWD is weighed and the weight in pounds. Any drastic changes in food consumption or weight gain or loss are reported to the veterinarian.

D-17. Enter the following information in the appropriate blocks:

- Enter the VCO's prescribed amount of food to be given daily directly under "Daily Feeding." Both blocks are used when an MWD is fed twice daily.
- Record the amount of food eaten per day, in the appropriate day's block under "Quantity of Food (by Weight)."
- Record the date and weight of the MWD in the respective blocks in the bottom right-hand corner.

### Remarks

D-18. The reverse side of *DA Form 2807-R* (*Figure D-1*) is used to annotate advanced techniques used during training and DFs with CAs and CRs listed. This section is very important for probable cause and shows the time line of events for training. Each day is listed with the action taken on that day. If no training is conducted due to company type training, the words "soldier development" are written on the line followed by the type of training conducted. "Soldier development" is only used when MWD handlers are conducting training in which the MWD is not needed or not taken. If no training is conducted due to leave, passes, or scheduled days off, no entry is needed; *DA Form 6 (Duty Roster)* can be used to verify these days.

## DEPARTMENT OF THE ARMY *FORM 3992-R* (FIRST PAGE)

D-19. *DA Form 3992-R (Narcotics or Explosives Detector Dog Training and Utilization Record)* (*Figure D-2*, page D-9) is a standard form used by all MWD personnel. It is used to record the use and training of detector dogs. Use the following instructions when preparing a *DA Form 3992-R:*

### Administrative Data

D-20. Enter the following information in the appropriate blocks:

- The date.
- The MWD's name and type, tattoo number, and age.
- The handler's name and grade.
- The organization and location.

## TRAINING

D-21. The training section of the form is used for recording all training of the detector dog team in the particular detection skill. The detection areas of buildings, containers, and vehicles are listed. There is one blank space for detector training in other areas. Examples of other areas may include—

- Mail facilities.
- Open or wooded areas.
- Warehouses.
- Aircraft.
- Areas considered significant or unique to the command to which the MWD team is assigned.

D-22. Although some of the examples may be included under one of the three prelisted categories, the others should be listed separately. For areas covered under the three general headings, the handler or trainer annotates the specific location under the appropriate day in the remarks (for example, Smith High School, NCO Club, or Corps Auditorium).

D-23. The top line of each training category is divided diagonally into two blocks. The top of each block is used to record the number of training aids planted for that day's detector training. The bottom of each block is used for the number of training aids found by the proper response of the detector dog. If the MWD responds on the same training aid more than once in the same training scenario, it will not be counted as a find unless the number of plants is changed as well. At no time will there be more finds annotated than there are plants. Continually allowing the MWD to return to the last aid found creates false response problems in the MWD and should not be encouraged. It is the handler's responsibility to keep the MWD from returning to the scent cone of the last found training aid, whether the MWD is being worked on or off leash. Once the MWD has entered the scent cone, he must be allowed to work to the source and final respond. At no time will an MWD be yanked off an aid once he is working the scent cone.

D-24. The bottom line of each training category is used to record the search time. Search time is the amount of time devoted to detector training that day. Detector training is the time from the moment the handler enters a problem and commands the MWD to "Seek" until the MWD has completed the search of that training problem, excluding handler or MWD breaks. Add the daily time entries in each training category, and enter the results in the last column, titled "Total Hours."

D-25. Record the following information beside the appropriate training location (building, containers, vehicles, or other):

- **Plants/Finds.** Record the number of plants in the top half of the block and the number of finds in the lower half.
- **Search Time.** Record the amount of time devoted for detector training that day.
- **Training Total Hours.** Record the sum of hours performing detection missions in the far right

## UTILIZATION

D-26. All operational detection missions are recorded in the utilization section. The detection missions are listed. Detection missions that are significantly different from these are listed separately to give an accurate record of the types of missions for which the detector dog team is used. There are two blocks under each day of the month for each utilization category. The top block is for recording the number of times the MWD detected a substance. The bottom block is for recording the total search or inspection time for the detection mission. The last column is the sum of the daily times. It gives a record of the total amount of time spent performing the detection missions for that month. The number of finds for the month can also be totaled and entered in the last column, above the time entry.

D-27. Record the following information beside the appropriate location (building, containers, vehicles, or other):

- **Finds.** Record the number of times the MWD finds the substance on a detection mission.
- **Search Time.** Record the total search time.

## Detector Dog Search Data

D-28. All relevant information about the productivity and success of each detection mission is recorded using the "Detector Dog Search Data" section. The location may be a building number, a unit designation, map grid coordinates, and/or any other information that helps identify where the detection mission was performed.

D-29. The MPR number is the number assigned to the case by military police to account for the custody and disposition of the substances found. The substance is the identification of the found material by a common name, such as heroin, marijuana, dynamite, or detonating cord.

D-30. Quantity is the measure of the amount of the substance found. Weight, volume, overall dimensions, length, or any other appropriate measure may be used.

D-31. The "Remarks" section may be used for adding any other relevant information about the substance found. This would include a field or laboratory verification of the type of substance, an EOD evaluation of an explosive device or explosive substance as live or inert, or the presence of other hazardous material in or around the substance found (such as razor blades, trip wires, or poisons). Record any information that may be useful in preparing the team for future searches or that may be applicable or useful to other detector dog teams.

D-32. Record the following information in the appropriate boxes:

- **Time.** Record the time the detector dog performed the mission.
- **Date.** Record the date the detector dog performed the mission.
- **Location.** Record the exact location of the detector mission (such as front gate, Fort Sill, Oklahoma; National Airport, Gabon, Africa; or building 442, Fort Bragg, North Carolina).
- **Military Police Report (MPR) Number.** Record the MPR number in this column, if applicable.
- **Substance.** Record the common name of the substance found, if applicable.
- **Quantity.** Record the quantity of the substance found, if applicable.
- **Remarks.** Use this column to record any other relevant information about the substance found (such as a positive response, no response, or a lab verification).

## Detector Dog Proficiency (Second Page)

D-33. A correct response occurs when the MWD detects the substance and responds with the proper final response, and the training aid or substance is found in the general location where the MWD has responded. A not-at-source response is a training tool used to help an MWD team move closer to the source location of a training aid plant. The distance away from the source in order for a not-at-source response to be given is six to ten feet.

D-34. A missed response is when the MWD fails to detect and respond to the presence of a training aid with a final response. A handler miss or error is when—

- The handler fails to search or clear an area with the MWD and a training aid plant is missed.
- The handler fails to read an MWD's change of behavior and pulls the MWD away from the area where a training aid is planted and a training aid plant is missed.
- The handler presents an area but fails to ensure that the MWD is sniffing and a training aid plant is missed.

Handler misses or errors are considered missed explosives or narcotics and are counted against the find rate percentages of the MWD team.

D-35. False responses are noted in the second to last section of *DA Form 3992-R*. A false response is when the MWD responds as if he has detected a substance and the handler accepts that response, but no training aid or substance can be found within a reasonable distance of the MWD's final response.

D-36. The detector dog team's proficiency is computed monthly by adding the total number of correct responses ("a" for the month) and adding the total number of false/missed (MWD or handler) responses ("b" for the month). These two numbers are used in the following formula to obtain the detector dog team's proficiency rating:

$$\left(\frac{a}{a+b}\right)100 = proficiency$$

D-37. For example, during the month the detector dog team made 93 correct responses on training aids, the MWD team also had 4 false responses, 2 MWD and 1 handler missed responses for a total of 7 false/missed responses. Applying these two numbers (a=93 and b=7) to the formula, the following is obtained:

$$\left(\frac{93}{93+7}\right)100 = 93\%$$

D-38. The computation above shows that the MWD team is working at a 93 percent proficiency rate. This rate is above the minimum for a PNDD team (90 percent), but it is below the minimum standard for a PEDD team (95 percent). A PNDD handler should continue training to maintain and possibly increase the team's proficiency. A PEDD handler needs to identify the causes of the team's substandard performance and immediately begin corrective training to bring the team up to and over the minimum 95 percent detection proficiency standard.

***Note:*** Responses on actual substances during actual searches are not included in this computation. Search data cannot be included because it is impossible to determine the number of false or missed responses during an actual search.

D-39. Record the following information on—

- **Line a.** Record the total correct alerts. The MWDMS makes these entries automatically.
- **Line b.** Record the total false or missed alerts. The MWDMS makes these entries automatically.

## Narcotics or Explosive Detector Dog Training and Utilization Record (Third and Fourth Pages)

D-40. The first part of section of the Narcotics or Explosive Detector Dog Training and Utilization Record section is used to record the number of each training aid used in training on each day of the month. The entry for each training aid type is divided into three rows to record the day of the month, correct alerts, and missed alerts, respectively. Record the appropriate information in the following sections.

- **Administrative Data.** Record the MWD's name and tattoo number, the month, and the year in the appropriate blocks.
- **Aid Type.** Enter the specific aid type in the upper left-hand corner of each "Detector Dog Proficiency" section.
- **Total Correct Alerts.** Record the number of correct alerts the MWD made during training.
- **Missed Alerts.** Record the number of missed and false alerts the MWD made during training.
- **Proficiency.** Record the actual proficiency in this block. The MWDMS makes this entry automatically.

## Remarks

D-41. Remarks for *DA Form 3992-R* are recorded on a continuation sheet. Each DF (for example, a missed aid or false sit) is listed by the date it occurred. The CA and CR are also listed. If more than one DF is noted per day, number the DF, CA, and CR series to avoid confusion.

| NARCOTICS OR EXPLOSIVES DETECTOR DOG TRAINING AND UTILIZATION RECORD<br>For use of this form see AR 190-12. The proponent is DCSPER. | | | | | | MONTH AND YEAR<br>SEP 2003 |
|---|---|---|---|---|---|---|
| NAME/TYPE OF DOG<br>NATZ/PEDD | TATTOO NUMBER<br>E309 | AGE<br>2 | NAME OF HANDLER<br>BOOKSTEIN, DAVID B. | GRADE<br>SGT | ORGANIZATION AND LOCATION<br>HHC 98TH ASG, BAMBERG, GE | |

| TRAINING | | 1 | 2 | 3 | 4 | 5 | 6 | 7 | 8 | 9 | 10 | 11 | 12 | 13 | 14 | 15 | 16 | 17 | 18 | 19 | 20 | 21 | 22 | 23 | 24 | 25 | 26 | 27 | 28 | 29 | 30 | 31 | TRAINING TOTAL HOURS 24.25 |
|---|---|---|---|---|---|---|---|---|---|---|---|---|---|---|---|---|---|---|---|---|---|---|---|---|---|---|---|---|---|---|---|---|---|
| BUILDINGS | PLANTS / | | | 7 | 8 | 4 | | | | 4 | 6 | | | | | | 11 | | 5 | | | | | | 4 | 8 | 6 | | | | 7 | | 70 |
| | FINDS | | | 7 | 5 | 4 | | | | 4 | 6 | | | | | | 10 | | 4 | | | | | | 4 | 8 | 6 | | | | 7 | | 65 |
| | SEARCH TIME | | | 120 | 80 | 60 | | | | 60 | 60 | | | | | | 130 | | 70 | | | | | | 60 | 120 | 120 | | | | 120 | | 16.67 HRS |
| CONTAINERS | PLANTS / | | | | | | | | | 19 | 57 | | | | | | 19 | | 19 | | | | | | | | | | | | | | 114 |
| | FINDS | | | | | | | | | 19 | 57 | | | | | | 19 | | 19 | | | | | | | | | | | | | | 114 |
| | SEARCH TIME | | | | | | | | | 45 | 120 | | | | | | 45 | | 45 | | | | | | | | | | | | | | 4.25 HRS |
| VEHICLES | PLANTS / | | | | | 4 | | | | | | | | | | | | | 5 | | | | | | | | | | | | | | 9 |
| | FINDS | | | | | 4 | | | | | | | | | | | | | 5 | | | | | | | | | | | | | | 9 |
| | SEARCH TIME | | | | | 60 | | | | | | | | | | | | | 80 | | | | | | | | | | | | | | 2.33 HRS |
| OTHER | PLANTS / | | | | | | | | | 4 | | | | | | | | | | | | | | | | | | | | | | | 4 |
| | FINDS | | | | | | | | | 4 | | | | | | | | | | | | | | | | | | | | | | | 4 |
| | SEARCH TIME | | | | | | | | | 60 | | | | | | | | | | | | | | | | | | | | | | | 1.00 HRS |

| UTILIZATION | | 1 | 2 | 3 | 4 | 5 | 6 | 7 | 8 | 9 | 10 | 11 | 12 | 13 | 14 | 15 | 16 | 17 | 18 | 19 | 20 | 21 | 22 | 23 | 24 | 25 | 26 | 27 | 28 | 29 | 30 | 31 | UTILIZATION TOTAL HOURS 0.00 |
|---|---|---|---|---|---|---|---|---|---|---|---|---|---|---|---|---|---|---|---|---|---|---|---|---|---|---|---|---|---|---|---|---|---|
| BUILDINGS | FINDS | | | | | | | | | | | | | | | | | | | | | | | | | | | | | | | | 0 |
| | SEARCH TIME | | | | | | | | | | | | | | | | | | | | | | | | | | | | | | | | 0.00 HRS |
| CONTAINERS | FINDS | | | | | | | | | | | | | | | | | | | | | | | | | | | | | | | | 0 |
| | SEARCH TIME | | | | | | | | | | | | | | | | | | | | | | | | | | | | | | | | 0.00 HRS |
| VEHICLES | FINDS | | | | | | | | | | | | | | | | | | | | | | | | | | | | | | | | 0 |
| | SEARCH TIME | | | | | | | | | | | | | | | | | | | | | | | | | | | | | | | | 0.00 HRS |
| OTHER | FINDS | | | | | | | | | | | | | | | | | | | | | | | | | | | | | | | | 0 |
| | SEARCH TIME | | | | | | | | | | | | | | | | | | | | | | | | | | | | | | | | 0.00 HRS |

DA FORM 3992-R, Oct 84 EDITION OF 1 DEC 72 IS OBSOLETE



Figure D-2. Sample DA Form 3992-R

DETECTOR DOG SEARCH DATA

| NO. | TIME | DATE | LOCATION | MPR NUMBER | SUBSTANCE | QUANTITY | REMARKS |
|---|---|---|---|---|---|---|---|
| 1 | | | | | | | |
| 2 | | | | | | | |
| 3 | | | | | | | |
| 4 | | | | | | | |
| 5 | | | | | | | |
| 6 | | | | | | | |
| 7 | | | | | | | |
| 8 | | | | | | | |
| 9 | | | | | | | |
| 10 | | | | | | | |
| 11 | | | | | | | |
| 12 | | | | | | | |
| 13 | | | | | | | |
| 14 | | | | | | | |
| 15 | | | | | | | |
| 16 | | | | | | | |
| 17 | | | | | | | |
| 18 | | | | | | | |
| 19 | | | | | | | |
| 20 | | | | | | | |
| 21 | | | | | | | |
| 22 | | | | | | | |
| 23 | | | | | | | |
| 24 | | | | | | | |
| 25 | | | | | | | |
| 26 | | | | | | | |
| 27 | | | | | | | |
| 28 | | | | | | | |
| 29 | | | | | | | |
| 30 | | | | | | | |
| 31 | | | | | | | |
| 32 | | | | | | | |
| 33 | | | | | | | |

Sample

DETECTOR DOG PROFICIENCY
(Alerts on training aids during training and actual searches)

| DAY OF MONTH | 1 | 2 | 3 | 4 | 5 | 6 | 7 | 8 | 9 | 10 | 11 | 12 | 13 | 14 | 15 | 16 | 17 | 18 | 19 | 20 | 21 | 22 | 23 | 24 | 25 | 26 | 27 | 28 | 29 | 30 | 31 |
|---|---|---|---|---|---|---|---|---|---|---|---|---|---|---|---|---|---|---|---|---|---|---|---|---|---|---|---|---|---|---|---|
| a. TOTAL CORRECT ALERTS | 0 | 0 | 7 | 5 | 8 | 0 | 0 | 0 | 26 | 63 | 0 | 0 | 0 | 0 | 0 | 29 | 0 | 28 | 0 | 0 | 0 | 0 | 0 | 4 | 8 | 6 | 0 | 0 | 0 | 7 | 0 |
| b. TOTAL FALSE/MISSED ALERTS | 0 | 0 | 0 | 3 | 0 | 0 | 0 | 0 | 1 | 0 | 0 | 0 | 0 | 0 | 0 | 3 | 0 | 1 | 0 | 0 | 0 | 0 | 0 | 0 | 0 | 0 | 0 | 0 | 0 | 0 | 0 |

PROFICIENCY = $\frac{a}{a+b} \times 100$

$\frac{191}{199}$ 96%

DA Form 3992-R, Oct 84

Figure D-2. Sample DA Form 3992-R (Continued)

## NARCOTICS OR EXPLOSIVES DETECTOR DOG TRAINING AND UTILIZATION RECORD

NATZ/PEDD ______ TATTOO: E309 MONTH: SEP YEAR: 2003

**DETONATING CORD** DETECTOR DOG PROFICIENCY (ALERTS ON TRAINING AIDS DURING TRAINING AND ACTUAL SEARCHES)

| DAY OF MONTH | 1 | 2 | 3 | 4 | 5 | 6 | 7 | 8 | 9 | 10 | 11 | 12 | 13 | 14 | 15 | 16 | 17 | 18 | 19 | 20 | 21 | 22 | 23 | 24 | 25 | 26 | 27 | 28 | 29 | 30 | 31 | PROFICIENCY A / A + B X 100 |
|---|---|---|---|---|---|---|---|---|---|---|---|---|---|---|---|---|---|---|---|---|---|---|---|---|---|---|---|---|---|---|---|---|
| a. TOTAL CORRECT ALERTS | | | 2 | 1 | | | | | | 20 | | | | | | 1 | | | | | | | | | 3 | | | | | | | 27 |
| b. TOTAL MISSED ALERTS | | | | 1 | | | | | | | | | | | | | | | | | | | | | | | | | | | | 28  96% |

**C-4 PLASTIC EXPLOSIVE** DETECTOR DOG PROFICIENCY (ALERTS ON TRAINING AIDS DURING TRAINING AND ACTUAL SEARCHES)

| DAY OF MONTH | 1 | 2 | 3 | 4 | 5 | 6 | 7 | 8 | 9 | 10 | 11 | 12 | 13 | 14 | 15 | 16 | 17 | 18 | 19 | 20 | 21 | 22 | 23 | 24 | 25 | 26 | 27 | 28 | 29 | 30 | 31 | PROFICIENCY A / A + B X 100 |
|---|---|---|---|---|---|---|---|---|---|---|---|---|---|---|---|---|---|---|---|---|---|---|---|---|---|---|---|---|---|---|---|---|
| a. TOTAL CORRECT ALERTS | | | | 1 | | | | | 22 | 1 | | | | | | 3 | | 2 | | | | | | | | 1 | | | | | | 30 |
| b. TOTAL MISSED ALERTS | | | | 1 | | | | | 1 | | | | | | | | | | | | | | | | | | | | | | | 32 94% |

**POTASSIUM CHLORATE** DETECTOR DOG PROFICIENCY (ALERTS ON TRAINING AIDS DURING TRAINING AND ACTUAL SEARCHES)

| DAY OF MONTH | 1 | 2 | 3 | 4 | 5 | 6 | 7 | 8 | 9 | 10 | 11 | 12 | 13 | 14 | 15 | 16 | 17 | 18 | 19 | 20 | 21 | 22 | 23 | 24 | 25 | 26 | 27 | 28 | 29 | 30 | 31 | PROFICIENCY A / A + B X 100 |
|---|---|---|---|---|---|---|---|---|---|---|---|---|---|---|---|---|---|---|---|---|---|---|---|---|---|---|---|---|---|---|---|---|
| a. TOTAL CORRECT ALERTS | | | | | | | | | 1 | | | | | | | 20 | | 2 | | | | | | 2 | 1 | | | | | | | 26 |
| b. TOTAL MISSED ALERTS | | | | | | | | | | | | | | | | 1 | | | | | | | | | | | | | | | | 27 96% |

**SODIUM CHLORATE** DETECTOR DOG PROFICIENCY (ALERTS ON TRAINING AIDS DURING TRAINING AND ACTUAL SEARCHES)

| DAY OF MONTH | 1 | 2 | 3 | 4 | 5 | 6 | 7 | 8 | 9 | 10 | 11 | 12 | 13 | 14 | 15 | 16 | 17 | 18 | 19 | 20 | 21 | 22 | 23 | 24 | 25 | 26 | 27 | 28 | 29 | 30 | 31 | PROFICIENCY A / A + B X 100 |
|---|---|---|---|---|---|---|---|---|---|---|---|---|---|---|---|---|---|---|---|---|---|---|---|---|---|---|---|---|---|---|---|---|
| a. TOTAL CORRECT ALERTS | | | | | 1 | | | | 2 | | | | | | | | | 21 | | | | | | 1 | 1 | | | | | | | 26 |
| b. TOTAL MISSED ALERTS | | | | | | | | | | | | | | | | | | 1 | | | | | | | | | | | | | | 27 96% |

**WATER GEL (TOVEX)** DETECTOR DOG PROFICIENCY (ALERTS ON TRAINING AIDS DURING TRAINING AND ACTUAL SEARCHES)

| DAY OF MONTH | 1 | 2 | 3 | 4 | 5 | 6 | 7 | 8 | 9 | 10 | 11 | 12 | 13 | 14 | 15 | 16 | 17 | 18 | 19 | 20 | 21 | 22 | 23 | 24 | 25 | 26 | 27 | 28 | 29 | 30 | 31 | PROFICIENCY A / A + B X 100 |
|---|---|---|---|---|---|---|---|---|---|---|---|---|---|---|---|---|---|---|---|---|---|---|---|---|---|---|---|---|---|---|---|---|
| a. TOTAL CORRECT ALERTS | | | | 1 | | | | | | 21 | | | | | | 1 | | 1 | | | | | | | | 2 | | | | 2 | | 28 |
| b. TOTAL MISSED ALERTS | | | | 1 | | | | | | | | | | | | | | | | | | | | | | | | | | | | 29 97% |

DA FORM 3992-R, OCT 84 (CONTINUATION SHEET)



Figure D-2. Sample DA Form 3992-R (Continued)

## NARCOTICS OR EXPLOSIVES DETECTOR DOG TRAINING AND UTILIZATION RECORD

NATZ/PEDD **TATTOO:** E309 **MONTH:** SEP **YEAR:** 2003

| | | | | | | | | | | | | | | | | | | | | | | | | | | | | | | | | |
|---|---|---|---|---|---|---|---|---|---|---|---|---|---|---|---|---|---|---|---|---|---|---|---|---|---|---|---|---|---|---|---|---|
| **SEMTEX** | DETECTOR DOG PROFICIENCY (ALERTS ON TRAINING AIDS DURING TRAINING AND ACTUAL SEARCHES) | | | | | | | | | | | | | | | | | | | | | | | | | | | | | | | PROFICIENCY A/(A + B) X 100 |
| DAY OF MONTH | 1 | 2 | 3 | 4 | 5 | 6 | 7 | 8 | 9 | 10 | 11 | 12 | 13 | 14 | 15 | 16 | 17 | 18 | 19 | 20 | 21 | 22 | 23 | 24 | 25 | 26 | 27 | 28 | 29 | 30 | 31 | |
| a. TOTAL CORRECT ALERTS | | | 1 | | 2 | | | | 1 | | | | | | | 2 | | 2 | | | | | | | 1 | | | | | 2 | | 11 |
| b. TOTAL MISSED ALERTS | | | | | | | | | | | | | | | | | | | | | | | | | | | | | | | | ~~11~~ 100% |
| **AMMONIUM NITRATE** | DETECTOR DOG PROFICIENCY (ALERTS ON TRAINING AIDS DURING TRAINING AND ACTUAL SEARCHES) | | | | | | | | | | | | | | | | | | | | | | | | | | | | | | | PROFICIENCY A/(A + B) X 100 |
| DAY OF MONTH | 1 | 2 | 3 | 4 | 5 | 6 | 7 | 8 | 9 | 10 | 11 | 12 | 13 | 14 | 15 | 16 | 17 | 18 | 19 | 20 | 21 | 22 | 23 | 24 | 25 | 26 | 27 | 28 | 29 | 30 | 31 | |
| a. TOTAL CORRECT ALERTS | | | | | | | | | | | | | | | | | | | | | | | | | | | | | | | | 0 |
| b. TOTAL MISSED ALERTS | | | | | | | | | | | | | | | | | | | | | | | | | | | | | | | | 0 #DIV/0! |
| **NITROGLYCERIN DYNAMITE** | DETECTOR DOG PROFICIENCY (ALERTS ON TRAINING AIDS DURING TRAINING AND ACTUAL SEARCHES) | | | | | | | | | | | | | | | | | | | | | | | | | | | | | | | PROFICIENCY A/(A + B) X 100 |
| DAY OF MONTH | 1 | 2 | 3 | 4 | 5 | 6 | 7 | 8 | 9 | 10 | 11 | 12 | 13 | 14 | 15 | 16 | 17 | 18 | 19 | 20 | 21 | 22 | 23 | 24 | 25 | 26 | 27 | 28 | 29 | 30 | 31 | |
| a. TOTAL CORRECT ALERTS | | | | 2 | | | | | | 21 | | | | | | | | | | | | | | | 1 | | | | | | | 24 |
| b. TOTAL MISSED ALERTS | | | | | | | | | | | | | | | | | | | | | | | | | | | | | | | | 24 100% |
| **TNT (Trinitrotoluene)** | DETECTOR DOG PROFICIENCY (ALERTS ON TRAINING AIDS DURING TRAINING AND ACTUAL SEARCHES) | | | | | | | | | | | | | | | | | | | | | | | | | | | | | | | PROFICIENCY A/(A + B) X 100 |
| DAY OF MONTH | 1 | 2 | 3 | 4 | 5 | 6 | 7 | 8 | 9 | 10 | 11 | 12 | 13 | 14 | 15 | 16 | 17 | 18 | 19 | 20 | 21 | 22 | 23 | 24 | 25 | 26 | 27 | 28 | 29 | 30 | 31 | |
| a. TOTAL CORRECT ALERTS | | | 3 | | 3 | | | | | | | | | | | 2 | | | | | | | | | | 1 | | | | 2 | | 11 |
| b. TOTAL MISSED ALERTS | | | | | | | | | | | | | | | | | | | | | | | | | | | | | | | | 11 100% |
| **SMOKELESS POWDER** | DETECTOR DOG PROFICIENCY (ALERTS ON TRAINING AIDS DURING TRAINING AND ACTUAL SEARCHES) | | | | | | | | | | | | | | | | | | | | | | | | | | | | | | | PROFICIENCY A/(A + B) X 100 |
| DAY OF MONTH | 1 | 2 | 3 | 4 | 5 | 6 | 7 | 8 | 9 | 10 | 11 | 12 | 13 | 14 | 15 | 16 | 17 | 18 | 19 | 20 | 21 | 22 | 23 | 24 | 25 | 26 | 27 | 28 | 29 | 30 | 31 | |
| a. TOTAL CORRECT ALERTS | | | 1 | | 2 | | | | | | | | | | | | | | | | | | | 1 | 1 | 2 | | | | 1 | | 8 |
| b. TOTAL MISSED ALERTS | | | | | | | | | | | | | | | | | | | | | | | | | | | | | | | | 8 100% |
| **Totals from TNG Aids (Cont. Sheet)** | | | | | | | | | | | | | | | | | | | | | | | | | | | | | | | | |
| TOTAL PLANTS | 0 | 0 | 7 | 8 | 8 | 0 | 0 | 0 | 27 | 63 | 0 | 0 | 0 | 0 | 0 | 30 | 0 | 29 | 0 | 0 | 0 | 0 | 0 | 4 | 8 | 6 | 0 | 0 | 0 | 7 | 0, | 197 |
| TOTAL FINDS | 0 | 0 | 7 | 5 | 8 | 0 | 0 | 0 | 26 | 63 | 0 | 0 | 0 | 0 | 0 | 29 | 0 | 28 | 0 | 0 | 0 | 0 | 0 | 4 | 8 | 6 | 0 | 0 | 0 | 7 | 0 | 191 97% |
| a. TOTAL FALSE ALERTS | | | | | | | | | | | | | | | | 2 | | | | | | | | | | | | | | | | 2 1% |
| **Totals from TNG locations (3992 Front)** | | | | | | | | | | | | | | | | | | | | | | | | | | | | | | | | |
| TOTAL PLANTS | 0 | 0 | 7 | 8 | 8 | 0 | 0 | 0 | 27 | 63 | 0 | 0 | 0 | 0 | 0 | 30 | 0 | 29 | 0 | 0 | 0 | 0 | 0 | 4 | 8 | 6 | 0 | 0 | 0 | 7 | 0 | 197 |
| TOTAL FINDS | 0 | 0 | 7 | 5 | 8 | 0 | 0 | 0 | 27 | 63 | 0 | 0 | 0 | 0 | 0 | 29 | 0 | 28 | 0 | 0 | 0 | 0 | 0 | 4 | 8 | 6 | 0 | 0 | 0 | 7 | 0 | 192 97% |

DA FORM 3992-R, OCT 84 (CONTINUATION SHEET)

Sample

**Figure D-2. Sample DA Form 3992-R (Continued)**

**DA Form 3992-R**
**Continuation Sheet**

| DATE | REMARKS |
|---|---|
| 3-Sep-03 | TRAINING AT SCHWEINFURT BOWLING ALLEY. (TRAINING AIDS PLANTED FROM 6 FT H x 5 FT D) |
| 4-Sep-03 | TRAINING AT SCHWEINFURT FMO. (TRAINING AIDS PLANTED FROM 8 FT H x 3 FT D)<br>**DF1:** MWD MISSED AID (1 STRAND DET CORD - 1 FT H x 1 FT D) DUE TO HANDLER ERROR.<br>**CA1:** HANDLER TOLD TO WORK A REVERSE PATTERN UNTIL THE TEAM REACHED THE AID LOCATION.<br>**CR1:** MWD FOUND AID AFTER HANDLER MADE PROPER PRESENTATION.<br>**DF2:** MWD MISSED AID. (1/4 BLOCK C4 2 FT H x 1 FT D)<br>**CA2:** MWD BROUGHT BACK FOR RE-PRESENTATION.<br>**CR2:** MWD FOUND AID AFTER HANDLER MADE SECOND PRESENTATION.<br>**DF3:** MWD MISSED AID. (1 STICK NITRO DYNAMITE- 7 FT H x 1 FT D)<br>**CA3:** MWD BROUGHT BACK FOR RE-PRESENTATION.<br>**CR3:** MWD FOUND AID AFTER HANDLER MADE HIGH PRESENTATION AND DRAGGED HAND UPWARD.<br>**DF4:** MWD MISSED AID. (1 STICK WATER GEL - 5 FT H x 1 FT D)<br>**CA4:** MWD BROUGHT BACK FOR RE-PRESENTATION.<br>**CR4:** MWD FOUND AID AFTER HANDLER MADE SECOND PRESENTATION.<br>MWD WILL BE PROTOCOLED ON ALL MISSED AIDS AT A LATER DATE DUE TO TIME CONSTRAINTS. |
| 5-Sep-03 | TRAINING AT 630TH MILITARY POLICE COMPANY MOTOR POOL. (TRAINING AIDS PLANTED FROM 3 FT H x 7 FT D) |
| 9-Sep-03 | TRAINING AT SCHWEINFURT OPEN AREA BY CONN CLUB & OUTDOOR REC. (TRAINING AIDS PLANTED FROM 7 FT H x 4 FT D)<br>**CA:** (REFERENCE DF2, 4 SEP) MWD WAS PROTOCOLED ON 1/2 BLOCK C4.<br>**CR:** COMPLETED 4 HV ASSISTED FOR 4 TRIALS AND 15 CONSECUTIVE UNASSISTED TRIALS. |
| 10-Sep-03 | TRAINING AT ANSBACH CMFO WAREHOUSES (TRAINING AIDS PLANTED FROM 4 FT H x 1 FT D).<br>**CA:** (REFERENCE DF2-4, 4 SEP) MWD WAS PROTOCOLED ON 1 STRAND DC, 1 STICK ND, AND 1 STICK WG.<br>**CR:** COMPLETED 4 HV ASSISTED FOR 4 TRIALS AND 15 CONSECUTIVE UNASSISTED TRIALS FOR EACH AID. |
| 11-Sep-03 | NO TRAINING DUE TO MWD GOING TO VET. MWD HAS BEEN DIAGNOSED WITH A POSSIBLE PULLED/STRAINED MUSCLE IN HIS RIGHT LEG WHICH HE IS FAVORING AS HE WALKS. X-RAYS SHOW NO BREAKS OR FRACTURES. MWD CANNOT DO ANYTHING STRENUOUS FOR THE NEXT WEEK, BUT IS CLEARED TO WORK DETECTION NEXT TUES (16 SEP). |
| 16-Sep-03 | TRAINING AT KITZIGEN TISA WAREHOUSE & LEIGHTON THEATER. (TRAINING AIDS PLANTED FROM 9 FT H x 6 FT D.)<br>**DF1:** MWD MISSED AID. (PC - 1 FT H x 6 FT D)<br>**CA1:** MWD BROUGHT BACK FOR REPRESENTATION. WILL CONDUCT PROTOCOL.<br>**CR1:** MWD RESPONDED ON AID AFTER HANDLER MADE SECOND PRESENTATION. MWD WAS PROTOCOLED ON PC COMPLETING 4 HV ASSISTED FOR 4 TRIALS AND 15 CONSECUTIVE UNASSISTED TRIALS.<br>**DF2:** MWD FALSE SAT IN THEATER LOBBY.<br>**CA2:** MWD WAS ESCAPED FROM AREA AND CONTINUED TO WORK.<br>**CR2:** MWD CONTINUED PATTERN WITHOUT FALSE SITTING.<br>**DF3:** MWD FALSE SAT IN THEATER BACK WALL.<br>**CA3:** MWD WAS ESCAPED FROM AREA AND CONTINUED TO WORK. MWD WILL BE WORKED ON BLANK PROBLEM.<br>**CR3:** MWD CONTINUED PATTERN WITHOUT FALSE SITTING. |
| 18-Sep-03 | TRAINING AT SCHWEINFURT STRIP LOT & OUTDOOR REC. (TRAINING AIDS PLANTED FROM 8 FT H x 6 FT D)<br>**CA1:** (REFERENCE DF2, 16 SEP) CONDUCTED BLANK PROBLEM.<br>**CR1:** MWD DID NOT ATTEMPT TO FALSE SIT.<br>**DF2:** MWD MISSED AID. (SC - 7 FT H x 3 FT D)<br>**CA2:** MWD BROUGHT BACK FOR RE-PRESENTATION. MWD WAS PROTOCOLED ON SC.<br>**CR2:** MWD FOUND AID AFTER HANDLER MADE SECOND PRESENTATION. COMPLETED 4 HV ASSISTED FOR 4 TRIALS AND 15 CONSECUTIVE UNASSISTED TRIALS. |
| 24-Sep-03 | TRAINING AT SCHWEINFURT FMO. (TRAINING AIDS PLANTED FROM 5 FT H x 5 FT D) |
| 25-Sep-03 | TRAINING AT THEATER. (TRAINING AIDS PLANTED FROM 2 FT H x 2 FT D) |
| 26-Sep-03 | TRAINING AT COMMISSARY. (TRAINING AIDS PLANTED FROM 9 FT H x 6 FT D) |
| 30-Sep-03 | TRAINING AT FURNITURE WAREHOUSE. (TRAINING AIDS PLANTED FROM 6 FT H x 7 FT D) |



**Figure D-2. Sample DA Form 3992-R (Continued)**

# Appendix E

# Deployment Equipment List

*Table E-1* is a list to be used as a general guide to aid leaders in the planning process prior to MWD deployments. Mission, enemy, terrain, troops, time available, and civilian considerations (METT-TC) should be considered when determining what equipment is necessary for mission success. Required and authorized equipment is determined by tables of organization and equipment (TOEs), modification tables of organization and equipment (MTOEs), and tables of distribution and allowances (TDAs).

**Table E-1. Deployment Equipment List**

| | | |
|---|---|---|
| | ***Worn or Carried*** | |
| M9 with LCE (LBV) | Identification card | Body armor |
| Canteen cup | Identification tag | Black gloves with liners |
| First aid pouch with dressing | Protective mask with a carrier | Military police brassard (subdued) |
| Kevlar® helmet | | |
| | ***MWD Equipment*** | |
| Choke chain | Grooming brush/comb | Feed pan |
| Leather or nylon collar | Reward ball/Kong | Kennel crate |
| Stake-out chain | 360-inch nylon leash | 16-foot leash |
| Plastic or leather muzzle | Water bucket | Harness |
| | ***Alice Pack*** | |
| BDU (2 sets) | Towels (2) | Sleeping bag with a cover |
| Brown T-shirt (2) | Gloves, black, cold weather | Shower shoes |
| Undergarments (2) | BDU Cap | PT uniform (complete set) |
| Green and black socks (2) | Boot polish kit | MOPP gear in an NBC bag |
| Poncho with a liner | Flashlight | Boots |
| Shaving kit | Lip balm | Personal hygiene gear |
| Waterproof bag | Foot powder | |
| | ***A/B Bag*** | |
| Duffel bag with a padlock | Brown T-shirt (5) | White socks (3) |
| Cold weather boots | BDU (2) | Green and black socks (5) |
| Cold weather gear | Towels (3) | Personal hygiene gear |
| Waterproof bag | Overboots | Undergarments (5) |
| Military police brassard | Laundry bag (2) | |
| | ***Additional Equipment*** | |
| 30-90 day supply MWD food | 50-gallon trash can (MWD food) | 5-gallon water can |
| MWD training records | Medical supplies | Engineer tape |
| MWD medical records | Carpenter kit/tools | Cots |
| MWD health certificate | Safe for narcotics aides | Fire extinguishers |

**Table E-1. Deployment Equipment List (Continued)**

| *Additional Equipment* | | |
|---|---|---|
| MWD first aid kit | Explosives and narcotics training aides | Axes, shovels, rakes, brooms |
| Arm protector, sleeve (hidden) | Parasite preventative mediation | Cooling vest (MWD) |
| Soldier first aid kit | Light set | Paw pad protectors (2 sets) |
| Bite suit (optional) | Generator | Foot lockers (2) or blood boxes |
| Dog shampoo | Transformer | Equipment duffel bag |
| Heaters and fans | Warning signs (MWD) | Camouflage netting with poles |

## Appendix F

# Force Protection Conditions

The FPCONs outlined below describe the progressive level of a terrorist threat to all US military facilities and personnel under *DOD Directive 2000.12*. As approved by the Joint Chiefs of Staff, the terminology and definitions in this appendix are recommended security measures designed to ease interservice coordination and to support US military antiterrorism activities. Each installation must localize these measures for its specific circumstances, accounting for service and commander requirements, local laws, and SOFAs as a minimum. See *AR 525-13* for further details concerning FPCONs. This appendix details what each condition is and the suggested actions of the MWD teams for each level and measure. Commanders are encouraged to employ MWD teams wherever an MWD team can enhance security and force protection measures that are already established.

## FORCE PROTECTION CONDITION NORMAL

F-1. FPCON normal applies when a general global threat of possible terrorist activity exists. It warrants a routine security posture.

## FORCE PROTECTION CONDITION ALPHA

F-2. FPCON Alpha applies when there is an increased general threat of possible terrorist activity against personnel or facilities, the nature and extent of which are unpredictable. Alpha measures must be capable of being maintained indefinitely. The following bullets explain the FPCON Alpha measures:

- **Alpha measure 1.** Inform personnel and family members of the general situation at regular intervals. Also, remind them to be alert for and report suspicious activities such as the presence of unfamiliar personnel and vehicles, suspicious/unattended parcels, and possible surveillance attempts.
- **Alpha measure 2.** Ensure that the duty officer or personnel with access to building plans as well as the plans for area evacuations are available at all times. Plans to execute access control procedures should be in place. Key personnel required to implement security plans should be on call and readily available.
- **Alpha measure 3.** Secure and randomly inspect buildings, rooms, and storage areas not in regular use.

---

***Note:*** MWD should randomly inspect buildings after normal duty hours. The KM should ensure that no inspection time or location patterns are ever set by MWD teams.

---

- **Alpha measure 4.** Increase security spot checks of vehicles and persons entering installations under the jurisdiction of the United States.

---

***Note:*** MWD should perform vehicle checks randomly at entry points.

---

- **Alpha measure 5.** Limit access points for vehicles and personnel commensurate with a reasonable flow of traffic.

- **Alpha measure 6.** Apply Alpha measures 4, 5, 7, or 8 either individually or in combination with each other as a deterrent.
- **Alpha measure 7.** Review intelligence, counterintelligence, and operations dissemination procedures.
- **Alpha measure 8.** Review and, if necessary, implement security measures for HRP.
- **Alpha measure 9.** Consult local authorities on the threat and mutual antiterrorism measures as appropriate.
- **Alpha measure 10.** Review all plans and be prepared to implement a higher FPCON.
- **Alpha measure 11 (spare).** Add additional measures as deemed appropriate by the commanders.

## FORCE PROTECTION CONDITION BRAVO

F-3. FPCON Bravo applies when an increased or more predictable threat of terrorist activity exists. Sustaining Bravo measures for a prolonged period may affect operational capabilities and impact relations with local authorities. The following bullets explain the FPCON Bravo measures:

- **Bravo measure 1.** Continue, or introduce, all measures in the previous FPCON level.
- **Bravo measure 2.** Check plans for the implementation of the next FPCON level.
- **Bravo measure 3.** Identify critical and high-occupancy buildings. Move cars and objects (such as crates and trash containers) away from buildings to reduce vulnerability to bomb attacks. Consider centralized parking.
- **Bravo measure 4.** Secure and inspect all buildings, rooms, and storage areas not in regular use.
- **Bravo measure 5.** Inspect the interior and exterior of buildings in regular use for suspicious packages at the beginning and end of each workday, as well as at random intervals. Patrol teams search the exterior of high-occupancy buildings at the beginning and end of each day.

---

***Note:*** Use PEDD to search buildings and areas.

---

- **Bravo measure 6.** Implement mail-screening procedures to identify suspicious letters and parcels.

---

***Note:*** Search the mail distribution center with both PEDD and PNDD teams randomly.

---

- **Bravo measure 7.** Inspect commercial deliveries randomly. Advise family members to check home deliveries.
- **Bravo measure 8.** Increase presence around and surveillance of domestic accommodations, schools, messes, clubs, and other soft targets to improve deterrence and defense and build confidence among staff and dependents.

---

***Note:*** Randomly patrol and perform walking patrols of populated areas with MWDs.

---

- **Bravo measure 9.** Implement plans to enhance off-installation security of DOD facilities. In areas with threat levels of moderate, significant, or high, coverage includes facilities (such as DOD schools and daycare centers) and transportation services and routes (such as bus routes) used by DOD employees and family members.
- **Bravo measure 10.** Inform local security committees of actions being taken.
- **Bravo measure 11.** Verify the identity of visitors and randomly inspect their suitcases, parcels, and other containers.
- **Bravo measure 12.** Conduct random patrols to check vehicles, people, and buildings.
- **Bravo measure 13.** Implement additional security measures for HRP.
- **Bravo measure 14.** Identify and brief personnel who may augment guard forces. Review specific rules of engagement including the use of deadly force.

- **Bravo measure 15.** Test the mass notification system.
- **Bravo measure 16.** Verify the identity of all personnel entering buildings as deemed appropriate.
- **Bravo measure 17.** Identify off-installation safe havens for use at the next FPCON.
- **Bravo measure 18.** Review and implement proper OPSEC, communications security (COMSEC), and information security (INFOSEC) status.
- **Bravo measure 19.** Consider using tactical deception techniques.
- **Bravo measure 20 (spare).** Add additional measures as deemed appropriate by the commanders.

## FORCE PROTECTION CONDITION CHARLIE

F-4. FPCON Charlie applies when an incident occurs or intelligence is received indicating that some form of terrorist action or targeting against personnel or facilities is likely. Implementing FPCON Charlie measures creates hardship and affects the activities of the unit and its personnel. The following bullets explain the FPCON Charlie measures:

- **Charlie measure 1.** Continue, or introduce, all measures in the previous FPCON.
- **Charlie measure 2.** Ensure that duty personnel are cognizant of actions required to implement antiterrorism and force protection plans. Recall required personnel.
- **Charlie measure 3.** Reduce access control to a minimum and strictly enforce entry. Randomly search vehicles.

***Note:*** Increase the PEDD and PNDD teams' search times at entry points.

- **Charlie measure 4.** Increase standoff from sensitive buildings based on the threat.
- **Charlie measure 5.** Issue weapons and appropriate communication equipment to guards according to SOFAs. Include specific orders on the issue of ammunition.
- **Charlie measure 6.** Increase patrols of the installation to include waterside perimeters, if appropriate.

***Note:*** Use MWD assets to expand the perimeter patrol area and detect intruders.

- **Charlie measure 7.** Implement a barrier plan to prevent a vehicle-borne attack.
- **Charlie measure 8.** Protect all designated vulnerable points. Give special attention to vulnerable points outside the military establishment.
- **Charlie measure 9.** Consult local authorities about closing public (and military) roads and facilities to reduce the vulnerability to attack.
- **Charlie measure 10.** Consider searching suitcases, briefcases, and packages being brought onto the installation through ACPs and consider randomly searching suitcases, briefcases, and packages leaving.

***Note:*** Use PEDD and PNDD teams to search the items above.

- **Charlie measure 11.** Review personnel policy procedures to determine a course of action for family members.
- **Charlie measure 12.** Review access procedures for all non-US personnel and adjust as appropriate.
- **Charlie measure 13.** Consider escorting children to and from DOD schools.
- **Charlie measure 14 (spare).** Add additional measures as deemed appropriate by the commanders.

# FORCE PROTECTION CONDITION DELTA

F-5. FPCON Delta applies in the immediate area where a terrorist attack has occurred or when intelligence has been received that terrorist action against a specific location or person is imminent. Normally, FPCON Delta is declared as a localized condition. The following bullets explain the FPCON Delta measures:

- **Delta measure 1.** Continue, or introduce, all measures in the previous FPCON.
- **Delta measure 2**. Augment guards as necessary.
- **Delta measure 3.** Identify all vehicles within operational or mission support areas.
- **Delta measure 4.** Search all vehicles and contents before allowing entrance to the installation.

---

***Note:*** Use PEDD and PNDD as needed to search vehicles and contents.

---

- **Delta measure 5.** Control access and implement positive identification of all personnel with no exceptions.
- **Delta measure 6.** Search all suitcases, briefcases, and packages brought into the installation.

---

***Note:*** Use PEDD and PNDD as needed to search the items above.

---

- **Delta measure 7.** Escort children to and from DOD schools or close the schools.
- **Delta measure 8.** Make frequent checks of the exterior of buildings and of parking areas. Use MWD teams to make frequent checks of the exterior of buildings and parking areas.
- **Delta measure 9.** Restrict all nonessential movement.
- **Delta measure 10.** Close public and military roads and facilities if permitted.
- **Delta measure (spare).** Add additional measures as deemed appropriate by the commanders.

# Appendix G
# After-Action Reviews

The AAR process is used to document and discuss MWD operations and training. AARs are focused on performance standards that enable other MWD handlers, training NCOs, KMs, program managers, and concerned agencies to discover what happened, why it happened, and how to sustain strengths and improve on weaknesses. Leaders, planners, and developers can use AARs to get the maximum benefit from every mission or task conducted during predeployment, deployment, and redeployment.

## AFTER-ACTION REVIEW PURPOSE

G-1. AARs must provide—

- Candid insights into specific soldier, leader, and MWD strengths and weaknesses from various perspectives.
- Feedback and insight critical to future successes.
- Details that evaluation reports alone often lack.

G-2. The USAMPS MWD Career Management NCO collects historical data, end-of-tour reports, and other organizational specific issues. With the submission of AARs, USAMPS can build a historical database to benefit all in the improvement of Army MWD team tactics, techniques, and procedures. Ultimately, information will be submitted to the Center for Army Lessons Learned (CALL) for distribution to the Army.

## AFTER-ACTION REVIEW SUBMISSION

G-3. After the completion of TDY missions, special operations, deployments or other significant events, MWD handlers submit AARs to their respective MACOMs. MACOMs forward the AARs to HQDA Office of the Provost Marshal General (OPMG) and USAMPS.

## POINT OF CONTACT

G-4. The point of contact for the MWD Career Management NCO is commercial: (573) 563-8040 and defense switched network (DSN): 676-8040/7946. Send hard-copy correspondence to: US Army Military Police School, ATTN: ATSJ-P, 401 MANSCEN Loop, Suite 1058, Fort Leonard Wood, MO 65473-8926.

## EXAMPLE AFTER-ACTION REVIEW

G-5. *Figure G-1*, page G-2, should be used as the format for AARs submitted to the USAMPS MWD Career Management NCO. Refer to *TC 25-20* for comprehensive guidance on AARs.

DEPARTMENT OF THE ARMY
$42^{D}$ Military Police Detachment
Ft. Bragg, NC 27380

OFFICE SYMBOL DATE

MEMORANDUM FOR RECORD

SUBJECT: After-Action Review, Task Force Iron Dog, Sep 03 – Nov 03

1. State what was supposed to happen: This paragraph contains a brief introduction of the mission that was being performed by the MWD team or section. Include operation objectives, the overall mission, and the intent from the task force commander as well as any other relevant information.

2. State what happened: This paragraph contains pertinent information about what actually happened during the mission. Was the mission accomplished as stated above or were other objectives established inside of the area of operation?

3. Determine what went right and wrong during the mission:

   a. Topic: What was right and wrong with what happened during the mission.

   b. Discussion: Give the who, what, when, where, and how for the mission. Give a good summary of what the issue was.

   c. Recommendation: Give recommendations to improve the issues that were identified as needing improvement.

4. Point of contact for this memorandum is the undersigned at DSN: 123-4567 or jon.doe@us.army.mil.



Jon Doe
SSG, USA
Kennel Master / PEDD Handler

***Note***: If you have a positive topic just use Topic and Discussion sections.

**Figure G-1. Sample AAR Format**

# Appendix H
# Health and Welfare Inspections Briefing Guide

Commanders frequently request MWD support during health and welfare inspections. Commanders should be made aware that the conduct and responsibility of health and welfare inspections lies within their purview and that the MWD team is not the authority for the mission. This appendix identifies the actions that need to be addressed with the commander to ensure a successful and lawful health and welfare inspection. These actions are needed to establish and act on probable cause determinations. Questions concerning these matters should be directed to the local Staff Judge Advocate Office for clarification.

## COORDINATION

H-1. The KM coordinates and approves each health and welfare inspection. After the KM receives the request for assistance, he assesses the needs of the mission and determines which detector dog team (if any) will assist the commander. The KM notifies the unit commander or 1SG and accepts or denies the mission. Once the mission is accepted, the KM briefs them on their roles and responsibilities and informs them that if his criteria are not met, the KM will terminate the search.

## HANDLERS BRIEFING

H-2. Prior to the inspection, the KM, plans NCO, or handler ensures that the commander requesting support is aware that the building being inspected must be unoccupied and that all doors are unlocked (or keys available on site). On the arrival of the MWD team at the location of the health and welfare inspection, the handler performs a commander's briefing and demonstration. The handler—

- Explains the probable cause folder to the commander.
- Shows the commander the training aids used for detection.
- Demonstrates the dog's ability to detect.

H-3. A briefing document shows handlers how to record the details of each health and welfare inspection (*Figure H-1*, page H-3). The commander is provided with a copy of the briefing document containing the following information:

- The MWD teams used for the inspection.
- The canines' names and tattoo numbers.
- The scope of the proficiency training of the MWD teams. Discuss with the commander the standards and scope of the detector training required to maintain the certification of assigned MWD teams.
- The MWDs' accuracy rates. Explain to the commander that the accuracy percentage rate is based on quarterly training.
- The MWDs' detection experience. Inform the commander of the amount of time that the MWDs have actually spent conducting detection missions. Include any possible finds in prior inspections.
- The MWDs' capabilities. Explain to the commander that the MWDs are capable of detecting either drugs or explosives.

- The MWDs' alert response. Describe what the MWDs will do when they indicate the presence or odor of found drugs or explosives.
- The unit representatives. Notify the commander that one SSG or above is required to escort each MWD team through out the inspection. The commander may not be a unit representative.

## UNIT REPRESENTATIVE DUTIES AND RESPONSIBILITIES

H-4. Ensure that the unit representatives are informed that they—

- Escort the MWD team at all times.
- Unlock doors or areas as needed.
- Carefully observe the MWD team.
- Follow the handler's instructions.
- Obey safety precautions.
- Spot areas that are missed by the MWD team.

## DOCUMENTATION

H-5. It is important that the MWD handler documents the commander's briefing on every occasion. *Figure H-1*, page H-3, gives an example for documenting the briefing given to the commander before a health and welfare inspection. Advise the commander to maintain the handlers briefing and other documentation concerning each response of the MWD. Advise the commander that a positive response by a narcotic detector dog provides probable cause for a command-directed urinalysis. Ensure that the commander signs the briefing form and then provide a copy for filing in case of a judicial action.

### INSPECTION

H-6. After the briefing has been conducted, the inspection may begin. Remember to advise the commander and unit representatives that this health and welfare inspection is the unit's mission and that the MWD team is assisting in the mission. The MWD team is not in charge of the health and welfare inspection but can be used as a tool by the commander.

ATZP-PM-K DATE: ____________

SUBJECT: Commander's Briefing for Health and Welfare Inspections Utilizing Detector Dogs

1. (commander's name) has requested narcotics detector dog support for a health and welfare inspection on (dd/mm/yyyy) .

2. Prior to canine(s) being utilized, the building must be unoccupied and all doors must be unlocked (or keys available).

3. The following provides the commander with information essential to a successful inspection:

_____ A. Team(s) to be utilized: (number of PEDD and PNDD to be used) .

_____ B. MWD name(s) and tattoo number(s): (name and tattoo number for each MWD)

_____ C. The scope of formal training of the canine team(s). Discuss detector training. Discuss the standards that must be successfully completed to maintain certification of detector dog team(s).

_____ D. The canine's accuracy rate(s) is/are (accuracy rates for each MWD) . This is based on the previous 3 months training records.

_____ E. The extent of actual MWD detection field experience. (experience for each MWD)

_____ F. Explain the types of drugs or explosives that the canine(s) is/are trained to detect.

_____ G. Describe the canine's specific behavior(s)/actions when it responds, indicating the presence or the odor of narcotics (detector response of each MWD) .

_____ H. Ensure that the commander(s) is/are aware that one representatives (SSG or above) is needed to act as an escort for each team. The commander may not be one of the individuals accompanying canine team(s).

_____ I. Ensure that the escorts are briefed on their duties and responsibilities. Discuss the safety precautions that they must follow, and ensure the safety of others.

_____ J. Ensure that the commander(s) is/are aware that the unit is responsible for maintaining proper documentation (date, time, room number, common or personal area and occupant[s] name[s]), concerning the response of the canine(s). This information should be kept on file with the commander's copy of this briefing sheet, in case of a judicial action.

_____K. Ensure that the commander is aware that a positive response by a narcotic detector dog is probable cause for a command-directed urinalysis on the occupant or occupants of that room.

I (commander's name) have received a personal briefing covering all the information stated above and now understand the capabilities of the canine team(s). I am satisfied that the canine team(s) have the training and experience necessary to provide reliable information upon which to authorize a search and/or perform a command-directed urinalysis. I authorize the canine team(s) to inspect the following area(s): (areas to be inspected)

Sample

**Figure H-1. Health and Welfare Briefing Document Sample**

**HANDLER(S) SIGNATURE**

1. ______________________ 2. ______________________

3. ______________________ 4. ______________________

**COMMANDER'S INFORMATION**

NAME: ______________________

RANK: ______________________

UNIT: ______________________

SIGNATURE: ______________________

**1SG'S INFORMATION**

NAME: ______________________

RANK: ______________________

UNIT: ______________________

SIGNATURE: ______________________

**UNIT REPRESENTATIVE**

NAME: ______________________

RANK: ______________________

UNIT: ______________________

SIGNATURE: ______________________

**UNIT REPRESENTATIVE**

NAME: ______________________

RANK: ______________________

UNIT: ______________________

SIGNATURE: ______________________



**Figure H-1. Health and Welfare Briefing Document Sample (Continued)**

# Appendix I
# Terrorist Bomb Threat Stand-off Distances

*Table I-1* contains information from the DOD technical support working group. Commanders, PMs, and MWD planners can use this information as a guide to determine the minimum distances necessary to evacuate personnel from suspected explosive threats.

**Table I-1. Minimum Distances for Personnel Evacuation**

| *Threat Description* | *Explosives Capacity* (TNT Equivalent)* | *Building Evacuation Distance*** | *Outdoor Evacuation Distance**** |
|---|---|---|---|
| Pipe bomb | 5 lbs/2.3 kg | 70 ft/21 m | 850 ft/259 m |
| Briefcase/suitcase bomb | 50 lbs/23 kg | 150 ft/46 m | 1,850 ft/564 m |
| Compact sedan | 500 lbs/227 kg | 320 ft/98 m | 1,500 ft/457 m |
| Sedan | 1,000 lbs/454 kg | 400 ft/122 m | 1,750 ft/534 m |
| Passenger/cargo van | 4,000 lbs/1,814 kg | 640 ft/195 m | 2,750 ft/838 m |
| Small moving van/delivery truck | 10,000 lbs/4,536 kg | 860 ft/263 m | 3,750 ft/1,143 m |
| Moving van/water truck | 30,000 lbs/13,608 kg | 1,240 ft/375 m | 6,500 ft/1,982 m |
| Semitrailer | 60,000 lbs/27,216 kg | 1,570 ft/475 m | 7,000 ft/2,134 m |

* Based on maximum volume or weight of explosives (TNT equivalent).
** Governed by the ability of an unstrengthened building to withstand severe damage or collapse.
*** Governed by either the fragment throw distance or glass breakage/falling glass hazard distance (whichever throws fragments farther than vehicle bombs).

NOTES:
1. All personnel must either seek shelter inside a building away from windows and exterior walls (with some risk) or move beyond the outdoor evacuation distance.
2. The preferred area is beyond the outdoor evacuation distance for the evacuation of personnel in buildings. This is mandatory for personnel outdoors.

# Glossary

## Acronyms and Abbreviations

| | |
|---|---|
| **1SG** | first sergeant |
| **AAR** | after-action review |
| **ACP** | access control point |
| **AF** | Air Force |
| **agg** | aggression |
| **AR** | Army regulation |
| **ASG** | area support group |
| **ASI** | additional skill identifier |
| **ASP** | ammunition supply point |
| **ATTN** | attention |
| **BDU** | battle dress uniform |
| **C2** | command and control |
| **C4** | composition 4 |
| **CA** | corrective action |
| **CALL** | Center for Army Lessons Learned |
| **CCTV** | Closed-circuit television |
| **cert** | certification |
| **CI** | civilian internee |
| **CID** | Criminal Investigation Division |
| **clg** | cleaning |
| **CMFO** | centralized furnishings management office |
| **comm** | communications |
| **COMSEC** | communications security |
| **CONEX** | container express |
| **cont** | continuted |
| **CP** | command post |
| **CPT** | captain |
| **CR** | corrective response |
| **DA** | Department of the Army |
| **d** | deep |
| **DARE** | Drug Abuse Resistance Education |
| **DCSPER** | Dputy Chief of Stagg for Personnel |
| **DC** | detonating cord |
| **DD** | Department of Defense |
| **DEA** | Drug Enforcement Administration |

| | |
|---|---|
| **Dec** | December |
| **det** | detonating |
| **DF** | deficiency |
| **DOD** | Department of Defense |
| **DODMWDVS** | Department of Defense Military Working Dog Veterinary Services |
| **DSN** | defense switched network |
| **DST** | drug suppression team |
| **EAC** | echelon above corps |
| **EOD** | explosive ordnance disposal |
| **EPW** | enemy prisoner of war |
| **eval** | evaluaiton |
| **FM** | field manual |
| **FMO** | Furnishing managemetn office |
| **FPCON** | force protection condition |
| **ft** | foot; feet |
| **GE** | Germany |
| **GOV** | government owned vehicle |
| **GSA** | General Services Administration |
| **h** | high |
| **HF** | hole, variable |
| **HRP** | high-risk personnel |
| **HRT** | high-risk target |
| **HVAC** | heating, ventilation, and air conditioning |
| **I/R** | internment and resettlement |
| **IDS** | Intrusion Detection System |
| **IED** | improvised explosive device |
| **INFOSEC** | information security |
| **IPB** | intelligence preparation of the battlefield |
| **iso** | isolation |
| **kg** | kilogram |
| **KM** | kennel master |
| **LAFB** | Lackland Air Force base |
| **LAN** | local area network |
| **LBS** | pounds |
| **LBV** | load-bearing vest |
| **LCE** | load-carrying equipment |
| **LED** | light-emitting diode |
| **LIN** | line item number |
| **LTC** | lieutenent colonel |
| **m** | meter |

| | |
|---|---|
| **MACOM** | major Army command |
| **MANSCEN** | Maneuver Support Center |
| **MCI** | military customs inspector |
| **MCIPM** | military customs inspector program manager |
| **MCM** | Manual for Courts-Martial |
| **mech** | mechanical |
| **METL** | mission-essential task list |
| **METT-TC** | mission, enemy, terrain, troops, time available, and civilian considerations |
| **MEVA** | mission-essential vulnerable areas |
| **MILSTRIP** | military standard transportation and issue procedures |
| **min** | minimum |
| **MMSO** | maneuver mobility support operations |
| **MOPP** | mission-oriented protective posture |
| **MPR** | military police report |
| **MSR** | main supply route |
| **MTOE** | modification table of organization and equipment |
| **MWD** | military working dog |
| **MWDMS** | Military Working Dog Management System |
| **NBC** | nuclear, biological, and chemical |
| **NCO** | noncommissioned officer |
| **NCOIC** | noncommissioned officer in charge |
| **no** | number |
| **Nov** | November |
| **NSN** | national stock number |
| **OB** | obstacle |
| **Oct** | October |
| **ODCSOPS** | Office of the Deputy Chief of Staff for Operaitons and Plans |
| **off/L** | off leash |
| **on/L** | on leash |
| **OIF** | Operation Iraqi Freedom |
| **OPMG** | Office of the Provost Marshal General |
| **OPSEC** | operational security |
| **pam** | pamphlet |
| **PC** | potassium chlorate |
| **PD** | patrol dog |
| **PEDD** | patrol explosive detector dog |
| **PIO** | police intelligence operations |
| **PM** | provost marshal |
| **PMO** | provost marshal office |

| | |
|---|---|
| **PNDD** | patrol narcotic detector dog |
| **POV** | privately owned vehicle |
| **PT** | physical training |
| **prep** | preparation |
| **RAMP** | random antiterrorism measures program |
| **rec** | recreation |
| **S2** | intelligence section |
| **SC** | sodium cholorate |
| **sec** | seconds |
| **Sep** | September |
| **SF** | standard form |
| **SFC** | sergeant first class |
| **SGT** | sergeant |
| **SOFA** | Status of Forces Agreement |
| **SOP** | standard operating procedures |
| **SP** | smokeless powder |
| **SRT** | special-reaction team |
| **SSG** | staff sergeant |
| **STX** | Semtex |
| **TDA** | table of distribution and allowancees |
| **TDY** | temporary duty |
| **TISA** | troop issue subsistence activity |
| **TMP** | transportation motor pool |
| **TNT** | trinitrotoluene |
| **TOE** | table of organization and equipment |
| **TRADOC** | United States Army Training and Doctrine Command |
| **Tues** | Tuesday |
| **US** | United States |
| **USA** | United States of America |
| **USC** | United States Code |
| **USCS** | United States Customs Service |
| **USSS** | United States Secret Service |
| **v** | version |
| **VCO** | veterinary corps officer |
| **vet** | veterinarian |
| **VIP** | very important person |
| **VTF** | Veterinary Treatment Facility |
| **VTR** | Veterinary Treatment Record |
| **WT** | weight |

# References

DA Forms are available on the Army Publishing Directorate web site <http://www.apd.army.mil>. DD forms are available from the Office of the Secretary of Defense web site <http://www.dtic.mil/whs/directives/infomgt/forms/formsprogram.htm>.

## SOURCES USED

These are the sources quoted or paraphrased in this publication.


AR 190-12. *Military Working Dogs.* 30 September 1993.

AR 190-14. *Carrying of Firearms and Use of Force for Law Enforcement and Security Duties.* 12 March 1993.

DA Pam 190-12. *Military Working Dog Program.* 30 September 1993.

DA Pam 611-21. *Military Occupational Classification and Structure.* 31 March 1999.


## DOCUMENTS NEEDED

These documents must be available to the intended users of this publication.


AR 700-81. *DOD Dog Program.* 5 May 1971.

DA Form 2807-R. *Military Working Dog Training and Utilization Record.*

DA Form 3992–R. *Narcotics or Explosives Detector Dog Training and Utilization Record.*

FM 3-19.15. *Civil Disturbance Operations.* 18 April 2005.


## READINGS RECOMMENDED

These sources contain relevant supplemental information.


AF Form 1256. *Certificate of Training.*

AR 190-47. *The Army Corrections System.* 5 April 2004.

AR 190-8. *Enemy Prisoners of War, Retained Personnel, Civilian Internees and Other Detainees.* 1 October 1997.

AR 350-1. *Army Training and Education.* 9 April 2003.

AR 40-1. *Composition, Mission, and Functions of the Army Medical Department.* 1 July 1983.

AR 40-3. *Medical, Dental, and Veterinary Care.* 12 November 2002.

AR 40-905. *Veterinary Health Services.* 16 August 1994.

AR 525-13. *Antiterrorism.* 4 January 2002.

AR 600-8-101. *Personnel Processing (In-, Out-, Soldier Readiness, Mobilization, and Deployment Processing).* 15 July 2003.

AR 600-9. *The Army Weight Control Program.* 10 June 1987.

AR 614-200. *Enlisted Assignments and Utilization Management.* 30 September 2004.

DA Form 2028. *Recommended Changes to Publications and Blank Forms.*

DA Form 3161. *Request for Issue or Turn-In.*

DA Form 5513-R. *Key Control Register and Inventory.*

DA Form 6. *Duty Roster*

DA Form 7281-R. *Command Oriented Arms, Ammunition, & Explosives Security Screening and Evaluation Record.*

DA Pam 710-2-1. *Using Unit Supply System (Manual Procedures).* 31 December 1997.

DD Form 1834. *Military Working Dog Service Record.*

DEA Form 223. *Controlled Substance Registration Certificate.*

DOD Directive 2000.12. *DoD Antiterrorism (AT) Program.* 18 August 2003.

DOD Directive 5030.49. *DoD Customs and Border Clearance Program.* 4 May 2004.

DOD Directive 6400.4. *DoD Veterinary Services Program.* 22 August 2003.

DOD Instruction 3025.12. *Military Assistance for Civil Disturbances (MACDIS).* 4 February 1994

FM 21-20. *Physical Fitness Training.* 30 September 1992.

FM 27-10. *The Law of Land Warfare.* 18 July 1956.

FM 3-19.40. *Military Policy Internment/Resettlement Operations.* 1 August 2001.

Lackland AFB Form 375. *Patrol Dog Certification.*

Lackland AFB Form 375a. *Detector Dog Certification.*

MCM, Part III Military Rules of Evidence; Rule 311, *Evidence obtained from unlawful searches and seizures.* 2000.

MCM, Part III Military Rules of Evidence; Rule 313, *Inspections and inventories in the armed forces.* 2000.

MCM, Part III Military Rules of Evidence; Rule 314, *Searches not requiring probable cause.* 2000.

MCM, Part III Military Rules of Evidence; Rule 315, *Probable cause searches.* 2000.

MCM, Part III Military Rules of Evidence; Rule 315(f)(2), *Probable cause determination.* 2000.

MCM, Part III Military Rules of Evidence; Rule 316, *Seizures.* 2000.

SF 702. *Security Container Check Sheet.*

TC 19-210. *Access Control Handbook.* 4 October 2004.

TC 25-20. *A Leader's Guide to After Action Reviews.* 30 September 1993.

USC Title 18, Crimes and Criminal Procedure; Part I, Crimes; Section 1385, *Use of Army and Air Force as posse comitatus.*

# Index

Made in the USA
Monee, IL
05 November 2026

67858293R00418